Lecture Notes in Computer Science 3172

Commenced Publication in 1973
Founding and Former Series Editors:
Gerhard Goos, Juris Hartmanis, and Jan van Leeuwen

Marco Dorigo   Mauro Birattari
Christian Blum   Luca M. Gambardella
Francesco Mondada   Thomas Stützle (Eds.)

Ant Colony Optimization and Swarm Intelligence

4th International Workshop, ANTS 2004
Brussels, Belgium, September 5 - 8, 2004
Proceedings

 Springer

Volume Editors

Marco Dorigo
Mauro Birattari
Christian Blum
Université Libre de Bruxelles, IRIDIA CP 194/6
Avenue Franklin Roosevelt 50, 1050 Bruxelles, Belgium
E-mail: {mdorigo, mbiro, cblum}@ulb.ac.be

Luca M. Gambardella
IDSIA, Instituto Dalle Molle di Studi sull'Intelligenza Artificiale
Galleria 2, 6928 Manno-Lugano, Switzerland
E-mail: luca@idsia.ch

Francesco Mondada
Swiss Federal Institute of Technology of Lausanne (EPFL)
Autonomous Systems Lab - LSA
LSA-I2S-EPFL
CH-1015 Lausanne, Switzerland
E-mail: francesco.mondada@epfl.ch

Thomas Stützle
Darmstadt University of Technology
Computer Science Department, Intellectics Group
Hochschulstr. 10, 64283 Darmstadt, Germany
E-mail: stuetzle@informatik.tu-darmstadt.de

Library of Congress Control Number: 2004109777

CR Subject Classification (1998): F.2.2, F.1.1, G.1, G.2, I.2, C.2.4, J.1

ISSN 0302-9743
ISBN 3-540-22672-9 Springer Berlin Heidelberg New York

Springer is a part of Springer Science+Business Media

springeronline.com

© Springer-Verlag Berlin Heidelberg 2004
Printed in Germany

Typesetting: Camera-ready by author, data conversion by Olgun Computergrafik
Printed on acid-free paper   SPIN: 11307815   06/3142   5 4 3 2 1 0

Preface

With its fourth edition, the *ANTS* series of workshops[1] has changed its name. The original *"ANTS – From Ant Colonies to Artificial Ants: International Workshop on Ant Algorithms"* has become *"ANTS – International Workshop on Ant Colony Optimization and Swarm Intelligence"*. This change is mainly due to the following reasons.

First, the term *"ant algorithms"* was slower in spreading in the research community than the term *"swarm intelligence"*, while at the same time research in so-called *swarm robotics* was the subject of increasing activity: it was therefore an obvious choice to substitute the term *ant algorithms* with the more accepted and used term *swarm intelligence*.

Second, although *swarm intelligence* research has undoubtedly produced a number of interesting and promising research directions[2], we think it is fair to say that its most successful strand is the one known as *"ant colony optimization"*. *Ant colony optimization*, first introduced in the early 1990s as a novel tool for the approximate solution of discrete optimization problems, has recently seen an explosion in the number of its applications, both to academic and real-world problems, and is currently being extended to the realm of continuous optimization (a few papers on this subject being published in these proceedings). It is therefore a reasonable choice to have the term *ant colony optimization* as part of the workshop name.

As mentioned above, this is the fourth edition of the *ANTS* workshops. The series started in 1998 with the organization of *ANTS'98*. On that occasion more than 50 researchers from around the world joined for the first time in Brussels, Belgium to discuss *swarm intelligence* related research, and a selection of the best papers presented at the workshop was published as a special issue of the *Future Generation Computer Systems* journal (Vol. 16, No. 8, 2000). Two years later the experience was repeated with the organization of *ANTS 2000*, which attracted more than 70 participants. The 41 extended abstracts presented as talks or posters at the workshop were collected in a booklet distributed to participants, and a selection of the best papers was published as a special section of the *IEEE Transactions on Evolutionary Computation* (Vol. 6, No. 4, 2002). After these first two successful editions, it was decided to make of *ANTS* a series of biannual events. Accordingly, the third edition was organized in September 2002, in Brussels, Belgium. The success of the workshop and the quality of the papers presented in the second edition had also made it clear that it was the right time to have an official workshop proceedings: the *ANTS 2002* proceedings was

---

[1] http://iridia.ulb.ac.be/~ants/

[2] Think, for example, in addition to the already mentioned swarm robotics, of algorithms for clustering and data mining inspired by the ants' cemetery building behavior, of dynamic task allocation algorithms inspired by the behavior of wasp colonies, of particle swarm optimization, and so on.

published by Springer as Volume 2463 of LNCS, and contained 36 contributions: 17 full papers, 11 short papers, and 8 extended abstracts, selected out of a total of 52 submissions.

The *Ant Colony Optimization and Swarm Intelligence* field is still growing, as testified, for example, by the success of the *1st IEEE Swarm Intelligence Symposium*, held in 2003 in Indianapolis, Indiana, US; or by the steady increase we are observing in the number of submissions to *ANTS* workshops, which resulted in the 79 papers submitted to *ANTS 2004*. This relatively high number of submissions allowed us to set the acceptance threshold for full and short papers at approximately 50%, which guaranteed a fairly high quality of the proceedings, and, at the same time, a reasonably dense workshop program[3]. We are sure that the readers of these proceedings will enjoy the quality of the papers collected in this volume, quality that somehow reflects the growing maturity of the *swarm intelligence* field.

We wish to conclude by saying that we are very grateful to the authors who submitted their works; to the members of the international program committee and to the additional referees for their detailed reviews; to the IRIDIA people for their enthusiasm in helping with organization matters; to the Université Libre de Bruxelles for providing rooms and logistic support; and, more generally, to all those contributing to the organization of the workshop. Finally, we would like to thank our sponsors, the company *AntOptima*[4] and the *Metaheuristics Network*[5], who financially supported the workshop.

June 2004

Marco Dorigo
Mauro Birattari
Christian Blum
Luca M. Gambardella
Francesco Mondada
Thomas Stützle

---

[3] In addition to the accepted papers, a small number of posters were selected for presentation: these are works that, although in a rather preliminary phase, show high potential and are therefore worth discussing at the workshop.
[4] More information available at www.antoptima.com
[5] A Marie Curie Research Training Network funded by the European Commission. More information available at www.metaheuristics.org

Organization

ANTS 2004 was organized by IRIDIA, Université Libre de Bruxelles, Belgium.

Workshop Chair

Marco Dorigo   IRIDIA, ULB, Brussels, Belgium

Technical Program Chairs

Luca M. Gambardella   IDSIA, USI-SUPSI, Manno-Lugano, Switzerland
Francesco Mondada   ASL, EPFL, Lausanne, Switzerland
Thomas Stützle   Intellektik, TUD, Darmstadt, Germany

Publication Chairs

Mauro Birattari   IRIDIA, ULB, Brussels, Belgium
Christian Blum   IRIDIA, ULB, Brussels, Belgium

Program Committee

Tucker Balch   Georgia Tech, Atlanta, GA, USA
Christian Blum   IRIDIA, ULB, Brussels, Belgium
Eric Bonabeau   Icosystem, Boston, MA, USA
Oscar Cordón   Universidad de Granada, Spain
David Corne   University of Reading, UK
Jean-Louis Deneubourg   CENOLI, ULB, Brussels, Belgium
Gianni di Caro   IDSIA, Manno-Lugano, Switzerland
Dario Floreano   ASL, EPFL, Lausanne, Switzerland
Michel Gendreau   Université de Montréal, Canada
Deborah Gordon   Stanford University, CA, USA
Walter Gutjahr   Universität Wien, Austria
Richard Hartl   Universität Wien, Austria
Owen Holland   University of Essex, Colchester, UK
Holger Hoos   University of British Columbia, Vancouver, Canada
Paul B. Kantor   Rutgers University, New Brunswick, NJ, USA
Joshua Knowles   MIB, UMIST, Manchester, UK
Sven Koenig   Georgia Tech, Atlanta, GA, USA

Vittorio Maniezzo	Università di Bologna, Italy
Alcherio Martinoli	EPFL, Lausanne, Switzerland
Chris Melhuish	University of the West of England, Bristol, UK
Ronaldo Menezes	Florida Tech, Melbourne, FL, USA
Daniel Merkle	Universität Karlsruhe, Germany
Peter Merz	Universität Tübingen, Germany
Martin Middendorf	Universität Leipzig, Germany
Stefano Nolfi	CNR, Rome, Italy
Ben Paechter	Napier University, Edinburgh, UK
Van Parunak	Altarum Institute, Ann Arbor, MI, USA
Andrea Roli	Università degli Studi G. D'Annunzio, Chieti, Italy
Erol Şahin	Middle East Technical University, Ankara, Turkey
Michael Sampels	IRIDIA, ULB, Brussels, Belgium
Guy Theraulaz	Université Paul Sabatier, Toulouse, France
Franco Zambonelli	Università di Modena, Italy
Mark Zlochin	Weizmann Institute, Rehovot, Israel

Publicity Chair

Andrea Roli	Università degli Studi G. D'Annunzio, Chieti, Italy

Local Arrangements

Max Manfrin	IRIDIA, ULB, Brussels, Belgium
Carlotta Piscopo	IRIDIA, ULB, Brussels, Belgium

Additional Referees

Ashraf Abdelbar	Julia Handl	Rubén Ruiz García
Christian Almeder	Stephane Magnenat	Jürgen Schmidhuber
Erkin Bahceci	Marco Mamei	Alena Shmygelska
Levent Bayindir	Roberto Montemanni	Kevin Smyth
Leonora Bianchi	Alberto Montresor	Krzysztof Socha
Gilles Caprari	Luís Paquete	Onur Soysal
Karl Doerner	Yves Piguet	Christine Strauss
Alberto V. Donati	Andrea Emilio Rizzoli	Emre Ugur
Frederick Ducatelle	Daniel Roggen	Markus Waibel
Michael Guntsch	Martin Romauch	

Sponsoring Institutions

AntOptima (www.antoptima.com), Lugano, Switzerland
Metaheuristics Network (www.metaheuristics.org), a Marie Curie Research
Training Network of the Improving Human Potential Programme funded by the
European Commission

Table of Contents

Short Papers

A Comparison Between ACO Algorithms for the Set Covering Problem

Lucas Lessing[1], Irina Dumitrescu[2], and Thomas Stützle[1]

[1] Intellectics Group, Darmstadt University of Technology
llessing@gmx.net, stuetzle@informatik.tu-darmstadt.de
[2] Canada Research Chair in Distribution Management, HEC Montreal
irina@crt.umontreal.ca

Abstract. In this paper we present a study of several Ant Colony Optimization (ACO) algorithms for the Set Covering Problem. In our computational study we emphasize the influence of different ways of defining the heuristic information on the performance of the ACO algorithms. Finally, we show that the best performing ACO algorithms we implemented, when combined with a fine-tuned local search procedure, reach excellent performance on a set of well known benchmark instances.

1 Introduction

The Set Covering Problem (SCP) is an $\mathcal{NP}$-hard combinatorial optimization problem that arises in a large variety of practical applications, for example in airline crew scheduling or vehicle routing and facility placement problems [9, 12, 18]. Two Ant Colony Optimization (ACO) approaches for the SCP based on Ant System have been proposed so far [8, 10]; however, their computational performance is relatively poor compared to the state-of-the-art in SCP solving.

In this paper we present the results of a computational study comprising several ACO algorithms including $\mathcal{MAX}$–$\mathcal{MIN}$ Ant System ($\mathcal{MMAS}$) [16], Ant Colony System (ACS) [3], a hybrid between $\mathcal{MMAS}$ and ACS [17], as well as Approximate Nondeterministic Tree-Search (ANTS) [14]. Many of the best ACO computational results have been achieved using these algorithms and therefore they serve as good candidates for tackling the SCP. (For details on the full study we refer to [11].) We study the influence of different ways of defining the heuristic information on the performance of the algorithms. We distinguish between the usage of *static* and *dynamic* heuristic information. In the static case, the heuristic information is computed when initializing the algorithm and it remains the same throughout the whole run of the algorithm. In the dynamic case, the heuristic information depends on the partial solution available, hence it has to be computed at each step of an ant's walk. Therefore, it results in a higher computational cost that may be compensated by the higher accuracy of the heuristic values.

The paper is structured as follows. In the next section we introduce the SCP and lower bounds to it. In Section 3, we present some details on the ACO

M. Dorigo et al. (Eds.): ANTS 2004, LNCS 3172, pp. 1–12, 2004.

algorithms we applied, while in Sections 4 and 5 we give details on the local search and the heuristic information. We present experimental results in Section 6 and conclude in Section 7.

2   Set Covering Problem

The SCP can be formulated as follows. Let $A = (a_{ij})$ be an $m \times n$ 0-1 matrix and $c = (c_j)$ a positive integer n-dimensional vector, where each element c_j of c gives the cost of selecting column j of matrix A. We say that row i is covered by column j, if a_{ij} is equal to 1. The SCP consists of finding a subset of columns of minimal total cost such that all the rows are covered. The SCP is usually formulated as an integer program. Let $N = \{1, \ldots, n\}$ be the index set of the columns and $M = \{1, \ldots, m\}$ be the index set of the rows. The integer programming (IP) formulation of the SCP can be given as follows:

$$z_{SCP} = \min \quad \sum_{j \in N} c_j x_j$$

$$\text{s.t.} \quad \sum_{j \in N} a_{ij} x_j \geq 1, \forall i \in M \tag{1}$$

$$x_j \in \{0, 1\}, \quad \forall j \in N. \tag{2}$$

We now introduce the Lagrangean relaxation of this problem. For every vector $u = (u_1, \ldots, u_m) \in R_+^m$ we define the Lagrangean relaxation to be:

$$z_{LR}(u) = \min_{x \in \{0,1\}^n} \left[\sum_{j \in N} c_j x_j + \sum_{i \in M} u_i (1 - \sum_{j \in N} a_{ij} x_j) \right] \tag{3}$$

$$= \min_{x \in \{0,1\}^n} \left[\sum_{j \in N} c_j(u) x_j + \sum_{i \in M} u_i \right], \tag{4}$$

where $c_j(u) = c_j - \sum_{i \in M} a_{ij} u_i$ are called *Lagrangean costs*. The Lagrangean relaxation provides a lower bound on the optimal solution of the SCP, i.e. $z_{LR}(u) \leq z_{SCP}, \forall u \geq 0$. The best such lower bound on the optimal solution of the SCP is obtained by solving the *Lagrangean dual* problem, $z_{LD} = \max_{u \geq 0} z_{LR}(u)$. In our experiments, the Lagrangean dual was solved by the subgradient method [7], stopped after at most $100 \cdot m$ iterations as proposed by Yagiura et al. [19].

When we consider the IP formulation, a solution to the SCP is a 0-1 vector. However, when we refer to a solution, we sometimes refer to the set of indices of the variables fixed to one. This should be clear from the context.

3   Ant Colony Optimization for the SCP

Artificial ants in ACO algorithms can be seen as probabilistic construction heuristics that generate solutions iteratively, taking into account accumulated past search experience: pheromone trails and heuristic information on the instance under solution. In the SCP case, each column j has associated a phero-

```
procedure ACOforSCP
  initializeParameters;
  while termination condition is not true do
    for k := 1 to m_a do
      while solution not complete do
        applyConstructionStep(k);
      endwhile
      eliminateRedundantColumns(k);
      applyLocalSearch(k);
    endfor
    updateStatistics;
    updatePheromones;
  endwhile
  return best solution found
endprocedure ACOforSCP
```

Fig. 1. High-level view of the applied ACO algorithms. (For details, see text.)

mone trail τ_j that indicates the learned desirability of including column j into an ants' solution; η_j indicates the heuristic desirability of choosing column j.

In all ACO algorithms an ant starts with an empty solution and constructs a complete solution by iteratively adding columns until all rows are covered. It does so by probabilistically preferring solution components (columns in the SCP case) with high associated pheromone trail and/or heuristic value. Once m_a solutions are constructed and improved by local search, the pheromone trails are updated.

The application of ACO algorithms to the SCP differs from applications to other problems, such as the TSP. First, the solution construction of the individual ants does not necessarily end after the same number of steps for each ant, but only when a cover is completed. Second, the order in which columns are added to a solution does not matter, while in many other applications the order in which solution components are added to a partial solution may be important. Third, in the case of the SCP the solution constructed by the ants may contain redundant solution components which are eliminated before fine-tuning by a local search procedure. This latter feature is also present in two earlier applications of ACO algorithms to the SCP [8, 10]. However, the computational results obtained in [8, 10] are relatively poor, which may be due to the type of ACO algorithm chosen (i.e., Ant System, which is known to perform poorly compared to more recent variants of ACO algorithms). Here, we present the application of more recent ACO variants that exploit a large variety of different types of heuristic information. All these algorithms follow the algorithmic outline given in Figure 1.

3.1 $\mathcal{MAX}$–$\mathcal{MIN}$ Ant System

$\mathcal{MAX}$–$\mathcal{MIN}$ Ant System ($\mathcal{MM}$AS) [16] is at the core of many successful ACO applications. In $\mathcal{MM}$AS solutions are constructed as follows: an ant k ($k = 1, \ldots, m_a$) chooses column j with probability

$$p_j^k = \begin{cases} \frac{\tau_j[\eta_j]^\beta}{\sum_{h \notin S_k} \tau_h[\eta_h]^\beta}, & \text{if } j \notin S_k \\ 0, & \text{otherwise,} \end{cases} \tag{5}$$

where the parameter $\beta \geq 0$ determines the relative influence of the heuristic information with respect to the pheromone and S_k is the partial solution of ant k (note that here we assume that $S_k \subseteq N$). Once all solutions are completed, the pheromone values are updated by first evaporating all pheromone trails, i.e., by setting $\tau_j := (1 - \rho) \cdot \tau_j$, $\forall j \in N$, then by adding an amount of $\Delta\tau = 1/z$ to the columns contained in the best-so-far (i.e., the best since the start of the algorithm) or the iteration-best (i.e., the best found in the current iteration) solution, where z is the cost of the solution used in the pheromone update.

As usual in $\mathcal{MMAS}$, the range of feasible pheromone values is limited to an interval $[\tau_{min}, \tau_{max}]$ to avoid premature stagnation of the search and pheromones are initialized to τ_{max}. To further increase the exploration of solutions with columns that have only a small probability of being chosen, the pheromone trails are occasionally re-initialized. The re-initialization is triggered when no improved solution is found for a given number of iterations.

3.2   Ant Colony System

The second ACO algorithm we tested is Ant Colony Systems (ACS) [3]. ACS exploits the so called pseudo-random proportional action choice rule in the solution construction: ant k chooses the next column to be

$$j = \begin{cases} \text{argmax}_{l \notin S_k} \left\{ \tau_l[\eta_l]^\beta \right\}, & \text{if } q \leq q_0 \\ \text{draw}(J), & \text{otherwise,} \end{cases} \tag{6}$$

where q is a random number uniformly distributed in $[0, 1]$, q_0 ($0 \leq q_0 \leq 1$) is a parameter that controls how strongly the ants exploit deterministically the combined past search experience and heuristic information, and $\text{draw}(J)$ is a random number generated according to the probability distribution defined in Equation 5.

In ACS, ants modify the pheromone trails also while constructing a solution. Immediately after ant k's adding of a column j to its partial solution S_k, the ant modifies τ_j according to: $\tau_j := (1 - \xi)\tau_j + \xi \cdot \tau_0$, where ξ, $0 \leq \xi \leq 1$, is a parameter and $\tau_0 = 1/(n \cdot z_{GR})$, (where z_{GR} is the cost of a greedy solution), is the initial value of the pheromone trails. The local pheromone update has the effect that each time a column is chosen, it is made less attractive for the other ants, thus increasing exploration.

Once all ants have constructed their solutions, pheromones are deposited according to $\tau_j := (1 - \rho)\tau_j + \rho \cdot \Delta\tau_j^*, \forall j \in S^*$, where S^* is the best-so-far solution and $\Delta\tau_j^* = 1/z^*$, where z^* is the cost of S^*.

3.3 $\mathcal{MMAS}$–ACS–Hybrid

$\mathcal{MMAS}$–ACS–Hybrid is an ACO algorithm that follows the rules of $\mathcal{MMAS}$ but uses the pseudo-random proportional rule of ACS to determine the next column to be added to an ants' partial solution. Because of this aggressive construction rule, the hybrid uses generally much tighter pheromone trail limits than $\mathcal{MMAS}$. $\mathcal{MMAS}$–ACS–Hybrid has shown promising performance in [13, 17].

3.4   Approximate Nondeterministic Tree Search

Approximate Nondeterministic Tree Search (ANTS) is a recent ACO algorithm developed by Maniezzo [14]. The main innovative feature of ANTS is the use of lower bounds on the completion of a solution as the heuristic information used in the solution construction (see also Section 5.2). The probability of choosing the next column j to be added by ant k is given by:

$$p_j^k = \frac{\zeta\tau_j + (1 - \zeta)\eta_j}{\sum_{l\notin S_k}[\zeta\tau_l + (1 - \zeta)\eta_l]}, \qquad \text{if } j \notin S_k, \tag{7}$$

where $0 \leq \zeta \leq 1$ is a parameter that determines the influence of the pheromone trails versus the heuristic information. The contribution of all ants is considered in the pheromone trail update; the amount $\Delta\tau_j^k$ is given by:

$$\Delta\tau_j^k := \begin{cases} \vartheta\left(1 - \frac{C(S_k)-LB}{L_{avg}-LB}\right), & \text{if } j \in S_k \\ 0, & \text{otherwise,} \end{cases} \tag{8}$$

where $\vartheta \geq 0$, LB is a lower bound on the optimal solution value, and L_{avg} is a moving average of the solution qualities of recently constructed solutions; care is taken to avoid that pheromones become negative. If an ant's solution is worse than the current average, the pheromone trail of the columns in the solution is decreased; if the ant's solution is better, the pheromone trail is increased.

4   Local Search: r-Flip

In most applications of ACO to combinatorial optimization problems improving the solutions constructed by the ants in a local search phase [4, 5, 16] was worth doing. As our computational results will show, this is also the case for many SCP instances. We use an efficient local search , based on the r-flip neighborhood. The r-flip neighborhood of a solution $x = (x_1, \ldots, x_n)$ consists of all the solutions that can be obtained by flipping (i.e., removing a column from or adding a new column to the solution) at most r variables in the solution; this means that a solution $x' = (x_1', \ldots, x_n')$ is a neighbor of x if the Hamming-Distance $d(x, x') = |\{j \in \{1, \ldots, n\}|x_j \neq x_j'\}|$ between x and x' is at most r. An efficient implementation of a local search algorithm based on the 3-flip is given by Yagiura et al., [19]. The code of this local search algorithm has been provided to us by M. Yagiura; we adapted and used it to the ACO algorithms described above.

5   Heuristic Information

It is widely acknowledged that the heuristic information can play a significant role in the performance of ACO algorithms. However, there are several possible ways of defining heuristic information. The types of heuristic information used are *static* or *dynamic*. In the static case, the heuristic information can be computed only once, when initializing the algorithm. Then it remains the same throughout the whole run of the algorithm. In the dynamic case, the heuristic information depends on the partial solution constructed and it is computed at each construction step of each ant (the partial solutions between the ants differ). The use of dynamic heuristic information usually results in higher computational costs but typically also in higher accuracy of the computed heuristic values. Next we explain the types of heuristic information we considered in our study.

5.1 Static Heuristic Information

Column Costs: A very simple heuristic is to use the column costs for the heuristic information and to set $\eta_j = 1/c_j$.

Lagrangean Costs: Instead of the column costs, the Lagrangean costs associated to each column may be used to define the heuristic information. In this case, since the probability of selecting a column cannot be negative, the Lagrangean costs need to be normalized. Therefore, let the normalized Lagrangean costs be: $C_j = c_j(u) - \min_{h \in N} c_h(u) + \epsilon$, where ϵ is a small positive number, $c_j(u)$ are the Lagrangean costs, and the vector u considered is the one obtained at the end of the subgradient procedure. The heuristic information in this case is: $\eta_j = 1/C_j$.

5.2 Dynamic Heuristic Information

Cover Costs: This is the most commonly used heuristic information in the previously defined ACOs for the SCP. The cover cost of a column j are defined as $c_j/card_j(S)$, where $card_j(S)$ is the number of rows covered by column j, but not covered by any column in the partial solution S. Hence, the heuristic information can be defined as:

$$\eta_j = card_j(S)/c_j \quad \forall j \notin S. \tag{9}$$

Lagrangean Cover Costs: In the definition of the cover costs it is possible to use the normalized Lagrangean costs C_j instead of the column costs c_j; the resulting cost values are called Lagrangean cover costs. The heuristic information is then obtained by replacing c_j in Equation 9 by C_j.

Marchiori & Steenbeck Cover Costs: In their Iterated Greedy algorithm [15] Marchiori and Steenbeck propose a variant of the cover costs. Let S be a current (partial) solution and $cov(S)$ the set of rows covered by the columns in S; $cov(j, S)$ is the set of rows covered by column j, but not covered by any column in $S \setminus \{j\}$. Let $c_{min}(i)$ be the minimum cost of all columns that cover row i. The cover value cv of a column j with respect to a (partial) solution S is

$$cv(j, S) = \sum_{i \in cov(j,S)} c_{min}(i). \tag{10}$$

Note that if $cv(j, S) = 0$, then column j is redundant with respect to S. This cover value is used to define the modified cover costs $cov\_val(j, S)$ as

$$cov\_val(j, S) = \begin{cases} \infty, & \text{if } cv(j, S) = 0 \\ c_j/cv(j, S), & \text{otherwise.} \end{cases} \tag{11}$$

The heuristic information is then again given by $\eta_j = 1/cov\_val(j, S)$.

Marchiori & Steenbeck Lagrangean Cover Costs with Normalized Costs: The column costs c_j used to define the cover costs of Marchiori & Steenbeck can be replaced by the normalized Lagrangean costs C_j; the resulting cost values are called Marchiori & Steenbeck Lagrangean cover costs and the heuristic information is obtained like in the previous case.

Lower Bounds: In this case we consider that we have a partial solution available. Let $N_1, N_0 \subseteq N$, where N_1 is the index set of the variables already fixed to 1 and N_0 the index set of the variables fixed to 0, ($N_0 \cap N_1 = \emptyset$; the set N_0 is obtained, e.g., in a preprocessing stage that eliminates redundant columns). Let $N_{free} = N \setminus (N_0 \cup N_1)$, the index set of the variables that are still free, and let $M' = \{i \in M : \sum_{j \in N_1} a_{ij} = 0\}$, the set of rows that are not yet covered. Since our goal is to complete the partial solution to a full one as well as we can, we need to solve a reduced SCP, considering only the variables that are still free.

However, the reduced problem might still be difficult to solve, especially if the number of fixed variables is not very large. Therefore, we will not attempt to solve it directly; instead we will calculate lower bounds on the solution of the reduced problem. The Lagrangean relaxation of the reduced SCP is

$$z_{LR}^{SCP(N_{free})}(u) = \min_{x \in \{0,1\}^{|N_{free}|}} \sum_{j \in N_{free}} \bar{c}_j(u) x_j + \sum_{i \in M'} u_i,$$

where $\bar{c}_j(u) = c_j - \sum_{i \in M'} a_{ij} u_i$ are the Lagrangean costs of the reduced problem. The Lagrangean relaxation of the reduced problem gives a lower bound on the optimal solution of the reduced problem, i.e. $z_{LR}^{SCP(N_{free})}(u) \leq z_{SCP(N_{free})}$, for any $u \geq 0$. Since our aim is to use information that is already available, we calculate the lower bound using the value of u output by the subgradient procedure already run for the full SCP, thus, avoiding to run the subgradient procedure many times. The heuristic information for column j is inversely proportional to the lower bound obtained by tentatively adding this column.

6   Tests and Results

We performed an experimental analysis of the ACO algorithms presented, with emphasis on the study of the influence of the heuristic information on the final performance. We ran each of the ACO algorithms ($\mathcal{MMAS}$, ACS, $\mathcal{MMAS}$-ACS-Hybrid, ANTS) together with each of the seven different versions of obtaining heuristic information. The algorithms were implemented in C++ and compiled with the GNU-C-Compiler using -O2 optimization settings. All the experiments were executed on a Intel 2.4 GHz Xeon processor with 2 GB of RAM.

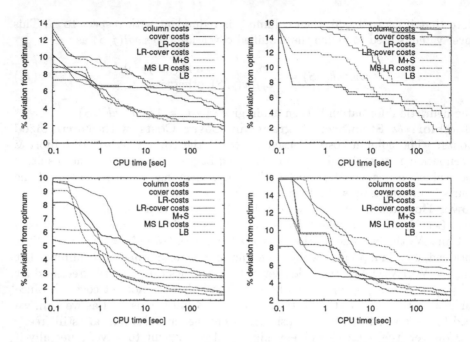

Fig. 2. Development of the average objective function value found by ACS using different types of heuristic information (without local search). Top left for instance SCPNRE2, top right SCPNRF5, bottom left SCPNRG2, and bottom right SCPNRH2

The algorithms were tested on a set of benchmark instances available from ORLIB. These instances are randomly generated instances, divided into 12 subsets identified as SCP4, SCP5, SCP6, SCPA, SCPB, SCPC, SCPD, SCPE, SCPNRE, SCPNRF, SCPNRG, SCPNRH. SCP4 and SCP5 have 10 instances each; all other classes have 5 instances. The instances differ in size and density of the matrices. In addition, we also tested the algorithms on the instances of the FASTER-competition, which stem from a real-world application of the SCP. The instance identifiers are RAIL507, RAIL516, RAIL582, RAIL2536, RAIL2586, RAIL4284, and RAIL4872, where the number in the identifier gives the number of rows; the number of columns in these instances ranges from 47,311 to 1,092,610.

In preliminary experiments we tried to find reasonable parameter settings for each of the algorithm/heuristic information combinations. We considered a set of parameters for each algorithm and then modifying one, while keeping the others fixed. The parameters tested include $m_a \in \{1, 5, 10, 20\}$, $\beta \in \{1, 3, 5\}$, $q_0 \in \{0.9, 0.95, 0.98, 0.99\}$, $\rho \in \{0.2, 0.5, 0.8\}$ (for ACS $\rho \in \{0.1, 0.2, 0.3\}$), $\xi \in \{0.1, 0.2, 0.3\}$, $\zeta \in \{0.0, 0.3, 0.5\}$, and $\vartheta \in \{0.1, 0.2, 0.3\}$ (the parameters apply only to the algorithms where they are actually used). We found that for $m_a = 5$, $\beta = 5$, and $\rho = 0.2$ the algorithms performed best; q_0 is set to 0.98, ξ to 0.2 in the algorithms that use these parameters; for ANTS the parameters are set $\zeta = 0.3$ and $\vartheta = 0.5$. In addition, each of the algorithms was tested with and without r-flip local search. The r-flip local search was stopped after a maximum of 200 iterations. Each algorithm was run 10 times with a computation time limit of 100 seconds for the ORLIB instances; for the large FASTER instances (RAIL2536 and larger) 500 seconds were allowed.

Figure 2 plots the development of the objective function values over computation times for ACS with the seven different types of heuristic information, without local search (without local search, ACS showed best performance with respect to the number of best known solutions found – see also Table 1 – and therefore we chose it here). It can be observed that dynamic heuristic information based on cover costs generally yields the best time–quality tradeoff with the cover costs defined by Marchiori & Steenbeck typically performing best. (In the latter case, it does not really matter whether the column costs or the Lagrangean costs are used; however, this makes a difference when using the standard cover costs). Static heuristic information typically shows quite poor performance with the only exception being the column cost on instance SCPNRH2. Hence, the higher computational cost of the dynamic heuristic information appears to be well spent. A notable exception is the usage of the lower bound information while constructing solutions, which typically results in rather poor performance. However, the ranking of the heuristic information is completely reversed when using local search, as shown in Figure 3. Here, the heuristic information based on the lower bounds (used in the ANTS algorithm) gives by far the best performance. Additionally, it is quite visible that on most of the instances the performance is rather similar, with the notable exception of the version using lower bounds as the heuristic information. Interestingly, on one of the largest instances, SCPNRH2, the heuristic information based on column costs is second best. Given that without local search these latter two kinds of heuristic information showed rather poor performance, one conclusion from these results may be that once a powerful local search such as the r-flip algorithm is available, it may be better that the starting solutions for the local search are not too good to leave the local search some room for improvement.

Table 1 shows for each possible ACO algorithm-heuristic information combination the total number of optimal solutions found and the sum of the ratios between the optimal (or best known) solution divided by the solution quality returned in each trial for the ORLIB-instances (proven optimal solutions are known for instances SCP4 through SCPE). The observations made above are confirmed. The combination of ANTS plus lower bound based heuristic information obtained in all 700 trials the best known or optimal solutions. The average computation times for classes SCP4 through SCPE are on average always less than one CPU second, while the hardest instance took on average 23.3 seconds. We note that this result slightly improves over that of the iterated r-flip heuristic of Yaguira et al. [19], which finds the best known solutions in 694 trials. Table 2 shows the same information for the FASTER instances. Here, occasionally, even the versions without local search perform better than the variants without r-flip. However, in part, this may be due to an inappropriate termination condition for the local search. In any case, on these instances the iterated r-flip algorithm of Yaguira et al. performs significantly better than the ants.

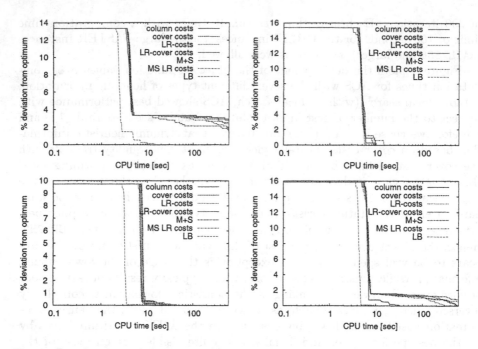

Fig. 3. Development of the average objective function value found by the best combination between ACO algorithm, local search, and heuristic information. Top left for instance SCPNRE2, top right SCPNRF5, bottom left SCPNRG2, and bottom right SCPNRH2

We tested ANTS plus LB heuristic information on a set of 16 benchmark instances taken from [1]; except for one instance, it found the optimal solutions for all instances, in every trial. The average computation times were typically below two seconds and only two instances took slightly longer (five and eight seconds, on average). Compared to other state-of-the-art algorithms than the iterated r-flip, our best performing variant appears to be competitive with the iterated greedy algorithm of Marchiori and Steenbeek [15]. The CFT heuristic of Caprara, Fischetti and Toth appears to have similar performance on the ORLIB instances, but it performs much better on the FASTER instances. Our best variant also outperforms earlier evolutionary algorithms like those presented in [2, 6] and two ACO algorithms based on Ant System that were proposed earlier in the literature [8, 10].

7   Conclusions

In this paper we studied the behavior of various ACO algorithms for the SCP and, in particular, the influence of various ways of defining the heuristic information on the final performance. When applying the ACO algorithms without local search, we noted that (i) for a fixed ACO algorithm, the heuristic information has significant influence on the final performance and that typically the best

Table 1. Summary statistics for the final computational results for 70 randomly generated ORLIB instances across 10 trials per instances. Given is n_{opt}, the number of best known (or optimal) solutions found in the 700 (= 70 · 10) trials and *rel*, the sum of the ratios of best known solution value divided by the solution quality returned by each trial

	c_j		cc		LR-c		LR-cc		MS-cc		MS-ccLR		LB	
	n_{opt}	*rel*	n_{opt}	*rel*	n_{opt}	*rel*	n_{opt}	*rel*	n_{opt}	*rel*	n_{opt}	*rel*	n_{opt}	*rel*
Without local search														
$\mathcal{MMAS}$	309	691.3	492	696.4	301	685.3	253	680.4	504	*696.5*	492	696.4	210	669.9
ACS	311	689.7	511	696.3	328	688.2	331	687.9	*536*	696.4	498	695.6	419	690.9
$\mathcal{MMAS}$-H.	192	684.0	496	695.4	239	680.4	238	681.9	515	695.9	475	694.8	194	675.8
ANTS	224	687.1	390	693.0	272	683.0	242	681.7	399	693.2	379	692.3	191	674.2
With *r*-Flip														
$\mathcal{MMAS}$	689	699.7	685	699.6	680	699.5	683	699.6	681	699.5	681	699.5	690	699.7
ACS	680	699.5	679	699.5	684	699.6	677	699.5	679	699.5	679	699.5	689	699.7
$\mathcal{MMAS}$-H.	688	699.6	684	699.6	683	699.6	678	699.5	686	699.6	683	699.6	687	699.6
ANTS	690	699.7	683	699.6	685	699.6	684	699.6	682	699.6	683	699.6	*700*	*700*

Table 2. Summary statistics for the final computational results for the seven FASTER instances. For details see caption of Table 1

	c_j		cc		LR-c		LR-cc		MS-cc		MS-ccLR		LB	
	n_{opt}	*rel*	n_{opt}	*rel*	n_{opt}	*rel*	n_{opt}	*rel*	n_{opt}	*rel*	n_{opt}	*rel*	n_{opt}	*rel*
Without local search														
$\mathcal{MMAS}$	0	60.4	0	67.0	0	62.6	0	64.4	0	*67.2*	0	65.4	0	58.8
ACS	0	60.9	0	65.7	0	62.3	0	65.4	0	65.9	0	64.2	0	66.4
$\mathcal{MMAS}$-H.	0	60.3	0	66.1	0	62.3	0	65.6	0	66.5	0	64.9	0	64.2
ANTS	0	58.8	0	62.6	0	61.8	0	63.2	0	62.7	0	62.7	0	60.1
With *r*-Flip														
$\mathcal{MMAS}$	3	63.7	1	64.3	0	64.3	0	64.7	0	64.3	0	64.3	0	63.8
ACS	0	63.8	0	66.1	1	64.1	0	65.5	0	66.3	0	64.8	1	66.3
$\mathcal{MMAS}$-H.	0	63.8	0	66.2	0	64.3	1	65.5	0	*66.5*	1	65.4	0	66.1
ANTS	5	63.9	2	63.7	4	63.8	1	64.1	3	63.7	0	63.7	*7*	63.8

performance is reached by using dynamic heuristic information based on cover costs or extensions thereof proposed by Marchiori and Steenbeek; (ii) the overall best performance with respect to the number of best known solutions found was obtained by ACS, which suggests that a rather aggressive construction policy, as implemented by ACS, is needed to achieve good performance. However, if local search is used, the heuristic information based on lower bound information, which performs as one of the 'worst' without local search, gives the overall best results in combination with the ANTS algorithm.

Comparisons with other algorithms have shown that on various instance classes, the best variants we tested reach state-of-the-art performance. However, this is not true for the FASTER benchmark set that stems from real-world SCP applications. However, experiments with optimized parameters for this instance class indicate that there is room for considerable improvement.

Acknowledgments

This work was supported by a European Community Marie Curie Fellowship, contract HPMF-CT-2001.

References

1. E. Balas and M. C. Carrera. A dynamic subgradient-based branch and bound procedure for set covering. *Operations Research*, 44(6):875–890, 1996.
2. J. E. Beasley and P. C. Chu. A genetic algorithm for the set covering problem. *European Journal of Operational Research*, 94(2):392–404, 1996.
3. M. Dorigo and L. M. Gambardella. Ant colony system: A cooperative learning approach to the traveling salesman problem. *IEEE Transactions on Evolutionary Computation*, 1(1):53–66, 1997.
4. M. Dorigo and T. Stützle. The ant colony optimization metaheuristic: Algorithms, applications and advances. In F. Glover and G. Kochenberger, editors, *Handbook of Metaheuristics*, pages 251–285. Kluwer, 2002.
5. M. Dorigo and T. Stützle. *Ant Colony Optimization*. MIT Press, USA, 2004.
6. A. V. Eremeev. A genetic algorithm with a non-binary representation for the set covering problem. In P. Kall and H.-J. Lüthi, editors, *Operations Research Proceedings 1998*, pages 175–181. Springer Verlag, 1999.
7. M. L. Fisher. The Lagrangean relaxation method for solving integer programming problems. *Management Science*, 27(1):1–17, 1981.
8. R. Hadji, M. Rahoual, E. Talbi, and V. Bachelet. Ant colonies for the set covering problem. In M. Dorigo et al., editors, *Proceedings of ANTS2000*, pages 63–66, 2000.
9. E. Housos and T. Elmoth. Automatic optimization of subproblems in scheduling airlines crews. *Interfaces*, 27(5):68–77, 1997.
10. G. Leguizamón and Z. Michalewicz. Ant Systems for subset problems. Unpublished manuscript, 2000.
11. L. Lessing. Ant colony optimization for the set covering problem. Master's thesis, Fachgebiet Intellektik, Fachbereich Informatik, TU Darmstadt, Germany, 2004.
12. H. R. Lourenço, R. Portugal, and J. P. Paixão. Multiobjective metaheuristics for the bus-driver scheduling problem. *Transportation Science*, 35(3):331–343, 2001.
13. H. R. Lourenço and D. Serra. Adaptive search heuristics for the generalized assignment problem. *Mathware & Soft Computing*, 9(2–3):209–234, 2002.
14. V. Maniezzo. Exact and approximate nondeterministic tree-search procedures for the quadratic assignment problem. *INFORMS Journal on Computing*, 11(4):358–369, 1999.
15. E. Marchiori and A. Steenbeek. An evolutionary algorithm for large scale set covering problems with application to airline crew scheduling. In S. Cagnoni et al., editors, *Proc. of EvoWorkshops 2000*, volume 1803 of *LNCS*, pages 367–381. Springer Verlag, 2000.
16. T. Stützle and H. H. Hoos. $\mathcal{MAX}$–$\mathcal{MIN}$ Ant System. *Future Generation Computer Systems*, 16(8):889–914, 2000.
17. T. Stützle and H.H. Hoos. $\mathcal{MAX}$–$\mathcal{MIN}$ Ant System and local search for combinatorial optimization problems. In S. Voss et al., editors, *Meta-Heuristics: Advances and Trends in Local Search Paradigms for Optimization*, pages 137–154. Kluwer, 1999.
18. F. J. Vasko and F. E. Wolf. Optimal selection of ingot sizes via set covering. *Operations Research*, 35:115–121, 1987.
19. M. Yagiura, M. Kishida, and T. Ibaraki. A 3-flip neighborhood local search for the set covering problem. Technical Report #2004-001, Department of Applied Mathematics and Physics, Graduate School of Informatics, Kyoto University, Kyoto, Japan, 2004.

A VLSI Multiplication-and-Add Scheme
Based on Swarm Intelligence Approaches

Danilo Pani and Luigi Raffo

DIEE - Department of Electrical and Electronic Engineering
University of Cagliari
P.zza d'Armi, 09123, Cagliari, Italy
{pani,luigi}@diee.unica.it

Abstract. The basic tasks in Digital Signal Processing systems can be expressed as summation of products. In this paper such operation is analyzed in terms of parallel distributed computation starting from an improvement of the Modified Booth Algorithm able to avoid useless sub-operation in the process of multiplication. We show how a such reformulation can take advantages from the cooperation between cells of a small colony statistically achieving shorter computation time. The interaction among cells, based on simple social rules typical of Swarm Intelligence systems, leads to a full exploitation of cells computational capabilities obtaining a more efficient usage of their computational resources in a so important task. In this paper a preliminary VLSI implementation and theoretical results are presented to show the feasibility of this approach.

1 Introduction

The Swarm Intelligence approach derives from observation of swarms, large sets of simple individuals that can perform complex tasks taking advantages by cooperation among themselves. There are some examples where collaboration among individuals accelerates the execution speed of an activity, and others where single individuals can't perform a too onerous task without the collaboration of other colony members. As a result of cooperation in many kind of micro-scale behavior, macro-scale complex behavior seems to emerge without any central or hierarchical control [4]. The self-organization theory simply tells that *at some level of description* it is possible to explain complex behavior by assuming that it is the result of the activities of simple interacting entities [1].

At the moment the application of such approaches have mainly concerned developing of computer programs to implement optimization and problem solving algorithms like the Travelling Salesman Problem (TSP) and other computational-intensive problems [3, 12]; the particular filed of swarm robotic has conducted to the creation of simple little robots realizing real world operative tasks [5, 7]. Also adaptive routing algorithms based on Swarm Intelligence approaches have been developed [11].

In this scenario we collocate our work, that applies the concepts of Swarm Intelligence and cooperative systems to the VLSI implementation of a typical

M. Dorigo et al. (Eds.): ANTS 2004, LNCS 3172, pp. 13–24, 2004.

Digital Signal Processing (DSP) task: the multiply-and-add operation. In this work we present a preliminary VLSI implementation of a system composed by 8 cells that cooperate by means of simple rules to perform the sum of 8 products. It is possible to conceive a system based on an improvement of the Modified Booth Algorithm where useless sub-operations in the sequential flow are skipped, making it possible the active cooperation among cells, and then allowing the full exploitation of the architectural capabilities. The overall system has been described in Verilog HDL, and synthesized using Synopsys Design Compiler on a CMOS 0.18μm standard cell technology.

It should be clear that the goal of this research is not to develop a new multiplier, but just to explore possible alternatives to the traditional approaches exploiting the Swarm Intelligence approach.

Section 2 introduces the multiply-and-add algorithm. Section 3 explores data complexity, a central issue in this work. Section 4 illustrates the adopted collaborative scheme. Section 5 shows the proposed architecture, whose implementation and performances evaluation are exposed in section 6. Conclusions are presented in section 7.

2   The Multiply-and-Add Algorithm

A broad range of DSP algorithms, can be described as a set of elementary summations of products. Typical examples can be found in digital filters, in particular in Finite Impulse Response (FIR) filters where the relationship between input and output takes the form (1):

$$y(n) = \sum_{k=0}^{M} x(n-k)b_k \qquad (1)$$

The impact of a single multiplier in terms of silicon area and critical path length is heavy, hence its hardware implementation has to cope with different cost/performance factors. With respect to their implementation, multipliers can be divided into sequential (multi-step) and combinatorial (single-step) ones. Sequential multipliers can be optimized to find the best solution for area and performance.

2.1   The Modified Booth Algorithm

There are many approaches to sequential multiplication implementation, and one of these is the Modified Booth Algorithm (MBA). MBA speeds up the multiplication reducing the number of partial products generated. The traditional Booth algorithm consumes N steps with N-bit operands, because examines one multiplier bit at a time, whereas the MBA examines two or more multiplier bits at a time [8]. Hence this approach reduces the latency of the operation. In particular we focus on the MBA with multiplier partitioning into 3-bit groups with one bit overlap [6].

Take a binary 2's complement number α, and let we say α_i its bits, with $i = 0, 1, ..., \omega_\alpha\text{-}1$. The decimal value of α obviously is:

$$\alpha = -\alpha_{\omega_\alpha-1} \cdot 2^{\omega_\alpha-1} + \sum_{i=0}^{\omega_\alpha-2} \alpha_i 2^i \tag{2}$$

Now let us consider α composed by triplets β of bits, with one bit overlap between adjacent triplets:

$$\beta_i = \{\alpha_{2i+1}, \alpha_{2i}, \alpha_{2i-1}\}, \qquad 0 < i < \frac{\omega_\alpha}{2} - 1 \tag{3}$$

where the overlap with the β_{i+1} triplet is on the bit α_{2i+1}. Every triplet in the number α has an associated weight, and each bit of the triplet has a relative weight. The triplet weight is the original weight in positional binary notation of its bit α_{2i}, whereas the bits inside the triplet have weights $w = \{-2, 1, 1\}$. Now the decimal interpretation of α can be written like in (4), taking $\alpha_{-1} = 0$ (such a bit is nonexistent in the number, and is only appended at the end of it).

$$\alpha = \sum_{i=0}^{\frac{\omega_\alpha}{2}-1} (-2\alpha_{2i+1} + \alpha_{2i} + \alpha_{2i-1}) \cdot 2^{2i} \tag{4}$$

Observe that the term between brackets can take only the values listed below, functions of $\{\alpha_{2i+1}, \alpha_{2i}, \alpha_{2i-1}\}$:

 0   for $\{0,0,0\}$ or $\{1,1,1\}$
 1   for $\{0,1,0\}$ or $\{0,0,1\}$
-1   for $\{1,1,0\}$ or $\{1,0,1\}$
 2   for $\{0,1,1\}$
-2   for $\{1,0,0\}$

If we coded using the above coding scheme the multiplier α in a multiplication between α and another number γ, the result may be obtained calculating only $\omega_\alpha/2$ partial products and then summating them. In a sequential scheme the summation is performed step-by-step in form of accumulation, and take $\omega_\alpha/2$ cycles. The result is:

$$\gamma \times \alpha = \sum_{i=0}^{\frac{\omega_\alpha}{2}-1} (-2\alpha_{2i+1} + \alpha_{2i} + \alpha_{2i-1}) \cdot 2^{2i} \cdot \gamma \tag{5}$$

Because all factors in the summation, with the only exception of the number γ, are powers of two, the partial products are simple left-shifted replicas of γ, and the shift amount corresponds to the exponent of two. Observing that the triplet may assume positive or negative values, the final adder has to be able to perform additions and subtractions as well.

3   Data Complexity

Typical variables that are taken into account for VLSI system designs are area, frequency (delay), latency and power consumption. Whereas silicon area is a relevant cost factor, the other variables are performance factors: frequency and latency are responsible for time performances, power consumption is very important for feasibility in embedded systems and portable devices.

In the design of VLSI systems, input data are considered only for their width, which affects the system dimension and therefore the silicon area and the operative frequency too. However in some cases input data value is an interesting design variable: among them we can comprise those systems that speculate on the distance between subsequent data or coefficients (i.e. in digital filters), adopting differential approaches (DCM [10], DCIM [2], DECOR [9], etc.). Nevertheless they exploit characteristics proper of a *data sequence* and not of *single data inputs*.

The proposed approach is completely different, assuming that in some cases the value of *each* bit forming the data word is relevant for computational latency reduction, that is the number of clock cycles required to accomplish the computation.

3.1   Data Complexity in the Modified Booth Algorithm

The data carry with itself natively the complexity concept: if an operation is decomposable into elementary steps whose number varies depending on data complexity, a Swarm Intelligence approach can conduce to a faster achievement of the result exploiting collective effort of colony members in the specific task.

Let us consider again the MBA described above. If we imagine a 2's complement number (i.e. a small number) made up of almost all $\{0,0,0\}$ and $\{1,1,1\}$ triplets (in the following named *null triplets*), implementing the system we find out that every time a such triplet is encountered the system wastes a cycle making nothing useful to obtain the result. It follows that the effective number of steps needed could be equal to the number of non-*null triplets* composing the multiplier. We define data complexity in the context of MBA, limitedly to the multiplier, accordingly.

Definition 1. *Given a ω_α-bit binary number α, described like an ordered set of triplets β_i, we define complexity of α the number $\mathbf{C}$ of triplets β_i that differs from null triplets $\{0,0,0\}$ and $\{1,1,1\}$.*

It follows that for a 32-bit wide binary number (which can be decomposed into 16 triplets) the complexity is $0 \leq C \leq 16$. An immediate outcome is that if the MBA could perform an overall number of steps equal to its multiplicand complexity C, the system should operate like a traditional system in the worst case, but statistically it could obtain a percentage clock cycles reduction exploiting data complexity fluctuations, which obviously are random.

3.2   Complexity Data Analysis

Figure 1 shows the histogram of data *complexity* for a dataset of 64,000 absolutely random numbers: it should be noted a shape similar (but a little skewed) to a Gaussian shape centered in 12 (the average value is 12.01). Almost the same

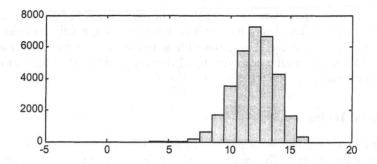

Fig. 1. Histogram of data complexity for 64,000 random numbers

shape can be obtained with arbitrarily large datasets. This analysis implies an immediate consequence: if the VLSI multiplier is able to skip *null triplets* (without wasting cycles), on the average it should implies 12 rather than 16 cycles to complete the operation, with a percentage reduction of 25% in latency.

4 Collaborative Scheme

If we restrict our analysis to a multiplication algorithm, a collaborative approach should be applied at a lower level. In fact, in a swarm system the exploitation of colony elements cooperation is oriented to perform a task valuable for the overall community, not for the single individual. Reminding that we were talking about multiply-and-add task, we can conceive that every *cell* in the colony is a kind of multiplier that potentially can perform a simple multiplication on its own. The algorithm we want to execute is this:

$$y = \sum_{k=0}^{M-1} \gamma_k \alpha_k = \sum_{k=0}^{M-1} \sum_{i=0}^{\frac{\omega_\alpha}{2}-1} (-2\alpha_{2i+1} + \alpha_{2i} + \alpha_{2i-1}) \cdot 2^{2i} \cdot \gamma_k \qquad (6)$$

In other words, we want to execute a sum of products, which is the sum of a sum of partial products. It is obvious that which cell register you accumulate in is not relevant to achieve the result. If we have a small colony of cell (i.e. 8) that are able to perform multiplications using the algorithm above, and we allow that when a cell terminate its own operation could help another cell accumulating partial products on its register, and at the end the final summation could take place, we have created a cooperative multiply-and-add. In this manner we obtain a *dense* computation.

Consider loading into an 8-cells system 8 data with complexities 4, 8, 6, 4, 1, 5, 1, and 0. Without cooperation among cells, even if our MBA multipliers skip *null triplets*, the overall latency equals the worst cell latency, that is 8. However the average complexity is 3.625, not 8, and one cell is never used, two cells are used only for one step, and so on. We are wasting computational resources and time too.

If we define *computational density* the sum over all cells of activity instants with respect to the sum of activity and inactivity instants, we obtain a density of 45% for the example above. A collaborative approach should potentially achieve a latency of 4 steps, with a computational density of 90% (that implies a good usage of hardware).

4.1   Social Rules

Swarm Intelligence is characterized by simple social rules that govern the interactions among individuals. The swarm's goal is to execute all the simple operations (obtainable after the triplets decomposition of every multiplier) as soon as possible, each cell starting with its own input data. If data complexities for all cells are almost the same, and no cooperation can take place, then every cell perform a multiply operation, and at the end all cells perform the final summation. If data complexity is different for some cells they can cooperate by means of the following simple rules:

- if a cell is working and its residual complexity differs from zero, than it asks for collaboration spreading a help request the request will be served during the next cycle if this cell is the most priority cell requesting. We'll call a such cell a H cell;
- if a cell residual complexity goes to zero, that cell indicates its availability to help other cells of the swarm, and it becomes bus slave if it is the most priority cell available. We'll call such cell a D cell;
- if a H cell receives an availability signal, it propagates its multiplicand and shift factor to the D cell (helper), and it decreases its own residual complexity monitor;
- if a D cell receives an help request, it processes the incoming data storing the result in its own register;
- when all cells declare to have finished the elaboration, they organize themselves to perform the last summation, at the end of which they indicate the end of elaboration (the elaboration time is complexity-dependent).

4.2   Cell Number Effect over Performances

A typical aspect of Swarm Intelligence based systems is that performances increase with the number of interacting cells. Starting from a hypothetical architectural partitioning between control and datapath, and assuming the control faster than datapath, interactions should take place during data elaboration. Due to hardware operative frequency limits, and preferring for this preliminary study a synchronous implementation, it arises the issue about the number of allowed interactions among cells. We have plotted a graph comparing a non-collaborative approach with a collaborative one, analyzing performances in terms of cycles speed-up with respect to the same implementation of MBA without skipping and cooperation, for different number of cooperations allowed during an algorithm step. Figure 2 depicts the curves.

It was predictable that increasing the number of allowed collaboration for a single algorithm step gets an increment in performances. The non-cooperative approach exhibits only a 10% in cycle speed-up with respect to a non-skipping implementation, whereas the cooperative one obtain a cycle speed-up near to 40%. In the following we'll show how these curves have influenced our architectural choices.

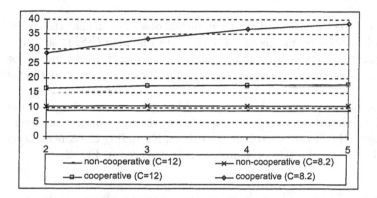

Fig. 2. Performances comparison (percentage clock cycles speed-up) with respect to a cycle-equivalent MBA without skipping and cooperation. The 2 pairs of lines have been obtained for 2 different datasets with average complexity 8.2 (random number generated starting from random complexity) and 12.0 (absolutely random number, generating the complexity distribution depicted in Figure 1). On the x-axis, the number of allowed cooperations during one algorithm step

5   The Swarm Multiply-and-Add

In this section we present the architectural strategies for the implementation, starting from the above considerations.

5.1   Architectural Definition

The architecture of the proposed Swarm Multiply-and-Add (SM&A) consists of 8 undifferentiated cells that form a colony of interacting *hardware agents* able to perform an operation of multiply-and-add based on an improvement of the sequential MBA. We state first that in this preliminary work some architectural choices have been adopted to maintain a position closest to swarm systems even if some of these choices are not much efficient in terms of area/performance for the implementation.

Even if the best result with 8 cells seems achievable allowing 5 time-slots for interactions among cells (between one elaboration step and the successive, see Fig. 2), performing a lot of synthesis we have found that the optimal solution considering the needed timing constraints is 3. To work with reasonable number complexities the input data size has been chosen equal to 32 bits, whereas the output size (full precision) is 67 bits (64 plus 3 guard bits for cells results accumulation). The communication network is bus-based and then allows interactions between any couple of cells. For the final summation, cells define an original pipelined structure requiring only one additional adder.

5.2   Cell Architecture

The cell architecture is based on the empirical observation that the control structure, implemented by means of a Finite State Machine (FSM) should be much faster than the datapath structure, allowing cooperation during an operative step of the algorithm, anything using a single clock structure and a synchronous traditional approach (to allow more useful comparisons).

Cell structure is divisible in layers. Mainly we can distinguish between 4 layers: the elaboration layer, the source layer, the interconnection layer and the control layer, the latest controlling the others. Excluding the control layer, which is a FSM, a block diagram of a single cell is depicted in Fig. 3.

Elaboration Layer. The elaboration layer accommodates the datapath which produces the partial result, taking as input the multiplicand and the shift amount. This operation requires 4 clock cycles (an algorithm step corresponds to 4 clock cycles), during which the cell can't elaborate other data. This layer is controlled from the control layer in terms of control points and register enables. The flexibility of this layer allows to reuse structural resources by means of run-time reconfiguration of the datapath, decided autonomously by each cell without a centralized controller.

Source Layer. The source layer of a cell implements the data source for that cell and for other cells during computation. If we intend cells like agents consuming spread resources, the source layer is the resources container. It consists of triplets decoding systems that allow *null triplets* skipping during elaboration. The control layer of its own cell controls it, even if another cell is sourcing from it (the controller of the cell works as an intermediary). The layer exports towards the other layers only multiplicand, shift amount for the actual triplet, and a signal that indicate data unavailable (when the number of non-*null triplet* not yet processed reaches zero). In a swarm system, this layer represents the environment.

Interconnection Layer. This layer allows communications among cells, and it passes data from source layer towards elaboration layer. For a total decentralized approach, even if this choice imply duplication of logic, the bus arbitrage is fully decentralized, so by means of this layer every cell knows if it can or cannot access the bus (like a bus master or slave). This layer operates without the explicit control of the control layer.

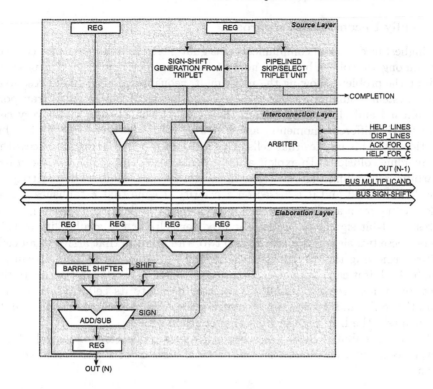

Fig. 3. Block diagram of a single cell. The control layer is not depicted because it is a simple FSM and the the excessive number of control signal may cause less readability

Control Layer. The cell controller has been implemented by means of a FSM, optimized to obtain the highest performances allowable. This is the core element of the cell, and operates passing through 4 modes: load mode, internal elaboration mode, helping mode and accumulation mode. The controller operates complying with the social rules exposed above.

- During load mode the controller govern only the source layer to load the operands and initialize the pipeline stages.
- During the internal elaboration mode, the cell executes its own single step of operation, during which other cells can help it. In this condition the controller govern the elaboration layer, and the source layer if one or more cells is available for helping.
- During the helping mode a cell that consumed its data executes partial products of other cells, accumulating the result into its own output register. In this mode the controller governs only the elaboration layer.
- Finally, in accumulation mode, the controller sets up its elaboration layer to perform the final accumulation. This operation is conducted using an algorithm that allows the exploitation of a pipelined binary tree structure of the adders/subtracters included in the elaboration layer.

5.3   Fully Decentralized Approach

The highest flexibility in communication is guaranteed by point-to-point connections among the colony, but this should cause a connectivity issue. It is possible to limit the problem allowing the non-simultaneous usage of a shared resource. In our system, communications take place by means of buses which transport multiplicand and shift amount from the H cell to the D cell. Since every cell could be (in different moments) an H or a D cell, cells access to the bus bidirectionally. This implies that cells need to adopt a bus arbitrage to access the bus avoiding collisions. To exploit a fully decentralized approach, every cell has an arbiter controlling the bus access in master or slave mode. The arbitrage is fully decentralized and it is based on a priority scheme that establishes a cells rank starting from a cell-ID assigned to cells at synthesis time. Every cell propagates two 1-bit signals to display the need for help, or the availability to help others: these two signals are putted into two wire bundles that arrive to all cells arbiters. Knowing the cell-ID assigned at synthesis time, every arbiter can resolve by itself if it can or cannot access to data buses. If a cell can't access to the buses for an help request, simply the arbiter masks to its cell any availability from other cells, and if a cell is not enough priority to help another cell, than its arbiter masks the help request from the requesting cell.

A such distributed arbitrage scheme involves replication of logical blocks in every cell, but this choice ensures the total absence of a centralized control system.

6   Implementation and Performances Evaluation

The overall system composed by 8 cells, with 32-bit input data and 67-bit output result, has been synthesized using Synopsys Design Compiler v2001.08 on standard cell CMOS technology (UMC18μ1P6M). Note that some architectural modules should take advantages by custom implementations both in terms of area and delay constraints.

To adapt a theoretically fully asynchronous system to a synchronous implementation we have chosen multicycle solutions for datapath implementation, preserving the synchrony property for implementation comparisons. In this first version we set at 3 the number of interactions allowable between two algorithm steps. This choice reduced the complexity of the system but its performances too.

In Tab. 1 we compare the result of the simulations for different architectures. In the first row our approach with a cooperative effort 3 (three interaction per algorithm step) is presented. The effect of cooperation is evidenced by the comparison with row 2 where the cooperative effort is reduced to 0. The improvement is more relevant when the dataset is characterized by a reduced complexity (C=8). Row 3 and 4 report standard sequential implementation (Booth and MBA), we can see how performances are independent from the complexity of the data and lower than the SM&A's ones. Areas (i.e. hardware complexity) are in the same order of magnitude. The last two rows report two top (combinatorial and order

Table 1. Comparison of performances for different architectures

Architecture	Area eq.gate	Frequenza [MHz]	C=12		C=8	
			Latency cycles	Elab. Time [ps]	Latency cycles	Elab. Time [ps]
SM&A coop3	36985.56	1428.57	65.96	46.18	53.9	37.73
SM&A coop0	36985.56	1428.57	70.13	49.09	71.29	49.9
Booth	32817.16	641.03	38	59.28	38	59.28
MBA	35731.65	507.61	22	43.34	22	43.34
Combinatorial	64526.47	84.75	1	11.8	1	11.8
DW_2_stages	81322.16	510.2	2	3.92	2	3.92

2 pipelined) implementations, in which no area constraint is considered and only the throughput is considered.

7 Conclusions

In this paper we have presented an efficient sequential multiplication-and-add architecture in which the computation is assigned to a number of simple units that recursively (on each operation) and parallely (on different data) perform the task. The control of such units is distributed, and the overall system expose a self-organizing behavior. We have shown how introducing cooperation rules typical of Swarm Intelligence the performances of the system can be improved, in particular when complexity (according to a proper definition of it) of the data stream is limited. It should be noted that reduced complexity of dataset is typical of stream processing.

In our SM&A interactions among cells take place by means of buses and signalling lines. These interactions are inherently local since the system is composed of only 8 cells, so all the interactions take place between neighbours. To exploit this *locality* (one important aspect of Swarm Intelligence) to reduce communication traffic, SM&A should grow replicating the presented structure. In this case, long distances interactions should take place moving inactive cells from low complexity zones towards high complexity zones, and very active cells from high complexity zones towards low complexity zones, in practice introducing *mobility*, another important aspect of Swarm Intelligence. In larger system the mechanism of mobility should be carefully considered with respect to the resources needed.

Also it should be noted that the Swarm Intelligence approach silently and automatically introduces the property of fault tolerance: in fact if a cell "die" the others can finish the operation in progress without loss of precision, even spending more time. Once again, with respect to fault tolerance, swarm systems exhibits an extraordinary robustness.

A SM&A has no limited application field (limitedly to the cases where it is useful a sequential multiplier, with its consequent limited speed), because the multiplication-and-add is an operation common to many DSP and non-DSP algorithms. Other possible application fields, different from DSP systems, are big-number crunching and vectorial operations.

The core of the system described has been implemented and synthesized on standard CMOS technology. Results of simulations are comparable to the state of the art implementations, showing the potentiality of the approach.

References

1. Bonabeau, E., Dorigo, M., Theraulaz, G. : Swarm Intelligence, From Natural To Artificial Systems. Oxford University Press (1999)
2. Chang, T. S., Chu, Y. H., Jen, C. W.: Low Power FIR Filter Realization with Differential Coefficients and Inputs. IEEE Trans. On Circuits and Systems - II **47** (2000) No. 2, 137–145
3. Dorigo, M., Maniezzo, V., Colorni, A.: The Ant System: Optimization by a Colony of Cooperating Agents. IEEE Trans. on Systems, Men and Cybernetics - B **26** (1996) 29–41
4. Kawamura, H., Yamamoto, M.: Multiple Ant Colonies Algorithm Based on Colony Level Interactions. IEICE Trans. Fundamentals **E83-A** (2000) No.2, 371–379
5. Kube, C. R.: Collective Robotics: From Local Perception to Global Action. Ph.D. Thesis, University of Alberta (1997)
6. Lee, H.R., Jen, C.W., Liu, C.M.: A New Hardware-Efficient Architecture for Programmable FIR Filters. IEEE Trans. on Circuits and Systems - II, **43** (1996), No. 9, 637–644
7. Martinoli, A. Yamamoto, M., Mondada, F.: On the Modelling of Bio-Inspired Collective Experiments with Real Robots. Proc. of the Fourth European Conference on Artificial Life (ECAL 97) Brighton, UK (1997)
8. Parhi, K.K.: VLSI Digital Signal Processing Systems - Design and Implementation. Wiley-Interscience, (1999), 484–489
9. Ramprasad, S., Shanbhag, N. R., Hajj, I. N.: Decorrelating (DECOR) Transformations for Low-Power Digital Filters. IEEE Trans. On Circuits and Systems - II **46** (1999), No. 6, 776–788
10. Sankarayya, N., Roy, K., Bhattacharya, D.: Algorithm for Low Power and High Speed FIR Filter Realization Using Differential Coefficients. IEEE Trans. on Circuits and Systems - II **44** (1997) 488–497
11. Schoonderwoerd, R., Holland, O., Bruten, J., Rothkrantz, L.: Ant-Based Load Balancing in Telecommunications Networks. Adaptive Behavior **5** (1996) 169–207
12. Taillard, E.: Ant Systems. Technical Report IDSIA-05-99, IDSIA, Lugano (1999)

ACO for Continuous
and Mixed-Variable Optimization

Krzysztof Socha

IRIDIA – Université Libre de Bruxelles – Brussels, Belgium
ksocha@ulb.ac.be

Abstract. This paper presents how the Ant Colony Optimization (ACO) metaheuristic can be extended to continuous search domains and applied to both continuous and mixed discrete-continuous optimization problems. The paper describes the general underlying idea, enumerates some possible design choices, presents a first implementation, and provides some preliminary results obtained on well-known benchmark problems. The proposed method is compared to other ant, as well as non-ant methods for continuous optimization.

1 Introduction

Optimization algorithms inspired by the ants' foraging behavior proposed by Dorigo in his PhD thesis in 1992 have been initially used for solving combinatorial optimization problems. They have been eventually formalized into the framework of the Ant Colony Optimization (ACO) metaheuristic [7]. ACO has proven to be an efficient and versatile tool for solving various combinatorial optimization problems. Several versions of ACO have been proposed, but they all follow the same basic ideas:

- search performed by a population of *individuals*, i.e. simple independent agents,
- incremental construction of solutions,
- probabilistic choice of solution components based on stigmergic information,
- no direct communication between the individuals.

Since the emergence of ant algorithms as an optimization tool, some attempts were also made to use them for tackling continuous optimization problems. However, at the first sight, applying the ACO metaheuristic to continuous domain was not straightforward. Hence, the methods proposed often drew inspiration from ACO, but did not follow exactly the same methodology.

Up to now, only a few ant approaches for continuous optimization have been proposed in the literature. The first method – called Continuous ACO (CACO) – was proposed by Bilchev and Parmee [2] in 1995, and also later used by some others [17, 12]. Other methods include the API algorithm by Monmarché [13], and Continuous Interacting Ant Colony (CIAC), proposed by Dréo and Siarry [9, 8].

M. Dorigo et al. (Eds.): ANTS 2004, LNCS 3172, pp. 25–36, 2004.

Although both CACO and CIAC claim to draw inspiration from the ACO metaheuristic, they do not follow it closely. All the algorithms add some additional mechanisms (e.g. direct communication – CIAC and API – or nest – CACO) that do not exist in regular ACO. They also disregard some other mechanisms that are otherwise characteristic of ACO (e.g. stygmergy – API – or incremental construction of solutions – all of them). CACO and CIAC are dedicated strictly to continuous optimization, while API may also be used for discrete problems.

Contrary to those earlier approaches, this paper presents a way to extend a generic ACO to continuous domains without the need to make any major conceptual changes. Such extended ACO, due to its closeness to the original formulation of ACO, provides an additional advantage – the possibility of tackling mixed discrete-continuous optimization problems. In other words, with ACO it should be now possible to consider problems where some variables are discrete and others are continuous.

The reminder of the paper is organized as follows. Section 2 presents the idea and enumerates the possible design choices. Section 3 provides a short discussion of the proposed solution with regard to other methods for continuous and mixed-variable optimization. Section 4 presents the choices made for the first implementation and compares some initial results with those obtained by competing methods. Finally, Sec. 5 presents the conclusions and future work plans.

2   ACO Extended to Continuous Domain

When ACO is used for combinatorial optimization problems, ants construct solutions incrementally. Each ant starts with an empty solution S^0 and at each construction step i a component of the solution is added. The definition of a *solution component* depends on the problem tackled. In case of the popular example of Traveling Salesman Problem (TSP), a component of the solution is a city that is added to a tour. For other problems the solution components may be defined differently.

In order to choose, which of the available solution components C^i should be added to the current partial solution S^i, a probabilistic choice is made. This decision is usually influenced by amount of *pheromone* τ associated with available choices, and by heuristic information about the problem. Without the loss of generality, we focus on a case when no heuristic information is used. The probability of choosing a solution component $c \in C^i$ at step i in iteration t, assuming that the partial solution constructed so far is S^i, is a normalized pheromone value associated with this component:

$$p_{S^i c}(t) = \frac{\tau_{S^i c}(t)}{\sum_{j \in C^i} \tau_{S^i j}(t)} \tag{1}$$

Hence, in case of combinatorial optimization problems, at each construction step the ants make a probabilistic decision according to some discrete probability distribution.

Algorithm 1 Ant Colony Optimization extended to continuous domain

---

input: An objective function $\mathbb{R} \ni f(x) : x \in \mathbb{R}^n$
$\tau^i \leftarrow$ initial probability distribution $P^i(x^i)$, $i \in \{1..n\}$
while (stop condition not met) **do**
  {iterate through all m ants}
  for $a = 1$ **to** m **do**
    {construction process of ant a}
    $s^0 \leftarrow \emptyset$
    for $i = 1$ **to** n **do**
      choose value x^i randomly according to probability distribution $P^i(x^i)$
      $s^i \leftarrow s^{i-1} \cup \{x^i\}$
    end for
  end for
  $s_{\text{I best}} \leftarrow$ iteration best solution
  $s_{\text{G best}} \leftarrow$ best of $s_{\text{I best}}$ and previous global best $s_{\text{G best}}$
  $\tau \leftarrow$ pheromone updated based on one or more solutions found
end while
output: Best solution found $s_{\text{G best}}$.

In case of continuous optimization problems, the domain changes from discrete to continuous. The logical adaptation would be to also move from using the discrete probability distribution to a continuous one – the Probability Density Function (PDF). Instead of choosing at step i a component $c \in C^i$, the ants would generate a random number according to a certain PDF $P^i(x^i)$.

More formally, the proposed ACO algorithm extended to continuous domain is presented in Alg. 1. Clearly, the general structure does not differ from a generic ACO, but it is extended to handle continuous variables. The probability distributions $P^i(x^i)$ may be either discrete or continuous.

2.1   Probability Density Function (PDF)

One of the most popular functions that is used as PDF for estimating distributions is the normal (or gaussian) function:

$$g(x, \mu, \sigma) = \frac{1}{\sigma\sqrt{2\pi}} e^{-\frac{(x-\mu)^2}{2\sigma^2}} \tag{2}$$

The normal PDF has some clear advantages, such as a fairly easy way of generating random numbers according to it, but it also has some disadvantages. A single normal PDF is not able to describe a situation where two disjoint areas of the search space are promising, as it only has one maximum.

Due to this fact, we used a distribution based on the normal PDF, but slightly enhanced – a mixture of normal kernels. Similar constructs have been used before [3], but not exactly in the same way. We define it as a weighted sum of several normal PDFs, and denote it as G:

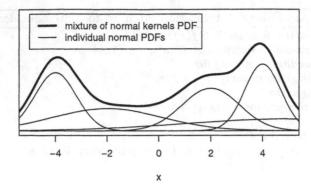

Fig. 1. Example of five normal PDFs and their sum – the resulting mixture of normal kernels – on the interval $(-5, 5)$

$$P(x) = G(x, \boldsymbol{\omega}, \boldsymbol{\mu}, \boldsymbol{\sigma}) = \sum_{j=1}^{k} \omega_j \cdot g(x, \mu_j, \sigma_j) \qquad (3)$$

where $\boldsymbol{\omega}$ is the vector of weights associated with the components of the mixture, $\boldsymbol{\mu}$ is the vector of means, and $\boldsymbol{\sigma}$ is the vector of standard deviations. The dimensions of all those vectors are equal to the number of normal PDFs constituting the mixture. For convenience we will use parameter k to describe this number $\dim \boldsymbol{\omega} = \dim \boldsymbol{\mu} = \dim \boldsymbol{\sigma} = k$.

Such a distribution allows for reasonably easy generation of random numbers accoding to it, and yet it provides a much increased flexibility in the possible shape. An example of how such a mixture may look like is presented in Fig. 1.

For the remaining part of this paper we will use the notation $P^i(x^i)$ to indicate the i-th mixture of normal kernel PDFs: $P^i(x^i) = G(x^i, \boldsymbol{\omega}^i, \boldsymbol{\mu}^i, \boldsymbol{\sigma}^i)$, and $P_j^i(x^i)$ to indicate the j-th single normal PDF: $P_j^i(x^i) = g(x^i, \mu_j^i, \sigma_j^i)$ being part of the i-th mixture. Therefore:

$$P^i(x^i) = \sum_{j=1}^{k^i} \omega_j^i \cdot P_j^i(x^i) \qquad (4)$$

Additionally, we will use simply the term *mixture* to refer to a mixture of normal kernel PDFs.

2.2 Solution Construction

The construction of solutions in ACO for continuous domain is done in principle in the same manner as in the case of regular ACO. At each step i, each of the ants chooses a component of the solution based on the probability distribution. If the component is to be chosen based on discrete probability distribution, it is done exactly the same way as in the case of regular ACO.

In case of continuous domain, a component is a value x^i of a single dimension of the solution $x \in \mathbb{R}^n$. Rather than using discrete distribution, the continuous PDF $P^i(x^i)$ is used as defined in (3). At step i, an ant generates a random number according to the i-th mixture $P^i(x^i)$. This is accomplished in two stages. First, an ant chooses probabilistically a single normal PDF $P_j^i(x)$ from the mixture $P^i(x^i)$ as in (4), with probability p_j^i proportional to ω_j^i:

$$p_j^i = \frac{\omega_j^i}{\sum_{l=1}^{k^i} \omega_l^i} \tag{5}$$

Following that, an ant generates a random number according to the chosen $P_j^i(x^i)$. This may be done using a random number generator that is able to generate random numbers according to a parameterized normal distribution, or using a uniform random generator and (for instance) the Box-Muller method [4].

2.3 Pheromone Maintenance

Each mixture $P^i(x^i)$ representing the i-th pheromone distribution is described by a triplet of vectors $(\omega^i, \mu^i, \sigma^i)$ of equal dimension k^i. The larger the k^i, the more complex distributions may be described by this PDF. In particular, such a PDF may have at the most k^i maxima.

Initialization. Initially the pheromone distribution $P^i(x^i)$ must consist of at least one (it could be more) normal PDF $P_j^i(x^i)$. A reasonable choice must be made. If no prior information is given about the problem, it is reasonable to start with a rather uniform distribution over the search domain (a, b). This is not quite achievable using the mixture of normal kernel PDFs, but reasonable examples include a single normal PDF[1] $P_1^i(x^i)$:

$$P^i(x^i) = P_1^i(x^i) = g\left(x^i, \frac{a+b}{2}, \frac{b-a}{2}\right) \tag{6}$$

or a set of k^i normal distributions $P_j^i(x^i)$ with uniformly distributed means:

$$P^i(x^i) = \sum_{j=1}^{k^i} \frac{1}{k^i} \cdot g\left(x^i, a + (2j-1)\frac{b-a}{2k^i}, \frac{b-a}{2k^i}\right) \tag{7}$$

In some cases, in particular when the algorithm uses frequent restarts, the distribution $P^i(x^i)$ may be initialized with a randomized version of the one presented in (7). In such a case, one or more parameters of the distribution may be randomized: (i) standard deviations, (ii) means, (iii) weights, or (iv) number k^i of the normal PDFs.

---

[1] Note that it is necessary in such a case that $\omega_1^i = 1$.

Update. Pheromone update is a process of modifying the probability distribution used by the ants during the construction process, so that it can guide the ants towards better solutions. This process traditionally consists of two actions: (i) reinforcing the probability of the choices that lead to good solutions – *positive update*, and (ii) decreasing probability of other choices (i.e. *forgetting* bad solutions) – *negative update*.

In ACO for continuous domain, the positive update may be accomplished by incorporating in the mixture $P^i(x^i)$, an additional normal PDF $P_j^i(x^i)$ for each solution used for the update. The mean μ_j^i of this new distribution should be equal to the value of the solution component x^i used for updating the probability distribution $P^i(x^i)$. The values of ω_j^i and σ_j^i may be chosen based on the state of the search, or the quality of the solution used for the update. The number of normal PDFs k^i constituting the mixture $P^i(x^i)$ should be appropriately increased: $k^i \leftarrow k^i + o$, where o is the number of solutions used for the update. For each dimension $i \in \{1..n\}$ the update is given by:

$$\begin{cases} P^i(x^i) \leftarrow P^i(x^i) + \sum_{j=1}^{o} P_j^i(x^i) \\ k^i \leftarrow k^i + o \end{cases} \tag{8}$$

Negative update is usually done in traditional ACO through pheromone evaporation [7]. However, also other alternative solutions have been presented in the literature. For instance, in Population-Based ACO [10]–the pheromone is added or removed from the pheromone matrix, as the individual solutions enter or are being removed from the population.

ACO for continuous domain allows for significant flexibility in the way the negative update is accomplished. In the following paragraphs we describe three example methods of achieving it. Others may also be possible.

One of the most obvious methods for negative update is the opposite of the positive update method presented in (8). Just as any new normal PDF $P_j^i(x^i)$ may be added to the mixture $P^i(x^i)$, in the same manner any existing one may be removed. More formally the process of removing a set of $O \subset \{1..k^i\}$ normal PDFs for each dimension $i \in \{1..n\}$ may be presented as follows:

$$\begin{cases} P^i(x^i) \leftarrow P^i(x^i) - \sum_{j \in O} P_j^i(x^i) \\ k^i \leftarrow k^i - |O| \end{cases} \tag{9}$$

The second method for negative update is inspired directly by the pheromone evaporation in case of generic ACO. In case of a mixture of normal kernels PDF $P^i(x^i)$, it is possible to evaporate the vector of weights ω^i, for each dimension $i \in \{1..n\}$ and each normal PDF $j \in \{1..k^i\}$ according to evaporation rate ρ:

$$\omega_j^i \leftarrow (1 - \rho) \cdot \omega_j^i \tag{10}$$

The third method that may be considered, exploits the properties of the probability distribution used – the mixture of normal kernel PDFs. Assuming that the positive update is done as presented in (8), each new normal PDF added to the mixture is based on some point that identifies a promising area of the search space. The idea of negative update is that as time passes, the old promising areas may become less promising. One of the ways of expressing this fact is, in the case of the normal PDF, the increase of the standard deviation σ. As this does not mean that the particular normal PDF will be less probable, but rather that the numbers generated according to this distribution will have more spread, we call this type of negative update *dissolving* instead of evaporation. The idea is then to increase the standard deviation σ_j^i of normal PDF $P_j^i(x^i)$ for each dimension $i \in \{1..n\}$ and for each normal PDF $j \in \{1..k^i\}$, with each iteration by:

$$\sigma_j^i \leftarrow \gamma \cdot \sigma_j^i \tag{11}$$

where γ is the parameter describing the rate of dissolving.

Obviously, the three methods proposed here for negative update of the probability distribution may be combined. In particular, the method presented in (9) may be combined with the other two in order to maintain a reasonable number of normal PDFs within the mixture. Otherwise, the size of the mixture may become too large to be handled easily. Ideas based on $\mathcal{MAX}$-$\mathcal{MIN}$ Ant System [16] may be used for removing for instance only those normal PDFs whose weight ω dropped below a certain minimal value, or whose standard deviation σ exceeded a certain maximal threshold.

3 Discussion

The ACO algorithm presented in this paper may be studied from at least two different points of view. It may be used either as a method of solving continuous (or *global*) optimization problems, or as a method of solving mixed-variable (discrete and continuous) optimization problems.

3.1 Continuous Domain

ACO as a continuous optimization algorithm is part of a rather large family of algorithms for continuous optimization. For these types of problems, a number of methods have been proposed in the literature. They include some ant-related methods [2, 13, 9] already briefly presented in Sec. 1, but also many others. Many optimization algorithms have been originally developed for combinatorial optimization and only afterwards adapted also to the continuous case. Examples include the Continuous Genetic Algorithm (CGA) [6], Enhanced Simulated Annealing (ESA) [15], or Enhanced Continuous Tabu Search (ECTS) [5].

There are also other methods that – similarly to ACO – explicitly use some notion of probability estimation. Examples of the latter include the Iterated Density Estimation Algorithm (IDEA), or PBIL–Population-Based Incremental Learning. Similarly to ACO, they have been initially used for combinatorial optimization, and only later also adapted to handle continuous domains [3, 18].

ACO for continuous domain is similar to continuous PBIL or IDEA in the same sense, as generic ACO is similar to their discrete counterparts. It explicitly uses an estimation of probability distribution in order to find promising areas for the search. ACO differs from almost all of the methods mentioned here in the sense that it does not simply choose and evaluate the solutions, but builds them incrementally. Only PBIL uses similar incremental approach.

3.2 Mixed Variable Domain

ACO as an algorithm for optimization of mixed-variable problems does not have too many competitors at the moment. None of the other ant-related methods allows the intuitive handling of mixed-variable problems. There are few examples in the literature of other types of algorithms that explicitly do that. They include the Mixed Bayesian Optimization Algorithm (MBOA) [14], Pattern Search Algorithms [1], and the Bell-Curve Genetic Algorithm [11].

Since the mixed-variable optimization is not yet a very popular subject, there are not many benchmark problems available that would allow to properly test such an algorithm. Mixed-variable optimization remains nonetheless an interesting area of research. The most obvious field where it may be used is optimization of parameters. This could mean parameters of any physical process, or an algorithm's parameters – any case where some of the parameters may only assume discrete and others continuous values from a certain range. A practical example of such a mixed variable problem is the design of a thermal insulation system, as presented in [1].

4   Initial Results

4.1   Implementation

The extended version of ACO algorithm presented here may be used for both continuous and mixed-variable optimization problems. As a proof of concept we have implemented ACO for continuous optimization problems. Many benchmarks are available for these types of problems, and it is easy to compare the results of ACO with those obtained by other methods.

We have implemented ACO allowing optimization of multidimensional continuous functions on a given range and to a given accuracy. As input, the algorithm takes the following: the function to be optimized $f(x) \in \mathbb{R}$, the search domain $x \in (a, b)^n \subset \mathbb{R}^n$ – where interval (a, b) is assumed to be the same for all n dimensions, the known optimum[2] s_o, the accuracy ϵ, and a set of parameters, such as the number of ants m and the size of the mixtures k. As output, the algorithm presents the best solution found and the number of function evaluations performed.

---

[2] As this is proof of concept algorithm, the known optimum was used to test the performance of the algorithm against other methods.

The algorithm uses a different mixture of normal kernel PDFs for each of the n dimensions. Each mixture consists of the same number $k = k^{1..n}$ of normal PDFs. The solution construction is done exactly as presented in Sec. 2.2. The pheromone update is done in the following way: At each iteration, the *iteration best* solution is used for pheromone update. Positive update is done according to (8), and negative update according to (9) – only one (the oldest) normal PDF is removed. Since both actions are performed at the same time, the number k of normal PDFs in the mixture does not change. An additional improvement in the algorithm's performance was achieved by adding a simple elitist strategy. The normal PDF associated with the *global best* solution is never removed. If it is also the oldest one, then the second oldest normal PDF is rather removed.

The pheromone distribution for each variable x^i is initialized with a slightly randomized version of the one presented in (7). The vector of means μ is randomly selected from the interval (a, b) using a uniform random number generator.

At each iteration of the algorithm, m ants construct the m solutions to the problem. A different mixture $P^i(x^i)$ is used when choosing each component x^i of the solution. Due to the nature of the PDF used, it is possible that some of the generated solutions S^c will fall outside of the search domain $x \in (a, b)^n \subset \mathbb{R}^n$. Those solutions are dropped before being evaluated. All other solutions are evaluated, and the *iteration best* solution $s_{I\,best}$ is chosen. The oldest component $P^i_j(x^i)$ of each mixture is removed, and a new component is added. The new component PDF takes its mean from the value of the respective solution variable: $\mu^i = x^i_{s_{I\,best}}$.

The standard deviation σ^i of the normal PDFs $P^i_j(x^i)$ used for the update, is chosen adaptively based on the index of current iteration c and the solutions found in this iteration $s_{1..m} = (x^1, .., x^n)_{1..m} \in S^c$:

$$\sigma^i = \max \left(\frac{\max(x^i_{1..m}) - \min(x^i_{1..m})}{\sqrt{c}}, \epsilon \right) \qquad (12)$$

At each iteration, if the $s_{I\,best}$ is better than the $s_{I\,best}$, the latter is replaced. The stopping criterion is the required accuracy. The new global best solution $s_{G\,best}$ is compared to the known optimum s_o. The algorithm stops if the following condition is met [6, 8]:

$$|s_{G\,best} - s_o| < \epsilon_{rel} \cdot s_o + \epsilon_{abs} \Rightarrow STOP \qquad (13)$$

where $\epsilon_{rel} = \epsilon_{abs} = \epsilon$.

4.2   Results

The preliminary results presented here are based on 100 independent runs of the ACO algorithm on each of the test functions. The parameters used for tackling each test function are presented in Tab. 1. In order to have comparable results, the accuracy ϵ was chosen based on the results of the other algorithms published in the literature. The other parameters were chosen empirically based on some limited experimentation – about 10 configurations of parameters were tried for

Table 1. ACO parameters used for solving different benchmark problems: m–number of ants, k–number of normal PDFs per each of n dimensions, and ϵ–required accuracy

Test Function	d	m	k	ϵ
Sphere Model (SM)	6	8	3	$1 \cdot 10^{-4}$
Goldstein and Price (GP)	2	6	4	$1 \cdot 10^{-4}$
Rosenbrock (R_2)	2	30	8	$3 \cdot 10^{-3}$
Zakharov (Z_2)	2	8	4	$1 \cdot 10^{-4}$
Hartmann ($H_{3,4}$)	3	12	5	$1 \cdot 10^{-3}$

Table 2. Comparison of average number of function evaluations until the algorithm stop, on different benchmark problems. Results for some of the algorithms were not available – hence some entries are missing. The brackets indicate that the results are based on the runs with a fixed number of evaluations

$f(x)$	ACO	Other Ant Methods			Non-Ant Methods		
		CACO [17]	API [13]	CIAC [8]	CGA [6]	ECTS [5]	ESA [15]
SM	695	22050	[10000]	50000	750	338	-
GP	364	5330	-	23391	410	231	783
R_2	2905	6842	[10000]	11797	960	480	796
Z_2	401	-	-	-	620	195	15820
$H_{3,4}$	457	-	-	-	582	548	698

each test function. The results obtained by the other methods found in the literature are presented in Tab. 2.

The comparison presented here may be used only as an indication of the potential of ACO, and not as an exhaustive comparison of continuous optimization methods. The results found in the literature have been obtained under different conditions (e.g. slightly different stopping criteria). Also, the simple average number of evaluations does not fully describe an algorithm's performance. Other measures of the performance may be important, such as variance, stability, accuracy, or even the ease of the implementation.

It is clear that when considering only the number of function evaluations, ACO (for the limited number of cases presented here) is better than any of the other ant-related methods. However, more detailed analysis on a wider sample of test functions will have to be performed in order to add statistical significance to this claim. Considering the other continuous optimization methods, ACO's performance is similar. Again, larger sets of benchmarks and unified evaluation methods would have to be employed to indicate particular advantages or disadvantages of ACO.

Due to the space limitations, the definitions of the test functions used for performance evaluation are not provided in the paper. They may be found in the literature [8, 6, 5], and also online[3], along with the source code (in R) of the ACO used to obtain the results presented in this paper.

---

[3] http://iridia.ulb.ac.be/~ksocha/extaco04.html

5 Conclusions

We have shown how Ant Colony Optimization may be extended to continuous and mixed-variable optimization domains. This can be done without any major conceptual change to the original definition of ACO.

We have shown how such a new ACO algorithm may be designed, and what are the possible design choices. We have reviewed other methods used for both continuous and mixed-variable optimization from the literature. We have positioned ACO among them and explained how it differs from all other ant-related approaches to continuous optimization.

Finally, we have presented the first implementation of ACO for continuous problems, and shown that the method is at least competitive with the others found in the literature.

Our future work plans are twofold. On one side we would like to fine-tune the algorithm, so that its performance on continuous benchmark functions may be further improved. Also, as the initial tests of the algorithm were performed on problems that only had few dimensions, we plan to test the algorithm on higher-dimensional problems.

The second direction of possible research is a practical application of ACO to mixed-variable optimization. We plan to either find or create benchmarks that would allow the evaluation the performance of the algorithm on those types of problems. Also, if possible we will look for practical examples, where mixed-variable optimization may be used.

Acknowledgments

This research was supported by the "Metaheuristics Network", a Marie Curie Research Training Network funded by the Improving Human Potential programme of the CEC, grant HPRN-CT-1999-00106, and by the "ANTS" project, an "Action de Recherche Concertée" funded by the Scientific Research Directorate of the French Community of Belgium.

References

1. C. Audet and J. J. E. Dennis. Pattern Search Algorithms for Mixed Variable Programming. *SIAM Journal on Optimization*, 11(3):573–594, 2001.
2. G. Bilchev and I. C. Parmee. The Ant Colony Metaphor for Searching Continuous Design Spaces. In T. C. Fogarty, editor, *Proceedings of the AISB Workshop on Evolutionary Computation*, volume 993 of *LNCS*, pages 25–39. Springer-Verlag, Berlin, Germany, 1995.
3. P. A. N. Bosman and D. Thierens. Continuous Iterated Density Estimation Evolutionary Algorithms within the IDEA Framework. In M. Pelikan, H. Mühlenbein, and A. O. Rodriguez, editors, *Proceedings of OBUPM Workshop at GECCO-2000*, pages 197–200. Morgan-Kaufmann Publishers, San Francisco, CA, USA, 2000.
4. G. E. P. Box and M. E. Muller. A note on the generation of random normal deviates. *Annals of Mathematical Statistics*, 29(2):610–611, 1958.

5. R. Chelouah and P. Siarry. Enhanced Continuous Tabu Search. In S. Voss, S. Martello, I. H. Osman, and C. Roucairol, editors, *Meta-Heuristics Advances and Trends in Local Search Paradigms for Optimization*, chapter 4, pages 49–61. Kluwer Academic Publishers, Boston, MA, USA, 1999.

6. R. Chelouah and P. Siarry. A Continuous Genetic Algorithm Designed for the Global Optimization of Multimodal Functions. *Journal of Heuristics*, 6:191–213, 2000.

7. M. Dorigo and G. Di Caro. The Ant Colony Optimization meta-heuristic. In D. Corne, M. Dorigo, and F. Glover, editors, *New Ideas in Optimization*. McGraw-Hill, New York, NY, USA, 1999.

8. J. Dréo and P. Siarry. Continuous Interacting Ant Colony Algorithm Based on Dense Heterarchy. *Future Generation Computer Systems*, to appear.

9. J. Dréo and P. Siarry. A New Ant Colony Algorithm Using the Heterarchical Concept Aimed at Optimization of Multiminima Continuous Functions. In M. Dorigo, G. D. Caro, and M. Sampels, editors, *Proceedings of the Third International Workshop on Ant Algorithms (ANTS'2002)*, volume 2463 of *LNCS*, pages 216–221. Springer-Verlag, Berlin, Germany, 2002.

10. M. Guntsch and M. Middendorf. A population based approach for ACO. In S. Cagnoni, J. Gottlieb, E. Hart, M. Middendorf, and G. Raidl, editors, *Applications of Evolutionary Computing, Proceedings of EvoWorkshops 2002: EvoCOP, EvoIASP, EvoSTim*, volume 2279, pages 71–80. Springer-Verlag, Berlin, Germany, 3-4 2002.

11. R. K. Kincaid, S. Griffith, R. Sykes, and J. Sobieszczanski-Sobieski. A Bell-Curve Genetic Algorithm for Mixed Continuous and Discrete Optimization Problems. In *Proceedings of 43rd AIAA Structures, Structural Dynamics, and Materials Conference*. AIAA, Denver, CO, USA, 2002.

12. M. Mathur, S. B. Karale, S. Priye, V. K. Jyaraman, and B. D. Kulkarni. Ant Colony Approach to Continuous Function Optimization. *Ind. Eng. Chem. Res.*, 39:3814–3822, 2000.

13. N. Monmarché, G. Venturini, and M. Slimane. On how *Pachycondyla apicalis* ants suggest a new search algorithm. *Future Generation Computer Systems*, 16:937–946, 2000.

14. J. Očenášek and J. Schwarz. Estimation Distribution Algorithm for Mixed Continuous-Discreet Optimization Problems. In *Proceedings of the 2nd Euro-Inernational Symposium on Computational Inteligence*, pages 227–232. IOS Press, Amsterdam, Netherlands, 2002.

15. P. Siarry, G. Berthiau, F. Durbin, and J. Haussy. Enhanced Simulated Annealing for Globally Minimizing Functions of Many Continuous Variables. *ACM Transactions on Mathematical Software*, 23(2):209–228, 1997.

16. T. Stützle and H. H. Hoos. $\mathcal{MAX}\text{-}\mathcal{MIN}$ Ant System. *Future Generation Computer Systems*, 16(8):889–914, 2000.

17. M. Wodrich and G. Bilchev. Cooperative distributed search: the ant's way. *Control & Cybernetics*, (3):413–446, 1997.

18. B. Yuan and M. Gallagher. Playing in Continuous Spaces: Some Analysis and Extension of Population-Based Incremental Learning. In Sarker, R. *et al.*, editor, *Proceedings of Congress of Evolutionary Computation (CEC)*, pages 443–450, 2003.

An Ant Approach
to Membership Overlay Design*
Results on the Dynamic Global Setting

Vittorio Maniezzo[1], Marco Boschetti[2], and Mark Jelasity[1,**]

[1] Department of Computer Science, University of Bologna, Italy
{maniezzo,jelasity}@cs.unibo.it
[2] Department of Mathematics, University of Bologna, Italy
boschett@csr.unibo.it

Abstract. Designing an optimal overlay communication network for a
set of processes on the Internet is a central problem of peer-to-peer (P2P)
computing. Such a network defines membership and allows for members
to disseminate information within the group. The network has to be
robust and the available bandwidth has to be utilized in an optimal
manner to allow for maximally efficient communication. This problem
can be formulated as a dynamic optimization problem where classical
combinatorial optimization techniques must face the further challenge of
time-varying input data. ACO systems appear to be particularly fit for
this class of problems, being able to construct an internal model of the
instance to face and to exploit it for fast adaptation to modified contexts.
This paper proposes to use elements resulting from mathematical techniques, in this case Lagrangean relaxation, in an ACO framework in order to achieve sound hot start states for fast response to varying network
structures.

1 Introduction

The rapid evolution of the Internet and related technology, the increasing bandwidth and the enormous number of network-enabled computing devices resulted
in a new field of distributed computing, the so called peer-to-peer (P2P) computing [21], which is an umbrella term for a wide range of areas which include file
sharing, grid computing, distributed search, distributed hash tables, etc. This
new field poses a large number of new challanges, in particular, optimization
problems, most of which are dynamic in nature as the defining elements change
over time. A recent research thread in ACO systems supports the intuition that
ant algorithms are particularly fit for dynamic optimization problems because
of their ability to construct an internal representation of the essential elements
of the instance to solve, representation which needs to be updated and not re-
constructed when the instance changes ([9, 23, 6, 12, 11]).

---

* This work was partially supported by the Future & Emerging Technologies unit of
the European Commission through Project BISON (IST-2001-38923).
** Also with MTA RGAI, SZTE, Szeged, Hungary.

M. Dorigo et al. (Eds.): ANTS 2004, LNCS 3172, pp. 37–48, 2004.

This work reports on our approach of using an ant algorithm for a dynamic network design problem: P2P membership overlay network design. This problem arises in all P2P applications that are large and fully decentralized, like popular file sharing networks [10, 24] or gossip-based protocols for information dissemination [4, 5] and information aggregation (data mining) [22, 15]. Lacking a central service, participating nodes talk to each other directly, typically using a relatively small list of peer nodes they are aware of. Given the heterogenious bandwidth and availability constraints at each node, it is crucial for scalability and performance that participating nodes share the costs of the application in a fair manner. The optimization problem arises through the fact that the peers have to intelligently select the other peers they communicate with so that no participants get overloaded but available bandwidth is utilized reliably in an optimal manner by all network nodes.

For the full specification of the problem the target application has to be specified as well. In this work we are focusing on overlay networks applied by gossip-based protocols. In this case, each node sends gossip messages periodically to its neighbors. The applied membership overlay (as defined by the set of neighbors used to send gossip messages to) is typically a random or pseudo-random topology [8] sometimes taking into account network locality [20, 16]. Random topologies have a number of advantages from a theoretical point of view, however they ignore bandwidth and other constraints.

Unlike our problem, other rigorously formulated network optimization problems that consider communication cost, like the optimum communication cost spanning tree (OCST) problem [13] and the Steiner tree problem [14] are formulated to find a specific topology (e.g., tree) and they target other applications like broadcasting or search. Besides, the extremely dynamic character of P2P environments represents a novel requirement that needed to be incorporated into the formulation of our problem as well.

While our long term objective is to obtain a local algorithm, that – when run at all nodes concurrently – solves the above problem in a distributed manner, in this work we report results on a global algorithm that can however handle dynamicity, our main focus of research. Unfortunately paper length constraints prevent us to provide a detailed account of the rationals and the techniques we used. However, we believe to have achieved a sufficient presentation clarity and we refer the interested reader to [2] for further details.

2   Problem Description

As described in the Introduction, we are focusing on overlay networks for gossip based protocols. We assume that we have a set of Internet nodes that wish to form a gossip group, that is, a group over which gossip-based protocols can be applied. Gossip protocols can disseminate information among the members [4, 5] or they can analyze some attributes of the nodes in a distributed fashion [22, 15]. In principle, each node can send a message to any other node applying the routing service of the Internet provided the target IP address is known (if we ignore the effect of firewalls). In practice, it is not feasible to store all member addresses at all nodes because there can be too many of them and membership can change dynamically, so scalability problems arise. The typical solution that is normally adopted is storing only a limited number of peer addresses and using only those to send gossip messages to, which still allows gossip to spread efficiently.

The set of known peers at each node defines the *overlay network* in which there is a directed link between nodes i and j if i has the IP address of j in its list of peers. This network defines the membership in the group: those nodes that are connected to the overlay network are members since they can participate in the gossiping. The structure of this network has a major impact on the performance of the communication. If nodes with limited bandwidth have to send or receive too many messages (i.e., have too many outgoing or incoming connections) then the network will not function properly. It is important that load is distributed in a fair manner so that the throughput of the network is maximized without any nodes being overloaded.

In the following we will use the term "connection" as a shorthand for "allocation of non-zero bandwidth for possible communication". That is, node j is connected to, or a neighbor of, i if i allocates non-zero bandwidth for communicating with j. This sense of connection is not identical to direct physical connection or even proximity at the network level, it expresses a logical connection in the overlay network.

Let us now formulate this networking problem, that we shall call the Membership Overlay Design Problem (MOP), in a mathematical model as follows. A graph G=(V,E) of n vertices is given, where the nodes correspond to peers that want to communicate with each other, that is, that want to form a gossip group. The edges correspond to possible communication, that is, if there is an edge (ij) then i can possibly send a message to j using the underlying routing infrastructure. Each node can dynamically enter and exit the network, and when it is connected it can make use of a limited bandwidth. Therefore, each node has two associated weights, p_i and w_i, $i = 0, \dots, n$, corresponding to its uptime (measured as the percentage of time that the peer is available and responding to traffic [25], normalized to 1) and to the available bandwidth of its connection to the Internet, respectively. The finer structure within the core Internet (the high performance routers and backbones) are not taken into account. In other words, as in P2P networks the bandwidth bottleneck is typically represented by the connection between the core Internet and the participating computer on the edge of the Internet, and only rarely by a main backbone within the core Internet, we approximate the bandwidth constraint locally using the bandwidth of the Internet connection of the participating node.

The MOP asks to find a subgraph $G'=(V, E')$ of G. The edges in the graph G' define the fact that two nodes actually decide to allocate some bandwidth to communicate with each other. In other words, when two nodes i and j establish a connection, each one must allocate part of its bandwidth. If b_i and b_j are the bandwidths which could be allocated by i and j, then the bandwidth of the connection can be at most $b_{ij} = min\{b_i, b_j\}$. The two values b_i and b_j could be equal to w_i and w_j or could be less than that, due to other connections already maintained by the peers. Moreover, there is a lower bound L on the bandwidth of acceptable connections and it is anyway possible to put a limit on the maximal value that b_{ij} can take.

The graph G' has to be such that

1. the expected network throughput is maximized
2. the diameter of G' is minimized.
3. the total bandwidth used by each node i is less than or equal to w_i.

Note that 2) implies that G' is connected.

As mentioned, the final algorithm for solving this problem should be local: no global knowledge of the network is provided, each node i can exchange informations only with the nodes in $\delta'(i)$, that is, with its neighbors in G'. However, at this stage of our research we present a solution algorithm working at a global level.

2.1   The Static Subproblem

First a mathematical formulation (P) of the static version of the problem will be presented, which will be later adapted to the dynamic case. Formulation P is a mixed integer formulation for which we will later derive a polynomial upper bound and a relaxation framework.

A MIP Formulation. A comprehensive mathematical analysis of the MOP can be found in [2], here we report only some results which are relevant for this work. Specifically, we present a mathematical formulation of MOP, with reference only to objective 1) and leave the other two objectives to be handled by the ants metaheuristic.

Two sets of decision variables are used: $[x_{ij}]$ and $[\xi_{ij}]$, $(ij) \in E$. The decision variables x_{ij} specify the bandwidth allocated to the connection between peers i and j. Therefore they are continuous variables $0 \leq x_{ij} \leq b_{ij}$, which will be further constrained when they are not 0 to be at least L. Decision variables ξ_{ij} are binary variables which are 1 if arc (ij) is used for a connection, 0 otherwise.

The formulation, denoted P, is the following.

$$z_P = max \sum_{(ij) \in E} p_{ij} x_{ij} \tag{1}$$

$$s.t. \sum_{j \in \delta(i)} x_{ij} \leq w_i \quad i \in V \tag{2}$$

$$x_{ij} \geq L\xi_{ij} \quad (ij) \in E \tag{3}$$

$$x_{ij} \leq b_{ij}\xi_{ij} \quad (ij) \in E \tag{4}$$

$$\xi_{ij} \in \{0, 1\} \quad (ij) \in E \tag{5}$$

where $p_{ij} = p_i * p_j$, for each edge $(ij) \in E$ and $\delta(i)$ represents the neighborhood of i in G (i.e., $V \setminus \{i\}$ if graph G is complete).

The complexity of this problem is under study, but no straight solution methodology is available.

Lagrangean Relaxation of P. Formulation P can be effectively solved by a sequence of successive relaxations. This is justified by the assumption, to be *a posteriori* verified, that the optimum of the relaxed problem is structurally sufficiently similar to a feasible solution to permit to obtain one with minor adjustments without loosing much in solution quality.

The first relaxation is a LP relaxation of constraints (5), and substitutes them with constraints in the form $0 \leq \xi_{ij} \leq 1$. This makes the problem equivalent to the following problem LP, where variables ξ_{ij} become unnecessary.

$$z_{LP} = max \sum_{(ij)\in E} p_{ij}x_{ij} \tag{6}$$

$$s.t. \sum_{(ij)\in\delta(i)} x_{ij} \leq w_i \qquad\qquad i \in V \quad (7)$$

$$0 \leq x_{ij} \leq b_{ij} \qquad\qquad (ij) \in E \quad (8)$$

Notice that constraints 3 have been removed because the LP solution has them always satisfied (with fractional variables), they will be later dealt with by the ants procedure (see section 3.2).

Problem LP can obviously be solved by a LP-solver. However, we preferred to further relax it for two reasons. The first one is that we are not interested directly in the optimal LP solution but in a good approximation of its optimal dual variables, to be later used by the ants metaheuristic (see section 3.2). The second one is that, as mentioned in section 1 our ultimate objective is the design of a fully local optimization procedure: this is better supported by a further relaxation, in a Lagrangean fashion, of problem LP than by a straight application of LP solution algorithms.

This further relaxion can be done by associating a positive Lagrangean penalty λ_i to each constraint 7, resulting in the following formulation, denoted LR.

$$z_{LR}(\lambda) = max \sum_{(ij)\in E} (p_{ij} - \lambda_i - \lambda_j)x_{ij} + \sum_{i\in V} w_i\lambda_i \tag{9}$$

$$s.t. \ 0 \leq x_{ij} \leq b_{ij} \qquad\qquad (ij) \in E \quad (10)$$

In order to solve problem LP we must now find the minimum over all feasible λ vectors of the $z_{LR}(\lambda)$ costs, i.e., we must solve the Lagrangean dual $min\,[z_{LR}(\lambda) : \lambda \geq 0]$.

2.2   The Dynamic Case

In actual practice of P2P networks it can be observed that nodes continuously enter and exit the network, some nodes spending more time in the network while others join only for a short time. The problem formulation is not affected by dynamicity, in the sense that at any moment in time the problem formulation is as described. The only effect is that the graph G and all related elements are time-varying.

Whenever the average uptime of a node in the network is higher than the optimization time, it becomes feasible to re-optimize the network, taking into account the new network conditions. This could be done periodically or whenever significant network topology changes are detected. In addition, re-optimizations should be sufficiently frequent to ensure smooth performance. This means that optimization time must be short with respect to average node permanence time, putting a stress on optimization efficiency.

3   Ants for the Static Case

We used an ants metaheuristic for solving the whole static problem and we considered LP as a subproblem. The upper bound provided by the LP solution can be infeasible because the resulting overlay topology could be disconnected and some connections could be allocated a bandwidth less then L. It will be the task of ants to construct a feasible, high-throughput, low diameter connected solution.

In the following we will first detail how to get a possibly disconnected solution but with feasible bandwidths, then we describe how to include this routine in an ants framework.

3.1   Disconnected Upper Bound

Let z_{LR}^* be the solution obtained by the subgradient optimization of problem LR using penalties λ_i^*, $i \in V$. The solution could be infeasible for problem P because of the relaxed constraints, thus it could contain arcs (ij) which have an allocated bandwidth less than L.

An heuristic solution is obtained by considering the optimized costs $c_{ij}^* = (p_{ij} - \lambda_i^* - \lambda_j^*)$, ranking all arcs $(ij) \in E$ by non increasing c_{ij}^* values and allocating all possible bandwidth to each successively considered connection. More in detail, the algorithm is the following.

LAGRHEURISTIC(c^*)
1   Order all arcs in E by decreasing c^*
2   initialize $s_i = b_i$ for each $i \in V$
3   **foreach** arc (ij) **in** E in nonincreasing c^* order
4   **do** $slack = min\{s_i, s_j\}$
5   **if** $slack \geq L$
6   **then** $x_{ij} = slack$
7               $\xi_{ij} = 1$
8               $s_i = s_i - slack; s_j = s_j - slack$

This approach is derived from the exact method for solving continuous knapsack problems [19]. In our case it is not guaranteed to be optimal but it consistently produces good quality solutions in time $O(n \log n)$, where n is the number of arcs, the highest cost operation being the ordering of the arcs.

3.2   The Ants Heuristic

The solution obtained by the procedure described in subsection 3.1 can be infeasible for problem P because of its disconnectedness. It is necessary to reduce the solution objective function value in order to introduce new arcs which ensure connectivity. This is the task of the ants algorithm. The objective of each ant is to construct a connected solution. This is done by first identifying all connected components, then all vertices which could be endpoints of a new solution arc and finally by searching in the space of these candidate arcs.

The connected components in the *LagrHeuristic* solution S^d are identified and maintained by means of up-trees [3]. It is assumed that in actual practice it is always possible to identify in each connected component at least one node which has an incoming connection whose allocated bandwidth can be decreased by L. To enter a new arc in the solution it will in fact be necessary to use bandwidth previously allocated by at least one of its endpoints to another, more profitable connection. The new arcs will therefore be introduced with the least feasible bandwidth (i.e., L) in order to disrupt the solution quality the least.

Let S_h^d be the subset of nodes belonging to the $h-th$ connected component which have at least one connection whose bandwidth can be decreased by L. Each ant will try to identify arcs (ij), with $i \in S_h^d$ and $j \in S_l^d$, $h \neq l$, so that the global solution is connected and the solution cost is maximized.

The implemented ants algorithm from a structural viewpoint is a very standard ants algorithm, whose main steps are as follows.

ANTSMOP(c^*, λ)
```
1   Initialize trails [τij] and solutions Sk = ∅, k = 1...n
2   repeat
3           repeat
4                   for ants k=1 to m
5                           do Choose the next arc (ij) ∈ E to append to Sk
6                                   Sk = Sk ∪(ij)
7                   until (all ants have their solution completed)
8           Update trails
9   until (Termination condition)
```

The choice of the next arc to append, at step 5, is obviously made in probability. The formulae used at step 5 and at step 8 are those proposed in [18], that is, each $k-th$ ant moves from its current state ι to a feasible state ψ in probability, coording to the formula:

$$p_{\iota\psi}^k = \frac{\alpha \cdot \tau_{\iota\psi} + (1-\alpha) \cdot \eta_{\iota\psi}}{\Sigma_{(\iota\nu) \notin tabu_k}(\alpha \cdot \tau_{\iota\nu} + (1-\alpha) \cdot \eta_{\iota\nu})} \qquad (11)$$

Whereas trails are updated by means of formula 12

$$\tau_{\iota\psi}(t) = \tau_{\iota\psi}(t-1) + \tau_{\iota\psi}(0) \cdot (1 - \frac{z_{curr} - LB}{\bar{z} - LB}) \tag{12}$$

where z_{curr} is the cost of each current solution, $\bar{z}$ is the average of the last computed solutions and LB is a lower bound to the optimal problem solution cost. Detailed descriptions in [18].

Parameters to tune are m, number of ants, α, relative importance of visibility w.r.t. trail (multiplicative coefficient), and the termination condition, in our case maximum CPU time allowed. The cost average to use in the trail update formula at step 8 was made over the costs of all solutions computed at each iteration of loop 3-7.

The elements to define for the MOP-specific algorithm are trail and visibility initialization, and the bound to use for trail update.

Trail Initialization. Trails were initialized by means of the optimized Lagrangean multipliers λ, which penalize relaxed total node load constraints thus are higher for a nodes i and j when it would be desirable to have connection (ij) in the solution (i.e., $x_{ij} > 0$) but there is not enough bandwidth available to allocate at least L to it. Therefore penalties quantify desirability other than the cost, and we used those values to initialize trails. Notice that this is not the first work proposing the use of non-uniform trail initialization, other relevant contributions to this topic can be found for example in [18], [1] and [17].

Visibility Initialization. Being MOP a maximization problem, it comes natural to set η_{ij} equal to the arc cost, for ll $(ij) \in E$. While this is true, we choose to use costs c_{ij}^* in order to take into consideration also the global information which comes from bound pricing.

3.3   ANTS in the Dynamic Setting

The dynamic setting introduced in section 2.2 has been tackled by means of a continuous application of the interwoven Lagrangean and ants procedures described for the static case.

Since variations of the network structure happen continuously, the optimization algorithm is run continuously. It is assumed that the speed of execution of iterations of the subgradient optimization procedure is higher than the rate of network changes, thus that a few subgradient iterations can be performed between consecutive network changes. The exact number of iterations is currently a parameter, which therefore implicitly quantifies the network variability.

Three algorithms are actually run concurrently and syncronously. The first is a standard subgradient optimization procedure which updates Lagrangean penalties [7]. The second is the lagrangean heuristic *LagrHeuristic* described in section 3.1, which is run every 5 iterations of the subgradient procedure.

The third procedure is *AntsMOP*, and we perform one ants iteration for each subgradient iteration. The full pseudocode, at a high abstraction level, is thus the following. It is assumed that network changes are handled programmatically as events, thus in separate threads.

DynAntsMOP(G)

```
1    Initialize subgradient step and penalties λᵢ, i ∈ V
2    Initialize trails [τᵢⱼ] and iteration counter ItCount
3    while (true)
4        do ItCount + +
5            Solve problems SP1 and SP2
6            Check for infeasibilities and update λ
7            if (ItCount mod 5 == 0) then call LagrHeuristic
8            Initialize solutions Sₖ = ∅, k = 1...n
9            repeat
10               for ants k=1 to m
11                   do Choose the next arc (ij) ∈ E to append to Sₖ
12                       Sₖ = Sₖ ⋃(ij)
13               until (all ants have their solution completed)
14           Update trails
```

4   Computational Results

The proposed algorithms were coded partly in c# and partly in c++ and run on a 1000 MHz Pentium III machine, under Windows XP. To validate the described techniques, we conducted a number of experiments on different simulated scenarios. We generated two sets of instance graphs; the first one (set A) has parameters which match those measured on real P2P networks as reported in [25], the second set (set B) has the nodes randomly generated on a x-y plane and p_{ij} inversely proportional to the Euclidean distance of nodes i and j, $\forall i, j \in V$, in order to produce meaningful visualizations.

More precisely, the parameters to be defined when constructing a testset are:
- n: number of network nodes.
- L: minimal acceptable connection bandwidth.
- p_i: uptime of node i.
- w_i: total bandwidth of node i.

For set A we have L=14 (Kbps), the uptime distribution (p_i) is derived from figure 6 of [25] and the bandwidth distribution (w_i is derived from from figure 4 of [25]. For set B we have L=2, uptime distribution such that $p_{ij} = M - dist[i, j]$, where M is a suitably big constant and $dist[i, j]$ is the euclidean distance of nodes i and j, bandwidth distribution uniform random in [2,11] and $L = 2$.

The computational testing was carried out separately for the static and the dynamic case. In the static case we wanted to determine the solution quality disruption, w.r.t. to upper bound z_P, due to the successive heuristics, while in the dynamic case the ease of adaptation to a mutated environment was of interest. Table 1 summarizes the results obtained. The columns show:

- *Id*: an instance identifier.
- *n*: number of nodes.
- z^*_{LR}: cost of best solution found for problem LR.
- t^*_{LR}: time to find the solution of cost z^*_{LR}.
- $\%z^*_d$: percentage decrease of cost for solution of algorithm *Lagr*Heuristic.
- t^*_d: time to find the solution of algorithm *Lagr*Heuristic.
- $\%z^*_{ants}$: percentage decrease of cost for solution of algorithm *Ants*MOP.
- t^*_{ants}: time to find the solution of algorithm *Ants*MOP.

The last two elements are reported for a number of ants / subgradient optimization iterations which is 10, 20 and 30, respectively. For the ants, parameters were $\alpha = 0.5$ and $num.ants = n/10$.

Table 1. Results on different MOP instances.

Problem		LP		LagrHeu		10 iter		20 iter		30 iter	
Id	n	z^*_{LR}	t^*_{LR}	$\%z^*_d$	t^*_d	$\%z^*_{ants}$	t^*_{ants}	$\%z^*_{ants}$	t^*_{ants}	$\%z^*_{ants}$	t^*_{ants}
A20	20	842	0.00	0.04	0.00	0.95	0.00	0.59	0.04	0.27	0.05
A50	50	2137	0.01	0.06	0.01	0.88	0.10	0.52	0.15	0.36	0.19
A100	100	8162	0.02	0.01	0.01	0.83	0.23	0.67	0.37	0.24	0.48
A500	500	20192	0.35	0.01	0.31	0.96	4.46	0.66	7.96	0.23	11.25
A1000	1000	53765	1.37	0.01	1.33	0.92	21.98	0.58	38.64	0.30	56.37
B20	20	53295	0.00	2.22	0.00	4.46	0.17	4.05	0.26	3.71	0.33
B50	50	131891	0.00	2.84	0.00	4.35	0.17	4.00	0.29	3.67	0.39
B100	100	281836	0.00	2.31	0.00	4.50	0.21	4.06	0.37	3.93	0.48
B500	500	1413301	0.33	2.81	0.44	4.59	4.05	4.15	8.14	3.95	10.72
B1000	1000	2940334	1.63	2.96	1.65	4.35	20.90	3.97	40.19	3.84	61.44

The results show how the proposed problem relaxation is highly effective on both, structurally very different, instance sets. The gap between the upper bound z^*_{LR} and the lower bound z^*_d is always reasonably small, and the feasible solution obtained by the ants does never disrupt excessively the bounds solution quality. Moreover, in the dynamic setting a fast adaptation is achieved, as testified by the solution quality obtained after different numbers of internal iterations. The results reported here are admittedly still incomplete, as computational testing is still going on. These results are to be read as a validation of the feasibility of the approach, but significant improvements are possible. We did not concentrate much on this as we are now focused on the possibility of designing a fully decentralized, asynchronous Lagrangean optimizer.

5   Conclusions

The work reported in this paper has three main foci of interest. First, it confirms the viability of ants approaches for dynamic problems and it does this in the specific case of a cogent problem in telecommunication network design. Second, it proposes a structural way to hybridize a fundamental mathematical technique

– Lagrangean optimization – with ants procedures. Both techniques implement a progressive refinement of an implicit model of the problem to solve and algorithm AntsMOP shows a way to intertwine their evolutions so that each one exploits current results of the other (lower bounds and Lagrangean penalties, respectively). Finally, the paper contains a mathematical analysis of the MOP, which can represent a foundation for further algorithmic advances.

Our final research objective is the definition of a fully distributed, local protocol for optimized overlay network design. The results and the methods reported in this work, besides being of interest in their own, also represent a promising base over which to continue research toward the final objective.

References

1. C. Blum. Aco applied to group shop scheduling: A case study on intensification and diversification. In *Proc. ANTS'02*, 2002.
2. M. Boschetti, M. Jelasity, and V. Maniezzo. A local approach to membership overlay design. Working paper, Department of Computer Science, University of Bologna, 2004.
3. T. Corman, C. Leiserson, and R. Rivest. *Introduction to Algorithms*. MIT Press, 1990.
4. Alan Demers, Dan Greene, Carl Hauser, Wes Irish, John Larson, Scott Shenker, Howard Sturgis, Dan Swinehart, and Doug Terry. Epidemic algorithms for replicated database management. In *Proceedings of the 6th Annual ACM Symposium on Principles of Distributed Computing (PODC'87)*, pages 1–12, Vancouver, August 1987. ACM.
5. Patrick Th. Eugster, Rachid Guerraoui, Anne-Marie Kermarrec, and Laurent Massoulié. From epidemics to distributed computing. *IEEE Computer*. to appear.
6. C.J. Eyckelhof and M. Snoek. Ant systems for a dynamic tsp: Ants caught in a traffic jam. In M. Dorigo, G. Di Caro, and M. Sampels, editors, *Ant Algorithms : Third International Workshop, ANTS 2002*, volume 2463 / 2002 of *Lecture Notes in Computer Science*. Springer-Verlag, Heidelberg, 2002.
7. M.L. Fisher. The lagrangean relaxation method for solving integer programming problems. *Management Science*, 27(1):1–18, 1981.
8. Ayalvadi J. Ganesh, Anne-Marie Kermarrec, and Laurent Massoulié. Peer-to-peer membership management for gossip-based protocols. *IEEE Transactions on Computers*, 52(2), February 2003.
9. R.M. Garlick and R.S. Barr. Dynamic wavelength routing in wdm networks via ant colony optimization. In M. Dorigo, G. Di Caro, and M. Sampels, editors, *Ant Algorithms : Third International Workshop, ANTS 2002*, volume 2463 / 2002 of *Lecture Notes in Computer Science*. Springer-Verlag, Heidelberg, 2002.
10. Gnutelliums. http://www.gnutelliums.com/.
11. M. Guntsch, J. Branke, M. Middendorf, and H. Schmeck. Aco strategies for dynamic tsp. In *ANTS'2000 - From Ant Colonies to Artificial Ants: Second International Workshop on Ant Algorithms*, 2000.
12. M. Guntsch and M. Middendorf. Applying population based aco to dynamic optimization problems. In M. Dorigo, G. Di Caro, and M. Sampels, editors, *Ant Algorithms: Third International Workshop, ANTS 2002*, volume 2463 / 2002 of *Lecture Notes in Computer Science*. Springer-Verlag, Heidelberg, 2002.

13. T. C. Hu. Optimum communication spanning trees. *SIAM Journal on Computing*, 3(3):188–195, September 1974.
14. Frank K. Hwang, Dana S. Richards, and Pawel Winter. *The Steiner Tree Problem*. North-Holland, 1992.
15. Márk Jelasity and Alberto Montresor. Epidemic-style proactive aggregation in large overlay networks. In *Proceedings of The 24th International Conference on Distributed Computing Systems (ICDCS 2004)*, pages 102–109, Tokyo, Japan, 2004. IEEE Computer Society.
16. Meng-Jang Lin and Keith Marzullo. Directional gossip: Gossip in a wide area network. In Jan Hlavička, Erik Maehle, and András Pataricza, editors, *Dependable Computing – EDCC-3*, volume 1667 of *Lecture Notes on Computer Science*, pages 364–379. Springer-Verlag, 1999.
17. H. Lourenço and D. Serra. Adaptive search heuristics for the generalized assignment problem. *Mathware and Soft Computing*, 9(2-3):209–234, 2002.
18. V. Maniezzo. Exact and approximate nondeterministic tree-search procedures for the quadratic assignment problem. *INFORMS J. on Computing*, 11(4):358–369, 1999.
19. S. Martello and P. Toth. *Knapsack problems: algorithms and computer implementations*. John Wiley & Sons, Inc., 1990.
20. Laurent Massoulié, Anne-Marie Kermarrec, and Ayalvadi J. Ganesh. Network awareness and failure resilience in self-organising overlays networks. In *Proceedings of the 22nd Symposium on Reliable Distributed Systems (SRDS 2003)*, pages 47–55, Florence, Italy, 2003.
21. Dejan S. Milojicic, Vana Kalogeraki, Rajan Lukose, Kiran Nagaraja, Jim Pruyne, Bruno Richard, Sami Rollins, and Zhichen Xu. Peer-to-peer computing. Technical Report HPL-2002-57, HP Labs, Palo Alto, 2002.
22. Alberto Montresor, Márk Jelasity, and Ozalp Babaoglu. Robust aggregation protocols for large-scale overlay networks. Technical Report UBLCS-2003-16, University of Bologna, Department of Computer Science, Bologna, Italy, December 2003. to appear in the proceedings of Distributed Systems and Networks (DSN 2004).
23. S. Nouyan. Agent-based approach to dynamic task allocation. In M. Dorigo, G. Di Caro, and M. Sampels, editors, *Ant Algorithms : Third International Workshop, ANTS 2002*, volume 2463 / 2002 of *Lecture Notes in Computer Science*. Springer-Verlag, Heidelberg, 2002.
24. FastTrack: Wikipedia page. http://en.wikipedia.org/wiki/FastTrack.
25. S. Saroiu, P. Krishna Gummadi, and S.D. Gribble. A measurement study of peer-to-peer file sharing systems. In *Proceedings of Multimedia Computing and Networking 2002 (MMCN'02)*, San Jose, CA, 2002.

An Ant Colony Optimisation Algorithm for the Set Packing Problem

Xavier Gandibleux[1], Xavier Delorme[1], and Vincent T'Kindt[2]

[1] LAMIH/ROI – UMR CNRS 8530
Université de Valenciennes, Campus "Le Mont Houy"
F-59313 Valenciennes cedex 9 - France
{Xavier.Gandibleux,Xavier.Delorme}@univ-valenciennes.fr
[2] Laboratoire d'Informatique
Polytech'Tours
64 avenue Jean Portalis
F-37200 Tours - France
Tkindt@univ-tours.fr

Abstract. In this paper we consider the application of an Ant Colony Optimisation (ACO) metaheuristic on the Set Packing Problem (SPP) which is a NP-hard optimisation problem. For the proposed algorithm, two solution construction strategies based on exploration and exploitation of solution space are designed. The main difference between both strategies concerns the use of pheromones during the solution construction. The selection of one strategy is driven automatically by the search process. A territory disturbance strategy is integrated in the algorithm and is triggered when the convergence of the ACO stagnates. A set of randomly generated numerical instances, involving from 100 to 1000 variables and 100 to 5000 constraints, was used to perform computational experiments. To the best of our knowledge, only one other metaheuristic (Greedy Randomized Adaptative Search Procedure, GRASP) has been previously applied to the SPP. Consequently, we report and discuss the effectiveness of ACO when compared to the best known solutions and including those provided by GRASP. Optimal solutions obtained with Cplex on the smaller instances (up to 200 variables) are indicated with the calculation times. These experiments show that our ACO heuristic outperforms the GRASP heuristic. It is remarkable that the ACO heuristic is made up of simple search techniques whilst the considered GRASP heuristic is more evolved.

1 Introduction

The set packing problem (SPP) is formulated as follows. Given a finite set $I = \{1, \ldots, n\}$ of items and $\{T_j\}, j \in J = \{1, \ldots, m\}$, a collection of m subsets of I, a packing is a subset $P \subseteq I$ such that $|T_j \cap P| \leq 1, \forall j \in J$. The set J can be also seen as a set of exclusive constraints between some items of I. Each item $i \in I$ has a positive weight denoted by c_i and the aim of the SPP is to calculate the packing which maximises the total weight. This problem can be formulated by integer programming as follows:

M. Dorigo et al. (Eds.): ANTS 2004, LNCS 3172, pp. 49–60, 2004.
© Springer-Verlag Berlin Heidelberg 2004

$$\left[\begin{array}{c} Max \ z = \sum_{i \in I} c_i x_i \\ \sum_{i \in I} t_{i,j} x_i \leq 1, \forall j \in J \\ x_i \in \{0, 1\} \quad , \forall i \in I \\ t_{i,j} \in \{0, 1\} \quad , \forall i \in I, \forall j \in J \end{array} \right] \qquad (1)$$

In the above model, the variables are the x_i's with $x_i = 1$ if item $i \in P$, and $x_i = 0$ otherwise. The data $t_{i,j}, \forall i \in I, \forall j \in J$, enable us to model the exclusive constraints with $t_{i,j} = 1$ if item i belongs to set T_j, and $t_{i,j} = 0$ otherwise.

Notice that the special case in which $\sum_{i \in I} t_{i,j} = 2, \forall j \in J$, is the node packing problem.

The SPP is known to be strongly NP-Hard, according to Garey and Johnson [6]. The most efficient exact method known for solving this problem (as suggested in [9]) is a Branch & Cut algorithm based on polyhedral theory and the works of Padberg [12] to obtain facets. However, only small-sized instances can be solved to optimality. To the best of our knowledge, and also according to Osman and Laporte [10], few metaheuristics have been applied to the solution of the SPP. Besides, few applications have been reported in the literature. Rönnqvist [13] worked on a cutting stock problem formulated as a SPP and solved it using a lagrangian relaxation combined with subgradient optimisation. Kim [7] modelled a ship scheduling problem as a SPP and used LINDO software to solve it. Mingozzi et al. [8] used a SPP formulation to calculate bounds for a resource constrained project scheduling problem. This SPP formulation is solved by a greedy algorithm. At last, Rossi [14] modelled a ground holding problem as a SPP and solved it with a Branch & Cut method.

The application of the SPP to a railway planning problem has been first studied by Zwaneveld et al. [18]. Railway infrastructure managers now have to deal with operators' requests for increased capacity. Planning the construction or reconstruction of infrastructures must be done very carefully due to the huge required investments. Usually, assessing the capacity of one component of a rail system is done by measuring the maximum number of trains that can be operated on this component within a certain time period. Measuring the capacity of junctions is a matter of solving an optimisation problem called the *feasibility problem*, and which can be formulated as a SPP. Zwaneveld et al. [18] proposed reduction tests and a Branch & Cut method to solved it to optimality.

More recently, Delorme et al. have taken up again this application of the SPP and proposed heuristic algorithms [1, 3]. Among these heuristics, the most efficient one is derived from the metaheuristic GRASP [2]. The designed algorithm integrates advanced strategies like a path-relinking technique, a learning process and a dynamic tuning of parameters. Besides, it should be noticed that basically GRASP is not a *population based* heuristic since at each iteration of the algorithm only one solution is considered.

In this paper, we consider an implementation of Ant Colony Optimisation (ACO) principles on the SPP. ACO algorithms are population based heuristics in which, at each iteration, ants build solutions by exploiting a common memory. Henceforth, ACO algorithms have the capability of learning "good and bad" decisions when building solutions. This motivated the design of the proposed ACO heuristic for the SPP, in which taking a decision consists in choosing if a given item belongs to a packing under construction. Since the last decade, ACO algorithms become more and more used in the field of Operations Research ([4]). Notably, they have been applied to the solution of the quadratic assignment problem ([5]) and the traveling salesman problem ([16]).

The remainder of the paper is organised as follows: Section 2 is devoted to the presentation of the designed ACO heuristic and its implementation. Section 3 deals with numericals experiments conducted on various kind of instances. Our ACO heuristic is compared to the GRASP heuristic presented in [1, 2] and to Cplex solver applied on Model 1. We conclude this section and the paper by providing an analysis of the behavior of our ACO heuristic.

2 The Ant Colony Optimisation Algorithm for the SPP

The basic idea of ACO metaheuristics comes from the capability of ants to find shortest paths from their nest to food locations. For combinatorial optimization problems this means that ants search how to build good solutions. At each iteration of this search, each ant builds a solution by applying a constructive procedure which uses the common memory of the colony. This memory, referred to as the *pheromone matrix*, corresponds to the pheromone trails for real ants. Rougthly speaking, the pheromone matrix contains for a combinatorial optimisation problem, the probabilities of building good solutions. Once a complete solution has been computed, pheromone trails are updated according to the quality of the best solution built. Hence, cooperation between ants is performed by exploiting the pheromone matrix. In this paper we use some principles of the ACO framework, denoted by SACO, proposed by Stuztle [15] and revised successfuly by T'kindt et al. [17] for a scheduling problem. Basically, when constructing a solution, an ant iteratively takes decisions either in the *exploration mode* or in the *exploitation mode*. In a basic ACO framework the choice of a mode is done following a fixed known probability. However, in the SACO framework we use an important feature of Simulated Annealing algorithms which is to allow more diversification at the beginning of the solution process and more intensification at the end. Henceforth, in the SACO framework the above mentionned probability evolved along the solution process. Besides, we also consider that after having built a solution an ant applies on it a local search. A territory disturance strategy is integrated in this framework. This strategy is inspired by a warming up strategy, well-known for the simulated annealing metaheuristic. We now describe more accurately the ACO heuristic designed for the SPP.

The general outline of the proposed ACO heuristic is given in Algorithm 1. (The arrows ↓, ↑ and ↕ specify the transmission mode of a parameter to a procedure; they correspond respectively to the mode IN, OUT and INOUT.) Initially, a greedy heuristic is applied to provide the initial best known solution. It works

---

Algorithm 1 The main procedure

```
--| Generate an initial solution using a greedy algorithm and a local search
elaborateSolutionGreedy( sol ↑ )
localSearch( sol ↕ )
copySolution( sol ↓ , bestSolKnown ↑ )

--| ACO Algorithm
initPheromones( φ ↑ ); iter ← 0
while not( isFinished?( iter ↓ ) ) do
  resetToZero( bestSolIter ↑ )
  for ant in 1... maxAnt do
    if isExploitation?(ant ↓, iter ↓, iterOnExploit ↓, maxIter ↓ ) then
      elaborateSolutionGreedyPhi( φ ↓ , solution ↑ )
    else
      elaborateSolutionSelectionMethod( φ ↓ , solution ↑ )
    end if
    localSearch( sol ↕ )
    if performance( sol ) > performance( bestSolIter ) then
      copySolution( sol ↓ , bestSolIter ↑ )
      if performance( sol ) > performance( bestSolKnown ) then
        copySolution( sol ↓ , bestSolKnown ↑ )
      end if
    end if
  end for
  managePheromones( φ ↕ , bestSolKnown ↓ , bestSolIter ↓ )
  iter++
end while

--| The best solution found
putLine( bestSolKnown ↓ )
```

---

as follows (procedure `elaborateSolutionGreedy`, see Algorithm 2). Iteratively
the candidate variable which involves a minimum number of constraints with
a maximum value is selected. This process is repeated until there is no more
candidate variable available.

Algorithm 2 The `elaborateSolutionGreedy` procedure

$I_t \leftarrow I$; $x_i \leftarrow 0, \forall i \in I_t$
$valuation_i \leftarrow c_i / \sum_{j \in J} t_{i,j}, \forall i \in I_t$
while $(I_t \neq \emptyset)$ do
  $i^* \leftarrow$ bestValue($valuation_i, i \in I_t$)
  $x_{i^*} \leftarrow 1$; $I_t \leftarrow I_t \setminus \{i^*\}$; $I_t \leftarrow I_t \setminus \{i : \exists j \in J, t_{i,j} + t_{i^*,j} > 1\}$
end while

The neighbourhood $\mathcal{N}$ used for the local search procedure is based on $k - p$
exchanges. The $k - p$ exchange neighbourhood of a solution x is the set of
solutions obtained from x by changing the value of k variables from 1 to 0,
and changing p variables from 0 to 1. Due to the combinatorial explosion of the
number of possible exchanges when k and p increase, we decided to implement
the $1 - 1$ exchanges. This exchange is valuable only for weighted instances, *i.e.*
instances in which there exists, at least, c_i and c_j such that $c_j \neq c_i$. Consequently,
no local search is applied for unicost instances, *i.e.* those instances with $c_i = c_j$,

Algorithm 3 The `elaborateSolutionSelectionMode` procedure

$I_t \leftarrow I$; $x_i \leftarrow 0, \forall i \in I_t$
$\mathcal{P} \leftarrow \log_{10}(\texttt{iter})/\log_{10}(\texttt{maxIter})$
while $(I_t \neq \emptyset)$ **do**
  if (randomValue(0,1) $> \mathcal{P}$) **then**
    $i^* \leftarrow$ rouletteWheel$(\phi_i, i \in I_t)$
  else
    $i^* \leftarrow$ bestValue$(\phi_i, i \in I_t)$
  end if
  $x_{i^*} \leftarrow 1$; $I_t \leftarrow I_t \setminus \{i^*\}$; $I_t \leftarrow I_t \setminus \{i : \exists j \in J, t_{i,j} + t_{i^*,j} > 1\}$
end while

$\forall i, j \in I$. Moreover, the search procedure was implemented using a non-iterative first-improving strategy (*i.e.* we selected the first neighbour whose value is better than the current solution).

Let ϕ be the pheromone matrix and ϕ_i be the probability of having item i in a good packing for the SPP. Initially, the pheromones are initialized (routine `initPheromones`) by assigning $\phi_i \leftarrow$ `phiInit` for all variables $i \in I$, with `phiInit` a given parameter. Each ant elaborates a feasible saturated solution (*i.e.* a solution in which it is impossible to add one more variable without violating the constraints set) starting from the trivial feasible solution, $x_i = 0, \forall i \in I$. Some variable are set to 1, as long as the solution is maintained feasible. Changes concern only one variable at each step and there is no more change when no variable can be fixed to 1 without losing feasibility. The choice of a variable x_i to be set to 1 is done either in the *exploration mode* or the *exploitation mode* according to the procedure `elaborateSolutionSelectionMode` described in Algorithm 3. In the exploration mode a roulette wheel is applied on the set of candidate variables whilst in the exploitation mode the candidate variable with the greatest value of pheromone is selected. The ceil probability $\mathcal{P}$ evolves along the solution process following a logarithmic curve regurlarly restarted. This mechanism enables the ants to periodically strongly diversify their search for a good solution.

Notice that for some ants when the predicate `isExploitation?` is true, a solution is built by applying the greedy strategy on the current level of pheromones (like in procedure `elaborateSolutionGreedy` but with *valuation$_i$* $= \phi_i$). The above predicate is true for each first ant of an iteration, every `iterOnExploit` iterations.

After all ants have built a solution the best one for the current iteration is retained and *evaporation* and *deposition* of pheromones is performed. It means that we increase the pheromones ϕ_i corresponding to the items selected in the best packing of the current iteration, whilst we decrease the other pheromones. This process is described in the procedure `managePheromones` (Algorithm 4). Besides a disturbance strategy has been integrated to this management procedure. This strategy is triggered when three conditions are true: (1) the convergence of the ACO is in stagnation (predicate `isStagnant?` is true), (2) at least one pheromone has its level set to zero (predicate `isExistsPhiNul?` is true), and (3) it remains enough iterations after the application of the disturbance to stabilize the pheromones (predicate `isRestartEnable?` is true). Finally, the procedure is

---

Algorithm 4 The `managePheromones` procedure

```
--| Pheromone evaporation
for i in 1 ... ncol do
    φ_i ← φ_i * rhoE
end for

--| Pheromone deposition
for i in 1... ncol do
    if  (bestSolIter.x[i] = 1)  then
        φ_i ← φ_i + rhoD
    end if
end for

--| Territory disturbance
if      isStagnant?( bestSolution , iterStagnant )
    and isExistsPhiNul?( φ )
    and isRestartEnable?( iter , lastIterRestart , maxIteration ) then
    --| Disturb the pheromones
    for i in 1... ncol do
        φ_i ← φ_i * 0.95 * log10(iter)/log10(maxIteration)
    end for
    for i in 1...random(0.0, 0.1 * ncol) do
        φ_{random(1,ncol)} ← random(0.05, (1.0 − iter/maxIteration) * 0.5)
    end for
    --| Offset on the pheromones with low level
    for i in 1... ncol do
        if φ_i < 0.1  then
            φ_i ← φ_i + random(0.05, (1.0 − iter/maxIteration) * 0.5)
        end if
    end for
end if
```

---

stopped after a predefined number of iterations (predicate `isFinished?` is true). The parameters used in the ACO heuristic are reported in Table 1.

3   Numerical Experiments and Analysis

This section presents the computational results obtained on all the randomly generated instances considered in our study. The solutions generated by GRASP and our ACO implementation are included. Our implementation of ACO has been performed with C language whilst GRASP has been developped with Ada Language. The results were obtained on a PC Pentium III at 800 MHz for GRASP and ACO. We also compare these heuristics to the optimal solutions calculated by Cplex solver for instances with up to 200 variables. As Cplex was not capable of solving all larger instances, we consider in this case the best known integer solution which is compared to the two heuristics. Roughly speaking, this best solution is taken, for a given instance, as the one returned by ACO, GRASP and Cplex (when time limited) which yields the highest value of the total cost.

Characteristics of the instances and results provided by Cplex and the heuristics are given in Tables 2 and 3. In each of these tables and for each instance, the column #*var* contains the number of variables and the column #*cst* contains the number of constraints. The *Density* column corresponds to the percentage of non-null elements in the constraint matrix. The two remaining columns describing the characteristics of the instances are *MaxOne* which provides the

Table 1. The parameters

maxIter	200	Number of iterations determined a priori
maxAnt	15	Number of ants for each iteration
phiInit	1.0	Initial pheromone assigned to a variable
rhoE	0.8	Pheromone evaporation rate
rhoD	phiInit * (1.0 - rhoE)	Pheromone deposition rate
phiNul	0.001	Level of pheromon considered as zero
iterOnExploit	0.750	Percentage of iterations when Exploitation mode is activated
iterStagnant	8	Declare the procedure stagnant when no improvement is observed

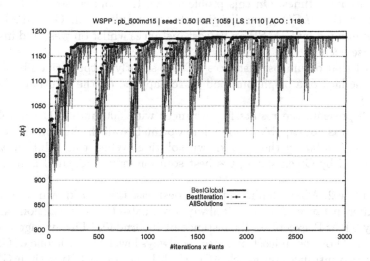

Fig. 1. Behavior of the ACO heuristic on the instance pb500rnd15

maximum number of data $t_{i,j}$ different from 0, and *Weight* which indicates the interval in which the costs c_i are comprised. Notice that instances for which the interval is $[1-1]$ are instances of the unicost SPP. For all the heuristics we report the average objective function value (column *Avgvalue*) found over 16 runs. We also give the maximum objective function value found (column *Bestvalue*) and the average required CPU time (column *CPUt*).

All the tested instances are freely available on the Internet at the address www3.inrets.fr/~delorme/Instances-fr.html.

We first focus on the behavior of the ACO heuristic on the instance pb500rnd15. Figure 1 reports the evolution of the best solution calculated (*BestGlobal* curve) as well as the best solution of each iteration (*BestIteration*). It appears that the evolution of the ceil probability $\mathcal{P}$ implies a quick convergence towards a good global solution value. However, the best solution is obtained thanks to the disturbance strategy. This is a typical behavior of the designed ACO.

Table 2 presents the results for instances with up to 200 variables. Cplex solver was capable of solving all these instances to optimality. Henceforth, column 0/1 *solution* reports, for each instance, the information about the optimal solution value as well as the CPU time required by Cplex to calculate it. Notice that on some instances this CPU time is prohibitive.

It appears that the ACO found in the best case the optimal solutions except for one instance. Besides, the best solution found by ACO always outperforms the best solution value obtained by GRASP. The comparison of average calculated solution values of ACO and GRASP show that on instances with 100 variables GRASP slightly outperforms ACO, and only on unicost instances. On instances with 200 variables ACO strictly outperforms GRASP 8 times whilst the opposite situation occurs 6 times. On this problem size, the comparison of the average values shows that ACO generally gives better results than GRASP. Besides, ACO is dominated by GRASP on unicost instances whilst on weighted instances the converse occurs.

The fact that on unicost instances ACO is slightly outperformed by GRASP is due to the lack of a neighbourhood search applied on the solution calculated by each ant.

Table 3 presents the results for instances with 500 and 1000 variables. As Cplex solver was not capable of solving to optimality all the instances, we report in column 0/1 *solution* the best known solution. When this solution was not shown optimal by Cplex solver, the best solution value indicated is marked by an asterix.

As in Table 2, ACO often found in the best case the best known solution (except for three instances). Besides, it always dominates the best solution value calculated by GRASP, except on the instance pb1000rnd400. The average value obtained by ACO on the unicost instances is always lower than the one of GRASP, except for two instances on which ACO slightly performs better than GRASP. On the weighted instances, GRASP strictly outperforms ACO 6 times whilst the opposite situation also occurs 6 times. From this viewpoint, the two algorithms seem to be complementary. However, it should be noticed that when GRASP provides better averaged results than ACO, it is always of at most 3 units. But when ACO improves over GRASP, the gap between the two averaged values can be up to 25 units.

Figure 2 provides a graphical overview of the performances of the ACO heuristic on all the instances. This figure shows that ACO generally calculates good solutions even if, as expected, the gap between the average solution value and the best solution value increases as the problem size increases.

The first conclusion of these experiments is that ACO is capable of finding the best known solutions and providing, at least, similar results than those given by GRASP heuristic. This is an interesting conclusion since the proposed ACO heuristic has a simple structure and does not use evolved search mechanisms as, for instance, the path-relinking process used in the GRASP heuristic. Henceforth, ACO is simpler than GRASP but provides similar or better results. A drawback

Table 2. Instances and results (1/2)

Instances	Characteristics					0/1 Solution		GRASP			ACO			
	#var	#cst	Density	Max One	Weight	Optimal value	CPUt (s)	Best value	Avg value	CPUt (s)	Best value	Avg value	CPUt (s)	Optimal found
pb100rnd01	100	500	2.0%	2	[1-20]	372	2.92	372	372.00	1.97	372	372.00	3.33	•
pb100rnd02	100	500	2.0%	2	[1-1]	34	0.60	34	34.00	1.31	34	34.00	2.00	•
pb100rnd03	100	500	3.0%	4	[1-20]	203	7.81	203	203.00	1.14	203	203.00	2.00	•
pb100rnd04	100	500	3.0%	4	[1-1]	16	52.86	16	16.00	1.29	16	15.56	0.67	•
pb100rnd05	100	100	2.0%	2	[1-20]	639	0.01	639	639.00	0.80	639	639.00	1.67	•
pb100rnd06	100	100	2.0%	2	[1-1]	64	0.01	64	64.00	0.69	64	64.00	1.00	•
pb100rnd07	100	100	2.9%	4	[1-20]	503	0.00	503	503.00	1.00	503	503.00	1.00	•
pb100rnd08	100	100	3.1%	4	[1-1]	39	0.02	39	38.75	0.57	39	38.68	0.67	•
pb100rnd09	100	300	2.0%	2	[1-20]	463	0.49	463	463.00	1.26	463	463.00	1.67	•
pb100rnd10	100	300	2.0%	2	[1-1]	40	1.13	40	40.00	1.28	40	39.62	1.00	•
pb100rnd11	100	300	3.1%	4	[1-20]	306	0.48	306	306.00	0.63	306	306.00	1.67	•
pb100rnd12	100	300	3.0%	4	[1-1]	23	6.80	23	23.00	1.13	23	22.93	0.33	•
pb200rnd01	200	1 000	1.5%	4	[1-20]	416	8 760.73	416	415.18	7.32	416	415.25	27.33	•
pb200rnd02	200	1 000	1.5%	4	[1-1]	32	156 109.36	32	32.00	7.35	32	31.56	14.67	•
pb200rnd03	200	1 000	1.0%	2	[1-20]	731	5 403.23	726	722.81	10.81	729	725.12	44.33	○
pb200rnd04	200	1 000	1.0%	2	[1-1]	64	63 970.91	63	63.00	9.12	64	62.93	24.33	○
pb200rnd05	200	1 000	2.5%	8	[1-20]	184	1 211.37	184	184.00	4.62	184	182.56	16.00	•
pb200rnd06	200	1 000	2.5%	8	[1-1]	14	8 068.20	14	13.37	3.48	14	12.87	4.00	•
pb200rnd07	200	200	1.5%	4	[1-20]	1 004	0.02	1 002	1 001.12	4.20	1 004	1 003.50	6.33	•
pb200rnd08	200	200	1.5%	4	[1-1]	83	0.04	83	82.87	2.7_	83	82.75	2.67	•
pb200rnd09	200	200	1.0%	2	[1-20]	1 324	0.01	1 324	1 324.00	3.75	1 324	1 324.00	7.33	•
pb200rnd10	200	200	1.0%	2	[1-1]	118	0.02	118	118.00	3.6_	118	118.00	4.00	•
pb200rnd11	200	200	2.5%	8	[1-20]	545	0.33	545	544.75	2.36	545	545.00	4.33	•
pb200rnd12	200	200	2.6%	8	[1-1]	43	1.70	43	43.00	1.01	43	43.00	1.33	•
pb200rnd13	200	600	1.5%	4	[1-20]	571	830.39	571	566.43	6.01	571	568.50	20.33	•
pb200rnd14	200	600	1.5%	4	[1-1]	45	10 066.91	45	45.00	3.92	45	44.43	8.67	•
pb200rnd15	200	600	1.0%	2	[1-20]	926	12.20	926	926.00	4.22	926	926.00	27.00	•
pb200rnd16	200	600	1.0%	2	[1-1]	79	14 372.85	79	78.31	6.80	79	78.37	15.33	•
pb200rnd17	200	600	2.5%	8	[1-20]	255	741.52	255	251.31	3.61	255	253.25	11.00	•
pb200rnd18	200	600	2.6%	8	[1-1]	19	19 285.06	19	18.06	2.35	19	18.12	3.00	•

Table 3. Instances and results (2/2)

Instances	Characteristics					0/1 Solution	GRASP			ACO			
	#var	#cst	Density	Max One	Weight	Best known value	Best value	Avg value	CPUt (s)	Best value	Avg value	CPUt (s)	Best known found
pb500rnd01	500	2 500	1.2%	10	[1-20]	323*	323	319.38	32.08	323	319.87	154.67	●
pb500rnd02	500	2 500	1.2%	10	[1-1]	24*	24	23.69	25.62	24	23.06	26.67	●
pb500rnd03	500	2 500	0.7%	5	[1-20]	776*	772	767.63	70.33	776	772.75	244.00	●
pb500rnd04	500	2 500	0.7%	5	[1-1]	61*	61	60.13	57.30	61	60.06	69.00	●
pb500rnd05	500	2 500	2.2%	20	[1-20]	122	122	121.50	15.48	122	120.62	71.00	●
pb500rnd06	500	2 500	2.2%	20	[1-1]	8*	8	8.00	12.08	8	7.87	9.67	●
pb500rnd07	500	2 500	1.2%	10	[1-20]	1 141	1 141	1 141.00	13.43	1 141	1 141.00	60.00	●
pb500rnd08	500	2 500	1.2%	10	[1-1]	89	89	88.25	15.80	89	88.12	21.67	●
pb500rnd09	500	2 500	0.7%	5	[1-20]	2 236	2 235	2 235.00	23.44	2 236	2 234.43	84.33	●
pb500rnd10	500	2 500	0.7%	5	[1-1]	179	179	178.06	18.20	179	178.31	44.67	●
pb500rnd11	500	2 500	2.3%	20	[1-20]	424	423	419.31	19.25	424	418.25	37.33	●
pb500rnd12	500	2 500	2.2%	20	[1-1]	33*	33	33.00	11.91	33	32.62	8.00	●
pb500rnd13	500	1 500	1.2%	10	[1-20]	474*	474	470.00	32.88	474	468.00	105.00	●
pb500rnd14	500	1 500	1.2%	10	[1-1]	37*	37	36.94	20.77	37	36.50	25.66	●
pb500rnd15	500	1 500	0.7%	5	[1-20]	1 196*	1 196	1 186.94	59.36	1 196	1 190.93	161.67	●
pb500rnd16	500	1 500	0.7%	5	[1-1]	88*	88	86.63	36.31	88	86.12	60.67	●
pb500rnd17	500	1 500	2.2%	20	[1-20]	192*	192	191.75	18.38	192	188.31	57.00	●
pb500rnd18	500	1 500	2.2%	20	[1-1]	13*	13	13.00	12.03	13	12.43	9.00	●
pb1000rnd100	1 000	5 000	2.60%	50	[1-20]	67	67	65.50	53.50	67	64.00	117.67	●
pb1000rnd200	1 000	5 000	2.59%	50	[1-1]	4	4	3.15	39.30	4	3.81	15.00	●
pb1000rnd300	1 000	5 000	0.60%	10	[1-20]	661*	649	639.50	221.20	661	640.50	700.67	●
pb1000rnd400	1 000	5 000	0.60%	10	[1-1]	48*	48	46.83	149.70	47	45.06	108.33	○
pb1000rnd500	1 000	1 000	2.60%	50	[1-20]	222*	222	217.98	64.80	222	219.62	85.67	●
pb1000rnd600	1 000	1 000	2.65%	50	[1-1]	15*	14	13.68	41.40	15	13.37	8.00	●
pb1000rnd700	1 000	1 000	0.58%	10	[1-20]	2 260	2 222	2 214.10	119.70	2 248	2 239.56	296.67	○
pb1000rnd800	1 000	1 000	0.60%	10	[1-1]	175*	172	170.81	82.60	173	170.50	94.00	○

* The asterisks indicate that we don't know if the best known solution is optimal

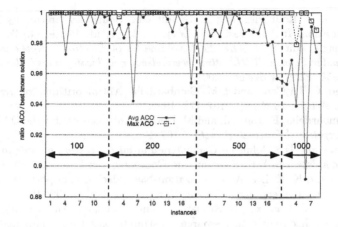

Fig. 2. Comparison between ACO and the best known solution for all instances

of ACO is related to the numerical "instability" of the output. In fact, on some instances, when performing 16 runs of this heuristic on the same instances it appears that the 16 returned solution values can be quite distinct. This leads us to the conclusion that more iterations should be allowed to the ACO heuristic to have a stronger convergence and increase the average calculated solution value. But, this would increase the required CPU time.

Besides, from the experiments we can derive that a local search should be applied in the ACO heuristic on unicost instances to provide as good results as on the weighted instances.

The conducted experiments presented in this section clearly show that the ACO heuristic proposed to solve the SPP performs well on the tested instances. Incorporating evolved search mechanisms, as such used in the GRASP heuristic with which we compared, may lead to a highly efficient ACO heuristic.

Acknowledgment

The authors would like to thank Coralie Paccou who worked during her Master thesis in 2001-2002 [11] on a first version of the ACO heuristic, under the supervision of Xavier Gandibleux and Xavier Delorme.

References

1. X. Delorme. *Modélisation et résolution de problèmes liés à l'exploitation d'infrastructures ferroviaires.* PhD thesis, Université de Valenciennes, Valenciennes, France, 2003.
2. X. Delorme, X. Gandibleux, and J. Rodriguez. GRASP for set packing problems. *European Journal of Operational Research*, 153 (3):564–580, 2004.

3. X. Delorme, J. Rodriguez, and X. Gandibleux. Heuristics for railway infrastructure saturation. In L. Baresi, J-J. Lévy, R. Mayr, M. Pezzè, G. Taentzer, and C. Zaroliagis, editors, *ICALP 2001*, volume 50 of *Electronic Notes in Theoretical Computer Science (URL: http://www.elsevier.nl/locate/entcs/volume50.html)*, pages 41–55. Elsevier Science, 2001.
4. M. Dorigo, G. Di Caro, and L.M. Gambardella. Ant algorithms for discrete optimization. *Artificial Life*, 5(3):137–172, 1999.
5. L.M. Gambardella, E. Taillard, and M. Dorigo. Ants colonies for the QAP. *Journal of the Operational Research Society*, 5:167–176, 1999.
6. M.R. Garey and D.S. Johnson. *Computers and intractability : a guide to the theory of NP-Completeness*. V.H. Freeman and Company, San Francisco, 1979.
7. S.-H. Kim and K.-K. Lee. An optimization-based decision support system for ship scheduling. *Computers and Industrial Engineering*, 33:689–692, 1997.
8. A. Mingozzi, V. Maniezzo, S. Ricciardelli, and L. Bianco. An exact algorithm for the project scheduling with ressource constraints based on a new mathematical formulation. *Management Science*, 44(5):714–729, mai 1998.
9. G.L. Nemhauser and L.A. Wolsey. *Integer and combinatorial optimization*. Willey-Interscience, New York, 1999.
10. I.H. Osman and G. Laporte. Metaheuristics : a bibliography. *Annals of Operations Research*, 63:513–623, 1996.
11. C. Paccou. L'algorithme des fourmis appliqué au Set Packing Problem. Master's thesis, Université de Valenciennes, Valenciennes, France, April 2002.
12. M.W. Padberg. On the facial structure of set packing polyhedra. *Mathematical Programming*, 5:199–215, 1973.
13. M. Rönnqvist. A method for the cutting stock problem with different qualities. *European Journal of Operational Research*, 83:57–68, 1995.
14. F. Rossi and S. Smriglio. A set packing model for the ground holding problem in congested networks. *European Journal of Operational Research*, 131:400–416, 2001.
15. T. Stützle. An ant approach for the flow shop problem. In *Proceedings of EUFIT'98, Aachen (Germany)*, pages 1560–1564, 1998.
16. T. Stützle and M. Dorigo. ACO algorithms for the traveling salesman problem. *in K. Miettinen, M. Makela, P. Neittaanmaki and J. Periaux (Eds.): Evolutionary Algorithms in Engineering and Computer Science: recent Advances in Genetic Algorithms, Evolution Strategies, Evolutionary Programming Genetic Programming and Industrial Applications, John Wiley & Sons*, pages 1560–1564, 1999.
17. V. T'kindt, N. Monmarché, F. Tercinet, and D. Laugt. An ant colony optimization algorithm to solve a 2-machine bicriteria flowshop scheduling problem. *European Journal of Operational Research*, 142(2):250–257, 2002.
18. P.J. Zwaneveld, L.G. Kroon, H.E. Romeijn, M. Salomon, S. Dauzère-Pérès, Stan P.M. Van Hoesel, and H.W. Ambergen. Routing trains through railway stations : Model formulation and algorithms. *Transportation Science*, 30(3):181–194, august 1996.

An Empirical Analysis of Multiple Objective Ant Colony Optimization Algorithms for the Bi-criteria TSP*

Carlos García-Martínez, Oscar Cordón, and Francisco Herrera

Dept. of Computer Science and Artificial Intelligence
University of Granada, 18071 - Granada, Spain
gcarlos@fedro.ugr.es, {ocordon,herrera}@decsai.ugr.es

Abstract. The difficulty to solve multiple objective combinatorial optimization problems with traditional techniques has urged researchers to look for alternative, better performing approaches for them. Recently, several algorithms have been proposed which are based on the Ant Colony Optimization metaheuristic. In this contribution, the existing algorithms of this kind are reviewed and experimentally tested in several instances of the bi-objective traveling salesman problem, comparing their performance with that of two well-known multi-objective genetic algorithms.

1 Introduction

Multi-criteria optimization problems are characterized by the fact that several objectives have to be simultaneously optimized, thus making especially difficult their solving. The existence of many multi-objective problems in the real world, their intrinsic complexity and the advantages of metaheuristic procedures to deal with them has strongly developed this research area in the last few years [11].

Ant colony optimization (ACO) is a metaheuristic inspired by the shortest path searching behavior of various ant species. Since the initial work of Dorigo, Maniezzo, and Colorni on the first ACO algorithm, the Ant System, several researchers have developed different ACO algorithms that performed succesfully when solving many different combinatorial problems [6].

Recently, some researchers have designed ACO algorithms to deal with multi-objective problems (MOACO algorithms) [1, 2, 4, 5, 7, 8, 10]. The most of them are specific proposals to solve a concrete multicriteria problem such as scheduling, vehicle routing, or portfolio selection, among others. The aim of the current contribution is to analyze the application of these proposals to the same benchmark problem, the multi-objective traveling salesman problem (TSP), and to compare their performance to two second generation, elitist multi-objective genetic algorithms (MOGAs) that currently represent the state-of-the-art of multi-objective evolutionary optimization: SPEA2 and NSGA-II [3].

* This work was partially supported by the Spanish Ministerio de Ciencia y Tecnología under project TIC2003-00877 (including FEDER fundings) and under Network HEUR TIC2002-10866-E.

M. Dorigo et al. (Eds.): ANTS 2004, LNCS 3172, pp. 61–72, 2004.

This paper is structured as follows. In Section 2, some basics of multi-objective optimization are reviewed. In Section 3, the existing MOACO algorithms are introduced, reporting their key characteristics. In Section 4, the performance of most of these algorithms is analyzed by applying them to the bi-objective TSP. Finally, some concluding remarks and proposals for future work are showed in Section 5.

2   Multi-objective Optimization

Multi-criteria optimization problems are characterized by the fact that several objectives have to by simultaneously optimized. Hence, there is not usually a single best solution solving the problem, but a set of solutions that are superior to the remainder when all the objectives are considered, the Pareto set. These solutions are known as Pareto-optimal or non-dominated solutions, while the remainder are known as dominated solutions. All of the former are equally acceptable as regards the satisfaction of all the objectives.

This way, the formal definition of the dominance concept is as follows. Let us consider, without loss of generality, a multiobjective minimization problem with m parameters (decision variables) and K objectives:

$$Min\ f(x) = (f_1(x), f_2(x), \dots, f_K(x)),\ \ with\ x = (x_1, x_2, \dots, x_m) \in X\ .$$

A decision vector $a \in X$ dominates another $b \in X$ ($a \succ b$) if, and only if:

$$\forall i \in 1, 2, \dots, K \mid f_i(a) \leq f_i(b) \quad \wedge \quad \exists j \in 1, 2,, \dots, K \mid f_j(a) < f_j(b)\ .$$

3   Multiple Objective Ant Colony Optimization Algorithms

In this section, the different existing proposals for Pareto-based MOACO algorithms that aim at obtaining set of non-dominated solutions for the problem being solved are reviewed.

3.1   Multiple Objective Ant-Q Algorithm

Multiple objective Ant-Q algorithm (MOAQ) is an MOACO algorithm that was proposed by Mariano and Morales in [10] to be applied to the design of water distribution irrigation networks. It was based on a distributed reinforcement learning algorithm called Ant-Q, a variant of the classical Ant Colony System (ACS). In Ant-Q, several autonomous agents learn an optimal policy $\pi : S \to A$, that outputs an appropriate action $a \in A$, given the current state $s \in S$, where A is the set of all possible actions in a state and S is the set of states. The available information to the agent is the sequence of immediate rewards $r(s_i, a_i)$ for all the possible actions and states $i = 0, 1, 2, \dots, Q(s, a)$, and a domain dependent heuristic value indicating how good is to select a particular action (a) being in the actual state (s), $HE(s, a)$. Each agent uses the following transition rule to select the action a according with the actual state s:

$$a = \begin{cases} arg \max_{a \in A}(HE^\alpha(s, a) \cdot Q^\beta(s, a)), & \text{if } q > q_0 \\ P, & \text{otherwise} \end{cases},$$

where α and β are parameters that weight the relative importance of pheromone trail and heuristic information, q is a random value in $[0,1]$, and q_0 $(0 \leq q_0 \leq 1)$ is calculated in every iteration as follows:

$$q_0 = \frac{q_0 \cdot \lambda}{q_{max}},$$

where $q_{max} \in [0, 1]$, and P is a random action selected according to the following probability distribution:

$$p(a) = \frac{HE^\alpha(s, a) \cdot Q^\beta(s, a)}{\sum_{b \in A} HE^\alpha(s, b) \cdot Q^\beta(s, b)},$$

with s being the current state.

The basic idea behind MOAQ is to perform an optimization algorithm with a family of agents (ants) for each objective. Each family k tries to optimize an objective considering the solutions found for the other objectives and its corresponding function HE^k. This way, all the agents from the different families act in the same environment proposing actions and expecting a reward value r which depends on how their actions helped to find trade-off solutions between the rest of the agents. The delayed reinforcement is computed as follows: $Q(s, a) = (1 - \rho) \cdot Q(s, a) + \rho \cdot [r(s, a) + \gamma \cdot Q(s', a')]$, where ρ is the learning step, s' and a' are the state and action in the next algorithm step, and γ is a discount factor.

Finally, MOAQ presents other three distinguishing characteristics. First, the j-th ant from the i-th family uses the solution found by the j-th ant of family i-1 while constructing its solution. Second, when a solution found is not feasible, the algorithm applies a punishment to its components on the Q values. And third, along the process, non-dominated solutions are stored in a external set, as usually done in elitist (second generation) MOGAs.

3.2   Ant Algorithm for Bi-criterion Optimization Problems

The so-called BicriterionAnt algorithm was designed by Iredi et al. in [8] to specifically solve a bi-criteria vehicle routing problem. To do so, it uses two different pheromone trail matrices, τ and τ', one for each of the criteria considered.

In every generation, each of m ants in the colony generates a solution to the problem. During its construction trip, the ant selects the next node j to be visited by means of the following probability distribution:

$$p(j) = \begin{cases} \dfrac{\tau_{ij}^{\lambda\alpha} \cdot \tau_{ij}'^{(1-\lambda)\alpha} \cdot \eta_{ij}^{\lambda\alpha} \cdot \eta_{ij}'^{(1-\lambda)\alpha}}{\sum_{h \in \Omega} \tau_{ih}^{\lambda\alpha} \cdot \tau_{ih}'^{(1-\lambda)\alpha} \cdot \eta_{ih}^{\lambda\alpha} \cdot \eta_{ij}'^{(1-\lambda)\alpha}}, & \text{if } j \in \Omega \\ 0, & \text{otherwise} \end{cases},$$

where α and β are the usual weighting parameters, η_{ij} and η'_{ij} are the heuristic values associated to edge (i,j) according to the first and the second objective, respectively, Ω is the current feasible neighborhood of the ant, and λ is computed for each ant t, $t \in \{1, \ldots, m\}$, as follows:

$$\lambda_t = \frac{t-1}{m-1} \; .$$

Once all the ants have generated their solutions, the pheromone trails are evaporated by applying the usual rule on every edge (i,j): $\tau_{ij} = (1 - \rho) \cdot \tau_{ij}$, with $\rho \in [0,1]$ being the pheromone evaporation rate.

Then, every ant that generated a solution in the non-dominated front at the current iteration is allowed to update both pheromone matrices, τ and τ', by laying down an amount equal to $\frac{1}{l}$, with l being the number of ants currently updating the pheromone trails. The non-dominated solutions generated along the algorithm run are kept in an external set, as it happened in MOAQ.

3.3 Multi Colony for Bi-criterion Optimization Problems

In the same contribution [8], Iredi et al. proposed another MOACO algorithm, BicriterionMC, very similar to the previous BicriterionAnt. The main difference between them is that each ant only updates one pheromone matrix in the new proposal. The authors introduce a general definition for the algorithm with p pheromone trail matrices and then make $p = 2$ to solve bi-criterion problems. To do so, they consider two different methods for the pheromone trail update:

- Method 1 - *Update by origin*: an ant only updates the pheromone trails in its own colony. This method enforces both colonies to search in different regions of the non-dominated front. The algorithm using method 1 is called UnsortBicriterion.
- Method 2 - *Update by region*: the sequence of solutions along the non-dominated front is split into p parts of equal size. Ants that have found solutions in the i-th part update the pheromone trails in colony i, $i \in [1, p]$. The aim is to explicitly guide the ant colonies to search in different regions of the Pareto front, each of them in one region. The algorithm using method 2 is called BicriterionMC.

The other difference is the way to compute the value of the transition parameter λ for the ant t. The authors describe three different rules and propose to use the third of them which gives better results. This rule overlaps the λ-interval of colony i in a 50% with the λ-interval of colony $i - 1$ and colony $i + 1$. Formally, colony i has ants with λ-values in $[\frac{i-1}{p+1}, \frac{i+1}{p+1}]$, with p being equal to the number of colonies.

3.4 Pareto Ant Colony Optimization

Pareto Ant Colony Optimization (P-ACO), proposed by Doerner et al. in [5], was originally applied to solve the multi-objective portfolio selection problem.

It considers the classical ACS as the underlying ACO algorithm but the global pheromone update is performed by using two different ants, the best and the second-best solutions generated in the current iteration for each objective k. In P-ACO, several pheromone matrices τ^k are considered, one for each objective k. At every algorithm iteration, each ant computes a set of weights $p = (p_1, \ldots, p_k)$, and uses them to combine the pheromone trail and heuristic information. When an ant has to select the next node to be visited, it uses the ACS transition rule considering the k pheromone matrices:

$$j = \begin{cases} arg\ \max_{j \in \Omega}([\sum_{k=1}^{K} p_k \cdot \tau_{ij}^k]^\alpha \cdot \eta_{ij}^\beta, & \text{if } q \leq q_0 \\ \hat{\imath}, & \text{otherwise} \end{cases},$$

where K is the number of objectives, η_{ij} is an aggregated value of attractiveness of edge (i, j) used as heuristic information, and $\hat{\imath}$ is a node selected according to the probability distribution given by:

$$p(j) = \begin{cases} \dfrac{\left[\sum_{k=1}^{K} p_k \cdot \tau_{ij}^k\right]^\alpha \cdot \eta_{ij}^\beta}{\sum_{h \in \Omega} \left[\sum_{k=1}^{K} p_k \cdot \tau_{ih}^k\right]^\alpha \cdot \eta_{ih}^\beta}, & \text{if } j \in \Omega \\ 0, & \text{otherwise} \end{cases}.$$

Every time an ant travels an edge (i, j), it performs the local pheromone update in each pheromone trail matrix, i.e., for each objective k, as follows: $\tau_{ij}^k = (1 - \rho) \cdot \tau_{ij}^k + \rho \cdot \tau_0$, with ρ being the pheromone evaporation rate, and τ_0 being the initial pheromone value.

The global pheromone trail information is updated once each ant of the population has constructed its solution. The rule applied for each objective k is as follows: $\tau_{ij}^k = (1 - \rho) \cdot \tau_{ij}^k + \rho \cdot \Delta\tau_{ij}^k$, where $\Delta\tau_{ij}^k$ has the following values:

$$\Delta\tau_{ij}^k = \begin{cases} 15, & \text{if edge } (i, j) \in \text{ best and second-best solutions} \\ 10, & \text{if edge } (i, j) \in \text{ best solution} \\ 5, & \text{if edge } (i, j) \in \text{ second-best solution} \\ 0, & \text{otherwise} \end{cases}.$$

Along the process, the non-dominated solutions found are stored in an external set, as in the previous MOACO algorithms.

3.5 Multiple Ant Colony System

Multiple Ant Colony System (MACS) [1] was proposed as a variation of the MACS-VRPTW introduced in [7]. So, it is also based in ACS but, contrary to its predecessor, MACS uses a single pheromone matrix, τ, and several heuristic information functions, η_k, initially two, η^0 and η^1. In this way, an ant moves from node i to node j by applying the following rule:

$$j = \begin{cases} arg\ \max_{j \in \Omega} \left(\tau_{ij} \cdot [\eta_{ij}^0]^{\lambda\beta} \cdot [\eta_{ij}^1]^{(1-\lambda)\beta}\right), & \text{if } q \leq q_0 \\ \hat{\imath}, & \text{otherwise} \end{cases},$$

where β weights the relative importance of the objectives with respect to the pheromone trail, λ is computed for each ant t as $\lambda = \frac{t}{m}$, with m being the number of ants, and $\hat{i}$ is a node selected according to the following probability distribution:

$$p(j) = \begin{cases} \dfrac{\tau_{ij} \cdot [\eta_{ij}^0]^{\lambda\beta} \cdot [\eta_{ij}^1]^{(1-\lambda)\beta}}{\sum_{h \in \Omega} \tau_{ih} \cdot [\eta_{ih}^0]^{\lambda\beta} \cdot [\eta_{ih}^1]^{(1-\lambda)\beta}}, & \text{if } j \in \Omega \\ 0, & \text{otherwise} \end{cases}.$$

Every time an ant crosses the edge (i, j), it performs the local pheromone update as follows: $\tau_{ij} = (1 - \rho) \cdot \tau_{ij} + \rho \cdot \tau_0$.

Initially, τ_0 is calculated from a set of heuristic solutions by taking their average costs in each of the two objective functions, f^0 and f^1, and applying the following expression:

$$\tau_0 = \frac{1}{\hat{f}^0 \cdot \hat{f}^1}.$$

However, the value of τ_0 is not fixed during the algorithm run, as usual in ACS, but it undergoes adaptation. Every time an ant t builds a complete solution, it is compared to the Pareto set P generated till now to check if the former is a non-dominated solution. At the end of each iteration, τ_0' is calculated by applying the previous formula with the average values of each objective function taken from the solutions currently included in the Pareto set.

Then, if $\tau_0' > \tau_0$, the current initial pheromone value, the pheromone trails are reinitialized to the new value $\tau_0 \leftarrow \tau_0'$.

Otherwise, the global update is performed with each solution x of the current Pareto optimal set P by applying the following rule on its composing edges (i, j):

$$\tau_{ij} = (1 - \rho) \cdot \tau_{ij} + \frac{\rho}{f^0(x) \cdot f^1(x)}.$$

3.6   Multi-objective Network ACO

Multi-objective Network ACO (MONACO) is quite different to the remaining algorithms reviewed in this section as it was designed to be applied to a dynamic problem, the optimization of the message traffic in a network [2]. Hence, the policy of the network changes according to the algorithm's steps, and it does not wait till the algorithm ends up. In the following, we present an adaptation of the original algorithm, developed by ourselves, to use MONACO in static problems. It requires several modifications such as the fact that the ants have to wait the cycle ends before updating the pheromone trails. The original algorithm takes the classical AS as a base but uses several pheromone trail matrices, τ^k. Each ant, which is defined as a message, uses the multi-pheromone trail and a single heuristic information to choose the next node to visit, according to the following probability distribution:

$$p(j) = \begin{cases} \dfrac{\eta_{ij}^{\beta} \cdot \prod_{k=1}^{K} (\tau_{ij}^k)^{\alpha_k}}{\sum_{h \in \Omega} \eta_{ih}^{\beta} \cdot \prod_{k=1}^{K} (\tau_{ih}^k)^{\alpha_k}}, & \text{if } j \in \Omega \\ 0, & \text{otherwise} \end{cases}.$$

In this formula, η_{ij} is the heuristic information for the edge (i, j), both β and the different α_k's weight the importance of each pheromone matrix value and the heuristic information, K is the number of objective functions and Ω is the feasible neighborhood of the ant at this step. After each cycle, the pheromone trails associated to every edge visited by at least one ant in the current iteration is evaporated in the usual way: $\tau_{ij}^k = (1 - \rho_k) \cdot \tau_{ij}^k$, with ρ_k being the pheromone evaporation rate for objective k (notice that a different one is considered for each pheromone trail matrix). Then, the pheromone trails of these edges are updated. Every ant lays down the following amount of pheromone in each edge (i, j) used by each ant and for every objective k:

$$\Delta\tau_{ij}^k = \frac{Q}{f^k(x)} \; ,$$

where Q is a constant related to the amount of pheromone laid by the ants, f^k is the objective function k, and x is the solution built by the ant. As said, the aim of the original algorithm is not to find a good set of non-dominated solutions but to make the network work efficiently. To apply it to static optimization problems, we have to store the non-dominated solutions generated in each run in an external set.

3.7 COMPETants

Initially, Doerner et al. introduced COMPETants to deal with bi-objective transportation problems [4]. The algorithm, based on the rankAS, used two ant colonies, each with its own pheromone trail matrix, τ^0 and τ^1, and its heuristic information, η^0 and η^1. An interesting point is that the number of ants in each population is not fixed but undergoes adaptation. When every ant has built its solution, the colony which has constructed better solutions gets more ants for the next iteration. The ants walk through the edges using the following probability distribution to select the next node to be visited (notice that it is adaptation of the AS transition rule to the case of multiple heuristic and pheromone trail values):

$$p(j) = \begin{cases} \dfrac{\tau_{ij}^{k\alpha} \cdot \eta_{ij}^{k\beta}}{\sum_{h \in \Omega} \tau_{ih}^{k\alpha} \cdot \eta_{ih}^{k\beta}}, & \text{if } j \in \Omega \\ 0, & \text{otherwise} \end{cases} .$$

Each ant uses the τ and η values of its colony k. Besides, some ants in every population, called spies, use another rule combining the information of both pheromone trails:

$$p(j) = \begin{cases} \dfrac{[0.5 \cdot \tau_{ij} + 0.5 \cdot \tau_{ij}']^\alpha \cdot \eta_{ij}^{k\beta}}{\sum_{h \in \Omega} [0.5 \cdot \tau_{ih} + 0.5 \cdot \tau_{ih}']^\alpha \cdot \eta_{ih}^{k\beta}}, & \text{if } j \in \Omega \\ 0, & \text{otherwise} \end{cases} .$$

In this new rule, τ is the ant's pheromone trail and τ' the pheromone trail of the other colony. η^k is the ant's heuristic information.

When every ant has built its solution, the pheromone trails of each edge (i, j) are evaporated: $\tau_{ij}^k = (1 - \rho) \cdot \tau_{ij}^k$.

Then, the Γ best ants of each population deposit pheromone on the edges visited using its own pheromone trail matrix and the following rule:

$$\Delta \tau_{ij}^\lambda = 1 - \frac{\lambda - 1}{\Gamma} \ ,$$

where λ is the position of the ant in the sorted list of the Γ best ants.

Before the end of the cycle, every ant is assigned to the first colony with the following probability:

$$\frac{\hat{f}^1}{\hat{f}^0 + \hat{f}^1} \ ,$$

with $\hat{f}^1$ being the average of the costs of the solutions in the second colony (that associated to the second objective function), and $\hat{f}^0$ being the average of the costs of the first colony (that corresponding to the first objective function). The remaining ants will be assigned to the second colony.

Finally, the number of spies in both colonies is randomly calculated with the following probability:

$$\frac{f(best)}{4 \cdot f'(best') + f(best)} \ ,$$

where f is the objective function of the current colony, f' is the objective function associated to the other colony, $best$ is the best ant of the current colony according to f and $best'$ is the best ant of the other according to f'.

4   Experimental Results

4.1   Experimental Setup

In our case, the K-objective symmetric TSP instances considered have been obtained from Jaszkiewicz's web page: http://www-idss.cs.put.poznan.pl/~jaszkiewicz/, where several instances usually tackled in evolutionary multi-objective optimization are collected. Each of these instances was constructed from $K = 2$ different single objective TSP instances having the same number of towns. The interested reader can refer to [9] for more information on them. In this first study, we will use four bi-criteria TSP instances: Kroab50, Krocd50, Krobc100, and Kroad100.

To perform the experimental comparison, besides the eight MOACO algorithms reviewed in Section 3, two of the most known, Pareto-based second generation MOGAs (which represent the state-of-the-art in multi-objective evolutionary optimization, see [3]) are considered as baselines for the MOACO algorithms performance. Hence, ten different algorithms will be run on the four bi-criteria TSP instances selected.

We have followed the same comparison methodology that Zitzler et al. in [12]. The comparison metric is thus based on comparing a pair of non-dominated sets by computing the fraction of each set that is covered by the other:

$$C(X', X'') = \frac{|\{a'' \in X''; \exists a' \in X' : a' \succ a''|}{|X''|} ,$$

where $a' \succ a''$ indicates that the solution a' dominates the solution a''.

Hence, the value $C(X', X'') = 1$ means that all the solutions in X'' are dominated by or equal to solutions in X'. The opposite, $C(X', X'') = 0$, represents the situation where none of the solutions in X'' are covered by the set X'. Note that both $C(X', X'')$ and $C(X'', X')$ have to be considered, since $C(X', X'')$ is not necessarily equal to $1 - C(X'', X')$.

In addition, we will compare the number of iterations and evaluations needed by each algorithm to generate its Pareto front.

As seen in Section 3, the eight MOACO algorithms considered are characterized by tackling very diverse applications. Hence, several changes have been done in our implementations to adapt them to solve the multi-objective TSP, moreover than making them manage an appropriate problem representation, the usual permutation. These changes are not reported due to the lack of space.

Each of the considered multi-objective algorithms (both MOACO and MO-GAs) has been run ten times for each of the four bi-criteria TSP instances during 300 seconds (small instances) and 900 seconds (large ones) in an Intel Celeron^{TM} 1200MHz computer with 256 MB of RAM. The generic parameter values considered are the usual ones when applying MOACO and MOGAs to the TSP problem: 20 ants, $(\alpha, \beta) = (1, 2)$, $\rho = 0.2$, $q_0 = 0.98$ (0.99 for MOAQ). To choose the initial pheromone trail value τ_0, we have employed the same method used by each algorithm to update the pheromone trails, obtaining it from the greedy solution of each of the K single-objective problems.

On the other hand, the values of the specific parameters, such as λ in MOAQ, have been obtained from the papers where the algorithms using them were defined ($\lambda = 0.9$, $\gamma = 0.4$). As regards the MOGAs, the population include 100 individuals (80 plus 20 in the elite population in SPEA2), and the crossover and mutation probabilities are respectively 0.8 and 0.1.

4.2   Results Obtained and Analysis of Results

All the experimental data are reported in the form of box-plots, where the minimum, maximum and median values as well as the upper and lower quartiles (upper and lower ends of the box) are graphically represented. Figures 1 and 2 show the statistics of the ten runs of each algorithm in each bi-criteria TSP instances (in a logarithmic scale and grouped in a single box-plot for each pair of instances of the same size) . In view of these graphics, we can see that the MOGAs, NSGA-II and SPEA2, can perform much more iterations and evaluations than the MOACO algorithms considered in the same fixed run time. Hence, this shows how MOACO algorithms are "slower" than MOGAs in the current

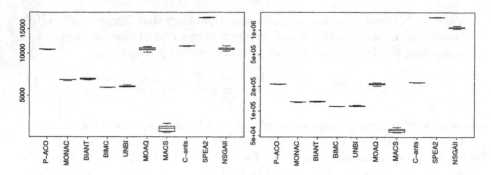

Fig. 1. Number of iterations and evaluations in the Kroab50 and Krocd50 runs

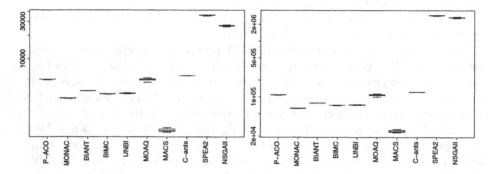

Fig. 2. Number of iterations and evaluations in the Krobc100 and Kroad100 runs

problem. Notice also that MACS is the algorithm which performs less iterations and evaluations while COMPETants is the quickest of the MOACO algorithms.

The graphics in Figure 3 are box-plots based on the C metric. Each rectangle contains four box-plots (from left to right, Kroab50, Krocd50, Krobc100, and Kroad100) representing the distribution of the C values for a certain ordered pair of algorithms. Each box refers to algorithm A associated with the corresponding row and algorithm B associated with the corresponding column and gives the fraction of B covered by A ($C(A, B)$). Consider, for instance, the top right box, which represents the fraction of solutions of SPEA2 covered by the non-dominated sets produced by the P-ACO algorithm.

In view of the box-plots of metric C, we can draw the conclusion that the MOACO algorithms considered are very competitive against the MOGAs implemented. The former offer good sets of non-dominated solutions which almost always dominate the solutions returned by NSGA-II and SPEA2. In addition, the Pareto fronts derived by the MOGAs do not dominate any of the fronts given by MOACO algorithms.

It is not easy to say which MOACO algorithm performs best, as all of them derive Pareto sets of similar quality. Maybe P-ACO could be considered as the algorithm with the best global performance as its Pareto fronts are few dominated by the remainder of the MOACO algorithms (see the box-plots in the

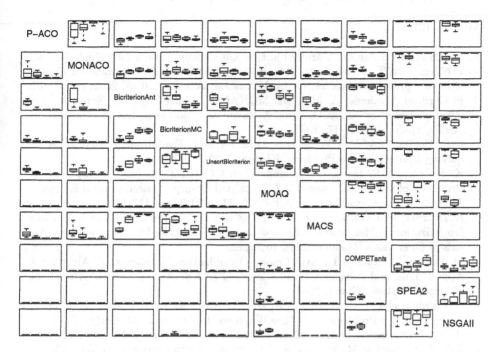

Fig. 3. Box-plots of the results obtained in the C metric

first column), while they dominate the remainder to some degree (see those in the first row). However, it is easier to identify the algorithms which usually return Pareto sets of bad quality for the problem. This is the case of SPEA2, NSGA-II, COMPETants, and MOAQ. Besides, we should notice that all of these non-dominated solution sets are very well dominated by those got by MACS.

5   Concluding Remarks and Future Works

In the current contribution, we have developed a experimental study comparing the performance of the different existing proposals for Pareto-based MOACO algorithms when applied to several classical instances of the bi-criteria TSP. From the results obtained, we have drawn the conclusion that the MOACO algorithms considered are more competitive in the current problem than two of the state-of-the-art MOGAs, SPEA-II and NSGA2.

Several ideas for future developments arise from this preliminary study: to analyze (i) the performance of the considered MOACO algorithms in other, larger instances of the multi-objective TSP; (ii) the influence of adding local search optimizers to the MOACO algorithms and to compare the performance of the resulting techniques against that of memetic algorithms such as Jaszkiewicz's MOGLS [9]; (iii) the performance of MOACO algorithms in other complex multi-objective combinatorial optimization problems.

References

1. B. Barán, M. Schaerer, A Multiobjective Ant Colony System for Vehicle Routing Problem with Time Windows, Proc. Twenty first IASTED International Conference on Applied Informatics, Insbruck, Austria, February 10-13, 2003, pp. 97-102.
2. P. Cardoso, M. Jesús, A. Márquez, MONACO - Multi-Objective Network Optimisation Based on an ACO, Proc. X Encuentros de Geometría Computacional, Seville, Spain, June 16-17, 2003.
3. C.A. Coello, D.A. Van Veldhuizen, G.B. Lamant, Evolutionary Algorithms for Solving Multi-objective Problems, Kluwer, 2002.
4. K. Doerner, R.F. Hartl, M. Teimann, Are COMPETants More Competent for Problem Solving? - The Case of Full Truckload Transportation, Central European Journal of Operations Research (CEJOR), 11:2, 2003, pp. 115-141.
5. K. Doerner, W.J. Gutjahr, R.F. Hartl, C. Strauss, C. Stummer, Pareto Ant Colony Optimization: A Metaheuristic Approach to Multiobjective Portfolio Selection, Annals of Operations Research, 2004, to appear.
6. M. Dorigo, T. Stützle, The Ant Colony Optimization Metaheuristic: Algorithms, Applications, and Advances, In: F. Glover, G.A. Kochenberger (Eds.), Handbook of Metaheuristics, Kluwer, 2003.
7. L. Gambardella, E. Taillard, G. Agazzi, MACS-VRPTW: A Multiple ACS for Vehicle Routing Problems with Time Windows, In: D. Corne, M. Dorigo, F. Glover (Eds.), New Ideas in Optimization, McGraw-Hill, 1999, pp. 73-76.
8. S. Iredi, D. Merkle, M. Middendorf, Bi-Criterion Optimization with Multi Colony Ant Algorithms, Proc. First International Conference on Evolutionary Multicriterion Optimization (EMO'01), LNCS 1993, 2001, pp. 359-372.
9. A. Jaszkiewicz. Genetic Local Search for Multi-objective Combinatorial Optimization, European Journal of Operational Research, 137:1, 2002, pp. 50-71.
10. C.E. Mariano, E. Morales, A Multiple Objective Ant-Q Algorithm for the Design of Water Distribution Irrigation Networks, Technical Report HC-9904, Instituto Mexicano de Tecnología del Agua, June 1999.
11. E.L. Ulungu, J. Teghem, Multi-objective Combinatorial Optimization: A Survey, Journal of Multi-Criteria Decision Analysis, 3, 1994, pp. 83-104.
12. E. Zitzler, K. Deb, L. Thiele, Comparison of Multiobjective Evolutionary Algorithms: Empirical Results, Evolutionary Computation, 8:2, 2000, pp. 173-195.

An External Memory Implementation in Ant Colony Optimization

Adnan Acan

Computer Engineering Dept., Eastern Mediterranean University, Gazimağusa
T.R.N.C. Mersin 10, Turkey
adnan.acan@emu.edu.tr

Abstract. An ant colony optimization algorithm using a library of partial solutions for knowledge incorporation from previous iterations is introduced. Initially, classical ant colony optimization algorithm runs for a small number of iterations and the library of partial solutions is initialized. In this library, variable size solution segments from a number of elite solutions are stored and each segment is associated with its parent's objective function value. There is no particular distribution of ants in the problem space and the starting point for an ant is the initial point of the segment it starts with. In order to construct a solution, a particular ant retrieves a segment from the library based on its goodness and completes the rest of the solution. Constructed solutions are also used to update the memory. The proposed approach is used for the solution of TSP and QAP for which the obtained results demonstrate that both the speed and solution quality are improved compared to conventional ACO algorithms.

1 Introduction

Ant colony optimization (ACO) is a general-purpose metaheuristic which can be applied to many kinds of optimization problems. This nature-inspired computation method is based on observations of the foraging behavior of natural ant colonies which choose the shortest path between their nests and the food sources. This collective problem solving ability is a result of a reinforcement process in which ants deploy a pheromone trail as they walk and this trail attracts other ants to take the path having the most pheromone concentration. Since ants following the shortest path will make higher number of tours between their nests and the food source, more pheromone will be deposited on the shortest path than on the longer ones. This ant-based optimization principle combined with a pheromone update strategy, to avoid premature convergence to locally optimal solutions, is transformed into a remarkable optimization algorithm, the ACO algorithm, and applied for the solution of a wide variety of combinatorial optimization problems [2], [4], [5], [17].

Among many efforts on the development of new variants of ACO algorithms toward improving their efficiency under different circumstances, recently the idea of knowledge incorporation from previous iterations became attractive and

M. Dorigo et al. (Eds.): ANTS 2004, LNCS 3172, pp. 73–82, 2004.

handled by a number of researchers. Mainly, these population- or memory-based approaches take their inspiration from studies in genetic algorithms. In fact, in genetic algorithms, knowledge incorporation from previous generations has been studied and a number powerful strategies were developed. In memory-based implementations, information stored within a memory is used to adapt the GAs behavior either in problematic cases where the solution quality is not improved over a number of iterations, or a change in the problem environment is detected, or to provide further directions of exploration and exploitation. Memory in GAs can be provided externally (outside the population) or internally (within the population) [6].

External memory implementations store specific information within a separate population (memory) and reintroduce that information into the main population at a later moment. In most cases, this means that individuals from memory are put into the initial population of a new or restarted GA. One of the earliest external memory approach is the work of Cobb that is based on a simple triggered hypermutation mechanism to deal with continuously changing environments [3]. In this approach, the quality of the best solution is monitored over time and a declination of this measure is taken as a sign of change in problem environment. In this method, knowledge incorporation into a GA is carried out in a blind fashion by creating random search directions.

Case-based memory approaches, which are actually a form of long term elitism, are the most typical form of external memory implemented in practice. In general, there are two kinds of case-based memory implementations: in one kind, case-based memory is used to re-seed the population with the best individuals from previous generations when a change in the problem domain takes place [14]. A different kind of case-based memory stores both problems and solutions. When a GA has to solve a problem similar to problems in its case-based memory, it uses the stored solutions to seed the initial population [11], [12]. Case-based memory aims to increase the diversity by reintroducing individuals from previous generations and achieves exploitation by reusing individuals from case-based memory when a restart from a good initial solution is required.

Other variants of external memory approaches are provided by several researchers for both specific and general purpose implementations. Simoes and Costa introduced an external memory method in which gene segments are stored instead of the complete genomes. In their so called transformation-based GA, TGA, there is no mutation operator and the crossover is replaced by an asexual transformation operator which retrieves gene segments from a gene pool (an external memory) and inserts them into randomly chosen regions within the selected parent individuals [15], [16]. The idea implemented within the proposed approach is very similar to what the authors did for genetic algorithms. Recently, Acan et al. proposed a novel external memory approach based on the reuse of insufficiently utilized promising chromosomes from previous generations for the production of current generation offspring individuals. In the implementation, a chromosome library is established to hold above-average quality solutions from previous generations and individuals in this library are associated with lifetimes that are updated after every generation. Chromosome library is combined with the current population in the generation of offspring individuals. Then, those library elements running out of their age limit are replaced by new promising and under-utilized individuals from the current population [1].

The most common approaches using internal memory are polyploidy structures. Polyploidy structures in combination with dominance mechanisms use redundancy in genetic material by having more than one copy of each gene. From biological point of view, polyploidy is a form of memory of previously useful adaptations which may be useful when a change in the environment occurs. When a chromosome is decoded to determine the corresponding phenotype, the dominant copy is chosen. This way, some genes can shield themselves from extinction by becoming recessive. Through switching between copies of genes, a GA can adapt faster to changing environments and recessive genes are used to provide information about fitness values from previous generations [7], [10].

In ACO the first internally implemented memory-based approach is the work of Montgomery et al. [13]. In their work, named as AEAC, they modified the characteristic element selection equations of ACO to incorporate a weighting term for the purpose of accumulated experience. This weighting is based on the characteristics of partial solutions generated within the current iteration. Elements that appear to lead better solutions are valued more highly, while those that lead to poorer solutions are made less desirable. They aim to provide, in addition to normal pheromone and heuristic costs, a more immediate and objective feedback on the quality of the choices made. Basically, considering the TSP, if a link (r, u) has been found to lead to longer paths after it has been incorporated into a solution, then its weight $w(r, u) < 1$. If the reverse is the case, then $w(r, u) > 1$. If the colony as a whole has never used the link (r, u), then its weight is selected as 1. The authors suggested simple weight update procedures and proposed two variations of their algorithm. They claimed that the achieved results for different TSP instances are either equally well or better than those achieved using normal ACS algorithm.

The work of Guntsch et al. [8] is the first example of an external memory implementation within ACO. Their approach, P-ACO, uses a population of previously best solutions from which the pheromone matrix can be derived. Initially the population is empty and, for the first k iteration of ants, the best solutions found in each iteration enters the population. After that, to update the population, the best solution of an iteration enters the population and the oldest one is removed. That is, the population is maintained like a FIFO-queue. This way, each solution in the population influences the decisions of ants over exactly k iterations. For every solution in the population, some amount of pheromone is added to the corresponding edges of the construction graph. The whole pheromone information of P-ACO depends only on the solutions in the population and the pheromone matrix is updated as follows: whenever a solution π enters the population, do a positive update as $\forall i \in [1, n] : \tau_{i\pi(i)} \rightarrow \tau_{i\pi(i)} + \Delta$ and whenever a solution σ leaves the population, do a negative update as $\forall i \in [1, n] : \tau_{i\sigma(i)} \rightarrow \tau_{i\sigma(i)} - \Delta$. These updates are added to the initial pheromone value τ_{init}. The authors also proposed a number of population update strategies in [9] to decide which solutions should enter the population and which should leave. In this respect, only the best solution generated during the past iteration is considered as a candidate to enter the population and the measures used in population update strategies are stated as age, quality, prob, age and prob, and elitism. In age strategy, the oldest solution is removed from the population. In quality strategy, the population keeps the best k solutions found over all past iterations, rather than the best solutions of the last k iterations. Prob strategy probabilistically chooses the element to be removed form the population and the aim is to reduce the number of identical copies that might be caused the quality strategy. Combination of age and prob strategies use prob for removal and age for insertion into the population. In elitist strategy, the best solution found by the algorithm so far is never replaced until a better solution is found.

This paper introduces another population based external memory approach where the population includes variable-size solution segments taken from elite individuals of previous iterations. Initially, the memory is empty and a classical ant colony optimization algorithm runs for a small number of iterations to initialize the partial solutions library. Each stored solution segment is associated with its parent's objective function value that will be used as measure for segment selection and in updating the memory. There is no particular distribution of ants in the problem space and the starting point for an ant is the initial point of the segment it starts with. In order to construct a solution, a particular ant retrieves a segment from the library based on its goodness and completes the rest of the solution. Constructed solutions are also used to update the memory. The details of the practical implementation are given in the following sections.

This paper is organized as follows. The basics solution construction procedures for the travelling salesman problem (TSP) and the quadratic assignment problem (QAP) are presented in Section 2. The proposed approach is described with its implementation details in Section 3. Section 4 covers the results and related discussions. Conclusions and future research directions are given in Section 5.

2   ACO for TSP and QAP Problems

In this section, the basic solution construction processes of ACO algorithms for the solution of TSP and QAP will be briefly described.

In its classical implementation, an ACO algorithm uses a number of artificial ants for the construction of solutions such that each ant starts from a particular assignment and builds a solution in an iterative manner. In this respect, the TSP and QAP are two most widely handled problems for the illustration and testing of ACO strategies.

In TSP, there are n cities which are assumed to be located on the nodes of a construction graph where edge weights are the pairwise distances between the cities. The goal is to find a minimum-length tour over the n cities such that the tour passes through each city only once, except the starting city that is also the termination point for the tour. To construct a solution, an ant starts from a particular city and proceeds over neighbor cities, with a list of already visited cities in its internal memory, until the tour is complete [17].

In QAP, there are n facilities, n locations, two $n \times n$ matrices, D and F, such that $d_{ij} \in D$ is the distance between location i and j and $f_{ij} \in F$ is the amount of flow between facilities i and j. The goal is to assign facilities to locations such that the cost function $\sum_{i=1}^{n} \sum_{j=1}^{n} d_{\pi(i)\pi(j)} f_{(i,j)}$ is minimized. An ant builds a solution by going to a randomly chosen unassigned location and placing one of the remaining facilities there, until no free locations are left [18].

In the construction of a solution, the decision of an ant for its next assignment is affected by two factors: the pheromone information and the heuristic information which both indicate how good it is to choose the item under consideration for its assignment to the current position.

3 The Proposed External Memory Approach

In our approach, there are m ants that build the solutions. An external memory of variable-size solution segments from a number of elite solutions of previous iterations is maintained. Initially the memory is empty and a number classical ACO iterations is performed to fill in the memory. In this respect, after every iteration, the top k-best solutions are considered and randomly positioned variable-size segments are cut from these individuals, one segment per solution, and stored in the memory. The memory size M is fixed and normal ACO iterations are repeated until the memory is full.

After the initial phase, ACO algorithm works in conjunction with the implemented external memory as follows: there is no particular assignment of ants over the problem space and, in order to construct a solution, each ant selects a solution segment from the memory using a tournament selection strategy and the starting point for the corresponding ant becomes the starting point of the partial solution. That is, the ant builds the rest of the solution starting from the end point of the segment. In the selection of solution segments from memory, each ant makes a tournament among Q randomly selected solution segments and the best segment is selected as the seed to start for construction. The solution construction procedure is the same as the one followed in Max-Min AS algorithm. After all ants complete their solution construction procedures, pheromone updates are carried in exactly the same way it is done in normal ACO algorithm. Based on these descriptions the proposed ACO algorithm can be put into an algorithmic form described in Figure 1.

In memory update procedure, the solutions constructed by all ants are sorted in ascending order and the top k-best are considered as candidate parents from which new segments will be inserted into memory. One randomly-positioned and random-length segment is cut from each elite solution and two strategies are followed in updating the memory elements. First of all, a new segment replaces the worst of all the segments having an associated cost larger than the cost of itself. If no such segment exists, all existing segments have equal or lower costs than the newly cut segment, then, for the purpose of creating diversity, the cut is concatenated with the highest cost memory element and the repeated data values removed from the resulting solution segment. The main purpose of this segment concatenation operation is to achieve additional diversification through the combination of two promising solution segments.

Algorithm: ACO with an external memory

1. Initialize the external memory.
2. Repeat
   2.1 Place m ants within the construction graph.
   2.2 Let all ants construct a solution using the neighborhood graph and the pheromone matrix.
   2.3 Sort the solutions in ascending order of their objective function values.
   2.4 Consider the top k-best and cut a randomly-positioned and randomly-sized segment from each solution and insert the segment into the memory. Also, store the length of the segment and the cost of its parent.
   2.5 Update the pheromone matrix.
3. Until the memory is full.
4. DONE=FALSE
5. While (NOT DONE)
   5.1 Let all ants select a solution segment from the library using tournament selection.
   5.2 Let all ants construct a solution starting from the partial solution they retrieved from the library.
   5.3 Update pheromone matrix.
   5.4 Update memory.
6. End While

Fig. 1. ACO with an external solution-segment library

4   Experimental Results

To study the performance of the proposed external memory ACO approach, it is compared with a well-known classical implementation of ACO algorithms, namely the Max-Min AS algorithm, for the solution of two widely known and difficult combinatorial optimization problems, the TSP and the QAP. The TSP problem instances considered are taken from the web-site http://www.iwr.uni-heidelberg.de/groups/comopt/software/TSPLIB95/, and the QAP problem instances used in evaluations are taken from the web-site http://www.opt.math.tu-graz.ac.at/qaplib. The TSP and QAP instances used in experimental evaluations are given in Table 1. These problem instances are selected for comparison, since they are representative instances of different problem groupes, they are also commonly handled by several researchers, and they reasonable problem sizes for experimental evaluations.

Table 1. TSP and QAP instances used in experimental evaluations

TSP Instances			QAP Instances		
Instance	Description	Best Known	Instance	Description	Best Known
KroA100	100 cities (sym)	21282	wil50	50 Locs (sym)	48816
eil101	101 cities (sym)	629	wil100	100 Locs (sym)	273038
kro124p	100 cities (asym)	36230	tai100a	100 Locs (sym)	21125314
ftv170	170 cities (asym)	2755	tai100b	100 Locs (asym)	1185996137
d198	198 cities (sym)	15780	tai35b	35 Locs (asym)	283315445
KroA200	200 cities (sym)	29368	tai40b	40 Locs (asym)	637250948
rd400	400 cities (sym)	15281	tai50b	50 Locs (asym)	458821517
pcb442	442 cities (sym)	50778	tai60b	60 Locs (asym)	608215054
att532	532 cities (sym)	27686	tai80b	80 Locs (asym)	818415043

The computing platform used for experiments is a 1 GHz PC with 256 MB memory and all implementations are prepared and run in Matlab. Each experiment is performed 10 times over 5000 iterations. The relative percentage deviation, RPD, is used to express the experimental results. RPD is a measure of the percentage difference between the best known solution and an actual experimental result. RPD is simply calculated by the following formula.

$$RPD = 100 \times (\frac{Cost_{Actaul} - Cost_{Best}}{Cost_{Best}}) \qquad (1)$$

In the implementation of the proposed approach, the parameter values used for both Max-Min AS and the proposed approach are $\rho = 0.1$, $\alpha = 1$, and $\beta = 5$, $\tau_{init} = 1/(n-1)$ for TSP, $\tau_{init} = 1/n$ for QAP, $\tau_{max} = 3.0$, and $q_0 = 0.9$. Population size $m = 25$ ants for all experiments. In the implementation of the proposed approach, the memory size M is taken equal to the number of ants, tournament size $Q = 5$, and the top $k = 5$ solutions are used in updating the memory. The following tables illustrate the improvements achieved by the proposed approach over Max-Min AS for the problem instances under consideration.

From Table 2 and Table 3, it can be observed that the proposed ACO strategy performs better than the Max-Min strategy in all problem instances. In addition, the effectiveness of the external memory and the implemented memory management approaches is observed much more clearly for larger problem instances. This is an expected result because partial solutions from potentially promising solutions provide more efficient intensification for more difficult problems.

5   Conclusions and Future Research Directions

In this paper a novel ACO strategy using an external library of solution segments from elite solutions of previous iterations is introduced. The stored segments are used in the construction of solutions in the current iteration to provide further intensification around potentially promising solutions. Using partially constructed solutions also improve time spent in the construction of solutions since part of

Table 2. Results for Max-Min AS and the proposed approach on TSP instances

Strategy	Instance	RPD		
		Min	Average	Max
Max-Min AS	KroA100	0.0	0.55	3.5
	eil101	0.9	2.96	4.3
	kro124p	1.89	3.13	6.4
	ftv170	3.8	5.19	8.35
	d198	1.3	2.1	3.5
	KroA200	1.1	3.2	5.2
	rd400	7.4	9.6	13.7
	pcb442	20.5	25.3	28.1
	att532	22.7	28.2	32.3
Proposed ACO	KroA100	0.0	0.132	12.964
	eil101	0.7	2.723	4.262
	kro124p	1.193	2.386	4.524
	ftv170	1.726	4.722	5.263
	d198	0.817	1.760	2.155
	KroA200	0.745	2.449	4.130
	rd400	5.124	8.582	11.433
	pcb442	11.50	18.891	22.892
	att532	12.654	21.43	26.429

Table 3. Results for Max-Min AS and the proposed approach on QAP instances

Strategy	Instance	RPD		
		Min	Average	Max
Max-Min AS	wil50	1.831	3.823	4.412
	wil100	3.512	5.711	7.3
	tai100a	8.910	13.36	18.44
	tai100b	10.117	15.19	19.5
	tai35b	4.218	9.3	14.5
	tai40b	4.887	6.1	11.7
	tai50b	5.628	7.6	13.6
	tai60b	7.917	9.5	17.32
	tai80b	13.226	17.5	28.49
Proposed ACO	wil50	1.21	3.32	4.4
	wil100	2.73	3.97	5.84
	tai100a	5.61	9.45	13.41
	tai100b	8.8	12.1	16.8
	tai35b	3.79	7.85	11.5
	tai40b	3.64	5.7	9.7
	tai50b	4.66	6.54	8.6
	tai60b	5.87	8.2	13.65
	tai80b	9.12	12.27	18.71

the solution is already available. Performance of the proposed external memory approach is tested using several instances of two well-known hard combinatorial optimization problems. For each particular experiment, the same parameters settings are used throughout in all trials.

From the results of case studies, it can easily be concluded that the proposed ACO strategy performs better than Max-Min AS algorithm in terms of the convergence speed and the solution quality. It can easily be observed that, even though the proposed strategy better than the Max-Min strategy, its real effectiveness is seen for larger size problems. This is an expected result because information incorporation form previous iterations should be helpful for more difficult problem instances. The proposed ACO strategy is very simple to implement and it does not bring significantly additional computational or storage cost to the existing ACO algorithms.

This work requires further investigation from following point of views:

- Using case-based external memory implementations in ACO
- Effects of different memory update strategies
- Use of distributed small-size external memories for individual ants
- Application to nonstationary optimization problems

References

1. Acan, A., Tekol, Y.: Chromosome reuse in genetic algorithms, in Cantu-Paz et al. (eds.): Genetic and Evolutionary Computation Conference GECCO 2003, Springer-Verlag, Chicago, (2003), 695-705.
2. Bonabeau, E., Dorigo, M., Theraluaz, G.: From Natural to Artificial Swarm Intelligence. Oxford University Press, (1999).
3. Cobb, H.G.: An investigation into the use of hypermutation as an adaptive operator in GAs having continuous, time-dependent nonstationary environment. NRL Memorandum Report 6760, (1990).
4. Dorigo, M., Caro, G.D., Gambardella, L.M.: Ant algorithms for distributed discrete optimization, Artificial Life, Vol. 5, (1999), 137-172.
5. Dorigo, M., Caro, G.D.,: The ant colony optimization metaheuristic, In Corne, D., Dorigo, M., Glover, F. (eds.): New ideas in optimzation, McGraw-Hill, London, (1999), 11-32.
6. Eggermont, J., Lenaerts, T.: Non-stationary function optimization using evolutionary algorithms with a case-based memory, Technical report, Leiden University Advanced Computer Sceice (LIACS) Technical Report 2001-11.
7. Goldberg, D. E., Smith, R. E.: Non-stationary function optimization using genetic algorithms and with dominance and diploidy, Genetic Algorithms and their Applications: Proceedings of the Second International Conference on Genetic Algorithms, (1987), 217-223.
8. Guntsch, M., Middendorf, M.: A population based approach for ACO, in S. Cagnoni et al., (eds.): Applications of Evolutionary Computing - EvoWorkshops2002, Lecture Notes in Computer Science, No:2279, Springer Verlag, (2002), 72-81.
9. Guntsch, M., Middendorf, M.: Applying population based ACO for dynamic optimization problems, in M. Dorigo et al., (eds.): Ant Algorithms - Third International Workshop ANTS2002, Lecture Notes in Computer Science, No:2463, Springer Verlag, (2002), 111-122.
10. Lewis, J., Hart, E., Ritchie, G.: A comparison of dominance mechanisms and simple mutation on non-stationary problems, in Eiben, A. E., Back, T., Schoenauer, M., Schwefel, H. (Editors): Parallel Problem Solving from Nature- PPSN V, Berlin, (1998), 139-148.

11. Louis, S., Li, G.: Augmenting genetic algorithms with memory to solve traveling salesman problem, Proceedings of the Joint Conference on Information Sciences, Duke University, (1997), 108-111.
12. Louis, S. J., Johnson, J.: Solving similar problems using genetic algorithms and case-based memory, in Back, T., (Editor):Proceedings of the Seventh International Conference on Genetic Algorithms, San Fransisco, CA, (1997), 84-91.
13. Montgomery, J., Randall, M: The accumulated experience ant colony for the travelling salesman problem, International Journal of Computational Intelligence and Applications, World Scientific Publishing Company, Vol. 3, No. 2, (2003), 189-198.
14. Ramsey, C.L., Grefenstette, J. J.: Case-based initialization of GAs, in Forest, S., (Editor): Proceedings of the Fifth International Conference on Genetic Algorithms, San Mateo, CA, (1993), 84-91.
15. Simoes, A., Costa, E.: Using genetic algorithms to deal with dynamic environments: comparative study of several approaches based on promoting diversity, in W. B. Langton et al. (eds.): Proceedings of the genetic and evolutionary computation conference GECCO'02, Morgan Kaufmann, New York, (2002), 698.
16. Simoes, A., Costa, E.: Using biological inspiration to deal with dynamic environments, Proceedings of the seventh international conference on soft computing MENDEL'2001, Czech Republic, (2001).
17. Stützle, T., Dorigo, M.: ACO Algorithms for the Traveling Salesman Problem. In: Miettinen, K., Neittaanmaki, P., Periaux, J. (eds.): Evolutionary Algorithms in Engineering and Computer Science, John Wiley & Sons, (1999), 163-184.
18. Stützle, T., Dorigo, M.: ACO Algorithms for the Quadratic Assignment Problem. In: Corne, D., Dorigo, M., Glover, F. (eds.): New Ideas in Optimization, McGraw-Hill, (1999), 33-50.

BeeHive: An Efficient Fault-Tolerant Routing Algorithm Inspired by Honey Bee Behavior

Horst F. Wedde, Muddassar Farooq, and Yue Zhang

Informatik III, University of Dortmund
44221, Dortmund, Germany
{wedde,farooq,zhang}@ls3.cs.uni-dortmund.de

Abstract. Bees organize their foraging activities as a social and communicative effort, indicating both the direction, distance and quality of food sources to their fellow foragers through a "dance" inside the bee hive (on the "dance floor"). In this paper we present a novel routing algorithm, *BeeHive*, which has been inspired by the communicative and evaluative methods and procedures of honey bees. In this algorithm, *bee agents* travel through network regions called *foraging zones*. On their way their information on the network state is delivered for updating the local routing tables. *BeeHive* is fault tolerant, scalable, and relies completely on local, or regional, information, respectively. We demonstrate through extensive simulations that *BeeHive* achieves a similar or better performance compared to state-of-the-art algorithms.

1 Introduction

Swarm Intelligence has been evolving as an active area of research over the past years. The major emphasis is *to design adaptive, decentralized, flexible and robust artificial systems, capable of solving problems through solutions inspired by the behavior of social insects* [2]. Research in the field has largely been focused on working principles of ant colonies and how to use them in the design of novel algorithms for efficiently solving combinatorial optimization problems. A major emphasis here is on ant colony optimization (ACO) techniques [2].

To our knowledge, little attention has been paid in utilizing the organizational principles in other swarms such as honey bees to solve real world problems although the study of honey bees has revealed a remarkable sophistication of the communication capabilities as compared to ants. Nobel laureate Karl von Frisch deciphered and structures these into a language, in his book *The Dance Language and Orientation of Bees* [13]. Upon their return from a foraging trip, bees communicate the distance, direction, and quality of a flower site to their fellow foragers by making waggle dances on a dance floor inside the hive. By dancing zealously for a good foraging site they recruit foragers for the site. In this way a good flower site is exploited, and the number of foragers at this site are reinforced. In our approach we model *bee agents* in packet switching networks, for the purpose of finding suitable paths between sites, by extensively borrowing from the principles behind the bee communication.

M. Dorigo et al. (Eds.): ANTS 2004, LNCS 3172, pp. 83–94, 2004.

Previous and Related Work. The focus of our research is on dynamic routing algorithms therefore we now provide an overview of two such algorithms, *AntNet* and *Distributed Genetic Algorithm (DGA)*, that have been inspired through ants swarm behavior. This will help in understanding not only the motivation for our work, but also assist the reader in understanding the different direction of our algorithm. However, we do use a classic static routing algorithm, OSPF (see [5]), in our comparative simulation for comprehensiveness.

AntNet was proposed by Di Caro and Dorigo in [3]. In *AntNet* the network state is monitored through two ant agents: *Forward_Ant and Backward_Ant*. A Forward_Ant agent is launched at regular intervals from a source to a certain destination. The authors proposed a model in [4] that enables a Forward_Ant agent to estimate queuing delay without waiting inside data packet queues. Forward_Ant agent is equipped with a stack memory on which the address and entrance time of each node on its path are pushed. Once the Forward_Ant agent reaches its destination it creates a Backward_Ant agent and transfers all information to it. Backward_Ant visits the same nodes as Forward_Ant in reverse order and modifies the entries in the routing tables based on the trip time from the nodes to the destination. At each node the average trip time, the best trip time, and the variance of the trip times for each destination are saved. The trip time values are calculated by taking the difference of entrance times of two subsequent nodes pushed onto the stack. Backward_Ant agent uses the system priority queues so that it quickly disseminates the information to the nodes. The interested reader may find more details in [3][4]. Later on the authors of [1] made significant improvements in the routing table initialization algorithm of *AntNet*, bounded the number of Forward_Ant agents during congestion, and proposed a mechanism to handle routing table entries at the neighbors of crashed routers.

The authors of [8] showed that the information needed by *AntNet* for each destination is difficult to obtain in real networks. Their idea of *global information* is that there is an entry in the routing table for each destination. This shortcoming motivated the authors to propose in [9] a *Distributed Genetic Algorithm (DGA)* that eliminates the need for having an entry for each destination node in the routing table. In this algorithm ants are asked to traverse a set of 6 nodes in a particular order, known as a *chromosome*. Once an agent visits the 6th node then it is converted into a backward agent that returns to its source node. In contrast to *AntNet* the backward agents only modify the routing tables at the source node. The source node also measures the fitness of this agent based on the trip time value, and then it generates a new population using single point cross over. New agents enter the network and evaluate the assigned paths. The routing table stores the agents' IDs, their fitness values and trip times to the visited nodes. Routing of a data packet is done through the path that has the shortest trip time to the destination. If no entry for a particular destination is found then a data packet is routed with the help of an agent that has the maximum fitness value. *DGA* was designed assuming that the routers could crash during network operations. The interested reader will find more details in [9].

In contrast to above-mentioned algorithms, we propose in this paper, an agent model in which agents need not be equipped with a stack to perform their duties. Moreover, our model requires only forward moving agents and they utilize an estimation model to calculate the trip time from their source to a given node. This model eliminates the need for global clock synchronization among routers, and it is expected that for very large networks routing information could be disseminated quickly with a small overhead as compared to *AntNet*. Our agent model does not require to store the average trip time, the variance of trip times, and the best trip time for each destination at a node to determine the goodness of a neighbor for a particular destination. Last but not the least, our algorithm works with a significantly smaller routing table as compared to *AntNet*.

Organization of the Paper. In section 2 we first develop the key ideas of the bee agent model underlying the *BeeHive* algorithm. On this basis we will present our *BeeHive* algorithm in section 2.1. Section 3 will describe the simulation and the network environment that were used to compare the performance of *BeeHive* with other state-of-the-art algorithms. In Section 4 we will discuss the results obtained from the extensive simulations. Finally we conclude on our findings, and provide an outlook to future research.

2   The *Bee Agent* Model

In this section, we will briefly describe the organizational principles of bee behavior that inspired us to transform them into an agent model. Honey bees evaluate the quality of each discovered food site and only perform *waggle dance* for it on the *dance floor* if the quality is above a certain threshold. So not each discovered site receives a reinforcement. As a result, quality flower sites are exploited quite extensively. Hence we abstract a *dance floor* into a routing table where *bee agents*, launched from the same source but arrived from different neighbors at a given node, could exchange routing information to model the network state at this node. The agent communication model of *BeeHive* could easily be realized as a *blackboard system* [10]. In comparison *AntNet* utilizes the *principle of stigmergy* [7] for communication among agents.

The majority of foragers exploit the food sources in the closer vicinity of their hive while a minority among them visit food sites faraway from their hive. We transformed this observation into an agent model that has two types of agents: *short distance bee agents* and *long distance bee agents*. Short distance bee agents collect and disseminate routing information in the neighborhood (upto a specific number of hops) of their source node while *long distance bee agents* collect and disseminate routing information to all nodes of a network. Informally, the *BeeHive* algorithm and its main characteristics could be summarized as follows:

1. The network is organized into fixed partitions called *foraging regions*. A partition results from particularities of the network topology. Each *foraging region* has one representative node. Currently the lowest IP address node in a *foraging region* is elected as the representative node. If this node crashes then the next higher IP address node takes over the job.

2. Each node also has a node specific *foraging zone* which consists of all nodes from whom *short distance bee agents* can reach this node.
3. Each non-representative node periodically sends a *short distance bee agent*, by broadcasting replicas of it to each neighbor site.
4. When a replica of a particular *bee agent* arrives at a site it updates routing information there, and the replica will be flooded again, however, it will not be sent to the neighbor from where it arrived. This process continues until the life time of the agent has expired, or if a replica of this *bee agent* had been received already at a site, the new replica will be killed there.
5. Representative nodes only launch *long distance bee agents* that would be received by the neighbors and propagated as in 4. However, their life time (number of hops) is limited by the *long distance limit*.
6. The idea is that each agent while traveling, collects and carries path information, and that it leaves, at each node visited, the trip time estimate for reaching its source node from this node over the incoming link. *Bee agents* use priority queues for quick dissemination of routing information.
7. Thus each node maintains current routing information for reaching nodes within its *foraging zone* and for reaching the *representative nodes* of *foraging regions*. This mechanism enables a node to route a data packet (whose destination is beyond the *foraging zone* of the given node) along a path toward the *representative node* of the *foraging region* containing the destination node.
8. The next hop for a data packet is selected in a probabilistic manner according to the quality measure of the neighbors, as a result, not all packets follow the best paths. This will help in *maximizing the system performance though a data packet may not follow the best path*, a concept directly borrowed from a principle of bee behavior: *A bee could only maximize her colony's profit if she refrains from broadly monitoring the dance floor to identify the single most desirable food* [11](In comparison *OSPF* always chooses a next hop on the shortest path).

Figure 1 provides an exemplary working of the flooding algorithm. *Short distance bee agents* can travel upto 3 hops in this example. Each replica of the shown *bee agent* (launched by Node 10) is specified with a different trail to identify its path unambiguously. The numbers on the paths show their costs. The flooding algorithm is a variant of breadth first search algorithm. Nodes 2,3,4,5,6,7,8,9,11 constitute the *foraging zone* of node 10.

Now we will briefly discuss the estimation model that *bee agents* utilize to approximate the trip time t_{is} that a packet will take in reaching their source node s from current node i (ignoring the protocol processing delays for a packet at node i and s).

$$t_{is} \approx \frac{ql_{in}}{b_{in}} + tx_{in} + pd_{in} + t_{ns} \tag{1}$$

where ql_{in} is the size of the queue (in bits) for neighbor n at node i, b_{in} is the bandwidth of the link between node i and neighbor n, tx_{in} and pd_{in} are transmission delay and propagation delay respectively of the link between node i and neighbor n, and t_{ns} is trip time from n to s. Bandwidth and propagation delays of all links of a node are approximated by transmitting *hello packets*.

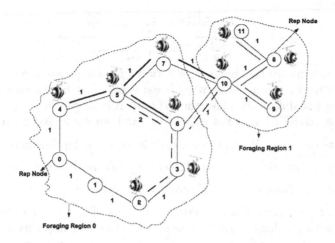

Fig. 1. Bee Agents Flooding Algorithm

Table 1. *Intra foraging zone* Routing Table

R_i	$D_1(i)$	$D_2(i)$	$\ldots$	$D_d(i)$
$N_1(i)$	(p_{11}, q_{11})	(p_{12}, q_{12})	$\cdots$	(p_{1d}, q_{1d})
$\vdots$	$\vdots$	$\vdots$	$\ddots$	$\vdots$
$N_n(i)$	(p_{n1}, q_{n1})	(p_{n2}, q_{n2})	$\cdots$	(p_{nd}, q_{nd})

2.1 Algorithm Design of BeeHive

In *BeeHive*, each node i maintains three types of routing tables: *Intra Foraging Zone* (IFZ), *Inter Foraging Region* (IFR) and *Foraging Region Membership* (FRM). *Intra Foraging Zone* routing table R_i is organized as a vector of size $|D(i)| \times (|N(i)|)$, where $D(i)$ is the set of destinations in *foraging zone* of node i and $N(i)$ is the set of neighbors of i. Each entry P_{jk} is a pair of queuing delay and propagation delay (q_{jk}, p_{jk}) that a packet will experience in reaching destination k via neighbor j. Table 1 shows an example of R_i. In the *Inter Foraging Region* routing table, the queuing delay and propagation delay values for reaching the *representative node* of each *foraging region* through the neighbors of a node are stored. The structure of the *Inter Foraging Region* routing table is similar to the one shown in Table 1 where destination is replaced by a pair of (representative, region). The *Foraging Region Membership* routing table provides the mapping of known destinations to a *foraging region*. In this way we eliminate the need to maintain $O(N \times D)$ (where D is total number of nodes in a network) entries in a routing table as done by *AntNet* and save a considerable amount of router memory needed to store this routing table.

Goodness of a Neighbor: The goodness of a neighbor j of node l (l has n neighbors) for reaching a destination d is g_{jd} and defined as follows

$$g_{jd} = \frac{\frac{1}{p_{jd}}(e^{-\frac{q_{jd}}{p_{jd}}}) + \frac{1}{q_{jd}}(1 - e^{-\frac{q_{jd}}{p_{jd}}})}{\sum_{k=1}^{n}(\frac{1}{p_{kd}}(e^{-\frac{q_{kd}}{p_{kd}}}) + \frac{1}{q_{kd}}(1 - e^{\frac{q_{kd}}{p_{kd}}}))} \qquad (2)$$

The fundamental motivation behind Definition 2 is to approximate the behavior of the real network. When the network is experiencing a heavy network traffic load then the queuing delay plays the primary role in the delay of a link. In this case it is trivial to say that $q_{jd} \gg p_{jd}$ and we could see from equation (2) that $g_{jd} \approx \frac{\frac{1}{q_{jd}}}{\sum_{k=1}^{n}\frac{1}{q_{kd}}}$. When the network is experiencing low network traffic then the propagation delay plays an important role in defining the delay of a link. As $q_{jd} \ll p_{jd}$, from equation (2) we get $g_{jd} \approx \frac{\frac{1}{p_{jd}}}{\sum_{k=1}^{n}\frac{1}{p_{kd}}}$. We use *stochastic sampling with replacement* [6] for selecting a neighbor. This principle ensures that a neighbor j with goodness g_{jd} will be selected as the next hop with at least the probability $\frac{g_{jd}}{\sum_{k=1}^{n}g_{kd}}$. Algorithm 1 provides the pseudo code of *BeeHive*.

Algorithm 1 (BeeHive)

$t:=$ *current_time*, $t_{end}:=$ *time to end simulation*
// *Short_Limit:= 7, Long_Limit:= 40, Bee_Generation_Interval:= 1*
// *i=current node, d=destination node, s=source node*
// *n=successor node of i, p=predecessor node of i*
// *z=Representative node of the foraging region containing s*
// *w=Representative node of the foraging region containing d*
// *q is queuing delay estimate of a bee agent from node p to s*
// *p is propagation delay estimate of a bee agent from node p to s*
$\Delta t:=$ *Bee_Generation_Interval*, $\Delta h:=$ *hello packet generation interval*
$b_{ip}:=estimated\_link\_band\_width\_to\_neighbor\_p$
$p_{ip}:=estimated\_propagation\_delay\_to\_neighbor\_p$
$h_i:=$ *hop limit for bees of i*, $l_{ip}:=size\_normal\_queue\_i\_to\_p$ *(bits)*
foreach *Node* // *concurrent activity over the network*
  while $(t \leq t_{end})$
    if $(t \bmod \Delta t = 0)$
      if(*i is representative node of the foraging region*)
        set $h_i:=$ *Long_Limit*, b_i is long distance bee agent
      else
        set $h_i:=$ *Short_Limit*, b_i is short distance bee agent
      endif
      launch a bee b_i to all neighbors of i
    endif
    foreach bee b_s received at i from p
      if(b_s *was launched by i or its hop limit reached*)
        kill bee b_s
      elseif(b_s *is inside foraging zone of node s*)
        $q:= q + \frac{l_{ip}}{b_{ip}}$ and $p:= p + p_{ip}$

$\quad$ *update IFZ routing table entries* $q_{ps} = q$ *and* $p_{ps} = p$

$\quad$ *update* q $(q := \sum\limits_{k \in N(i)} (q_{ks} \times g_{ks}))$ *and* p $(p := \sum\limits_{k \in N(i)} (p_{ks} \times g_{ks}))$

else

$\quad$ $q := q + \frac{l_{ip}}{b_{ip}}$ *and* $p := p + p_{ip}$

$\quad$ *update IFR routing table entries* $q_{pz} = q$ *and* $p_{pz} = p$

$\quad$ *update* q $(q := \sum\limits_{k \in N(i)} (q_{kz} \times g_{kz}))$ *and* p $(p := \sum\limits_{k \in N(i)} (p_{kz} \times g_{kz}))$

endif

if(b_s already visited node i)

$\quad$ *kill bee b_s*

else

$\quad$ *use priority queues to forward b_s to all neighbors of i except p*

endif

endfor

foreach data packet d_{sd} received at i from p

$\quad$ *if (node d is within* foraging zone *of node i)*

$\quad\quad$ *consult IFZ routing table of node i to find delays to node d*

$\quad\quad$ *calculate goodness of all neighbors for reaching d using equation 2*

$\quad$ *else*

$\quad\quad$ *consult FRM routing table of node i to find node w*

$\quad\quad$ *consult IFR routing table of node i to find delays to node w*

$\quad\quad$ *calculate goodness of all neighbors for reaching w using equation 2*

$\quad$ *endif*

$\quad$ *probabilistically select a neighbor n (n $\neq$ p) as per goodness*

$\quad$ *enqueue data packet d_{sd} in normal queue for neighbor n*

endfor

if (t mod $\Delta h = 0$)

$\quad$ *send a hello packet to all neighbors*

$\quad$ *if (time out before a response from neighbor) (4th time)*

$\quad\quad$ *neighbor is down*

$\quad\quad$ *update the routing table and launch bees to inform other nodes*

$\quad$ *endif*

endif

endwhile

endfor

3   Simulation Environment for BeeHive

In order to evaluate our algorithm *BeeHive* in comparison with *AntNet, DGA* and *OSPF*, we implemented all of them in the OMNeT++ simulator [12]. For *OSFP* we implemented a static link state routing that implements the deterministic Dijkstra Algorithm [5] which selects the next hop according to the shortest path from a source to a destination. For *AntNet* and *DGA* we used the same parameters that were reported by the authors in [3] and [9], respectively. The network instance that we used in our simulation framework is the Japanese Internet Backbone (NTTNET)(see Figure 2). It is a 57 node, 162 bidirectional links network. The link bandwidth is 6 Mbits/sec and propagation delay is from 2 to 5 milliseconds. Traffic is defined in terms of open sessions between two different nodes. Each session is characterized completely by sessionSize (2 Mbits), inter-arrival time between two sessions, source, destination, and packet inter-arrival time during a session. The size of data packet is 512 bytes, the size of a *bee agent* is 48 bytes. The queue size is limited to 500 Kbytes in all experiments. To inject dynamically changing data traffic patterns we have defined two states: uniform and weighted. Each state lasts 10 seconds and then a state transition to another state occurs. In *Uniform* state (U) a destination is selected from a uniform distribution. While in *Weighted* state (W), a destination selected in *Uniform* state is favored over other destinations. This approach provides a more challenging experimental environment than the one in which *AntNet* was evaluated.

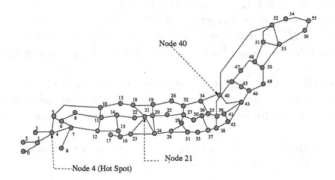

Fig. 2. Japanese Backbone NTTNet

4 Experimental Findings

Now we will report our results obtained from the extensive simulations. MSIA is the mean of session inter-arrival times and MPIA is the mean of packet inter-arrival times during a session. The session inter-arrival and packet inter-arrival times are taken from negative exponential distributions with mean MSIA and MPIA, respectively. All reported values are an average of the values obtained from ten independent runs. % Packets Delivered (on top of bars) report the percentage of deliverable packets that were actually delivered. By deliverable packet we mean a packet whose destination router is up.

Saturating Loads. The purpose of the experiments was to study the behavior of the algorithms by gradually increasing the traffic load, through decreasing MSIA from 4.7 to 1.7 seconds. MPIA is 0.005 seconds during these experiments. Figure 3 shows the average throughput and 90th percentile of the packet delay distribution. It is obvious from Figure 3 that *BeeHive* delivered approximately

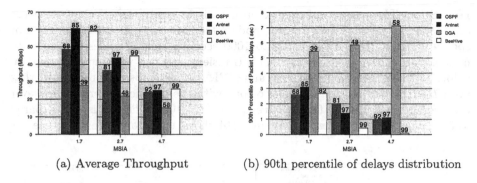

(a) Average Throughput   (b) 90th percentile of delays distribution

Fig. 3. Behavior under saturating traffic loads

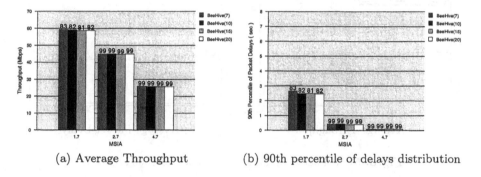

(a) Average Throughput   (b) 90th percentile of delays distribution

Fig. 4. Effect of *foraging zones* sizes

the same number of data packets as that of *AntNet* but with lesser packet delay. Both *OSFP* and *DGA* are unable to cope with a saturated load yet the performance of *DGA* is the poorest.

Size of Foraging Zones. Next we analyzed the effect of the size of a *foraging zone* in which a *bee agent* updates the routing table. We report the results for sizes of 7, 10, 15 and 20 hops in Figure 4. Figure 4 shows that increasing the size of *foraging zone* after 10 does not bring significant performance gains. This shows the power of *BeeHive* that it converges to an optimum solution with a size of just 7 hops.

Size of the Routing Table. The fundamental motivation of the *foraging zone* concept was not only to eliminate the requirement for global knowledge but also to reduce memory needed to store a routing table. *BeeHive* requires 88, 94 and 104 entries, on the average, in the routing tables for *foraging zone* sizes of 7, 10 and 20 hops respectively. *OSFP* needs just 57 entries while *AntNet* needs 162 entries on the average. Hence *BeeHive* achieves similar performance as that of *AntNet* but size of the routing table is of the order of *OSPF*.

Hot Spot. The purpose of this experiment is to study the effect of transient overloads in a network. We selected node 4 (see Figure 2) as hot spot. The hot spot was active from 500 seconds to 1000 seconds and all nodes sent data to node 4 with MPIA=0.05 seconds. This transient overload was superimposed on a normal load of MSIA=2.7 seconds and MPIA=0.3 seconds. Figure 5 shows that both *BeeHive* and *AntNet* are able to cope with the transient overload, however the average packet delay for *BeeHive* is less than 100 milliseconds as compared to 500 milliseconds for *AntNet*. Again *DGA* shows the poorest performance.

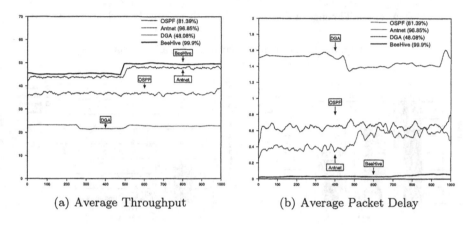

(a) Average Throughput   (b) Average Packet Delay

Fig. 5. Node 4 acted as hot spot from 500 to 1000 seconds

Router Crash. The purpose of this experiment was to analyze the fault tolerant behavior of *BeeHive* so we took MSIA = 4.7 seconds and MPIA = 0.005 seconds to ensure that no packets are dropped because of the congestion. We simulated a scenario in which Router 21 crashed at 300 seconds, and Router 40 crashed at 500 seconds and then both were repaired at 800 seconds. Figure 6 shows the results. *BeeHive* is able to deliver 97% of deliverable packets as compared to 89% by *AntNet*. Please observe that from 300 to 500 seconds (just Router 21 is down), *BeeHive* has a superior throughput and lesser packet delay but once Router 40 crashes, the packet delay of *BeeHive* increases because of higher load at Router 43. From Figure 2 it is obvious that the only path to the upper part of the network is via Router 43 once Router 40 crashed. Since *BeeHive* is able to deliver more packets the queue length at Router 43 increased and this led to relatively poorer packet delay as compared to *AntNet*. Please also observe that Router 21 is critical but in case of its crash still multiple paths exist to the middle and upper part of the topology via 15,18,19,20,24.

Overhead of BeeHive. *Routing overhead is defined as the ratio between the bandwidth occupied by the routing packets and the total available network bandwidth* [3]. The overhead of *BeeHive* is 0.00633 as compared to 0.00285 of *AntNet* [3]. *BeeHive* has a bit more overhead as compared to *AntNet*, but this overhead remains constant for a given topology.

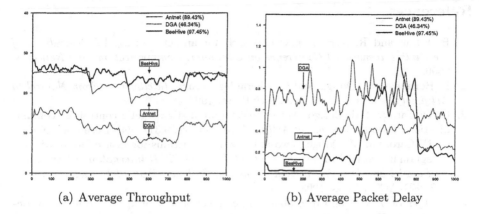

(a) Average Throughput   (b) Average Packet Delay

Fig. 6. Router 21 is down at 300 and Router 40 at 500 and both repaired at 800

5   Conclusion and Future Work

Swarm intelligence is evolving as a new discipline that is transforming the traditional algorithmic design paradigm of the developers by enabling them to contemplate the working principles of systems in Nature and then use them in design and development of the algorithms. In this paper we introduced a fault-tolerant, adaptive and robust routing protocol inspired from dance language and foraging behavior of honey bees. The algorithm does not need any global information such as the structure of the topology and cost of links among routers, rather it works with the local information that a *short distance bee agent* collects in a *foraging zone*. It works without the need of global clock synchronization which not only simplifies its installation on real routers but also enhances fault tolerance. In contrast to *AntNet* our algorithm utilizes only forward moving *bee agents* that help in disseminating the state of the network to the routers in real-time. The *bee agents* take less than 1% of the available bandwidth but provide significant enhancements in throughput and packet delay.

We implemented two state-of-the-art adaptive algorithms (*AntNet* and *DGA*) for the OMNeT++ simulator and then compared our *BeeHive* algorithm with them. Through extensive simulations representing dynamically changing operating network environments we have demonstrated that *BeeHive* achieves a better or similar performance as compared to *AntNet*. However, this enhancement in performance is achieved with a routing table whose size is of the order of *OSPF*. In the near future we will have implemented *BeeHive* inside the network stack of the Linux kernel in order to then test the algorithm on real network topologies. Our work during this phase will be subject of forthcoming publications.

Acknowledgments

We pay special thanks to Gianni Di Caro who provided valuable feedback on our implementation of *AntNet* on the OMNeT++ simulator. We would also like to thank Suihong Liang who helped us in understanding *DGA*.

References

1. B. Barán and R. Sosa. A new approach for antnet routing. In *Proceedings of the Ninth International Conference on Computer, Communications and Networks*, 2000.
2. E. Bonabeau, M. Dorigo, and G. Theraulaz. *Swarm Intelligence: From Natural to Artificial Systems*. Oxford University Press, 1999.
3. G. Di Caro and M. Dorigo. AntNet: Distributed stigmergetic control for communication networks. *Journal of Artificial Intelligence*, 9:317–365, December 1998.
4. G. Di Caro and M. Dorigo. Two ant colony algorithms for best-effort routing in datagram networks. In *Proceedings of the Tenth IASTED International Conference on Parallel and Distributed Computing and Systems (PDCS'98)*, pages 541–546. IASTED/ACTA Press, 1998.
5. E. Dijkstra. A note on two problems in connection with graphs. *Numerical Mathematics*, 1:269–271, 1959.
6. D. Goldberg. *Genetic algorithms in search, optimization and machine learning*. Addison Wesley, Reading, MA, 1989.
7. P. Grassé. La reconstruction du nid et les coordinations interindividuelles chez bellicositermes natalensis et cubitermes sp. la théorie de la stigmergie: essai d'interprétation du comportement des termites constructeurs. *Insectes Sociaux*, 6:41–81, 1959.
8. S. Liang, A. Zincir-Heywood, and M. Heywood. The effect of routing under local information using a social insect metaphor. In *Proceedings of IEEE Congress on Evolutionary Computing*, May 2002.
9. S. Liang, A. Zincir-Heywood, and M. Heywood. Intelligent packets for dynamic network routing using distributed genetic algorithm. In *Proceedings of Genetic and Evolutionary Computation Conference*. GECCO, July 2002.
10. P. Nii. The blackboard model of problem solving. *AI Mag*, 7(2):38–53, 1986.
11. T. Seeley. *The Wisdom of the Hive*. Harvard University Press, London, 1995.
12. A. Varga. OMNeT++: Discrete event simulation system: User manual. http://www.omnetpp.org.
13. K. von Frisch. *The Dance Language and Orientation of Bees*. Harvard University Press, Cambridge, 1967.

Competition Controlled Pheromone Update for Ant Colony Optimization

Daniel Merkle and Martin Middendorf

Department of Computer Science, University of Leipzig
Augustusplatz 10-11, D-04109 Leipzig, Germany
{merkle,middendorf}@informatik.uni-leipzig.de

Abstract. Pheromone information is used in Ant Colony Optimization (ACO) to guide the search process and to transfer knowledge from one iteration of the optimization algorithm to the next. Typically, in ACO all decisions that lead an ant to a good solution are considered as of equal importance and receive the same amount of pheromone from this ant (assuming the ant is allowed to update the pheromone information). In this paper we show that the decisions of an ant are usually made under situations with different strength of competition. Thus, the decisions of an ant do not have the same value for the optimization process and strong pheromone update should be prevented when competition is weak. We propose a measure for the strength of competition that is based on Kullback-Leibler distances. This measure is used to control the update of the pheromone information so that solutions components that correspond to decisions that were made under stronger competition receive more pheromone. We call this update procedure competition controlled pheromone update. The potential usefulness of competition controlled pheromone update is shown first on simple test problems for a deterministic model of ACO. Then we show how the new update method can be applied for ACO algorithms.

1 Introduction

A characteristic feature of Ant Colony Optimization (ACO) is the use of so called pheromone information to guide the search of simple agents, called ants, for good solutions (e.g., [4]). The pheromone information is also used to transfer knowledge from one iteration of an ACO algorithm to the next. Due to its central role for the optimization behavior of ACO algorithms several aspects of the use of pheromone information have been studied in the literature in some detail:

a. the modelling of an optimization problem so that the construction of a solution corresponds to a sequence of decisions where typically each decision provides a generic solution component which corresponds to a pheromone parameter (e.g., [1, 4]),
b. different methods to be used by the ants to evaluate the pheromone information for solution construction (e.g., [8–10]),

M. Dorigo et al. (Eds.): ANTS 2004, LNCS 3172, pp. 95–105, 2004.

c. the pheromone update, e.g., i) how the amount of pheromone update for a solution should depend on its relative or absolute quality (e.g. [4]), ii) which ants are allowed to update (e.g. [2, 4]), iii) whether pheromone update is done according to elitist solutions (e.g. [4]), iv) the use of stochastic gradient search ([6, 7]), or v) whether the amount of pheromone update should change with the state of the optimization process (see e.g. [4, 13–15]).

So far the research on pheromone update focused on the amount of pheromone that is added for the different solutions that have been found by the ants. The specific decisions that lead an ant to a good solution have been considered as of equal importance and receive the same amount of pheromone (assuming this ant is allowed to update the pheromone information). In this paper we show that the decisions of the ants are usually made under situations with different strength of competition between the ants. Thus, the decisions that have led an ant to a solution do not have the same value for the optimization process and therefore should be handled differently during pheromone update. In general a strong pheromone update should be prevented if the strength of competition is low. Therefore, we propose to use a measure for the strength of competition in order to determine the amount of pheromone update. The measure that we introduce here is based on the Kullback-Leibler distance between the estimated distribution of the ants decisions and the estimated distribution of the decisions of the better ants. This measure is then used to control the update of the pheromone information. The corresponding pheromone update procedure is called competition controlled pheromone update.

We first define and investigate competition controlled pheromone update on a deterministic model for ACO that was introduced in [11]. The advantage of using this model is that it allows to abstract from random effects and more clearly shows some effects of competition controlled update on the optimization process. It was shown in [11] that parts of the dynamics of ACO might be driven mostly by pure selection and that this forces the system to approach a fixed point state. Such kind of dynamic behavior is not always wanted because it can bring the system in a less favorable state. A method that was proposed to hinder the system to change the pheromone values for such parts of the problem instance where competition is too weak is pheromone update masking. This method masks the corresponding pheromone values so that they remain unchanged until competition becomes stronger. A problem with the masking method is that it is difficult to decide when masking should be applied to a pheromone value. The competition controlled pheromone update method that is proposed in this paper overcomes this problem. It can be seen as an advanced masking method that uses different degrees of masking and automatically modifies this degree according to the actual strength of competition. As a proof of concept we show for a simple example problem that competition controlled pheromone can significantly improve the optimization behavior of the ACO model.

Since the ACO model can map only some aspects of ACO algorithms behavior we also investigate how the principle of competition controlled pheromone update can be applied to (real) ACO algorithms. Since the effects of randomness in an ACO algorithm can not be neglected we extend the competition controlled pheromone update method as it is introduced for the ACO model by several heuristic features to make it suitable also for ACO algorithms. The results of test runs on selected test problems give hints when competition controlled pheromone update can be advantageous ACO algorithms.

In Section 2 we describe the simple ACO algorithm that is used in this paper. The ACO model is decribed in Section 3. In Section 4 we introduce a measure for competition and the competition based pheromone update method for the ACO model. Moreover results on a test problem are presented. In Section 5 we introduce extensions of the competition controlled pheromone update method so that is can be applied to ACO algorithms and discuss the results for several test problems. Conclusions are given in Section 6.

2 ACO Algorithm

A simple ACO algorithm that is used in this paper is described in the following. Clearly, for applying ACO to real-world problems several extensions and improvements have been proposed in the literature (e.g. usually ants use problem specific heuristic information or local search). The optimization problems that are considered in this paper are permutation problems where given are a set S of n items $1, 2, \ldots, n$ and an $n \times n$ cost matrix $C = [c(ij)]$ with integer costs $c(i, j) \geq 0$. The problem is to find a permutation π in set $\mathcal{P}_n$ of permutations of $(1, 2, \ldots, n)$ that has minimal costs $c(\pi) := \sum_{i=1}^{n} c(i, \pi(i))$, i.e., a permutation with $c(\pi) = \min\{c(\pi') \mid \pi' \in \mathcal{P}_n\}$.

The ACO algorithm consists of several iterations where in every iteration each of m ants constructs a solution for the optimization problem. For the construction of a solution (here a permutation) every ant selects the items one after the other. For its decisions the ant uses pheromone information τ_{ij}, $i, j \in [1 : n]$ where τ_{ij} is an indicator of how good it seems to have item j at place i of the permutation. The matrix $M = (\tau_{ij})_{i,j \in [1:n]}$ is called the pheromone matrix.

An ant always chooses the next item from the set S of items, that have not been placed so far, according to the following probability distribution that depends on the pheromone values in row i of the pheromone matrix: $p_{ij} = \tau_{ij} / \sum_{h \in S} \tau_{ih}$, $j \in S$ (see e.g. [5]). Before the pheromone update is done a certain percentage of the old pheromone evaporates according to the formula $\tau_{ij} = (1 - \rho) \cdot \tau_{ij}$. Parameter ρ allows to determine how strongly old pheromone influences future decisions. Then, for every item j of the best permutation found so far some amount Δ of pheromone is added to element τ_{ij} of the pheromone matrix. The algorithm stops when some stopping criterion is met, e.g. a certain number of iterations has been done. For ease of description we assume that the sum of the pheromone values in every row of the matrix is always one, i.e., $\sum_{j=1}^{n} \tau_{ij} = 1$ for $i \in [1 : n]$.

3 ACO Model

In this section we describe the deterministic ACO model that was proposed in [11]. In this model the pheromone update at an iteration is done by adding to each pheromone value the expected update value. Formally, the pheromone update that is done at the end of an iteration is defined by $\tau_{ij} = (1 - \rho) \cdot \tau_{ij} + \rho \cdot \sigma_{ij}^{(m)}$ where $\sigma_{ij}^{(m)}$ is the probability that the best of m ants in a generation selects item j for place i. $\sigma_{ij}^{(m)}$ can be computed as described in the following. Let C be the set of possible cost values for a permutation (i.e., the set of possible solution qualities). Let $\xi^{(m,x)}$ be the probability that the best of m ants in a generation finds a solution with quality $x \in C$. Let $\omega_{ij}^{(x)}$ be the probability that an ant which found a solution with quality $x \in C$ has selected item i for place j. Then

$$\sigma_{ij}^{(m)} = \sum_{x \in C} \xi^{(m,x)} \cdot \omega_{ij}^{(x)}$$

The formula shows that the pheromone update can be described as a weighted sum over the possible solution qualities. For each (possible) solution quality the update value is determined by the probabilities for the decisions of a single ant when it chooses between all possible solutions with that same quality. The effect of the number m of ants is only that the weight of the different qualities in this sum changes. The more ants per iteration, the higher becomes the weight of the optimal quality.

Since the update values in the simple ACO algorithm are always only zero or $\Delta = \rho$ the ACO model only approximates the average behaviour of the algorithm.

3.1   ACO Model for Composed Permutation Problems

A certain type of permutation problems, so called composed permutation problems, were introduced in [12] as idealized versions of real-world problems that are composed of a set of more or less independent subproblems. As an example consider a composed permutation problem $P_1 P_2$ with cost matrix $C^{(2)}$ that is composed from two elementary problems P_1 and P_2 of size $n = 3$ with cost matrices C_1, respectively C_2.

$$C_1 = \begin{pmatrix} 0\ 1\ 2 \\ 1\ 0\ 1 \\ 2\ 1\ 0 \end{pmatrix} \qquad C_2 = \begin{pmatrix} 0\ 2\ 4 \\ 2\ 0\ 2 \\ 4\ 2\ 0 \end{pmatrix} \qquad C^{(2)} = \begin{pmatrix} 0 & 1 & 2 & \infty & \infty & \infty \\ 1 & 0 & 1 & \infty & \infty & \infty \\ 2 & 1 & 0 & \infty & \infty & \infty \\ \infty & \infty & \infty & 0 & 2 & 4 \\ \infty & \infty & \infty & 2 & 0 & 2 \\ \infty & \infty & \infty & 4 & 2 & 0 \end{pmatrix}$$

Formally, define for a permutation problem P of size n a composed permutation problem P^q of size $q \cdot n$ such that for an instance of P with cost matrix $C = (c'_{ij})_{i,j \in [1:n]}$ the corresponding instance of P^q consists of q independent of these instances of P. Let $C^{(q)} = (c_{ij})_{i,j \in [1:qn]}$ be the corresponding cost matrix of the instance of problem P^q where $c_{(l-1) \cdot n+i,(l-1) \cdot n+j} = c'_{ij}$ for $i, j \in [1 : n], l \in [1 : q]$ and $c_{ij} = \infty$ otherwise. Note, that our definition of composed permutation problems does not allow an ant to make a decision with cost ∞. We call P the elementary subproblem of P^q. Since all subproblems of P^q are the same it is called homogeneous composed permutation problem. Let P_1, P_2 be two permutation problems. Then similar as above $P_1^q P_2^r$ denotes the heterogeneous composed permutation problem that consists of q instances of P_1 and r instances of P_2.

4   Competition Controlled Pheromone Update for the ACO Model

In this section we introduce a measure for competition between the ants and based on this measure the competition controlled pheromone update for the ACO model.

4.1   Measure of Competition

In order to to measure the strength (or importance) of competition for each single decision that an ants make for constructing a solution at an iteration we introduce a competition measure. To simply use the number of ants per iteration as a measure is not suitable because it can not differentiate between the effect of competition for different decisions and it is also only a static measure.

Competition can be considered as important for a decision when there is a difference between the outcomes of this decision with and without competition. For permutation problems that means that, considering a row in the pheromone matrix (which corresponds to a single decision) there should be a difference between the expected decisions of a single ant for that row and the expected decisions of the best of several ants for the same row. Thus, for a row in the pheromone matrix we compare the distribution of the selection probabilities of a single ant to the distribution of the selection probabilities of the best of m ants. As a measure for the distance between two probability distributions we use the Kullback-Leibler distance. Formally, for $i \in [1 : n]$

$$d_i = \sum_j \sigma_{ij}^{(1)} \cdot log_2(\frac{\sigma_{ij}^{(1)}}{\sigma_{ij}^{(m)}})$$

is the Kullback-Leibler distance between the probability distributions $\sigma_{i*}^{(1)}$ and $\sigma_{i*}^{(m)}$.

4.2   Competition Controlled Pheromone Update

The idea of the competition controlled pheromone update method (cc-pheromone update) is to let the amount of pheromone update in a row correspond to the relative strength of competition in this row. Hence the more important competition is in a row the stronger is the update. Moreover, the overall amount of pheromone update in the pheromone matrix should be the same as for the standard update. Formally, the first step of cc-pheromone update is defined by

$$\tau_{ij} = (1 - \rho \cdot w_i) \cdot \tau_{ij} + \rho \cdot w_i \cdot \sigma_{ij}^{(m)}$$

where $w_i := (n \cdot d_i)/(\sum_{j=1}^{n} d_j)$ measures the relative strength of competition. Note, that compared to standard pheromone update we have introduced additional weights w_i such that the total amount of pheromone update remains the same, i.e., $n \cdot \rho = \rho \cdot \sum_{i=1}^{n} w_i$.

A problem with the first step of cc-pheromone update is that the reduced amount of pheromone update in some rows can cause problems. Assume that in some row i with small weight w_i there is a large pheromone value τ_{ih}. If in a row $j > i$ which has a large weight w_j the pheromone value τ_{jh} increases this will not cause the value τ_{ih} to decrease. Instead τ_{ih} remains nearly unchanged and since $j > i$ and h in row i will often be chosen by the ants instead of in row j (simply because the decision for row j is made later). Therefore we add a second step to cc-pheromone update procedure to overcome this problem. In each column we evaporate all pheromone values by a certain amount $\delta \geq 0$. The amount of pheromone that was evaporated in a column is then added again to the elements in the column but so that elements in a row with large weight w_i receive more than those in rows with small weight. Formally, the second step of cc-pheromone update is defined by

$$\tau_{ij} = (1 - \delta) \cdot \tau_{ij} + \frac{w_i}{n} \cdot \sum_{i=1}^{n} (\delta \cdot \tau_{ij})$$

4.3 Results

For a proof of concept we apply the cc-pheromone update to a composed permutation test problem from [12]. Consider problem $P_1^{20} P_y^{20}$ where P_y has the cost matrix $y \cdot C_1$, i.e., all cost are y times higher than for P_1. In the test we used $y = 20$. The initial pheromone matrix for P_y was determined by $\tau_{11} = 0.1$, $\tau_{12} = 0.3$, $\tau_{21} = 0.6$, $\tau_{22} = 0.1$ and the initial pheromone matrix for P_1 was determined by $\tau_{11} = 0.9$, $\tau_{12} = 0.09$, $\tau_{21} = 0.05$, and $\tau_{22} = 0.9$ (test initialization). For this problem it was shown that for the ACO model with standard pheromone update the average solution quality on subproblem P_1^{20} becomes increasingly worse during the first 230 iterations when $m = 2$ ants are used (see [12] for more explanations). To circumvent this it was proposed to mask the pheromone values temporarily, i.e., to allow no pheromone update for P_1^{20} during the first 230 iterations. It was shown that with masking the model converges much earlier to the optimal solution than without masking.

As mentioned in the Introduction one motivation for the cc-pheromone update method is that it should work as an "automatic masking method". Figure 1 compares the solution qualities obtained with cc-pheromone update with those obtained with the standard pheromone update. The results show that the ACO model with cc-pheromone update performs very well and converges much earlier to the optimal solution than the ACO model with standard update. Moreover, the solution quality on subproblem P_1^{20} does (nearly) never decrease.

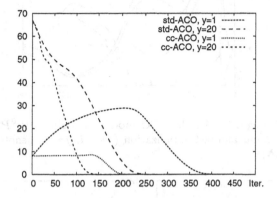

Fig. 1. Comparison of ACO model with standard (std-ACO) and cc-pheromone update (cc-ACO) for the on problem $P_1^{20} P_{20}^{20}$ with $\rho = 0.1$, $\gamma = 1$, $\delta = 0$, $m = 2$ ants and test initialization, curves show the solution qualities on subproblems P_1^{20} ($y = 1$) and P_{20}^{20} ($y = 20$) (where the solution qualities for P_{20}^{20} are divided by 20 for better comparison)

Figure 2 shows the change of the weights w_i which measure the relative strength of competition in the rows. It can be seen that the weights for rows of subproblem P_{20}^{20} are as expected much higher during the first about 140 iterations. Note, that the ACO model uses the expected behaviour of ants ants there fore behaves the same on each of the 20 subproblems P_1, respectively $P_2 0$. Therefore is does not matter from which of these 20 subproblem the 3 rows are taken. Then the weights for the rows of subproblem P_1^{20} increase and become larger than the weights for the rows of P_{20}^{20}. When the ACO model converges all weights converge to 1 for which the cc-pheromone update equals the standard pheromone update.

In order to show that the cc-pheromone update does work not only for the special initialization of the pheromone values we also measured the solution quality when standard initialization is used, i.e., where all pheromone values of initial pheromone matrix for P_y and for P_1 are 1/3. Figure 3 clearly shows that the ACO model with cc-pheromone update performs clearly better on P_{20}^{20} than the ACO model with standard pheromone update. On the subproblem P_1^{20} it performs slightly worse during the first about 120 iterations. But when the pheromone update starts to focus on this subproblem the ACO model with cc-pheromone update becomes clearly better. Overall the solution quality with cc-pheromone update is better over the whole range of iterations.

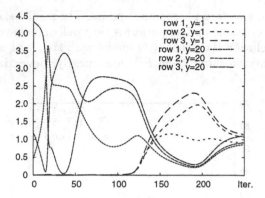

Fig. 2. Change of weights w_i for the ACO model on problem $P_1^{20} P_{20}^{20}$ with $\rho = 0.1$, $\gamma = 1$, $\delta = 0$, $m = 2$ ants and test initialization ("row i, $y = z$" denotes row i of one of the subproblems P_z)

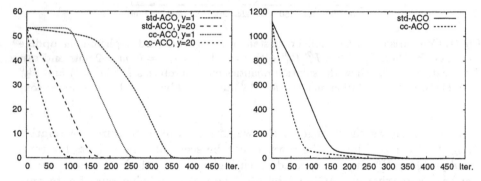

Fig. 3. Comparison of standard and cc-pheromone update for the ACO model on problem $P_1^{20} P_{20}^{20}$ with $\rho = 0.1$ and $m = 2$ ants and standard initialization; curves in the left part show the solution qualities on subproblems P_1^{20} ($y = 1$) and P_{20}^{20} ($y = 20$) (where the solution qualities for P_{20}^{20} are divided by 20 for better comparison); curves in the right part show the total solution qualities

5   Competition Controlled Pheromone Update for the ACO Algorithm

Since the ACO model does not exactly model the ACO algorithm it is necessary to carefully check whether the cc-pheromone update method has to be adapted when used for the ACO algorithm. A problem that occurs is that the pheromone update in the algorithm with standard pheromone update is for each pheromone value either 0 or Δ. Thus during the convergence phase of the algorithm it can happen by chance that in a row of the pheromone that has nearly converged there is a large absolute pheromone update for a very small pheromone value (Note, that this can not happen in the model where always the expected pheromone update is applied). When using the cc-pheromone update method this will usually

Table 1. Results of ACO with standard pheromone update (std-ACO) and cc-pheromone update (cc-ACO) for different problem instances; results are averages over 10 runs; the average time, when the best solution was found, is given in brackets

Instance	std-ACO	cc- ACO
$P_{6,15}^{20}$	264.2 (1062.1)	203.7 (1081.8)
$P_{11,20}^{5}$	73.8 (1402.1)	64.4 (1411.7)
$P_{11,20}^{10}$	235.4 (1556.9)	124.2 (1335.8)
P_{odd}^{20}	248.9 (1182.5)	195.0 (1058.9)
P_{rand}	47.7 (936.2)	49.4 (1097.1)

drastically increase the importance of competition in the corresponding row and leads to a situation that hinders the algorithm to converge. Therefor we propose to use a mixture of cc-pheromone update and standard pheromone update so that the relative influence of cc-pheromone update is high at the start of the algorithm and then slowly decreases so that it becomes nearly zero in the convergence phase. This is done by decreasing the weights w_i when the maximum pheromone value in row i becomes large which indicates that the algorithm converges in this row. Thus the following modified weights w_i' are used for the ACO algorithm

$$w_i' = ((1 - \max\{\tau_{ij} \mid j \in [1:n]\}) \cdot w_i)^\gamma$$

where γ is a parameter that is introduced in order to be able to adapt the influence of the weights.

Another problem for cc-pheromone update with ACO algorithms is that the determination of the weights w_i' requires the determination of the Kullback-Leibler distance between $\sigma_{i*}^{(1)}$ and $\sigma_{i*}^{(m)}$. Clearly this is computationally too expensive for large problem instances. Hence this distance has to be estimated. In this paper we intend to provide a proof of concept and therefore used a simple estimation method. For each iteration of the ACO algorithm we used $m = 10$ ants. To estimate the Kullback-Leibler distances at every iteration the solutions of 500 ants where used to estimate $\sigma_{i*}^{(1)}$ and 500 times the best solutions of 10 ants where determined to estimate $\sigma_{i*}^{(m)}$. Clearly, this estimation method is still computationally too expensive for practical purposes. A cheaper estimation method could be to add the the decisions of the ants over several iterations within a sliding window.

For the test we used parameters $\rho = 0.1$, $\gamma = 2$, $\delta = 1$. The test instances of permutation problems for the empirical investigations were defined as follows. A problem instance P was used where all costs values equal their distance to the main diagonal plus 1, i.e. $c(i, j) := |i - j| + 1$. Problem instance $P_{(i,j)}^s$ is similar to P but the costs in rows k, $i \leq k \leq j$ were scaled by a factor of s. Problem instance P_{odd}^s has all odd rows scaled by s when compared to P. For problem instance P_{rand} the costs $c(i, j)$ are randomly chosen from $[1:50]$.

Table 1 compares the results of the ACO algorithm with cc-pheromone updated and standard pheromone update. For instances of type $P^s_{(i,j)}$ cc-pheromone update performs significantly better results than standard pheromone update. In particular when there is a large scaling factor, i.e., a large difference between the rows with respect to their influence on the solution quality (compare the results for $P^{10}_{11,20}$ and $P^5_{11,20}$). It is interesting that cc-pheromone is also better on P_{odd} which indicates that it successfully can find our the rows with high competition when they occur not in blocks. On P_{rand} cc-pheromone update performs slightly worse than standard pheromone update. The reason why cc-pheromone can not profit on this instances needs further investigation.

The change of the weights w'_i that are a combined measure of relative competition and state of convergence in the rows is shown in Figure 2. The figure shows the weights for several rows of instance $P^5_{11,20}$. It can be seen that the weights for rows 13, 15, 18 which have scaling factor 5 are much larger during the first 1150 iterations than the weights for rows 3, 5, 8 which are not scaled. The weights for the lats rows become nearly zero during iteration 800-1000. But from iteration 1100 the weights for these rows increase and become significantly larger than the weights for rows 13, 15, 18. When the ACO model converges all weights converge to 1 for which the cc-pheromone update equals the standard pheromone update. Overall it can be seen that the weights change quite dynamically and adapt to the actual state of the optimization process.

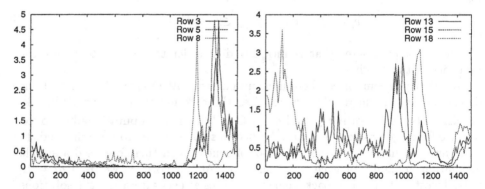

Fig. 4. Change of weights w'_i for a typical run of ACO algorithm on problem $P^5_{11,20}$; rows 3, 5, and 8 are not scaled (left), rows 13, 15, and 18 are scaled with 5 (right)

6   Conclusion

In this paper we have proposed a new type of pheromone update for ACO called competition controlled pheromone update (cc-pheromone update). For this pheromone update the amount of pheromone that is added for a decision of an ant depends on the strength of competition under which the decision was made. We have proposed a measure for the strength of competition. It was shown that cc-pheromone update clearly outperformed the standard pheromone update on several simple test problems. It is future work to investigate for which type of real world application problems cc-pheromone is suitable and how it might be combined best with standard pheromone update. It is interesting to study several aspects of cc-pheromone update in more detail, e.g., the estimation method for the Kullback-Leibler distance or alternative measures for competition.

References

1. Blum, C. and Sampels, M. (2002). Ant Colony Optimization for FOP Shop scheduling: A case study on different pheromone representations Proc. of the 2002 Congress on Evolutionary Computation (CEC'02), 1558-1563.
2. Bullnheimer, B., Hartl, R.F., and Strauss, C. (1999). A new rank based version of the ant system - a computational study. Central Europ. J. Oper. Res. 7(1): 25-38.
3. Dorigo, M. (1992). Optimization, Learning and Natural Algorithms *(in Italian)*. PhD thesis, Dipartimento di Elettronica, Politecnico di Milano, Italy.
4. Dorigo, M., and Di Caro, G. (1999). The ant colony optimization meta-heuristic. In Corne, D., Dorigo, M., and Glover, F., editors, *New Ideas in Optimization*, 11–32, McGraw-Hill, London.
5. Dorigo, M., Maniezzo, V., and Colorni, A. (1996). The Ant System: Optimization by a Colony of Cooperating Agents. *IEEE Trans. Systems, Man, and Cybernetics – Part B*, 26: 29–41.
6. Dorigo, M., Zlochin, M., Meuleau, N., and Birattari, M. (2002). Updating ACO Pheromones Using Stochastic Gradient Ascent and Cross-Entropy Methods. Proc. EvoWorkshops, LNCS 2279, 21-30.
7. Meuleau, N. and Dorigo, M. (2002). Ant colony optimization and stochastic gradient descent. Artificial Life, 8(2):103-121.
8. Merkle, D. and Middendorf, M. (2000). An Ant Algorithm with a new Pheromone Evaluation Rule for Total Tardiness Problems. In Cagnoni, S., et al. (Eds.) *Real-World Applications of Evolutionary Computing*, LNCS 1803, 287–296, Springer.
9. Merkle, D. and Middendorf, M. (2001). A New Approach to Solve Permutation Scheduling Problems with Ant Colony Optimization. In Boers, E. J. W., et al. (Eds.) *Applications of Evolutionary Computing*, LNCS 2037, 213–222, Springer.
10. Merkle, D. and Middendorf, M. (2002). Ant Colony Optimization with the Relative Pheromone Evaluation Method. Proc. 3rd European Workshop on Scheduling and Timetabling and 3rd European Workshop on Evolutionary Methods for AI Planning (EvoSTIM/EvoPLAN-2002), LNCS 2279, 325-333.
11. Merkle, D. and Middendorf, M. (2002). Modelling the Dynamics of Ant Colony Optimization Algorithms. Evolutionary Computation 10(3): 235-262, 2002.
12. Merkle, D. and Middendorf, M. (2002). Modelling ACO: Composed Permutation Problems. Proceedings of Third International Workshop ANTS 2002, Springer, LNCS 2463, 149-162.
13. Merkle, D., Middendorf, M., and Schmeck, H. (2002): Ant Colony Optimization for Resource-Constrained Project Scheduling. *IEEE Transactions on Evolutionary Computation*, 6(4): 333-346.
14. Randall, M. and Tonkes, E. (2000). Intensification and Diversification Strategies in Ant Colony Optimisation. *TR00-02*, School of Inf. Technology, Bond University.
15. Stützle, T. and Hoos, H. H. (2000). MAX-MIN Ant System. Future Generation Computer Systems. 16(8):889–914.

Cooperative Transport of Objects of Different Shapes and Sizes

Roderich Groß and Marco Dorigo

IRIDIA - Université Libre de Bruxelles - Brussels, Belgium
{rgross,mdorigo}@ulb.ac.be

Abstract. This paper addresses the design of control policies for groups of up to 16 simple autonomous mobile robots (called *s-bots*) for the cooperative transport of heavy objects of different shapes and sizes. The s-bots are capable of establishing physical connections with each other and with the object (called *prey*). We want the s-bots to self-assemble into structures which pull or push the prey towards a target location. The s-bots are controlled by neural networks that are shaped by artificial evolution. The evolved controllers perform quite well, independently of the shape and size of the prey, and allow the group to transport the prey towards a moving target. Additionally, the controllers evolved for a relatively small group can be applied to larger groups, making possible the transport of heavier prey. Experiments are carried out using a physics simulator, which provides a realistic simulation of real robots, which are currently under construction.

1 Introduction

The transport of heavy objects by groups of robots can be motivated by low cost of manufacture, high robustness, high failure tolerance, or high flexibility – all desirable properties for a robotic system to have. Instead of using one robot powerful enough to transport the object (also called *prey* hereafter) on its own, the task is accomplished in cooperation by a group of simpler robots.

In the literature, several works can be found that consider the transport of an object by a team of robots. Deneubourg et al. (1990) proposed the use of self-organized approaches for the collection and transport of objects by robots in unpredictable environments. Each robot unit could be simple and inefficient in itself, but a group of robots could exhibit complex and efficient behaviors. Cooperation could be achieved without any direct communication among robots (Grassé, 1959; Deneubourg and Goss, 1989).

Kube and Zhang (1993a,b) studied a distributed approach to let a group of simple robots find and push a prey towards a light. Kube and Bonabeau (2000), on a follow-up of Kube and Zhang's research, evaluated the sensitivity of their robotic system to the prey's geometry by comparing the performance on prey of different shapes. They report that when the prey was a small cuboid, "as the number of robots increased the task took longer to complete as the robot interference was high since the limited box side space created competition

M. Dorigo et al. (Eds.): ANTS 2004, LNCS 3172, pp. 106–117, 2004.

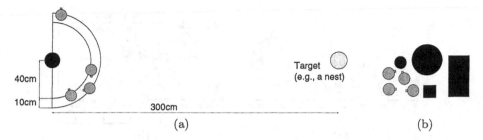

(a)   (b)

Fig. 1. (a) Example of initial placement: the prey (black) has to be transported towards a light-emitting beacon acting as a target (e.g., a robot nest). We assume that in the environments in which the robotic system will be applied, the s-bots will approach the prey from the half space on the side of the nest. To simulate this, the s-bots (gray) are placed randomly within a semi-circle 40 to 50 cm away from the prey. (b) Top view of four s-bots (gray) and of those four prey (black) having the most extreme dimensions. The mass of each prey is independent of its shape and size.

among the robots". If the robot's diameter was reduced while keeping the box dimensions fixed, the performance increased.

A similar observation has been reported by Groß and Dorigo (2004): they evolved neural networks to control two simple robots in order to pull or push a cylindrical prey as far as possible in an arbitrary direction. Increasing the number of robots while using heavier prey with the same cylindrical shape and dimension, the performance of the evolved strategies greatly decreased. However, if the diameter of the prey was multiplied by the same factor as the number of robots was increased, the controllers could reach again a satisfactory level of performance. In the latter case the density of robots around the prey is comparable with the one in the most successful setup used by Kube and Bonabeau (2000).

In this paper, we aim at a robotic system that is appropriate for the transport of prey of different shapes and sizes. We consider prey of cuboid as well as cylindrical shape, with footprints that may differ in size by factors up to 6.25. In addition, the prey can be of two different heights. The weight of the prey is independent of its geometry and may vary up to a factor of 2.

2   Experimental Framework

2.1   Environment and Task

The simulated environment consists of a flat plane, a prey with characteristics that change across simulations (shape, size and mass), and a light-emitting beacon acting as a target and placed 300 cm away from the initial position of the prey. A simulation lasts 20 simulated seconds. Initially, the prey is put with random orientation on the plane and four s-bots are put at random positions and with random orientations in the neighborhood of the prey (see Figure 1 (a)). Since the s-bots are placed in the vicinity of the prey, it is not necessary for the s-bot to deploy any elaborate exploration of the environment in order to find the

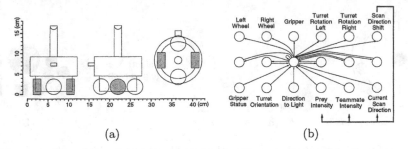

(a)   (b)

Fig. 2. The s-bot: (a) front, side and top view of the s-bot model and (b) the controller network. For details see Sections 2.3 and 2.4.

prey. The s-bot controllers are supposed to let the s-bots localize and approach the prey, and self-assemble into structures physically linked to the prey, in order to pull or push it towards the beacon.

2.2 Prey

The prey is modeled as a cylinder or as a cuboid of height 12 or 20 cm. Tall prey are of special interest since they can prevent s-bots from localizing the target location. In case of a cylinder, the radius (in cm) is chosen in the range $[6, 15]$. If the prey is modeled as a cuboid, the length l of one horizontal side (in cm) is chosen in the range $[12, 40]$, while the other horizontal side length is set in the range $[12, \min(l, 20)]$. The mass of the prey (in grams) is in the range $[500, 1000]$. Depending on its weight, the cooperative behavior of at least 2 or 3 s-bots is necessary to move the prey. Objects of height 12 cm have a uniform distribution of mass, while the taller ones are modeled as a stack of two geometrically identical objects of different masses: the mass of the object on top of the stack is 25% of the total mass. This prevents tall prey from toppling down when pushed or pulled by the s-bots. Every potential geometry has the same probability to be chosen in each simulation. Figure 1 (b) illustrates the dimensions of different prey with respect to the s-bot's size.

2.3 S-Bot Model

The s-bot model is illustrated in Figure 2 (a). It is an approximation of a real s-bot, currently under construction within the SWARM-BOTS project (Mondada et al., 2004; Dorigo et al., 2004, see also www.swarm-bots.org). The model provides a degree of freedom enabling the rotation of the s-bot's upper part (the turret) with respect to the lower part (the chassis) by means of a motorized axis. The turret is composed of a cylindrical torso, a *gripper* element fixed on the torso's edge heading outwards, and of a pillar placed on the top to support a mirror providing a camera system with an omni-directional view. The chassis is composed of a cylindrical element and of four wheels attached to it: two motorized wheels on the left and right, and two passive wheels, one in the front, and one in the back. The s-bot has a height of 16.85 cm and a mass of 660 g.

In the following, the actuators and sensors are described. Both the actuators and the sensors are affected by noise (for details see Groß (2003)).

Actuators. The s-bot is equipped with two motorized wheels on the left and right, a rotational base, a gripper element and an LED ring. Each motorized wheel can be controlled by setting the desired angular speed (in rad/s) in the range $[-8, 8]$. The rotational base actuator enables the s-bot to align its turret with respect to its chassis to any angular offset in $[-\pi, \pi]$. The gripper element is modeled as a box heading forward. If the element is in *gripping* mode, a connection will be established as soon as the gripper element touches the turret of another s-bot or the prey. If the gripper is set to the *open* mode, any gripped object is released. The gripper of the real s-bot can establish a physical link to another s-bot or to the prey by grasping the surrounding ring this object is equipped with. The LED ring can be activated with different colors. In this study it is permanently activated with a red color to facilitate the s-bots detection.

Sensors. The simulated s-bot is equipped with a gripper status sensor, a rotational base sensor, and an omni-directional camera. These sensors have their counterparts on the real s-bot. The gripper status sensor is able to detect whether the gripper element is connected to another object or not. The rotational base sensor measures the angular offset between the turret and the chassis.

The simulated camera provides data based not on recorded images but on information available in the simulator, e.g., if up to a pre-defined distance there is a green object in front of the s-bot or not. The implicit assumption is that when working with the real s-bot, such information can be obtained from the camera using feature extraction algorithms. Although the real camera is attached to the turret (that can be rotated with respect to the chassis), it is possible to simulate a camera that is aligned with the chassis, since it is possible to measure the angular mismatch between the turret and the chassis.

Using the camera, the s-bot is able to scan its surrounding for teammates or for the prey. Given any horizontal direction, the s-bot can detect, up to a distance of $R = 50$ cm in the specified direction, the presence of a prey (green object) or (if not shadowed) of an s-bot (red object). In both cases, the s-bot can compute a rough estimate of the object distance. In addition, the camera can localize the light-emitting beacon (horizontal angle), as long as it is not shadowed by the prey or by a teammate.

2.4 Control Policy

The architecture used in order to control the group of s-bots is a simple recurrent neural network (Elman, 1990) with six hidden nodes and whose weights are synthesized by an evolutionary algorithm. The same network is cloned so that it is identical on each s-bot. We hypothesize that Elman networks might be more appropriate than fully reactive networks, since the proper selection of sensory patterns by motor actions might be facilitated by the use of memory.

The neural network is illustrated in Figure 2 (b). It is activated every 100 ms. The six input neurons correspond to the five, current sensor readings and to the horizontal direction in which the camera scans for the teammates and prey. The activations of the output neurons are used to control the motorized wheels, the gripper and the angular offset of the rotational base. Furthermore, the neural network may shift the horizontal direction in which the camera scans the environment of any angle (in degrees) in the range $[-18, 18]$. Given such a controllable sensor, the s-bot can select sensory patterns by its motor actions[1].

In average the s-bot's detection of the prey at the beginning of the simulation would require at least several seconds since the perception is restricted to a single horizontal direction per control step. To facilitate the detection, the scan direction of the camera is manually aligned approximately towards the prey for the first second of simulation time.

2.5 Evolutionary Algorithm

The evolutionary algorithm used is a self-adaptive $(\mu + \lambda)$ evolution strategy (Rechenberg, 1965; Schwefel, 1975). The implementation is almost identical to the one utilized by Groß and Dorigo (2004). In this paper the initial population is not biased and purely random, while in the previous study it was constructed by a Monte Carlo search.

The number of offspring is $\lambda = 60$ and the number of parents is $\mu = 20$.

2.6 Fitness

Each individual represents a common controller for a group of s-bots. The objective is to let a group of $N = 4$ s-bots approach the prey, and self-assemble into structures physically linked to the prey, in order to pull or push it towards the beacon (the faster the better). The simulation lasts $T = 20$ seconds.

To favor the evolution of solutions that make use of the gripper element, the fitness function takes the assembling performance into account. Since the s-bots start disassembled up to $R = 50$ cm away from the prey, some time is required to approach the prey and to establish a connection. Therefore, the recording of the assembling performance A starts after 10 simulation seconds have elapsed:

$$A = \frac{1}{(T - 10 + 1)N} \sum_{k=10}^{T} \sum_{j=1}^{N} A_j(t = k), \tag{1}$$

where $A_j(t = k) \in [0, 1]$ is defined by

$$A_j(t = k) = \begin{cases} 1 & \text{if } j \in M(t = k); \\ 0 & \text{if } d_j(t = k) \geq R \wedge j \notin M(t = k); \\ 0.75 & \text{if } d_j(t = k) < \frac{R}{2} \wedge j \notin M(t = k); \\ 0.65 \left(\frac{R - d_j(t=k)}{R/2} \right) + 0.1 & \text{otherwise.} \end{cases} \tag{2}$$

---

[1] This is called *active perception* and, within evolutionary robotics, it has already been studied, for instance, by Kato and Floreano (2001) and Marocco and Floreano (2002).

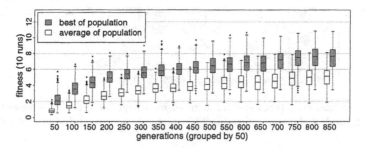

Fig. 3. Box-and-whisker plot providing information about the evolutionary progress in ten evolutionary runs (4 s-bots; prey of various dimensions and weights, see Section 2.2). Each box comprises observations ranging from the first to the third quartile. The median is indicated by a bar, dividing the box into the upper and lower part. The whiskers extend to the the farthest data points that are within 1.5 times the interquartile range. Outliers are indicated as circles. Characteristics about two types of observations for each population are displayed: the best fitness value (gray boxes) and the average fitness (white boxes) of a population.

The value $d_j(t = k)$ denotes the distance between s-bot j and the prey at time k. $M(t = k)$ is the set of s-bots that are physically linked to the prey at time t. This comprises s-bots that directly grip the prey, but also those that are connected to a chain of s-bots of which the foremost is gripped to the prey. The assembling performance A rewards s-bots for being close to the prey and even more for being physically linked to the prey.

The distance gain D is given by $D = D(t = 0) - D(t = T)$, where $D(t = k)$ denotes the horizontal distance (in cm) between the prey and the beacon after k seconds of simulation time.

Finally, the quality measure q is defined as

$$q = \begin{cases} A & \text{if } D \leq 0; \\ 1 + (1 + \sqrt{D})A^2 & \text{otherwise.} \end{cases} \tag{3}$$

The fitness value (to be maximized) is given by the average quality observed in five tests with different initial configurations and different prey. The sample of tests is changed at each generation.

3   Results

The experimental setup described above has been used in ten independent evolutionary runs of 850 generations each. Every run lasted about three weeks on a machine equipped with an AMD Athlon XP^{TM} 2400+ processor. Figure 3 shows the development of the best fitness (gray boxes) and the average fitness (white boxes) for all ten runs. Altogether there is a continuous increase of the best and average fitness for the entire range of 850 generations. The fitness values are noisy and obtained running five tests per individual so that they are only a very rough estimate of the attained quality level.

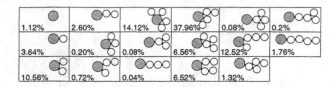

Fig. 4. Connection structures and frequencies by which the structures have been observed at the end of the simulation (in 2,500 tests of the best individual of run 10).

For each evolutionary run, we measure the transport quality of the $\mu = 20$ parent individuals of the last generation on a sample of 500 different tests. This set of μ parent individuals comprises all the genetic material that would be exploited in subsequent generations in case the evolution were continued. For every evolutionary run, the individual exhibiting the highest average of the observed transport quality values is selected. These individuals are post-evaluated for a second time, on a new sample of 2,500 tests.

The median assembling performance $A \in [0, 1]$ for the best individuals for all ten runs is in the range $[0.92, 1.0]$. The corresponding average values range from 0.79 to 0.92. In the following, we focus on the individuals with the highest average distance gain of the two most successful evolutions (here labeled as run 9 and run 10). Figure 4 illustrates all the possible connection structures in which s-bots that are connected to the prey can be organized. Each of these structures has appeared during the post-evaluation of the best individual of run 10 (frequencies indicated in the figure).

Figure 5 presents a box-and-whisker plot of the distance the prey has gained with respect to the beacon as observed in the 2,500 tests for the two best individuals. To evaluate the quality of these results, we proceed as follows.

First we experimentally measure the maximum distance that can be covered in 20 seconds by a single s-bot connected from simulation start (call it s-bot$_{con}$) to a prey of mass $125\,g$ $(250\,g)$: this distance turns out to be $222\,cm$ $(141\,cm)$. Then, we assume that within the same time interval (20 s) a group of n s-bot$_{con}$ connected from start to a prey of weight $n \cdot 125\,g$ $(n \cdot 250\,g)$, will transport it for the same maximum distance (i.e., $222\,cm$ $(141\,cm)$). Last, we experimentally observe that the average time for a group of 4 s-bots to self-assemble is 9.9 seconds (6.2 seconds) for the best controller of run 9 (run 10).

Therefore, we can use the distance covered by a group of n s-bot$_{con}$ within the average time available to the n s-bots for transport as a reference to the distance achievable on average by our group of n s-bots. For the best evolved controller of run 9 (run 10), the time available to a group of 4 s-bots for tranporting the prey is 10.1 seconds (13.8 seconds) and the reference distance is $112\,cm$ $(153\,cm)$ for the $n \cdot 125\,g$ prey, and $71\,cm$ $(97\,cm)$ for the $n \cdot 250\,g$ prey (on average).

Observing Figure 5, it is clear that, although in some tests the performance achieved is very good, the average is still rather low and there is surely room for future improvements.

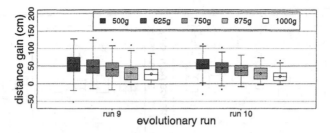

Fig. 5. Distance gain (in cm) within 20 seconds of simulation time by the two best individuals for different prey weights (4 s-bots; 500 observations per box). Average values are indicated by diamonds.

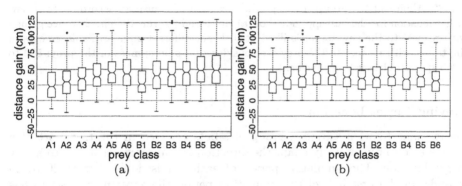

Fig. 6. Sensitivity of performance to different shapes and sizes of prey (4 s-bots; mass of the prey: 500 to 1,000 g): distance gain within 20 seconds by the best controller of (a) run 9 (188-215 observations per box) and (b) run 10 (205-229 observations per box).

Sensitivity to Different Shapes and Sizes of Prey

To assess the flexibility of the best two evolved controllers with respect to the shape and the size of the prey, the data concerning the 2,500 tests performed is partitioned according to different classes of prey (all measures in cm):

- The sets $A1, A2, \ldots, A6$ comprise the data concerning cylindrical prey. Ai corresponds to tests in which the cylinders' radii reside in the range $[6 + 1.5(i - 1), 6 + 1.5i)$.
- The sets $B1, B2, \ldots, B6$ comprise the data concerning prey of cuboid shape. $B1, B2, \ldots, B6$ refer to tests in which the length of the longer side is in the ranges $[12, 16)$, $[16, 20)$, $[20, 25)$, $[25, 30)$, $[30, 35)$ and $[35, 40)$. The length of the shorter side is always in the range $[12, 20]$.

Figure 6 shows the distance gain values observed for different classes of prey. The best individual of run 10 exhibits a decrease in performance if the prey is very small (i.e. for A1, A2 and B1), while the other performs robustly with respect to all combinations of size and shape.

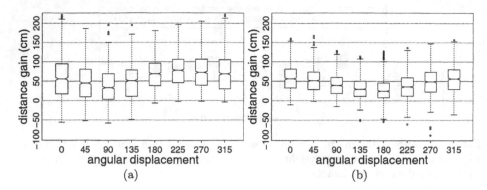

(a)   (b)

Fig. 7. Ability of the group to adapt the direction of transport according to a new target location for the best individual of (a) run 9 and (b) run 10. The distance gained with respect to the new target is illustrated. Eight different angular displacements of the target with respect to the prey are considered (4 s-bots; mass of the prey: 500 to 1,000 g; 1,250 observations per box).

Moving Target Location

We examine to what extent a group of s-bots engaged in the transport of a prey is able to dynamically adapt the direction of transport according to a new target location. The simulation period of each test is doubled from 20 to 40 seconds. In the first half of the simulation period the experimental setup has been retained unchanged. As soon as the first half of the simulation period is elapsed, the beacon (i.e., the target) is moved to a new location. The (horizontal) angular displacement (radian measure) of the beacon with respect to the prey is chosen randomly in $\{\frac{i}{8}2\pi \mid i \in \{0, 1, 2, \ldots, 7\}\}$. For each angular displacement the distance gained during the second part of the simulation is recorded for both individuals on 1,250 tests (see Figure 7). If the target direction does not change, the median distance gain of both individuals is 56 cm. Depending on the angular displacement, the median distance gain of the best individual of run 9 (run 10) may decrease to 17 cm (32 cm).

Scalability

In this section, we study the applicability of evolved controllers to larger groups of s-bots in order to transport prey of bigger weight. We consider groups of 4, 8, 12 and 16 robots. Along with the group size, the mass of the prey is also increased proportionally from 500 to 1,000, 1,500 and 2,000 g respectively.

The two best individuals have been evaluated on 250 tests per group size. Since for larger groups more time is necessary for assembling, the simulation period has been extended to 30 seconds. To ensure a non-overlapping placement of up to 16 s-bots in the immediate vicinity of the prey, the s-bots are initially put randomly within a full circle 25 to 50 cm away from the prey. Given the locality of sensing, it is not possible to position many more s-bots in the environment

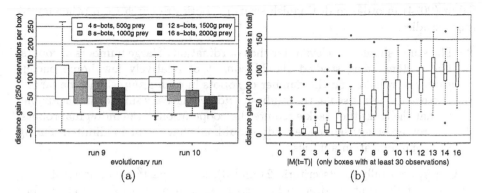

Fig. 8. Distance gain (cm) within 30 seconds of simulation time (a) with 4, 8, 12 and 16 robots transporting prey of mass 500, 1,000, 1,500 and 2,000 g and (b) with respect to $|M(t = T)|$ (i.e. the number of robots physically linked to the prey at time T) for 16 robots transporting prey of mass 2,000 g. The boxes are drawn with widths proportional to the square-roots of the number of observations in the groups.

(without having contacts among them) while ensuring at the same time that each one can locate the prey from its starting point.

For all group sizes the evolved individuals are able to cooperatively move the object. Overall, the performance decreases with group size (see Figure 8 (a)). We observed that a too high density of robots makes it difficult for the robots to self-assemble into structures connected to the prey. However, once the majority of s-bots is assembled, they exhibit a high performance even for group size 16 (see Figure 8 (b)).

4 Conclusions

In this paper we addressed the design of control policies for a group of simple autonomous mobile robots (the s-bots) for the cooperative transport of heavy objects called prey. We used artificial evolution to synthesize neural network controllers in order to let the s-bots localize and approach a prey, self-assemble into structures and pull or push the prey towards a target location. We consider this to be a challenging task; the evolved controllers are responsible for selecting sensory patterns by motor actions, and for controlling the wheels, a gripper device and the orientation of the s-bot's turret with respect to the chassis. The group has to cope with prey of different shapes and sizes. Moreover, for an s-bot the target can be shadowed by teammates or by the prey.

We have considered prey of cuboid and cylindrical shape with footprints that may differ in size by factors up to 6.25. One of the best individuals has shown to perform robustly with respect to all combinations of size and shape. Furthermore, the evolved controllers can be applied to let a group of robots pull or push a prey towards a target whose location changes in time.

As a final point, we wish to discuss to what extent our study can be considered a swarm robotics work. Recently, four criteria have been identified by Dorigo and

Şahin (2004) in order to distinguish swarm robotics research from other multi-robot studies:

1. The study should be relevant for the coordination of large numbers of robots.
2. The system being studied should consist of relatively few homogeneous groups or robots, and the number of robots in each group should be large.
3. The robots being used in the study should be relatively simple and incapable, so that the tasks they tackle require the cooperation of the individual robots.
4. The robots being used in the study should only have local and limited sensing and communication abilities.

Our system fully meets criteria 2) and 3), and in part[2] also criteria 4). For testing the fulfillment of criterion 1) further experiments with a greater number of s-bots need to be carried out. In this paper we have shown that for groups of size 4, 8, 12 and 16 robots, the best evolved controllers are able to cooperatively transport prey of mass 500, 1,000, 1,500 and 2,000 g respectively towards a target location. Once the majority of s-bots is physically linked to the prey, also for group size 16 a high performance level is reached. However, the assembling performance seems to decrease as the number of robots is increased. This is likely to be caused by the very high density of robots that are positioned in the immediate vicinity of the prey.

In the future, we want to extend this work by the integration of exploration and recruitment strategies aiming at a complete and scalable prey retrieval solution. In addition, we plan to transfer the solutions from simulation to the real s-bots.

Acknowledgments

This work was supported by the Belgian FNRS, of which Marco Dorigo is a Senior Research Associate, via the grant "Virtual Swarm-bots" (contract no. 9.4515.03), by the Scientific Research Directorate of the French Community of Belgium via an "Action de Recherche Concertée" (the "ANTS" project), and by the Future and Emerging Technologies programme (IST-FET) of the European Commission, via the "Swarm-bots" project (grant IST-2000-31010). The information provided is the sole responsibility of the authors and does not reflect the Community's opinion. The Community is not responsible for any use that might be made of data appearing in this publication. The authors also wish to thank all the members of the SWARM-BOTS project for their support, suggestions and comments.

References

J.-L. Deneubourg and S. Goss. Collective patterns and decision-making. *Ethology, Ecology and Evolution*, 1:295–311, 1989.

---

[2] The target of transport is visible for several meters (conflict with locality of sensing).

J.-L. Deneubourg, S. Goss, G. Sandini, F. Ferrari, and P. Dario. Self-organizing collection and transport of objects in unpredictable environments. In *Proceedings of Japan – U.S.A Symposium on Flexible Automation*, pages 1093–1098. ISCIE, Kyoto, Japan, 1990.

M. Dorigo and E. Şahin. Swarm robotics – special issue editorial. *Autonomous Robots*, 17(2–3):111–113, 2004.

M. Dorigo, V. Trianni, E. Şahin, R. Groß, T. H. Labella, G. Baldassarre, S. Nolfi, J.-L. Deneubourg, F. Mondada, D. Floreano, and L. M. Gambardella. Evolving self-organizing behaviors for a swarm-bot. *Autonomous Robots*, 17(2–3):223–245, 2004.

J. Elman. Finding structure in time. *Cognitive Science*, 14:179–211, 1990.

P. Grassé. La reconstruction du nid et les coordinations inter-individuelles chez *Bellicositermes natalensis et Cubitermes sp.* La théorie de la stigmergie : essai d'interprétation du comportement des termites constructeurs. *Insectes Sociaux*, 6: 41–81, 1959.

R. Groß. Swarm-intelligent robotics in prey retrieval tasks. Technical Report TR/IRIDIA/2003-27, IRIDIA, Université Libre de Bruxelles, Belgium, 2003. DEA thesis.

R. Groß and M. Dorigo. Evolving a cooperative transport behavior for two simple robots. In *Artificial Evolution – 6th International Conference, Evolution Artificielle (EA 2003)*, volume 2936 of *Lecture Notes in Computer Science*, pages 305–317. Springer Verlag, Berlin, Germany, 2004.

T. Kato and D. Floreano. An evolutionary active-vision system. In *Proceedings of the 2001 Congress on Evolutionary Computation (CEC'01)*, pages 107–114. IEEE Computer Society Press, Los Alamitos, CA, 2001.

C. Kube and E. Bonabeau. Cooperative transport by ants and robots. *Robotics and Autonomous Systems*, 30(1-2):85–101, 2000.

C. Kube and H. Zhang. Collective robotic intelligence. In *From Animals to Animats 2. Proceedings of the Second International Conference on Simulation of Adaptive Behavior*, pages 460–468. MIT Press, Cambridge, MA, 1993a.

C. Kube and H. Zhang. Collective robotics: from social insects to robots. *Adaptive Behaviour*, 2(2):189–218, 1993b.

D. Marocco and D. Floreano. Active vision and feature selection in evolutionary behavioral systems. In *From Animals to Animats 7: Proceedings of the Seventh International Conference on Simulation of Adaptive Behavior*, pages 247–255. MIT Press, Cambridge, MA, 2002.

F. Mondada, G. C. Pettinaro, A. Guignard, I. V. Kwee, D. Floreano, J.-L. Deneubourg, S. Nolfi, L. M. Gambardella, and M. Dorigo. SWARM-BOT: A new distributed robotic concept. *Autonomous Robots*, 17(2–3):193–221, 2004.

I. Rechenberg. *Cybernetic Solution Path of an Experimental Problem*. PhD thesis, Royal Aircraft Establishment, Hants Farnborough, UK, 1965.

H.-P. Schwefel. *Evolutionsstrategie und numerische Optimierung*. PhD thesis, Technische Universität Berlin, Germany, 1975.

Deception in Ant Colony Optimization

Christian Blum and Marco Dorigo

IRIDIA – Université Libre de Bruxelles – Brussels, Belgium
{cblum,mdorigo}@ulb.ac.be

Abstract. The search process of a metaheuristic is sometimes misled. This may be caused by features of the tackled problem instance, by features of the algorithm, or by the chosen solution representation. In the field of evolutionary computation, the first case is called *deception* and the second case is referred to as *bias*. In this work we formalize the notions of deception and bias for ant colony optimization. We formally define *first order deception* in ant colony optimization, which corresponds to deception as being described in evolutionary computation. Furthermore, we formally define *second order deception* in ant colony optimization, which corresponds to the bias introduced by components of the algorithm in evolutionary computation. We show by means of an example that second order deception is a potential problem in ant colony optimization algorithms.

1   Introduction

Deception and bias are well-studied subjects in evolutionary computation algorithms for the application to combinatorial optimization (CO) problems. The term deception was introduced by Goldberg in [12] with the aim of describing problems that are misleading for genetic algorithms (GAs). Well-known examples of deceptive problems are n-bit trap functions [7]. These functions are characterized by fixpoints with large basins of attraction that correspond to sub-optimal solutions, and by fixpoints with relatively small basins of attraction that correspond to optimal solutions. Therefore, for these problems a GA will – in most cases – not find an optimal solution. Except for deception, other efforts in the field of evolutionary computation were aimed at studying the effects of the bias that is sometimes introduced by the solution representation and the genetic operators. In some cases this bias was shown to have a negative impact on the search process (see, for example, [16]).

In the early 90's, ant colony optimization (ACO) [8–10] emerged as a novel nature-inspired metaheuristic method for the solution of combinatorial optimization problems. The inspiring source of ACO is the foraging behavior of real ants. When searching for food, ants initially explore the area surrounding their nest in a random manner. As soon as an ant finds a food source, it evaluates quantity and quality of the food and carries some of the food found to the nest. During the return trip, the ant deposits a chemical pheromone trail on the ground. The quantity of pheromone deposited, which may depend on the quantity and quality

M. Dorigo et al. (Eds.): ANTS 2004, LNCS 3172, pp. 118–129, 2004.

of the food, will guide other ants to the food source. The indirect communication between the ants via the pheromone trails allows them to find shortest paths between their nest and food sources. This behaviour of real ant colonies is exploited in artificial ant colonies in order to solve discrete optimization problems.

Research on bias in ACO algorithms is largely restricted to the work by Merkle and Middendorf [14, 13], and the work by Blum and Sampels [3, 4]. Merkle and Middendorf proposed and introduced the use of *models of ACO algorithms*. They studied the behaviour of a simple ACO algorithm by studying the dynamics of its model when applied to permutation problems. It was shown that the behaviour of ACO algorithms is strongly influenced by the pheromone model (and the way of using it) and by the competition between the ants. Moreover, it was shown that the performance of the model of an ACO algorithm may decrease during a run, which is clearly undesirable, because in general this worsens the probability of finding better and better solutions. Independently, Blum and Sampels showed on the example of an ACO algorithm for a general shop scheduling problem that ACO algorithms may fail depending on the chosen pheromone model.

Our Contribution. In this work we formalize the notions of deception and bias in ant colony optimization. We formally define *first order deception* in ant colony optimization, which corresponds to deception as being described in evolutionary computation. Then, we formally define *second order deception* in ant colony optimization, which corresponds to the bias introduced by components of the algorithm in evolutionary computation. Finally, we give an example of second order deception in order to show that the bias that leads to the occurrence of second order deception can make an ACO algorithm fail.

2   The Framework of a Basic ACO Algorithm

ACO algorithms are metaheuristic methods for tackling combinatorial optimization problems (see [11]). The central component of an ACO algorithm is the pheromone model, which is used to probabilistically sample the search space. As outlined in [2], the pheromone model can be derived from a *model* of the CO problem under consideration. A model of a CO problem can be stated as follows.

Definition 1. *A model $\mathcal{P} = (\mathcal{S}, \Omega, f)$ of a CO problem consists of:*

- *a **search (or solution) space** $\mathcal{S}$ defined over a finite set of discrete decision variables and a set Ω of **constraints** among the variables;*
- *an **objective function** $f : \mathcal{S} \to \mathbb{R}^+$ to be minimized.*

*The search space $\mathcal{S}$ is defined as follows: Given is a set of n **discrete variables** X_i with domains $D_i = \{v_i^1, \ldots, v_i^{|D_i|}\}$, $i = 1, \ldots, n$. A variable instantiation, that is, the assignment of a value v_i^j to a variable X_i, is denoted by $X_i = v_i^j$. A feasible solution $s \in \mathcal{S}$ is a complete assignment (i.e., an assignment in which each decision variable has a domain value assigned) that satisfies the constraints. If the set of constraints Ω is empty, then each decision variable can take any value*

Algorithm 1 The framework of a basic ACO algorithm

---

input: An instance P of a CO problem model $\mathcal{P} = (\mathcal{S}, f, \Omega)$.
InitializePheromoneValues($\mathcal{T}$)
$\mathfrak{s}_{bs} \leftarrow$ NULL
while termination conditions not met **do**
    $\mathfrak{S}_{iter} \leftarrow \emptyset$
    for $j = 1, \ldots, n_a$ **do**
        $\mathfrak{s} \leftarrow$ ConstructSolution($\mathcal{T}$)
        $\mathfrak{s} \leftarrow$ LocalSearch($\mathfrak{s}$)   {optional}
        if $(f(\mathfrak{s}) < f(\mathfrak{s}_{bs}))$ or $(\mathfrak{s}_{bs} =$ NULL) **then** $\mathfrak{s}_{bs} \leftarrow \mathfrak{s}$
        $\mathfrak{S}_{iter} \leftarrow \mathfrak{S}_{iter} \cup \{\mathfrak{s}\}$
    end for
    ApplyPheromoneUpdate($\mathcal{T}, \mathfrak{S}_{iter}, \mathfrak{s}_{bs}$)
end while
output: The best-so-far solution $\mathfrak{s}_{bs}$

---

*from its domain independently of the values of the other decision variables. In this case we call $\mathcal{P}$ an **unconstrained** problem model, otherwise a **constrained** problem model. A feasible solution $s^* \in \mathcal{S}$ is called a **globally optimal solution**, if $f(s^*) \leq f(s) \; \forall s \in \mathcal{S}$. The set of globally optimal solutions is denoted by $\mathcal{S}^* \subseteq \mathcal{S}$. To solve a CO problem one has to find a solution $s^* \in \mathcal{S}^*$.*

A model of the CO problem under consideration implies the finite set of solution components and the pheromone model as follows. First, we call the combination of a decision variable X_i and one of its domain values v_i^j a *solution component* denoted by c_i^j. Then, the pheromone model consists of a *pheromone trail parameter* T_i^j for every solution component c_i^j. The value of a pheromone trail parameter T_i^j – called *pheromone value* – is denoted by τ_i^j. The set of all pheromone trail parameters is denoted by $\mathcal{T}$. As a CO problem can be modelled in different ways, different models of the CO problem can be used to define different pheromone models.

Algorithm 1 captures the framework of a basic ACO algorithm. It works as follows. At each iteration, n_a ants probabilistically construct solutions to the combinatorial optimization problem under consideration, exploiting a given pheromone model. Then, optionally, a local search procedure is applied to the constructed solutions. Finally, before the next iteration starts, some of the solutions are used for performing a pheromone update. The details of this framework are explained in more detail in the following.

InitializePheromoneValues($\mathcal{T}$). At the start of the algorithm the pheromone values are all initialized to a constant value $c > 0$.

ConstructSolution($\mathcal{T}$). The basic ingredient of any ACO algorithm is a constructive heuristic for probabilistically constructing solutions. A constructive heuristic assembles solutions as sequences of elements from the finite set of solution components $\mathfrak{C}$. A solution construction starts with an empty partial solution $\mathfrak{s}^p = \langle \rangle$. Then, at each construction step the current partial solution $\mathfrak{s}^p$ is extended by adding a feasible solution component from the set $\mathfrak{N}(\mathfrak{s}^p) \subseteq \mathfrak{C} \setminus \mathfrak{s}^p$, which is defined by the solution construction mechanism. The process of constructing solutions can be regarded as a walk (or a path) on the so-called *construction graph* $\mathcal{G}_C = (\mathfrak{C}, \mathfrak{L})$, which is a fully connected graph whose vertices are the solution components $\mathfrak{C}$ and the set $\mathfrak{L}$ are the connections. The allowed walks on $\mathcal{G}_C$ are implicitly defined by the solution construction mechanism that defines the set $\mathfrak{N}(\mathfrak{s}^p)$ with respect to a partial solution $\mathfrak{s}^p$. We denote the set of all solution that may be constructed in this way by $\mathfrak{S}$. The choice of a solution component from $\mathfrak{N}(\mathfrak{s}^p)$ is, at each construction step, done probabilistically. In most ACO algorithms the probabilities for choosing the next solution component – also called the *transition probabilities* – are defined as follows:

$$\mathbf{p}(c_i^j \mid \mathfrak{s}^p) = \frac{\tau_i^{j\,\alpha} \cdot \eta(c_i^j)^{\beta}}{\sum\limits_{c_k^l \in \mathfrak{N}(\mathfrak{s}^p)} \tau_k^{l\,\alpha} \cdot \eta(c_k^l)^{\beta}} \,, \quad \forall\, c_i^j \in \mathfrak{N}(\mathfrak{s}^p) \,, \tag{1}$$

where η is a weighting function that assigns, at each construction step, a heuristic value $\eta(c_i^j)$ to each feasible solution component $c_i^j \in \mathfrak{N}(\mathfrak{s}^p)$. The values that are given by the weighting function are commonly called the *heuristic information*. α and β are positive parameters whose values determine the relative importance of pheromone and heuristic information.

ApplyPheromoneUpdate($\mathcal{T}, \mathfrak{S}_{iter}, \mathfrak{s}_{bs}$). Most ACO algorithms use the following pheromone value update rule:

$$\tau_i^j \leftarrow (1-\rho) \cdot \tau_i^j + \frac{\rho}{n_a} \cdot \sum_{\{\mathfrak{s} \in \mathfrak{S}_{upd} \mid c_i^j \in \mathfrak{s}\}} F(\mathfrak{s}) \,, \tag{2}$$

for $i = 1, \ldots, n$, and $j \in \{1, \ldots, |D_i|\}$ [1]. $\rho \in (0,1]$ is a parameter called *evaporation rate*. $F : \mathfrak{S} \mapsto \mathbb{R}^+$ is a function such that $f(\mathfrak{s}) < f(\mathfrak{s}') \Rightarrow F(\mathfrak{s}) \geq F(\mathfrak{s}')$, $\forall \mathfrak{s} \neq \mathfrak{s}' \in \mathfrak{S}$. $F(\cdot)$ is commonly called the *quality function*. Instantiations of this update rule are obtained by different specifications of $\mathfrak{S}_{upd}$, which – in all cases that we consider in this paper – is a subset of $\mathfrak{S}_{iter} \cup \{\mathfrak{s}_{bs}\}$, where $\mathfrak{S}_{iter}$ is the set of solutions that were constructed in the current iteration, and $\mathfrak{s}_{bs}$ is the best-so-far solution. A well-known example of update rule (2) is the AS-update rule (i.e., the update rule of Ant System (AS) [10]) which is obtained from (2) by setting $\mathfrak{S}_{upd} \leftarrow \mathfrak{S}_{iter}$. The goal of the pheromone value update rule is to increase the pheromone values on solution components that have been found in high quality solutions.

---

[1] Note that the factor $\frac{1}{n_a}$ is usually not used. We introduce it for the mathematical purpose of studying the expected update of the pheromone values. However, as the factor is constant it does not change the qualitative behaviour of an ACO algorithm.

3   Deception in Ant Colony Optimization

In the following, we first define and study first order deception in ACO, which corresponds to deception in EC. Then, we introduce second order deception in ACO, which corresponds to bias in EC.

3.1   First Order Deception

The desired behaviour of an ACO algorithm can be stated as follows: The average quality of the generated solutions should grow over time. This is desirable, as it usually increases the probability of finding better solutions over time. For studying the behaviour of an ACO algorithm, we study in the following its *model* as proposed by Merkle and Middendorf in [14]. The model of an ACO algorithm is obtained by applying the expected pheromone update instead of the real pheromone update. Therefore, models of ACO algorithms are deterministic and can be considered discrete dynamical systems. The behaviour of ACO algorithm models is characterized by the evolution of the expected quality of the solutions that are generated per iteration. We denote this *expected iteration quality* in the following by $W_F(\mathcal{T})$, or by $W_F(\mathcal{T} \mid t)$, where $t > 0$ is the iteration counter.

As an example, a simplified model of an ACO algorithm is obtained by assuming an infinite number of solution constructions (i.e., ants) per iteration. In this case, the expected iteration quality is given by

$$W_F(\mathcal{T}) = \sum_{\mathfrak{s} \in \mathfrak{S}} F(\mathfrak{s}) \cdot \mathbf{p}(\mathfrak{s} \mid \mathcal{T}) \ . \tag{3}$$

The expected pheromone update of the AS algorithm (i.e., Algorithm 1 using the AS-update rule) based on this simplified model can then be stated as follows:

$$\tau_i^j(t+1) \leftarrow (1 - \rho) \cdot \tau_i^j(t) + \rho \cdot \sum_{\{\mathfrak{s} \in \mathfrak{S} \mid c_i^j \in \mathfrak{s}\}} F(\mathfrak{s}) \cdot \mathbf{p}(\mathfrak{s} \mid \mathcal{T}) \ , \tag{4}$$

for $i = 1, \ldots, n$, $j = 1, \ldots, |D_i|$ (note that j depends on i), and where t is the iteration counter. However, this simplified model is only an example, and more accurate models of ACO algorithms are possible (see for example [14]).

Definition 2. *Given a model $\mathcal{P}$ of a CO problem, we call a model of an ACO algorithm applied to any instance P of $\mathcal{P}$ a **local optimizer** if for any initial setting of the pheromone values the expected update of the pheromone values is such that $W_F(\mathcal{T} \mid t + 1) \geq W_F(\mathcal{T} \mid t)$, $\forall t \geq 0$. In other words, the expected quality of the generated solutions per iteration must increase monotonically.*

Note that an increase in expected iteration quality does – due to the rather loose definition of the relation between quality function $F(\cdot)$ and objective function $f(\cdot)$ – not necessarily imply an increase in expected objective function values. However, in most cases this can be assumed. Based on this definition we introduce the definition of first order deceptive systems.

Definition 3. *Given a model $\mathcal{P}$ of a CO problem, we call a local optimizer* **applied to** *instance P of $\mathcal{P}$ a* **first order deceptive system (FODS)** *if there exists an initial setting of the pheromone values such that the algorithm does in expectation not converge to a globally optimal solution.*

This means that even if the model of an ACO algorithm is a local optimizer it is a first order deceptive system when for example applied to problem instances that are characterized by the fact that they induce more than one stable fixpoint of which at least one corresponds to a local minimum[2].

Ant System Applied to Unconstrained Problems. In [2] was proposed the hyper-cube framework (HCF) for ACO. The HCF is a framework for ACO algorithms that applies a normalized pheromone update at each iteration. For example, the AS-update rule in the HCF is

$$\tau_i^j \leftarrow (1 - \rho) \cdot \tau_i^j + \rho \cdot \sum_{\{\mathfrak{s} \in \mathfrak{S}_{iter} | c_i^j \in \mathfrak{s}\}} \frac{F(\mathfrak{s})}{\sum_{\{\mathfrak{s}' \in \mathfrak{S}_{iter}\}} F(\mathfrak{s}')} \ , \tag{5}$$

for $i = 1, \ldots, n$, $j = 1, \ldots, |D_i|$. The difference between pheromone update rules in the HCF and the ones that are used in standard ACO algorithms consists in the normalization of the added amount of pheromone. In [2] it was shown that the simplified model of the AS algorithm in the HCF (i.e., assuming an infinite number of solution constructions per iteration) possesses the property that the expected iteration quality monotonically increases over time when applied to unconstrained problems. This can be regarded as an indicator that the average quality of the generated solutions in empirical applications (i.e., using a finite number of ants per iteration) is likely to grow from iteration to iteration. In this section we show that the simplified model of the AS algorithm (not implemented in the HCF) also possesses this property when applied to unconstrained problems, which can be stated as follows: given are n decision variables X_i, $i = 1, \ldots, n$, with domains $D_i = \{v_i^1, \ldots, v_i^{|D_i|}\}$. To construct a solution we do n construction steps as follows. At construction step i, where $i \in \{1, \ldots, n\}$, we add one of the solution components c_i^j, where $j \in \{1, \ldots, |D_i|\}$, to the current partial solution $\mathfrak{s}^p$ under construction. This corresponds to assigning a value to decision variable X_i. The transition probabilities are as follows:

$$\mathbf{p}(c_i^j \mid \mathcal{T}) = \frac{\tau_i^j}{\sum_{k=1}^{|D_i|} \tau_i^k} \ , \quad i = 1, \ldots, n \ . \tag{6}$$

---

[2] The term of a stable fixpoint stems from the field of discrete dynamical systems. It denotes the state of a system that is characterized by the fact that when the system has entered this state, which is characterized by a an open basin of attraction, it will never leave it again.

Theorem 1. *The simplified model of AS (i.e., assuming an infinite number of solution constructions per iteration) is a local optimizer when applied to unconstrained problems.*

Proof. This theorem is a simple extension of Theorem 3 in [2], which proves that the simplified model of AS in the HCF is a local optimizer when applied to unconstrained problems. For distinguishing between the pheromone values of (a) AS and (b) AS in the HCF, we denote the pheromone values of AS by τ^a and the pheromone values of AS in the HCF by τ^b. We use the same notational convention for all the probabilities. Let us assume that at iteration $t \geq 0$ the pheromone values of AS and AS in the HCF are such that $\mathbf{p}^a(c_i^j \mid t) = \mathbf{p}^b(c_i^j \mid t)$, for $i = 1, \ldots, n$, and $j = 1, \ldots, |D_i|$. Assuming that $\rho = 1$, the expected update of the pheromone values of AS can, according to Equation 4, be stated as

$$\tau^{a j}_{\ i}(t+1) \leftarrow \sum_{\{s \in \mathfrak{S} \mid c_i^j \in s\}} F(s) \cdot \mathbf{p}(s \mid \mathcal{T}) \ . \tag{7}$$

In the same way, according to Equation 5, the expected update of the pheromone values of AS in the HCF can be stated as

$$\tau^{b j}_{\ i}(t+1) \leftarrow \sum_{\{s \in \mathfrak{S} \mid c_i^j \in s\}} \frac{F(s) \cdot \mathbf{p}(s \mid \mathcal{T})}{W_F(\mathcal{T})} \ . \tag{8}$$

Theorem 1 in [2] shows that $\mathbf{p}^b(c_i^j \mid t+1) = \tau^{b j}_{\ i}(t+1)$, for $i = 1, \ldots, n$, and $j = 1, \ldots, |D_i|$. Therefore, with Equation 8 it holds that

$$\mathbf{p}^b(c_i^j \mid t+1) = \sum_{\{s \in \mathfrak{S} \mid c_i^j \in s\}} \frac{F(s) \cdot \mathbf{p}(s \mid \mathcal{T})}{W_F(\mathcal{T})} \ . \tag{9}$$

Furthermore, from Equations 6 and 7 it follows that

$$\mathbf{p}^a(c_i^j \mid t+1) = \frac{\tau^{a j}_{\ i}(t+1)}{\sum_{k=1}^{|D_i|} \tau^{a k}_{\ i}(t+1)} = \sum_{\{s \in \mathfrak{S} \mid c_i^j \in s\}} \frac{F(s) \cdot \mathbf{p}(s \mid \mathcal{T})}{W_F(\mathcal{T})} \ . \tag{10}$$

From Equations 9 and 10 it follows that $\mathbf{p}^a(c_i^j \mid t+1) = \mathbf{p}^b(c_i^j \mid t+1)$, for $i = 1, \ldots, n$, and $j = 1, \ldots, |D_i|$. As the simplified model of AS in the HCF is a local optimizer as shown in Theorem 3 in [2], also the simplified model of AS is a local optimizer when $\rho = 1$. The general case follows from the fact that the new pheromone vector for $\rho < 1$ for AS in the HCF is on the line segment between the old pheromone vector and the pheromone vector that would be the result of the setting $\rho = 1$. The same holds for AS. $\square$

A consequence of this result is that the simplified model of AS is a FODS when applied to instances of unconstrained problems that induce more than one stable attractor such as for example n-bit trap functions. This is confirmed by

Table 1. A 4-bit trap function. Each of the 16 columns shows the 4 bits of a solution and its objective function value

1st bit	0 1 0 0 0 1 1 1 0 0 0 1 1 1 0 1
2nd bit	0 0 1 0 0 1 0 0 1 1 0 1 1 0 1 1
3rd bit	0 0 0 1 0 0 1 0 1 0 1 1 0 1 1 1
4th bit	0 0 0 0 1 0 0 1 0 1 1 0 1 1 1 1
$f(\cdot)$	5 1 1 1 1 2 2 2 2 2 2 3 3 3 3 4

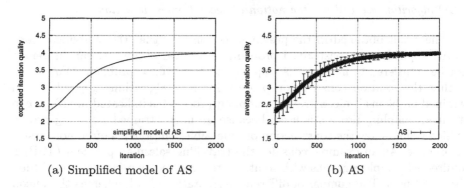

(a) Simplified model of AS   (b) AS

Fig. 1. (a) Evolution of the expected iteration quality W_F of the simplified model of AS when applied to a 4-bit trap function as shown in Table 1. As in this case we are considering a maximization problem, we choose $F(\cdot) = f(\cdot)$. The expected iteration quality continuously increases and the system converges to the sub-optimal solution with quality 4 (the optimal solution is characterized by $X_i = 0$, $i = 1, \ldots, 4$, and quality 5). (b) Evolution of the average iteration quality of AS (i.e., Algorithm 1 using the AS-update rule) when applied to the 4-bit trap function. The graph is the result of 100 runs of the algorithms (the vertical bars show the standard deviation). For (a) and (b) we chose the following settings: $\rho = 0.01$ (evaporation rate), and $c = 0.5$ (initial setting for the pheromone values). Furthermore, for (b) we chose $n_a = 10$ (number of ants per iteration)

Figure 1 which shows the evolution of the expected quality W_F of the simplified model of AS when applied to a 4-bit trap function (see Table 1) over time. The result shows that ACO algorithms suffer from the same type of deception as EC algorithms.

3.2 Second Order Deception

In most cases, for example when NP-hard problems are considered, we can not do better than having a local optimizer. Therefore, first order deception can not be considered a problem. In contrast, the empirical behaviour of an algorithm whose model is not a local optimizer might be very difficult to predict. In the following, we define a second order deceptive system (SODS) as a system that does not have the property of being a local optimizer.

Definition 4. *Given a model $\mathcal{P}$ of a CO problem, we call a model of an ACO algorithm* **applied to** *instance P of $\mathcal{P}$ a* **second order deceptive system (SODS)**, *if the evolution of the expected iteration quality contains time windows $[i, i+l]$ (where $i > 0$, $l > 0$) with $W_F(\mathcal{T} \mid t+1) < W_F(\mathcal{T} \mid t)$, $\forall\, t \in \{i, \ldots, i+l-1\}$. This means that the combination of a model of an ACO algorithm and an instance P of $\mathcal{P}$ may be a SODS, if the ACO algorithm model is not a local optimizer. We henceforth refer to the above mentioned time windows as* **second order deception effects**. *Also with respect to the empirical behaviour of an ACO algorithm we will use the notion of second order deception effects.*

The result obtained in the previous section, i.e., that the simplified model of AS applied to unconstrained problems is a local optimizer, can in general not be transfered to models of AS applied to constrained problems. The reason is that the solution construction process in ACO algorithms applied to constrained problems is a constrained sampling process (as indicated in [2]). A constrained sampling process is used, because an unconstrained sampling process (which would allow the construction of unfeasible solutions) is often not feasible in practice. As an example consider the travelling salesman problem (TSP) in undirected complete graphs with solutions represented as permutations of the n city identifiers. The number of different permutations is $n!$, whereas the unconstrained search space, which is the set of strings of length n over the alphabet $\{1, \ldots, n\}$, is of size n^n. Using Stirling's approximation (i.e., $n! \approx n^n e^{-n}\sqrt{2\pi n}$) we obtain $\frac{n^n}{n!} \approx \frac{e^n}{\sqrt{2\pi n}}$, which shows that the size of the unconstrained search space grows exponentially in comparison to the size of the constrained search space. Therefore, the probability of generating feasible solutions by an unconstrained sampling process might be extremely low.

4   Examples of Second Order Deception

Two examples of second order deception can be found in the literature. In [5], Blum and Sampels showed second order deception in the context of an ACO algorithm applied to the node-weighted k-cardinality tree (KCT) problem. However, in this case second order deception can only be detected when the AS-update rule is applied without the use of local search for improving the solutions constructed by the ants. A more serious case of second order deception is reported by Blum et. al in [3, 4] for the job shop scheduling (JSS) problem, in which is given a finite set of operations $\mathcal{O} = \{o_1, \ldots, o_n\}$ that have to be processed by machines. Each operation has a processing time assigned. The goal is to find a permutation of all operations that satisfies some precedence constraints and which is minimal in some function of the completion time of the operations. In order to apply ACO to the JSS problem, Colorni et al. [6] proposed the following CO problem model of the JSS problem: First, the set of operations $\mathcal{O}$ is augmented by two dummy operations o_0 and o_{n+1} with processing time zero. Operation o_0 serves as a source operation and o_{n+1} as a destination operation. The augmented set of operations is $\mathcal{O} = \{o_0, o_1, \ldots, o_n, o_{n+1}\}$. Then, for each operation o_i, where $i \in \{0, \ldots, n\}$, a decision variable X_i is introduced. The domain for decision variable X_0 is $D_0 = \{1, \ldots, n\}$, and for every other decision variable X_i the domain is $D_i = \{1, \ldots, n+1\} \setminus \{i\}$. The meaning of a domain value $j \in D_i$ for a decision variable X_i is that operation o_j is placed immediately after operation o_i in the permutation of all the operations to be constructed by the algorithm. We denote this CO problem model of the JSS problem by $\mathcal{P}_{\text{JSS}}^{\text{suc}}$. In order to derive the pheromone model we again introduce for each combination of a decision variable X_i and a domain value $j \in D_i$ a solution component c_i^j. The pheromone model then consists of a pheromone trail parameter $\mathcal{T}_i^j$ for each solution component c_i^j.

The ants' mechanism for constructing feasible solutions builds feasible permutations of all the operations from left to right. It works as follows. Let $\mathcal{I}$ denote the set of indices of the decision variables that have already assigned a value and the decision variable that receives a value in the current construction step. Furthermore, let i_c denote the index of the decision variable that receives a value in the current construction step. The solution construction starts with an empty partial solution $s^p = \langle \rangle$, with $i_c = 0$, and with $\mathcal{I} = \{0\}$. Then, at each of n construction steps $t = 1, \ldots, n$ a solution component $c_{i_c}^j \in \mathfrak{N}(s^p)$ is added to the current partial solution, where

$$\mathfrak{N}(s^p) = \left\{ c_{i_c}^j \mid o_j \in \mathcal{O}_t \right\} . \tag{11}$$

In this context, $\mathcal{O}_t$ is the set of allowed operations at construction step t. This means that at each construction step we decide a domain value for the decision variable with the index i_c. When adding the solution component $c_{i_c}^j$ to s^p we also set i_c to j and add j to $\mathcal{I}$. In the $(n+1)$-th construction step, the value of the last unassigned decision variable X_{i_c} is set to $n+1$. Each construction step is done according to the following probability distribution:

$$p(c_{i_c}^j \mid \mathcal{T}) = \begin{cases} \dfrac{\mathcal{T}_{i_c}^j}{\sum_{c_{i_c}^k \in \mathfrak{N}(s^p)} \mathcal{T}_{i_c}^k} & \text{if } c_{i_c}^j \in \mathfrak{N}(s^p) \\ 0 & \text{otherwise} . \end{cases} \tag{12}$$

Figure 2 shows the evolution of the average iteration quality over time for several versions of Algorithm 1 based on the above described pheromone model and construction mechanism (the parameter settings of the algorithm are given in the caption of the figure). The strongest second order deception effects are obtained with the AS-update rule. However, even when using the more common IB-update rule, that is, only the best of the solutions constructed per iteration is used for updating the pheromone values, the algorithm obtains solutions at the end of a run whose average quality is lower than the average quality of the solutions in the first iteration. This shows that the harmful bias that leads to the occurrence of second order deception can be a serious problem and may cause the failure of an algorithm. The reasons for the existence of such a bias are discussed in [1].

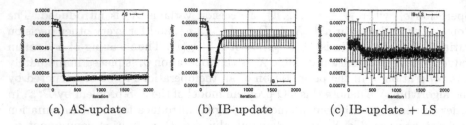

128   Christian Blum and Marco Dorigo

(a) AS-update   (b) IB-update   (c) IB-update + LS

Fig. 2. The three graphics show the evolution of Algorithm 1 applied to the JSS problem as outlined in the text for different pheromone update rules as well as with and without the application of steepest descent local search (based on the neighborhood structure introduced by Nowicki and Smutnicki in [15]). The following settings were chosen: $n_a = 10$ (number of ants), $\rho = 0.05$ (evaporation rate), and $c = 0.5$ (initial setting of the pheromone values). The graphics show the result of 100 runs of the algorithm. The vertical bars show the standard deviation

5   Conclusions

In this paper, we first have introduced the notion of first order deception in ant colony optimization. We showed that ACO algorithms suffer from this type of deception in the same way as evolutionary algorithms do. Then, we introduced second order deception, which is a bias introduced by algorithmic components. We presented an example of second order deception from the literature. The example of the job shop scheduling problem shows that second order deception is a major issue in ACO algorithms and should be taken into account by researchers that develop ACO algorithms. This is because an ACO algorithm that suffers from strong second order deception effects might fail. Therefore, research efforts have to be undertaken towards methods for avoiding second order deception. Possibilities include choosing different solution construction mechanisms, different pheromone models, and different pheromone update rules.

Acknowledgements

This work was supported by the "Metaheuristics Network", a Research Training Network funded by the Improving Human Potential program of the CEC, grant HPRN-CT-1999-00106. The information provided is the sole responsibility of the authors and does not reflect the Community's opinion. The Community is not responsible for any use that might be made of data appearing in this publication. Marco Dorigo acknowledges support from the Belgian FNRS, of which he is a Senior Research Associate, and from the "ANTS" project, an "Action de Recherche Concertée" funded by the Scientific Research Directorate of the French Community of Belgium.

References

1. C. Blum. *Theoretical and practical aspects of ant colony optimization*. PhD thesis, IRIDIA, Université Libre de Bruxelles, Belgium, 2004.
2. C. Blum and M. Dorigo. The hyper-cube framework for ant colony optimization. *IEEE Trans. on Systems, Man, and Cybernetics – Part B*, 34(2):1161–1172, 2004.
3. C. Blum and M. Sampels. Ant Colony Optimization for FOP shop scheduling: A case study on different pheromone representations. In *Proceedings of the 2002 Congress on Evolutionary Computation (CEC'02)*, volume 2, pages 1558–1563. IEEE Computer Society Press, Los Alamitos, CA, 2002.
4. C. Blum and M. Sampels. When model bias is stronger than selection pressure. In J. J. Merelo Guervós et al., editors, *Proceedings of PPSN-VII, Seventh Int. Conference on Parallel Problem Solving from Nature*, volume 2439 of *Lecture Notes in Computer Science*, pages 893–902. Springer, Berlin, Germany, 2002.
5. C. Blum, M. Sampels, and M. Zlochin. On a particularity in model-based search. In W. B. Langdon et al., editors, *Proceedings of the Genetic and Evolutionary Computation Conference (GECCO-2002)*, pages 35–42. Morgan Kaufmann Publishers, San Francisco, CA, 2002.
6. A. Colorni, M. Dorigo, V. Maniezzo, and M. Trubian. Ant System for job-shop scheduling. *JORBEL – Belgian Journal of Operations Research, Statistics and Computer Science*, 34(1):39–53, 1994.
7. K. Deb and D. E. Goldberg. Analyzing deception in trap functions. In L. D. Whitley, editor, *Foundations of Genetic Algorithms 2*, pages 93–108. Morgan Kaufmann, San Mateo, CA, 1993.
8. M. Dorigo. *Optimization, Learning and Natural Algorithms* (in Italian). PhD thesis, Dip. di Elettronica, Politecnico di Milano, Italy, 1992.
9. M. Dorigo, V. Maniezzo, and A. Colorni. Positive feedback as a search strategy. Technical Report 91-016, Dip. di Elettronica, Politecnico di Milano, Italy, 1991.
10. M. Dorigo, V. Maniezzo, and A. Colorni. Ant System: Optimization by a colony of cooperating agents. *IEEE Transactions on Systems, Man, and Cybernetics – Part B*, 26(1):29–41, 1996.
11. M. Dorigo and T. Stützle. *Ant Colony Optimization*. MIT Press, Cambridge, MA, 2004.
12. D. E. Goldberg. Simple genetic algorithms and the minimal deceptive problem. In L. Davis, editor, *Genetic algorithms and simulated annealing*, pages 74–88. Pitman, London, UK, 1987.
13. D. Merkle and M. Middendorf. Modelling ACO: Composed permutation problems. In M. Dorigo, G. Di Caro, and M. Sampels, editors, *Proceedings of ANTS 2002 – From Ant Colonies to Artificial Ants: Third International Workshop on Ant Algorithms*, volume 2463 of *Lecture Notes in Computer Science*, pages 149–162. Springer Verlag, Berlin, Germany, 2002.
14. D. Merkle and M. Middendorf. Modelling the dynamics of ant colony optimization algorithms. *Evolutionary Computation*, 10(3):235–262, 2002.
15. E. Nowicki and C. Smutnicki. A fast taboo search algorithm for the job-shop problem. *Management Science*, 42(2):797–813, 1996.
16. F. Rothlauf and D. E. Goldberg. Prüfer numbers and genetic algorithms: A lesson on how the low locality of an encoding can harm the performance of GAs. In *Proceedings of PPSN-VI, Sixth International Conference on Parallel Problem Solving from Nature*, volume 1917 of *Lecture Notes in Computer Science*, pages 395–404, Springer Verlag, Berlin, Germany, 2000.

Evolution of Direct Communication
for a *Swarm-bot* Performing Hole Avoidance

Vito Trianni, Thomas H. Labella, and Marco Dorigo

IRIDIA - Université Libre de Bruxelles - Brussels, Belgium
{vtrianni,hlabella,mdorigo}@ulb.ac.be

Abstract. Communication is often required for coordination of collective behaviours. Social insects like ants, termites or bees make use of different forms of communication, which can be roughly classified in three classes: indirect (*stigmergic*) communication, direct interaction and direct communication. The use of stigmergic communication is predominant in social insects (e.g., the pheromone trails in ants), but also direct interactions (e.g., antennation in ants) and direct communication can be observed (e.g., the waggle dance of honey bee workers). Direct communication may be beneficial when a fast reaction is expected, as for instance, when a danger is detected and countermeasures must be taken. This is the case of hole avoidance, the task studied in this paper: a group of self-assembled robots – called *swarm-bot* – coordinately explores an arena containing holes, avoiding to fall into them. In particular, we study the use of direct communication in order to achieve a reaction to the detection of a hole faster than with the sole use of direct interactions through physical links. We rely on artificial evolution for the synthesis of neural network controllers, showing that evolving behaviours that make use of direct communication is more effective than exploiting direct interactions only.

Keywords: evolutionary robotics, swarm robotics, communication.

1 Introduction

In collective robotics research, the coordination of the activities in a group of robots requires the definition of communication strategies and protocols among the individuals. These strategies and protocols need not, however, be particularly complex. In many cases, simple forms of communication – or no explicit communication at all – are enough to obtain the coordination of the activities of the group [11]. This is the case of *swarm robotics*, that, drawing inspiration from social insects such as ants, termites or bees, focuses on distributed robotic systems characterised by limited communication abilities among robots.

Communication in social insects has been thoroughly studied, identifying different modalities used for the regulation of the colony's activities. The study of the nest building behaviour of termites of the genus *Macrotermes* led Grassé to the introduction of the concept of *stigmergy* [9]. Impressed by the complexity of termites' nests and by their dimension with respect to an individual, Grassé

M. Dorigo et al. (Eds.): ANTS 2004, LNCS 3172, pp. 130–141, 2004.

suggested that the cooperation among termites in their building activities was not the result of either some direct interactions among individuals, nor some other form of complex communication. On the contrary, cooperation could be explained as the result of environmental stimuli provided by the work already done – i.e., the nest itself. Another example of stigmergic communication has been observed in the foraging behaviour of many ant species, which lay a trail of pheromone, thus modifying the environment in a way that can inform other individuals of the colony about the path to follow to reach a profitable foraging area [8]. The concept of stigmergy describes an indirect communication among individuals, which is mediated by the environment [4].

Stigmergy is not the only way of communication that can be observed in social insects. *Direct interactions* – such as antennation, mandibular contact, trophallaxis – account for various social phenomena. For example, in many species of ants such as *Œcophilla longinoda*, recruitment of nest-mates for the exploitation of a food source is performed with a mix of antennation and trophallaxis: when an ant returning from a food source encounters another worker, it stimulates the other ant to follow the laid pheromone trail touching the nest-mate with the antennas and regurgitating a sample of the food source [10].

Some forms of *direct communication* within insect societies have been studied, a well-known example being the waggle dance of honey bees. A bee is able to indicate to the unemployed workers the direction and distance from the hive of a patch of flowers, using a "dance" that gives also information on the quality and the richness of the food source [16]. Direct communication in ants has been reported by Hölldobler and Wilson [10]: ants may use sound signals – called *stridulation* – for recruiting or for help requests. In presence of a big prey, ants of the genus *Aphaenogaster* recruit nest-mates using stridulation. Here, the sound signal does not attract ants, but it serves as a reinforcement of the usual chemical and tactile attractors, resulting in a faster response of the nest-mates.

The above examples suggest a possible taxonomy of different forms of communications in insect societies that can be borrowed for characterising a collective robotic system. Defining what communication is and classifying its different forms is not trivial, as confirmed by the number of different taxonomies that can be found in the literature [1, 3, 6, 12]. In [12], Matarić distinguishes between indirect or stigmergic, direct and directed communication, on the base of the communication modality (through the environment versus through a "speech act") and of the receiver (unknown versus specified). In [3], Cao et al. introduce three "communication structures" specific for a robotic system: interaction via environment, via sensing and via communication. Defining yet another taxonomy for different communication modalities is out of the scope of this paper. Thus, we borrow the taxonomy introduced in [3], adapting it to the natural examples introduced above. In doing this, we will use the above mentioned terminology, partly borrowed by [12]. Summarising, we will talk of:

Indirect or Stigmergic Communication. A form of communication that takes place through the environment, as a result of the actions performed by some individuals, which indirectly influence someone else's behaviour.

Direct Interaction. A form of communication that implies a non-mediated transmission of information, as a result of the actions performed by some individuals, which directly influence someone else's behaviour.

Direct Communication. A form of communication that implies a non-mediated transmission of information, without the need of any physical interaction.

As described above, all these forms of communication can be observed in biological systems, and in particular in social insects: research in swarm robotics focuses on the application of these simple forms of communication to artificial, autonomous systems. Referring to the above taxonomy, in this paper we will show how direct communication can be beneficial for reinforcing direct interactions. In our work, we study a swarm robotic system composed of a swarm of autonomous mobile robots, called *s-bots*, which have the ability to connect one to the other forming a physical structure – called *swarm-bot* – that can solve problems the single *s-bots* are not able to cope with[1] [5, 13]. The physical connections provide direct interactions among *s-bots* that can be exploited for coordination. Additionally, *s-bots* are provided with a sound signalling system, which can be used for direct communication. In this paper, we show that, using the sound signalling system, *s-bots* can reinforce the information passed through the physical connections, thus achieving a faster reaction.

The rest of this paper is organised as follows. Section 2 describes the problem we are interested in, that is, the hole avoidance task. Section 3 details the experimental setup used to perform the experiments. Finally, Section 4 is dedicated to the obtained results and Section 5 concludes the paper.

2   The Hole Avoidance Task

The hole avoidance task has been defined for studying collective navigation strategies. It can be considered an instance of a broader family of tasks, aimed at the study of all-terrain navigation. This family of tasks includes scenarios in which the robotic system has to face challenges such as avoiding holes or obstacles, passing through narrow passages or over a trough, climbing steep slopes and coping with rough terrain. With respect to these scenarios, the single robot approach may fail due to physical constraints or to limited individual capabilities. Our approach consists in relying on a swarm of robots, that can cooperate to overcome the individual limitations. Here, we address the all-terrain navigation problems making use of self-assembled structures – i.e., the *swarm-bots*.

The hole avoidance task represents a relatively simple problem compared to others in the all-terrain navigation family, but it is still very interesting for the study of collective navigation behaviours for a *swarm-bot*. The *s-bots* are placed in an arena presenting open borders and holes, in which the *swarm-bot* could

---

[1] This research is carried out within the SWARM-BOTS project, funded by the Future and Emerging Technologies Programme (IST-FET) of the European Community, under grant IST-2000-31010.

Fig. 1. The hole avoidance task. The picture shows the arena used, which presents open borders and contains two large rectangular holes. A *swarm-bot* formed by four linearly connected *s-bots* is shown.

fall (see Fig. 1). Four *s-bots* are rigidly connected in a linear formation. Their goal is to efficiently explore the arena, avoiding to fall into the holes or out of the borders of the arena.

The control of the *swarm-bot* is completely distributed, and *s-bots* can only rely on local information. The problem consists in how to coordinate the activity of the *s-bots*. In particular, the difficulty of the collective navigation is twofold: (i) coordinated motion must be performed in order to obtain a coherent navigation of the *swarm-bot* as a whole, as a result of the motion of its components; (ii) holes are not perceived by all the *s-bots* at the same time. Thus, the presence of an hazard, once detected, must be communicated to the entire group, in order to trigger a change in the direction of motion.

The complexity of the task justifies the use of evolutionary robotics techniques for the synthesis of the *s-bots'* controller [15]. In a previous work, we studied the hole avoidance problem evolving simple neural controllers that were able to perform coordinated motion and hole avoidance, relying only on direct interactions among *s-bots* [17]. In this paper, we focus on the use of direct communication, in order to reinforce the direct interactions and therefore to obtain more efficient behaviours. In fact, direct communication among *s-bots* speeds up the reaction to the detection of a hole, thus it is beneficial for the efficiency of the navigation.

3   Experimental Setup

As already mentioned, we studied hole avoidance in a previous work [17], obtaining interesting results. In this paper, we aim at improving the obtained results modifying the experimental setup as follows: (i) the simulation model of the *s-bot* is modified, as described in Sec. 3.1; (ii) the controllers include the possibility to actuate the speaker, thus enabling direct communication among *s-bots* (see also Sec. 3.2); (iii) the fitness computation is simplified, taking into account only

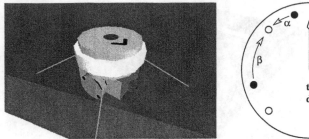

Fig. 2. Left: The simulated model of an *s-bot*. The light rays represent the position of the ground sensors, mounted on the rotating turret. The dark circle painted on the turret indicates that the *s-bot* is emitting a sound signal. Right: Description of the encoding of the sensors integral with the turret to the virtual sensors integral with the chassis. The filled circles indicate the position of the real sensors, while the empty circles refer to the position of the virtual sensor with respect to the direction of the chassis.

variables directly available to each *s-bot*, such as sensor readings or internal state variables (see Sec. 3.3).

In order to test the effectiveness of the use of direct communication among *s-bots*, we performed two sets of experiments: in the first setting only direct interactions were used, while in the second direct communication capabilities were added.

3.1   The Simulation Model

We developed a simulation software based on Vortex™, a 3D rigid body dynamics simulator that provides primitives for the implementation of detailed and realistic physics-based simulations (see [13] for more details about the simulator). We have defined a simple *s-bot* model that at the same time allows fast simulations and preserves those features of the real *s-bot* that were important for the experiments (see Figure 2 left). This model matches more closely the real *s-bot* than the one used in the previous work [17], both in the geometries and in the sensing abilities.

The *s-bot* has a differential drive motion provided by a traction system composed of four wheels: two lateral, motorised wheels and two spherical, passive wheels placed in the front and in the back, which serve as support. The four wheels are fixed to the chassis, which also holds the cylindrical rotating turret. The turret can rotate around its axis and it holds many sensory systems. Connections among *s-bots* can be made using a virtual gripper, which is modelled by dynamically creating a joint between two *s-bots*. The position of the virtual gripper is represented by an arrow painted on the turret. Finally, the turret also carries a loudspeaker that can be controlled to produce a tone that can be perceived by the other *s-bots*.

Each *s-bot* is provided with a *traction sensor*, which detects the forces that are applied to the junction between the chassis and the rotating turret. Four variables encode the traction force information from four different preferential orientations with respect to the chassis (front, right, back and left, see [2] for more details). Traction sensors are responsible for the detection of the direct interactions among *s-bots*. In fact, an *s-bot* can generate a traction force that is felt by the other *s-bots* connected through their grippers. This force mediates the communication among *s-bots*, and it can be exploited for coordinating the activities of the group: it proved to be important to evolve coordinated motion strategies in a *swarm-bot* and for collective obstacle and hole avoidance [2, 17].

The presence of holes is detected using four *ground sensors* – infrared proximity sensors pointing to the ground – that are integral with the rotating turret. In order to account for the rotation of the turret, we encode the information coming from the ground sensors in four virtual sensors integral with the chassis. As pictured in the right part of Fig. 2, the value taken by the virtual sensors is computed as the weighted average of the two closest ground sensors. In particular, if α and β are the angular differences from the two closest ground sensors, then $cos^2(\alpha)$ and $cos^2(\beta)$ are the weights for the average. Noise is simulated for all sensors, adding a random value uniformly distributed within the 5% of the sensor saturation value.

Each *s-bot* is also equipped with a loudspeaker and three directional microphones, used to detect the tone emitted by other *s-bots*. Also directional microphones, being integral with the turret, are encoded in three virtual sound sensors integral with the chassis following a procedure similar to the one used for ground sensors. The loudspeaker can be switched on, simulating the emission of a continuous tone, or it can be turned off. Exploiting this equipment, *s-bots* have direct communication capabilities.

S-bots can control the two wheels, independently setting their speed in the range $[-6.5, 6.5]$ *rad/s*. The virtual gripper is used to connect to another *s-bot*. However, in this work, the *s-bots* stay always assembled in a *swarm-bot* formation, thus connection and disconnection procedures have not been simulated. Finally, the motor controlling the rotation of the turret is actuated setting its desired angular speed proportionally to the difference between the desired angular speed of the left and right wheels. This setting helps the rotation of the chassis with respect to the turret also when one or both wheels of the *s-bot* do not touch the ground [2].

The *swarm-bot* is composed of four *s-bots* rigidly connected to form a chain. It is placed in a square arena of 4 meters side, that presents open borders and two rectangular holes (80×240 cm, see Fig. 1). The dimensions have been chosen to create passages that can be navigated by the *swarm-bot*, no matter its orientation.

3.2   The Controller and the Genetic Algorithm

The *s-bots* are controlled by artificial neural networks, whose parameters are set by an evolutionary algorithm. A single genotype is used to create a group of *s-bots* with an identical control structure – a homogeneous group. Each *s-bot* is controlled by a fully connected, single layer feed-forward neural network – a perceptron. Each input is associated with a single sensor, receiving a real value in the range [0.0, 1.0], which is a simple linear scaling of the reading taken from its associated sensor. Additionally, the network is provided with a bias unit – an input unit whose activation state is always 1.0 – and two output neurons that control the motors of the *s-bot*.

As mentioned above, we performed two sets of experiments, which differ in the form of communication used. Thus, the neural networks controlling the *s-bots* change depending on which sensors and actuators are employed. In all the experiments, traction and ground sensors have been used. Specifically, 4 inputs of the perceptron are dedicated to the traction sensors and 4 other inputs are dedicated to the virtual ground sensors (see Sec. 3.1). If direct communication is used, three more sensors are used, corresponding to the three microphones with which an *s-bot* is endowed. These sensors are connected to three additional neural inputs. Concerning the actuators, the two outputs of the perceptron are used to control the left and the right wheel. Additionally, the same two outputs control the turret-chassis motor, as described in Sec. 3.1. When direct communication is used, the activation of the loudspeaker has been handcrafted, simulating a sort of reflex action: an *s-bot* activates the loudspeaker whenever one of its ground sensors detects the presence of a hole. Thus, the neural network does not control the emission of a sound signal. However, it receives the information coming from the three directional microphones, and evolution is responsible for shaping the correct reaction to the perceived signals.

The weights of the perceptron's connections are genetically encoded parameters. A simple generational genetic algorithm (GA) is used [7]. Initially, a random population of 100 genotypes is generated. Each genotype is a vector of binary values – 8 bits for each parameters. The genotype is composed of 144 bits in the first setting (using direct interactions only) and 192 bits in the second setting (using direct communication). Subsequent generations are produced by a combination of selection with elitism and mutation. Recombination is not used. At every generation, the best 20 genotypes are selected for reproduction, and each generates 4 offspring. The genotype of the selected parents is copied in the subsequent generation; the genotype of their 4 offspring is mutated with a 3% probability of flipping each bit. One evolutionary run lasts 150 generations.

3.3   The Fitness Computation

During the evolution, a genotype is mapped into a control structure that is cloned and downloaded in all the *s-bots* taking part in the experiment (i.e., we make use of a homogeneous group of *s-bots*). Each genotype is evaluated 5 times – i.e., 5 trials. Each trial differs from the others in the initialisation of the random number generator, which influences both the initial position of the *swarm-bot* within the arena and the initial orientation of each *s-bot*'s chassis. Each trial lasts $T = 200$ simulation cycles, which correspond to 20 seconds of real time.

The behaviour produced by the evolved controller is evaluated according to a fitness function that takes into account only variables accessible to the *s-bots* (see [15], page 73). In each simulation cycle t, for each *s-bot* s belonging to the *swarm-bot* S, the individual fitness $f_s(t)$ is computed as the product of three components:

$$f_s(t) = \omega_s(t) \cdot \Delta\omega_s(t) \cdot \gamma_s(t), \tag{1}$$

where:

– $\omega_s(t)$ accounts for fast motion of an *s-bot*. It is computed as the sum of the absolute values of the angular speed of the right and left wheels, linearly scaled in the interval $[0, 1]$:

$$\omega_s(t) - \frac{|\omega_{s,l}(t)| + |\omega_{s,r}(t)|}{2 \cdot \omega_m}, \tag{2}$$

where $\omega_{s,l}(t)$ and $\omega_{s,r}(t)$ are respectively the angular speed of the left and right wheel of *s-bot* s at cycle t, and ω_m is the maximum angular speed achievable.

– $\Delta\omega_s(t)$ accounts for the straightness of the motion of the *s-bot*. It is computed as the difference between the angular speed of the wheels, as follows:

$$\Delta\omega_s(t) = \begin{cases} 0 & \text{if } \omega_{s,l}(t) \cdot \omega_{s,r}(t) < 0 \\ 1 - \sqrt{\frac{|\omega_{s,l}(t) - \omega_{s,r}(t)|}{\omega_m}} & \text{otherwise} \end{cases}, \tag{3}$$

where the difference is computed only if the wheels rotate in the same direction, in order to penalise more any turning-on-the-spot behaviour. The square root is useful to emphasise small speed differences.

– $\gamma_s(t)$ accounts for coordinated motion and hole avoidance. It is computed as follows:

$$\gamma_s(t) = 1 - \max\left(I_s(t), G_s(t)\right), \tag{4}$$

where $I_s(t)$ is the intensity of the traction force perceived by the *s-bot* s at time t, $G_s(t)$ is the maximum activation among the ground sensors. Both can take values in the interval $[0, 1]$. This component favours coordinated motion as it is maximised when the perceived traction is minimised, which corresponds to a coherent motion of the *swarm-bot*. It also favours hole avoidance because it is maximised if the *s-bots* stay away from the holes.

Given the individual fitness $f_s(t)$, the fitness F_θ of a trial θ is computed as follows:

$$F_\theta = \frac{1}{T} \sum_{t=1}^{T_f} \min_{s \in S} f_s(t), \tag{5}$$

where T is the maximum number of cycles and $T_f \leq T$ is the cycle at which the simulation ended, which may be smaller than the maximum allowed if the *swarm-bot* happens to fall into a hole. Averaging the individual components on T rather than on T_f simulation cycles puts an additional selective pressure for the evolution of hole avoidance. Additionally, at each simulation cycle t we select the minimum among the individual fitnesses $f_s(t)$, which refers to the worst-behaving *s-bot*, therefore obtaining a robust overall fitness computation.

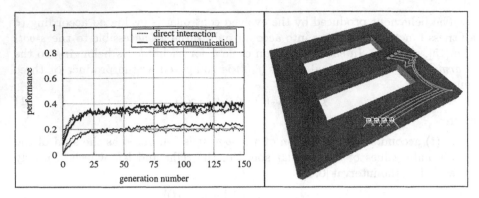

Fig. 3. Left: The average performance of the 10 replications is plotted against the generation number for each experimental setting. Thick lines refer to the best individual of the population, while thin lines refer to the population average. Right: trajectories of the *s-bots* while performing hole avoidance. Movies of this behaviour are available at http://www.swarm-bots.org/hole-avoidance.html.

4   Results

For both settings – using only direct interactions (hereafter indicated as DI) and complementing them with direct communication (hereafter indicated as DC) – the evolutionary experiments were replicated 10 times. The average fitness values, computed over all the replications, are shown in Figure 3 left. The average performance of the best individual and of the population are plotted against the generation number. All evolutionary runs were successful. The average fitness value of the best individuals reaches 0.4, where a value of 1 should be understood as a loose upper-bound to the maximum value the fitness can achieve[2]. It is worth noting that the average fitness of DC is slightly higher than in the case of DI. This suggests that the use of direct communication among *s-bots* is beneficial for the hole avoidance task.

A qualitative analysis of the behaviours produced by the two settings reveals no particular differences in the initial coordination phase that leads to a coherent motion of the *swarm-bot* (see Fig. 3 right). In both cases, the *s-bots* start to move in the direction they were positioned, resulting in a rather disordered overall motion. Within a few simulation cycles, the physical connections transform this disordered motion into traction forces, that are exploited to coordinate the group. When an *s-bot* feels a traction force, it rotates its chassis in order to cancel this force. Once the chassis of all the *s-bots* are oriented in the same direction, the traction forces disappear and the coordinated motion of the *swarm-bot* starts (see also [17, 2]).

---

[2] This maximum value could be achieved only by a *swarm-bot* coordinately moving in a flat environment, without holes. In the arena shown in Fig. 3 right, the narrow passages result in frequent activations of the ground sensors, and therefore in frequent re-organisations of the *swarm-bot*.

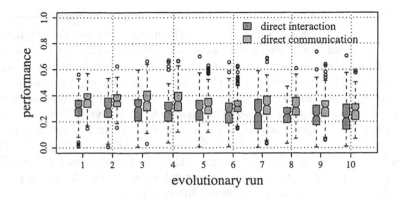

Fig. 4. Post-evaluation analysis performed evaluating 200 times the best individuals obtained from each replication of the experiment. Boxes represent the inter-quartile range of the data, while the horizontal bars inside the boxes mark the median values. The whiskers extend to the most extreme data points within 1.5 of the inter-quartile range from the box. The empty circles mark the outliers.

The differences between the two settings DC and DI are evident once the hole avoidance behaviour is considered. When an *s-bot* detects an edge, it rotates the chassis and changes the direction of motion in order to avoid falling. When using only direct interactions, this change in direction produces a traction force for the other *s-bots*, which triggers a new coordination phase that ends up in a new direction of motion that leads the *swarm-bot* away from the edge. This simple behaviour exploits the direct interactions among *s-bots* – shaped as traction forces – to communicate the presence of an hazard – the hole to be avoided. However, this strategy may fail as communication via traction is sometimes too weak to be perceived by the whole *swarm-bot*. On the contrary, the evolved controllers that makes use of direct communication react faster to the detection of a hole: the *s-bot* that detects the hole emits a sound signal that is immediately perceived by the rest of the group. Thus, the whole *swarm-bot* starts turning away from the hole, without waiting to perceive a strong traction force. Traction is then exploited again in order to perform coordinated motion.

From the qualitative analysis, the use of direct communication seems to confirm our expectations: direct communication provides a faster reaction to the detection of a hole and therefore a more efficient avoidance behaviour. In order to quantitatively assess the difference in performance between DC and DI, we performed a post-evaluation analysis and compared the results obtained with the two settings. For each evolutionary run, we selected the 20 best individuals of the final population and we re-evaluated them in 200 trials, each characterised by a different random initialisation. All individuals were tested using the same set of trials. The performance of the re-evaluations was measured using Eq. (5). We selected the individual with best mean performance in the post-evaluations and discarded the other nineteen individuals. A box-plot summarising the performance of these individuals is shown in Fig. 4. It is possible to notice that DC generally performs better than DI.

On the base of these data, we performed a two-way analysis of variance to test if there is a significant difference in performance between the settings [14]. The analysis considers 2 factors (the settings), 200 blocks (the testing trials) and 10 replications for each combination of factor/block (the evolutionary runs). The applicability of the method was checked looking at the residuals coming from the linear regression modelling of the data: no violation of the hypothesis to use the analysis of variance was found. The result of the analysis allows us to reject the null hypothesis that there is no difference among the two settings (p-value < 0.0001). On the base of the mean performance of the two settings – 0.3316 for DC and 0.2708 for DI – we can conclude that, in the experimental conditions considered, a system that uses direct communication among the *s-bots* performs better than one that exploits only direct interactions.

5   Conclusions

In this paper, we have shown how the use of direct communication in a *swarm-bot* performing hole avoidance can be beneficial for the effectiveness of the group behaviour. Comparing the use of direct communication with the case in which only direct interactions among *s-bots* were possible, we found that the former setting performs statistically better than the latter. It is worth noting that direct communication acts here as a reinforcement of the direct interaction among *s-bots*. In fact, *s-bots* react faster to the detection of the hole when they receive a sound signal, without waiting to perceive a traction strong enough to trigger the hole avoidance behaviour. However, traction is still necessary for avoiding the hole and coordinating the motion of the *swarm-bot* as a whole. Finally, it is important to remark that all controllers were synthesised by artificial evolution, which proved to be an efficient mean for automatically developing behaviours for homogeneous groups of robots.

Acknowledgements

This work was supported by the "SWARM-BOTS" project and by the "ECagents" project, two projects funded by the Future and Emerging Technologies programme (IST-FET) of the European Commission, under grant IST-2000-31010 and 001940 respectively. The information provided is the sole responsibility of the authors and does not reflect the Community's opinion. The Community is not responsible for any use that might be made of data appearing in this publication. Marco Dorigo acknowledges support from the Belgian FNRS, of which he is a Senior Research Associate, through the grant "Virtual Swarm-bots", contract no. 9.4515.03, and from the "ANTS" project, an "Action de Recherche Concertée" funded by the Scientific Research Directorate of the French Community of Belgium.

References

1. T. Balch and R. C. Arkin. Communication in reactive multiagent robotic systems. *Autonomous Robots*, 1(1):27–52, 1995.
2. G. Baldassarre, S. Nolfi, and D. Parisi. Evolution of collective behavior in a team of physically linked robots. In R. Gunther, A. Guillot, and J.-A. Meyer, editors, *Applications of Evolutionary Computing - Proceedings of the Second European Workshop on Evolutionary Robotics (EvoWorkshops2003: EvoROB)*, pages 581–592. Springer-Verlag, Berlin, Germany, 2003.
3. Y. U. Cao, A. S. Fukunaga, and A. B. Kahng. Cooperative mobile robotics: Antecedents and directions. *Autonomous Robots*, 4:1–23, 1997.
4. M. Dorigo, E. Bonabeau, and G. Theraulaz. Ant algorithms and stigmergy. *Future Generation Computer Systems*, 16(8):851–871, 2000.
5. M. Dorigo, V. Trianni, E. Şahin, R. Groß, T. H. Labella, G. Baldassarre, S. Nolfi, J.-L. Deneubourg, F. Mondada, D. Floreano, and L. M. Gambardella. Evolving self-organizing behaviors for a swarm-bot. *Autonomous Robots*, 17(2–3):223–245, 2004.
6. G. Dudek, M. Jenkin, and E. Milios. A taxonomy of multirobot systems. In T. Balch and L. E. Parker, editors, *Robot Teams: From Diversity to Polymorphism*. A K Peters Ltd., Wellesley, MA, 2002.
7. D. E. Goldberg. *Genetic Algorithms in Search, Optimization and Machine Learning*. Addison-Wesley, Reading, MA, 1989.
8. S. Goss, S. Aron, J.-L. Deneubourg, and J. M. Pasteels. Self-organized shortcuts in the argentine ant. *Naturwissenchaften*, 76:579–581, 1989.
9. P. P. Grassé. La reconstruction du nid et les coordinations interindividuelles chez *Bellicositermes natalensis et Cubitermes sp.* La théorie de la stigmergie: Essai d'interprétation du comportement des termites constructeurs. *Insectes Sociaux*, 6:41–81, 1959.
10. B. Hölldobler and E. O. Wilson. *Journey to the Ants: A Story of Scientific Exploration*. Bellknap Press/Harvard University Press, Cambridge, MA, 1994.
11. C. R. Kube and H. Zhang. Task modelling in collective robotics. *Autonomous Robots*, 4:53–72, 1997.
12. M. J. Matarić. Using communication to reduce locality in distributed multiagent learning. *Journal of Experimental and Theoretical Artificial Intelligence, Special Issue on Learning in DAI Systems*, 10(3):357–369, 1998.
13. F. Mondada, G. C. Pettinaro, A. Guignard, I. V. Kwee, D. Floreano, J.-L. Deneubourg, S. Nolfi, L. M. Gambardella, and M. Dorigo. SWARM-BOT: A new distributed robotic concept. *Autonomous Robots*, 17(2–3):193–221, 2004.
14. D. C. Montgomery. *Design and Analysis of Experiments*. John Wiley & Sons, Inc., New York, NY, 5th edition, 1997.
15. S. Nolfi and D. Floreano. *Evolutionary Robotics: The Biology, Intelligence, and Technology of Self-Organizing Machines*. MIT Press/Bradford Books, Cambridge, MA, 2000.
16. T. Seeley. *The Wisdom of the Hive*. Harvard University Press, Cambridge, MA, 1995.
17. V. Trianni, S. Nolfi, and M. Dorigo. Hole avoidance: Experiments in coordinated motion on rough terrain. In F. Groen, N. Amato, A. Bonarini, E. Yoshida, and B. Kröse, editors, *Intelligent Autonomous Systems 8*, pages 29–36. IOS Press, Amsterdam, The Netherlands, 2004.

Gathering Multiple Robotic A(ge)nts
with Limited Sensing Capabilities

Noam Gordon, Israel A. Wagner, and Alfred M. Bruckstein

Center for Intelligent Systems, CS Department
Technion – Israel Institute of Technology, 32000 Haifa, Israel
{ngordon,wagner,freddy}@cs.techniohn.ac.il

Abstract. We consider a swarm of simple ant-robots (or *a(ge)nts*) on
the plane, which are anonymous, homogeneous, memoryless and lack
communication capabilities. Their sensors are range-limited and they are
unable to measure distances. Rather, they can only acquire the directions
to their neighbors. We propose a simple algorithm, which makes them
gather in a small region or a point. We discuss three variants of the
problem: A continuous-space discrete-time problem, a continuous-time
limit of that problem, and a discrete-space discrete-time analog. Using
both analysis and simulations, we show that, interestingly, the system's
global behavior in the continuous-time limit is fundamentally different
from that of the discrete-time case, due to hidden "Zenoness" in it.

1 Introduction

The problem of gathering a swarm of robots in a small region or a point on the
plane is a fundamental one. From a practical standpoint, it may be useful for
collecting multiple ant robots after they have performed a task in the field, for enabling them to start a mission, after being initially dispersed (e.g., parachuted),
or even for aggregating many nano-robots in a self-assembly task. From a theoretical standpoint, the gathering problem is linked to *agreement* problems (as it
may imply or be implied by agreement on a reference location) and is the most
basic instance of the *formation* problem, i.e., the problem of arranging multiple
robots in a certain spatial configuration. The problem is most challenging when
the robots are ant-like – having very limited abilities, e.g. myopic, disoriented
and lacking explicit communication capabilities.

Several theoretical works on this subject exist. Suzuki and Yamashita suggested an agreement procedure, where the agents communicate data through
their movements, in order to agree on a meeting point [16]. Schlude suggested
the use of a *contraction point*, which is invariant to their moves toward it [14].
Ando et al. suggested an algorithm for myopic robots, which move according to
the exact locations of nearby robots [1]. Lin et al. also provided an algorithm for
mypoic robots and related to the case of a limited field of view [11]. Prencipe
et al. suggested an algorithm for myopic robots, which relies on a common com-
pass [7], as well as an algorithm which creates a unique point of multiplicity
(i.e., which contains several agents) to which all agents move [6]. Gordon et al.

M. Dorigo et al. (Eds.): ANTS 2004, LNCS 3172, pp. 142–153, 2004.
© Springer-Verlag Berlin Heidelberg 2004

suggested a simple gathering algorithm on the grid [8]. Others explored cyclic pursuit behaviors (where robots are cyclicly ordered and pursue each other accordingly), which, in some cases, lead to gathering [2, 3, 12]. Sugihara et al. suggested a simple behavior which makes robots fill any convex shape and evenly distribute inside it [15]. Also related is the work of Melhuish et al. [13], who considered aggregation of robots around a beacon in a noisy environment, and demonstrated a way to control the swarm size using minimal communication.

These works rely on some strong assumption about the robots (or *agents* as we shall call them henceforth): Some rely on labelling (e.g., pursuit), some on common orientation, and many on infinite range visibility. Furthermore, all works rely on the agents' ability to measure their mutual distances (except for a few pursuit strategies).

In this work, we suggest a simple gathering algorithm, which relies on very few capabilities: Our agents are both anonymous, homogenous, memoryless, asynchronous, myopic and are *incapable of measuring mutual distances*. It is similar in idea to the polygon-filling algorithm of Sugihara et al. [15]. However, they did not consider visibility limitations at all, so their algorithm could not be used as is, under our imposed limitations.

The inspiration and motivation for this work comes from experiments with real robots in our lab [5], made from LEGO parts and very simple sensors, which are range-limited and do not provide usable distance measurements.

We consider three flavors of the problem, differing mostly in the way time and space are modelled (continuous vs. discrete). Using simulations and analysis, we discuss some interesting implications of these differences on the resulting swarm behaviors. Due to limited space considerations, we have omitted or abridged most of our proofs. These will appear in a forthcoming extended paper.

2 An Asynchronous Gathering Algorithm on the Plane

In this section we present the continuous-space discrete-time case. We begin with the model of the world and its inhabitants.

The *world* consists of the infinite plane $\mathbb{R}^2$ and n point *agents* living in it. We adapt Suzuki and Yamashita's convenient way of modelling a system of asynchronous agents [16]: *Time* is a discrete series of *time steps* $t = 0, 1, \ldots$. At each time step, each agent may be either *awake* or *asleep*, having no control over the scheduling of its waking times. A sleeping agent does nothing and sees nothing, i.e., it is unaware of the world's state. When an agent wakes up, it is able to move instantly to a point on the plane within a distance σ (the *maximum step length*) according to its algorithm. The agent is able to see only the agents within distance V (the *visibility radius* or *range*). However, it cannot measure its *distance* from them. It only knows the *directions* in which the nearby agents are found, i.e., the input is a cyclic list of *angles* $\theta_1, \ldots \theta_m$ (relative to some arbitrary direction, e.g., the agent's heading). There are no collisions. Several agents may occupy the same point[1]. All agents are *memoryless, anonymous* (they cannot be

---

[1] In this case, they have undefined relative directions and are simply mutually ignored.

distinguished by their appearance) and *homogenous* (they lack any individuality, such as a name or ID, and perform the same algorithm).

Regarding the waking times of the agents, we make the following assumption. We say that the agents are *strongly asynchronous*: For any subset G of the agents and in each time step, the probability that G will be the set of waking agents is bounded from below by some constant $\varepsilon > 0$. This implies that each agent will always wake up again in finite expected time.

Define the *mutual visibility graph* of the world as an undirected graph with n vertices, representing the agents, and an edge between each pair of agents, if and only if they can see each other, i.e., the distance between them is at most V. Unless noted otherwise, we assume that this graph is initially connected.

2.1   Maintaining Visibility

We now present a sufficient condition on any movement algorithm for maintaining mutual visibility between agents. In what follows, denote a disc of radius r and center a (where a may signify the location of agent a) by $B_r(a)$.

Let a and b be two agents at some distance $d \leq V$ apart. If their next movement is confined to $B_{V/2}\left(\frac{a+b}{2}\right)$, they will remain visible, by definition. However, since the agents *cannot* measure d, they must consider all possible values of $d \in [0, V]$. Therefore, each agent's next move must rely within the intersection of all discs of diameter V, centered at all possible midpoints between a and b. It is easy to show that, for agent a, this is equivalent to $B_{V/2}(a) \cap B_{V/2}(r)$, where r is a point at a distance $V/2$ from a, in b's direction (cf. Fig. 1(a)). More generally, when a sees several other agents $b_1, \ldots b_m$, it is allowed to move to any point within the intersection of $m + 1$ discs of diameter V, one is $B_{V/2}(a)$, and the other m discs centered at a distance $V/2$ from a, in the directions of $b_1, \ldots b_m$ (cf. Fig. 1).

Denote by ξ the *largest angle between consecutive agents* in agent a's (cyclic) list of input angles $\theta_1, \ldots \theta_m$. It is straightforward to show that the allowable movement region is empty if and only if $\xi < \pi$, i.e., when the agent is "sur-

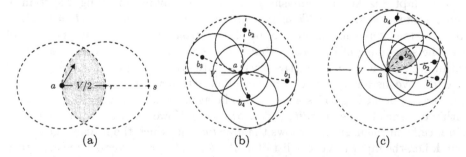

(a)   (b)   (c)

Fig. 1. Maintaining visibility. (a) a must remain within the shaded area to maintain visibility with b, which is somewhere on the line segment $\overline{as}$, as far as a knows. (b) a is surrounded and cannot move. (c) a can move only within the shaded area.

rounded" by other agents (cf. Fig. 1(b)). Otherwise, the allowable region is not empty, and is calculated as follows: Let θ_k and θ_{k+1} be the directions which form the angle $\xi = \theta_{k+1} - \theta_k$. Then the allowable region is the intersection of three discs of diameter V, centered around a and around the two points at a distance $V/2$ in directions θ_k and θ_{k+1} from a, respectively (cf. Fig. 1(c), 2). The following lemma follows from the above geometric arguments.

Lemma 1. *If each agent confines its movements to the allowable region defined above, then existing visibility will be maintained.*

A direct corollary to Lemma 1 is that a connected visibility graph will remain connected forever.

2.2   The Gathering Algorithm

Denote by ψ the complementary angle of ξ defined above (i.e., ψ is the angle of the *smallest wedge* containing all visible agents). The algorithm works as follows: Agents which are surrounded by other agents do not move, while other agents move as far as they can on the *bisector* of ψ. The idea is that outermost agents generally move *inside* the region containing the agents, gradually making it shrink.

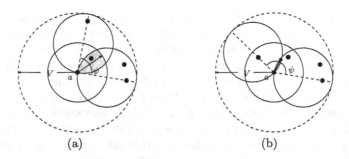

(a)   (b)

Fig. 2. Agent a's movement (assuming $\sigma > V/2$). (a) $\psi < 2\pi/3$. Travelling distance is $V/2$. (b) $\psi > 2\pi/3$. Travelling distance is $V\cos(\psi/2)$.

The exact movement rule is as follows: *If* $\psi \geq \pi$, *then do not move. Otherwise, move along the bisector of ψ, a distance*

$$\mu_\psi = \min\left(V/2, V\cos\left(\psi/2\right), \sigma\right) \ . \tag{1}$$

σ is the physical constraint, while the first two constraints express the maximum possible travelling distance, easily derived from the definition of the allowed movement region stated above (See Fig. 2). For convenience, we denote the *maximum possible step length* by $\mu = \min\left(V/2, \sigma\right)$.

It is worth noting that the allowable region is symmetrical about the bisector of ψ, and that the farthest point on this region is on the bisector. In other words, the bisector can be seen as the "natural" direction to move along, just for the sake of maintaining visibility. Interestingly, this movement direction was originally chosen by Sugihara et al., even though they were not at all concerned with range-limited visibility or step size.

We defer the discussion of the system's behavior to Sect. 5, after the other variants of the model are presented.

3   The Continuous-Time Limit

In this section we discuss the continuous-time limit behavior of the system described above, as follows. Assume that the agents are synchronous, i.e., active at all times. Denote the physical duration of each time step by Δt and let $\sigma = v \Delta t$ for some constant v. It follows from (1) that, for $\sigma < V/2$, $\mu_\psi = \sigma$ for $0 \le \psi < 2 \cos^{-1}(\sigma/V)$. Thus, in the limit $\Delta t \to 0$, we get $\mu_\psi = \sigma = v \Delta t$ for all $0 \le \psi < \pi$, and the movement rule becomes: *If $\psi < \pi$, move along the bisector of ψ at a (constant) speed v. Otherwise, do not move.*

Lemma 1 still holds. In fact, it is not hard to see that an even stronger result holds here: For any two mutually visible agents, their distance is non-increasing.

3.1   Collinearity, Varying Speeds and Zenoness

This simple movement rule (either move at a constant speed v or stand), exhibits a seemingly paradoxical behavior: Agents will sometimes move at *varying* speeds! To see this, consider the following scenario. Let a, b and c be three collinear agents and denote their wedge angles by ψ_a, ψ_b and ψ_c, respectively. b is the middle one, and $\psi_b = \pi$ is determined by the locations of a and c. Assume that $\psi_a, \psi_c < \pi$ (so that a and c move) at time $t = 0$. After an *arbitrarily short* time, as a result of their displacement, ψ_b slightly decreases and, therefore, b starts moving as well, in a normal direction to the segment $\overline{ac}$. Collinearity is now seemingly broken. However, within an arbitrarily small time, b necessarily moves farther enough so that it is "ahead" of $\overline{ac}$, and the bisector of ψ_b now flips and points backwards. As a result, b returns backwards until it crosses $\overline{ac}$ again. This process repeats itself over time, while a and c keep moving. Now, since the deviations of b from $\overline{ac}$ are also arbitrarily small, b effectively remains collinear with a and c. In summary, the following claim holds:

Proposition 1. *Let a, b and c be three collinear agents, where b is the middle one. They will remain collinear as long as in ψ_b is determined by a and c.*

When integrated over time, this "chattering" movement back and forth (at a constant speed v) becomes a smooth movement of b, always on $\overline{ac}$, at a speed generally not equal to v. Take, for instance, the symmetric case, where both a and c move at an angle $\pi/4$ relative to $\overline{ac}$. Then b will move in a straight line, normal to $\overline{ac}$, at a speed of $v/\sqrt{2}$.

This seemingly paradoxical behavior stems from the fact that agent b performs an infinite number of discrete direction switches over a finite period of time. This phenomenon is known as *Zenoness* in hybrid systems theory (See, e.g., [9]), and it shows that, in the continuous-time limit, our system is ill-posed and physically unrealizable. However, had we introduced a slight delay in the agents' responses, we would get a consistent behavior where agent b "oscillates" around $\overline{ac}$ at a finite rate. Indeed, we observed such a behavior in simulations of the discrete-time system of Sect. 2 (For a related discussion, cf. [4]).

3.2 Correctness and Termination

We now present a formal proof of correctness and termination in finite time of the algorithm. Denote the *convex hull* of the configuration (i.e., of the positions of all agents) at time t by $CH(t)$, and the number of its (strictly convex) corners by m. Denote the corner agents by a_i ($i = 1, \ldots, m$), and the inner angle at each corner a_i by ϕ_i. Denote the angle formed between a_i's movement direction and one of the adjacent edges of $CH(t)$ by α_i (Clearly, $0 \le \alpha_i \le \phi_i$). Denote the length of the edge adjacent to a_i by P_i, and the total perimeter of $CH(t)$ by $P = \sum_{i=1}^{m} P_i$.

Proposition 2. $\dot{P}$ *is negative and bounded away (by a constant) from zero, as long as* $P > 0$.

Proof. (abridged) Since the corner agents do not necessarily see each other, a corner agent may not necessarily move along the bisector of ϕ_i. However, since there are no agents outside $CH(t)$, and according to Lemma 1 the visibility graph is connected, it is guaranteed that each corner agent observes at least one other agent and therefore moves inside $CH(t)$.

Observe a single edge of $CH(t)$, connecting a_i and a_{i+1}. During an arbitrarily short period of time dt, these agents move into $CH(t)$ a distance $v\, dt$, at angles α_i and $\phi_{i+1} - \alpha_{i+1}$ relative to the edge, respectively. Any other agents on the edge, or arriving at the edge, will remain on it, according to Prop. 1. Therefore, they have no effect on it. It can be shown that, as a result, the edge will be shortened by the following amount:

$$P_i(t + dt) - P_i(t) = -v\, dt \, (\cos \alpha_i + \cos (\phi_{i+1} - \alpha_{i+1})) + \mathbf{o}\, (dt) \; . \qquad (2)$$

Summing over all m corners, dividing by dt and letting $dt \to 0$, we get:

$$
\begin{aligned}
\dot{P} &= -v \sum_{i=1}^{m} [\cos \alpha_i + \cos (\phi_i - \alpha_i)] \\
&\le -v \sum_{i=1}^{m} [1 + \cos (\phi_i)] \\
&\le -v \, [1 + \cos (\pi \, (1 - 2/n))] \; .
\end{aligned}
\qquad (3)
$$

The first inequality holds because, for any fixed $\phi_i < \pi$, the expression $\cos \alpha_i + \cos (\phi_i - \alpha_i)$ is minimal when $\alpha_i = 0$. The second inequality is straightforward, and the resulting expression on the right side is a negative constant, dependent only on n, which is finite. $\qquad \square$

Theorem 1. *Beginning with any initial configuration, whose visibility graph is connected, all agents will gather in a single point, in finite time.*

Proof. This is a direct corollary of Prop. 2. $\qquad\qquad\qquad\qquad\qquad$ □

Another obvious property from the above analysis is that for any two time instants t_1, t_2, if $t_1 < t_2$ then $CH(t_1) \supset CH(t_2)$. A corollary of this is that if the visibility graph is initially *not* connected, none of its connected components will ever merge (or split, of course), and therefore each group of agents, corresponding to a connected component, will gather in exactly one separate point.

4   A Discrete Analog

This section presents a discrete-space discrete-time analog of the problem presented in Sect. 2. The asynchronous operation of the agents is retained, with the world now being the infinite rectangular grid ($\mathbb{Z}^2$).

On the grid, we measure distances with *infinity norms*. Formally, the distance between two points $p_1 = (x_1, y_1)$ and $p_2 = (x_2, y_2)$ on the grid is

$$\|p_1 - p_2\|_\infty = \max(x_1 - x_2, y_1 - y_2) .$$

Accordingly, let $V \geq 1$ be the visibility range of the agents. Then the visible area of an agent is a $(2V + 1) \times (2V + 1)$ square, centered at the agent. An agent cannot measure the exact distance from a visible agent. Rather, it can only measure the signs (positive, negative or zero) of $x_1 - x_2$ and $y_1 - y_2$. In the grid world, an agent may move only to one of the four neighboring cells.

Define a move as *allowable* if and only if there are *no* visible agents "behind" the agent and there *exists* a visible agent "before" it, as it looks in that direction (e.g., the agent is allowed to move in the *positive x* direction if and only if there is *no* visible agent with a *smaller x coordinate* and there *exists* a visible agent with a *larger x coordinate*. See Fig. 3).

The movement rule is as follows: *The agent randomly picks an allowable move, if there is any, and performs it* (There can be 0 to 2 allowable moves. In case of two allowable moves, the agent tosses a coin). Intuitively, this rule makes agents get closer to, but not move away from, each other. Thus, visibility is maintained, and we expect the area occupied by all agents to shrink. We now present a proof for this.

Define the *bounding box BB(t)* of the agents as the smallest enclosing rectangle (oriented with the grid's axes) which contains all of the agents.

Proposition 3. *When following the movement rule defined above, mutually visible agents remain visible.*

Proposition 4. *The bounding box of the agents is monotonically non-inflating, i.e., $BB(t + 1) \subseteq BB(t)$ for all t.*

Proposition 5. *At least one of the bounding box's sides eventually move inwards (as long as BB is not a single cell).*

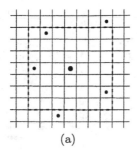   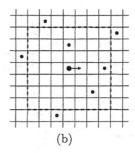

(a)   (b)

Fig. 3. The proposed gathering algorithm on the grid ($V = 3$). (a) The center agent is surrounded and cannot move. (b) The center agent can move only to the right.

Proof. (abridged) Observe w.l.o.g. all agents which reside on the *upper side* of the bounding box (i.e., all agents with the maximum y coordinate), and denote their number by $X(t)$. If the bounding box's height is 1 (i.e., $X = n$), then clearly the leftmost and rightmost agents move inwards (when they eventually wake up), narrowing the bounding box. Otherwise, due to the connectedness of the visibility graph (Prop. 3), at least one agent on the upper side observes another agent, which resides *below* the upper side. Therefore, that agent is allowed to move downwards and, from the strong asynchronicity assumption, there is a probability of at least $\varepsilon/2$ that it indeed wakes up and chooses to move in that direction (and no other agent wakes up and moves up), lowering the number of agents on the upper side. Thus, all states $X = k$, $0 < k < n$ are transitional and connected to either one of the states $X = 0$ and $X = n$, which are obviously trapping (due to Prop. 4). Therefore, eventually either $X = 0$ or $X = n$ will occur, meaning that either the upper side has moved down or the lower side has moved up all the way to merge with the upper side, respectively. ◻

Theorem 2. *Beginning with any initial configuration, whose visibility graph is connected, all agents will eventually gather in a single cell.*

Proof. The proof follows immediately by applying Proposition 5 repeatedly. ◻

5   Discussion

We performed extensive simulations of the problem described in Sect. 2. Not only did the simulations validate the correctness of the algorithm, but they also revealed some interesting global behaviors of the swarm. We discuss them qualitatively, and compare them to the continuous-time limit behavior (which apparently cannot be simulated, due to its Zenoness). We argue that, in the limit, the system's behavior is *fundamentally* different.

The system's evolution is clearly divided into two phases (1) A *contraction phase*, where the area occupied by the agents contracts until all agents become a small, dense cluster, whose diameter is in the order of μ (the maximum step

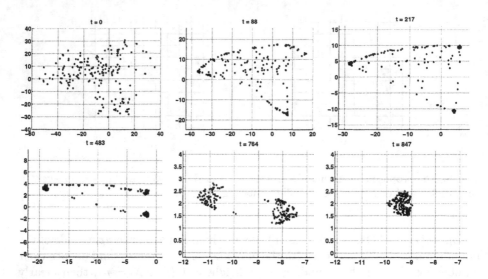

Fig. 4. A typical run of the continuous-space discrete-time algorithm. Here $n =$ 200, $\sigma = 1, V = 10$, and each agent wakes up in each time step with probability $p = 0.6$. Note that the scale changes between frames.

size); (2) A *wandering phase*, where the cluster stops contracting and begins to wander indefinitely in the plane.

5.1   The Contraction Phase

Observe Fig. 4, which shows several snapshots of one particular run of the system, beginning with a random configuration with a connected visibility graph. Consider the area occupied or "guarded" by the agents (informally speaking, the area enclosed by laying a "fence" between each pair of mutually visible agents). It is intuitively expected that the agents on the convex segments of its boundary would move inside, those on the concave segments would either move outside or stand, while the interior agents would generally stay inside[2]. What intrigues us is the evolution of that boundary over time. Evidently, the occupied region shrinks and its boundary contracts. As expected, the moving boundary "sweeps" more and more agents, and becomes a "belt" which accumulates most of the "mass" in the system. Moreover, a peculiar and less obvious phenomenon is evident from the simulations: The build-up of mass on the boundary belt is *not* uniform. Segments with high curvature (i.e., where the boundary's course bends sharply) tend to absorb more agents than (and from) segments with lower curvature. In addition, the former segments' curvature becomes even higher, while the latter segments tend to straighten slightly. This interaction between *mass* and *curvature* along the belt creates a positive feedback process: The large-scale

---

[2] We say "generally", since it is indeed possible that agents move a short distance outside the occupied region, if they are close enough to the boundary.

shape of the occupied region becomes an approximate polygon, where the curved segments turn into large clusters which form the corners, while the less curved segments become nearly straight edges between the corners. As the polygon contracts, edges gradually merge, until a triangle or a two-cluster "dipole" remains, ultimately collapsing into one dense cluster.

What causes this positive feedback process? An informal explanation for one direction can be illustrated if we consider the following approximated behavior: Replace the boundary with a smooth contour with equally spaced agents on it, and let each agent move a short distance μ along its normal. Clearly, the higher the curvature, the closer adjacent agents will become.

The other direction's possible explanation is easier to observe in one dimension first. Consider the following "leaping frogs" game: Imagine m frogs on the real axis $\mathbb{R}$. Their movement rule is that, at each time step, only the leftmost frogs leap a distance σ to the right. What is the average speed of the pack? Assuming that no frog lands exactly on another frog, then it is clearly σ/m, as only one of the m frogs move at each time step. Back in two dimensions, we notice that the boundary belt is especially *thick* in the denser segments, and the "cross sections" of these segments typically contain more agents. However, only the outermost agents (analogous to the leftmost frogs) move in each step, lowering the average contraction speeds of these segments, which, as a result, "lag behind" and become more corner-like.

If we set $\sigma = v \Delta t$, where Δt is the physical duration of each time step, the average speed of the frog pack should be v/m, regardless of the size of Δt. However, when we take $\Delta t \to 0$, a strange thing happens. The frogs' behavior changes from "leaping over" to "sweeping" – The leftmost frog simply moves at a constant speed v, eventually joined by all frogs. Analogously, in the continuous-time system of Sect. 3, due to Prop. 1, when a moving segment of the boundary meets an internal agent, it sweeps it along, without slowing at all. Thus, in this model, the "mass" of the boundary (which has no thickness in this case) does not affect its contraction speed, and the mass–curvature feedback link is broken. Therefore, we conjecture that the shape evolution of the occupied region's boundary is fundamentally different than that of the discrete-time case – No sharp corners and straight edges will be formed. Rather, convex segments will quickly contract, and the region will have a much rounder and smoother shape.

5.2 The Wandering Phase

Since the agents do not measure distances, their steps' sizes are invariant to the configuration's scale. The smaller the region they occupy, the relatively longer their steps become. Once the diameter of the occupied region is in the order of μ, the moving agents tend to "leap" over the region, rather than enter it. As a result, the region drifts rather than contracts[3]. Due to the random nature of the agents' activity schedules, the drift direction is also random. Thus, the

---

[3] Formally, we may choose to define the cluster's location as its center of mass, the center of the smallest enclosing circle, etc.

movement of the cluster is a random walk. We call it a *composite random walk*, as it is composed of the deterministic (yet randomly scheduled) movements of many agents. Just as with the boundary evolution case above, here more mass means slower wandering.

Random walks in two dimensions are known to be recurrent, which implies that two random walkers are bound to meet eventually (See, e.g., [10]). This has an important implication on our problem: If the visibility graph is initially *not* connected, then each connected component becomes an independent composite random walker. However, due to their recurrent nature, these clusters will eventually meet and merge. Thus, the proposed algorithm may eventually gather all agents, *even though the visibility graph is not connected initially.* Of course, we do not mean that it will work for any initial condition. The composite random walkers must be able to meet. This may be false for clusters of one agent (which doesn't move at all) or two agents (which always remain on one line). We conjecture that for clusters of three or more non-collinear agents, this is guaranteed, and a sufficient condition for the eventual success of the algorithm would be the existence of such a cluster. It should be noted that, as our simulations show, the merging process is generally agonizingly slow. Still, from a theoretical standpoint, merging should occur eventually.

Yet again, we see a significant difference between this behavior and the continuous-time limit behavior, where, as we showed in Sect. 3, gathering in a single point occurs, rather than wandering, and there is no hope of ever merging two connected components of the visibility graph.

6   Conclusion

The algorithm proposed in this paper is an example of how very simple individual behaviors can yield complex global behaviors of the swarm. In each of the three variants we presented, we showed how different global processes lead to the ultimate goal of gathering. In the continuous-space discrete-time model, the global shape of the swarm becomes an approximate polygon and converges to a wandering cluster. In the continuous-time limit, the swarm shape is much smoother, and the agents converge to a static point. Our main contribution is that we consider the gathering problem with such severe limits on the sensory abilities of the agents (being both myopic and unable to measure distances), in addition to being anonymous and memoryless.

In forthcoming papers, we shall provide an analytic proof for the convergence of the first variant of the algorithm, and a more rigorous analysis of the swarm's shape evolution. We will also explore the application of the proposed algorithm to row-straightening and formation of other convex shapes. Further work should relate to the effect of noise and error, both in sensing and movement, on the resulting global behavior. Clearly, the proposed algorithm is sensitive to such errors, as the connectivity of the visibility graph may be broken. It would be interesting to analyze the effects, and devise methods to overcome them.

References

1. H. Ando, Y. Oasa, I. Suzuki, and M. Yamashita. A distributed memoryless point convergence algorithm for mobile robots with limited visibility. *IEEE Trans. on Robotics and Automation*, 15(5):818–828, 1999.
2. A. M. Bruckstein, N. Cohen, and A. Efrat. Ants, crickets and frogs in cyclic pursuit. Technical Report CIS-9105, Technion – IIT, 1991.
3. A. M. Bruckstein, C. L. Mallows, and I. A. Wagner. Probabilistic pursuits on the grid. *American Mathematical Monthly*, 104(4):323–343, April 1997.
4. A. M. Bruckstein and O. Zeitouni. A puzzling feedback quantizer. Technical Report 879, EE Department, Technion IIT, May 1993.
5. The Center of Intelligent Systems, Technion IIT web site: *http://www.cs.technion.ac.il/Labs/Isl/index.html.*
6. M. Cieliebak, P. Flocchini, G. Prencipe, and N. Santoro. Solving the robots gathering problem. In *Proc. of ICALP 2003*, 2003.
7. P. Flocchini, G. Prencipe, N. Santoro, and P. Widmayer. Gathering of autonomous mobile robots with limited visibility. In *Proc. of STACS 2001*, 2001.
8. N. Gordon, I. A. Wagner, and A. M. Bruckstein. Discrete bee dance algorithms for pattern formation on a grid. In *Proc. of IEEE Intl. Conf. on Intelligent Agent Technology (IAT03)*, pages 545–549, October 2003.
9. T. A. Henzinger. The theory of hybrid automata. In *Proc. of LICS*, pages 278–292, 1996.
10. B. D. Hughes. *Random Walks and Random Environments*, volume 1. Oxford University Press, 1995.
11. Z. Lin, M. E. Broucke, and B. A. Francis. Local control strategies for groups of mobile autonomous agents. *IEEE Trans. on Automatic Control*, 49(4):622–629, April 2004.
12. J. A. Marshall, M. E. Broucke, and B. A. Francis. A pursuit strategy for wheeled-vehicle formations. In *Proc. of CDC03*, pages 2555–2560, 2003.
13. C. R. Melhuish, O. Holland, and S. Hoddell. Convoying: using chorusing to form travelling groups of minimal agents. *Robotics and Autonomous Systems*, 28:207–216, August 1999.
14. K. Schlude. From robotics to facility location: Contraction functions, weber point, convex core. Technical Report 403, CS, ETHZ, 2003.
15. K. Sugihara and I. Suzuki. Distributed algorithms for formation of geometric patterns with many mobile robots. *Journal of Robotic Systems*, 13(3):127–139, 1996.
16. I. Suzuki and M. Yamashita. Distributed anonymous mobile robots: Formation of geometric patterns. *SIAM Journal on Computing*, 28(4):1347–1363, 1999.

Improvements on Ant Routing
for Sensor Networks

Ying Zhang[1], Lukas D. Kuhn[2], and Markus P.J. Fromherz[1]

[1] Palo Alto Research Center, Palo Alto, CA, USA
{yzhang,fromherz}@parc.com
[2] Ludwig Maximilian University, Munich, Germany

Abstract. Ad-hoc wireless sensor networks have been an active research
topic for the last several years. Sensor networks are distinguished from
traditional networks by characteristics such as deeply embedded routers,
highly dynamic networks, resource-constrained nodes, and unreliable and
asymmetric links. Ant routing has shown good performance for communication networks; in this paper, we show why the existing ant-routing
algorithms do not work well for sensor networks. Three new ant-routing
algorithms are proposed and performance evaluations for these algorithms on a real application are conducted on a routing simulator for
sensor networks.

1 Motivation

Large-scale ad-hoc networks of wireless sensors have become an active topic of
research [6]. Such networks share the following properties:

- *embedded routers* – each sensor node acts as a router in addition to sensing
  the environment;
- *dynamic networks* – nodes in the network may turn on or off during operation
  due to unexpected failure, battery life, or power management [2]; attributes
  associated with those nodes (locations, sensor readings, load, etc.) may also
  vary over time;
- *resource constrained nodes* – each sensor node tends to have small memory
  and limited computational power;
- *dense connectivity* – the sensing range in general is much smaller than the
  radio range, and thus the density required for sensing coverage results in a
  dense network;
- *asymmetric links* – the communication links are not reversible in general.

Applications of sensor networks include environment monitoring, traffic control,
building management, and object tracking. Routing in sensor networks, however, has very different characteristics than that in traditional communication
networks. First of all, address-based destination specification is replaced by a
more general feature-based specification, such as geographic location [7] or information gain [3]. Secondly, routing metrics are not just shortest delay, but

M. Dorigo et al. (Eds.): ANTS 2004, LNCS 3172, pp. 154–165, 2004.
© Springer-Verlag Berlin Heidelberg 2004

also energy usage and information density. Thirdly, in addition to peer-to-peer communication, multicast (one-to-many) and converge-cast (many-to-one) are major traffic patterns in sensor networks. Even for peer-to-peer communication, routing is more likely to be source or destination driven than table-based [9], and source/destination pairs often are dynamic (changing from time to time) or mobile (moving during routing).

The characteristics of the *Ant System* [5] – positive feedback, distributed computation, and constructive greediness – seem to fit the domain of sensor networks very well. Various ant-routing algorithms have also been proposed for telecommunication networks [1, 4, 10], which demonstrated good performance for those networks [1]. We started by implementing an ant-routing algorithm with the basic ideas from AntNet [1], which we'll call the *basic* ant routing. Due to the properties of highly dynamic nodes and asymmetric links in sensor networks, the basic ant routing did not perform well. We then developed three improved versions of ant routing based on the message-initiated constraint-based routing framework [12]:

- *Sensor-driven and cost-aware ant routing (SC)*: One of the problems of the basic ant-routing algorithm is that the forward ants normally take a long time to find the destination, even when a tabu list is used (i.e., no repeating nodes if possible). That happens because ants initially have no idea where the destination is. Only after one ant finds the destination and traverses back along the links will the link probabilities of those links change. In SC, we assume that ants have sensors so that they can smell where the food is even at the beginning. That is not an unrealistic assumption for sensor networks, since feature-based routing dominates address-based routing in that space. Some features, such as geographic location, have a natural *potential field*. If the destination does not have a clear hint, pre-building the feature potential is sometimes still efficient. Cost awareness generalizes the objective of shortest path length so that ants can apply other routing metrics as well, e.g., energy-aware routing.
- *Flooded forward ant routing (FF)*: Even augmented with sensors, forward ants can be missguided due to obstacles or moving destinations. Flooded forward ant routing exploits the broadcast channel in wireless sensor networks. When a forward ant starts at the source, it tells all its neighbors to look for the food, and neighbors tell neighbors and so on, until the destination is found. Ants then traverse backward to the source and leave pheromone trails on those links. To control the flooding, only those ants who are "closer" to the food will join the food searching process.
- *Flooded piggybacked ant routing (FP)*: Single path routing tends to have high loss rates, due to dynamic and asymmetric properties of sensor networks. Multi-path routing such as flooding is very robust and has high success rate. In FP, we combine forward ants and data ants, using constrained flooding in the previous algorithm to route the data and to discover good paths at the same time.

The rest of the paper is organized as follows. Section 2 presents the basic ant-routing algorithm. Section 3 develops three improved ant-routing algorithms. Section 4 evaluates all four ant-routing algorithms in a routing simulator for sensor networks using a real application scenario. Section 5 concludes the paper.

2   Basic Ant Routing

Informally, the basic ant routing can be described as follows (cf. [1]):

- At some intervals, which may vary with time, a forward ant is launched from the source node toward the destination node.
- Each forward ant searches for the destination by selecting the next hop node according to the link probability distribution. Initially all the links have equal probability.
- While moving forward, each forward ant remembers the list of nodes it has visited and tries to avoid traversing the same node.
- Once a forward ant finds the destination, a backward ant is created, which moves back along the links that the forward ant had traversed.
- During the backward travel, the cost from the destination to each node in the path is recorded; rewards are then given according to the relative goodness of the path. Probabilities of the nodes in the path are updated according to the rewards.
- Once a backward ant arrives at the source, the next launch interval is calculated according to the relative goodness of the whole path.

At each node, like AntNet [1], the link probability distribution p_n is maintained for each neighbor node n, with $\sum_{n \in N} p_n = 1$, where N is the set of neighbor nodes. Initially, $p_n = 1/|N|$. In addition, the average cost (e.g., the number of hops) μ and the variance σ^2 from the current node to the destination is updated by the backward ants as follows. Let C be the current cost of the path from the destination to the current node:

$$\mu \leftarrow \mu + \eta(C - \mu), \sigma^2 \leftarrow \sigma^2 + \eta((C - \mu)^2 - \sigma^2) \qquad (1)$$

An observation window W of size M is kept for storing the cost of past M paths, so that the minimum cost within the window W can be obtained. In this case, we set $\eta = \min(5/M, 1)$ as in [1].

Given a reward $r \in [0, 1]$, the probability distribution on the links is updated as follows. Assuming the backward ant is coming from node $m \in N$, then[1]:

$$p_m \leftarrow p_m + r(1 - p_m), p_n \leftarrow p_n - r p_n, n \in N, n \neq m \qquad (2)$$

The reward r can be simply a constant, e.g., 0.5; however, it works better when r is cost-sensitive. For example, we use

$$r \leftarrow k_1 \left(\frac{C_{inf}}{C} \right) + k_2 \left(\frac{C_{sup} - C_{inf}}{(C_{sup} - C_{inf}) + (C - C_{inf})} \right) \qquad (3)$$

---

[1] Another update rule, $p_m \leftarrow \frac{p_m + r}{1 + r}$ and $p_n \leftarrow \frac{p_n}{1 + r}$, can be used as well. According to our experiments, there is no significant difference from (2).

where $C_{inf} = \min(W)$, $C_{sup} = \mu + z(\sigma/\sqrt{M})$, $k_1 + k_2 = 1$, and $z \in R^+$. This formula guarantees that $r \in (0, 1]$. To slow down the learning effect in (2), one may use $r \leftarrow \alpha r$, where $\alpha \in (0, 1)$ is a learning rate.

When a backward ant arrives at the source, the forward ant launch interval I is updated by $I \leftarrow e^{r-0.5}I$, where r is obtained from (3). The intuition is to increase I if the path has relatively lower cost (or higher reward with $r > 0.5$), and to reduce I otherwise.

The pseudo code for the basic ant routing is presented in Figure 1. Each forward ant carries a tabu list to avoid loops if possible. If the loop size is larger than half of the list size, the forward ant dies; otherwise the loop is removed from the list and the ant continues. Each backward ant follows the tabu list and updates the statistics and the probability distribution on the way back to the source. The cost is the number of hops in this basic algorithm. The forward ant interval is updated according to the cost of the whole path at the source. Like AntNet [1], the data ants are prevented from choosing links with very low probability by remapping p to p^β for $\beta > 1$.

The algorithm assumes that a neighborhood structure is established initially. To establish a neighborhood structure, each node sends out "hello" packets a few times at the beginning. A node u is a neighbor of node w if w can hear packets from u. For asymmetric links, u being a neighbor of w does not imply that w is a neighbor of u. When the ant routing starts, the probabilities are uniformly distributed among the neighbors.

The problem with the basic ant routing is that the forward ants initially have a hard time to find the destination. It is equivalent to random walks in a maze with no hint. Because of the density property (large neighborhood), the problem becomes more severe. The probability distributions are not modified until the first forward ant arrives at the destination and traverses back. Due to asymmetric links, ants who successfully reach the destination may not be able to move back to the source, causing the updates to happen even more slowly. Collisions and failure nodes contribute to the low performance as well.

3   Improved Ant-Routing Algorithms

3.1   Sensor-Driven Cost-Aware Ant Routing (SC)

To improve the performance of the forward ants, we equip ants with sensors to sense the best direction to go even initially. In addition to storing the probability distribution, each node estimates and stores the cost to the destination from each of its neighbors. Assume that the cost estimation is Q_n for neighbor n. The cost from the current node to the destination is 0 if it is the destination, otherwise $C = \min_{n \in N} (c_n + Q_n)$, where c_n is the local cost function. The initial probability distribution is calculated according to

$$p_n \leftarrow \frac{e^{(C-Q_n)^\beta}}{\sum_{n \in N} e^{(C-Q_n)^\beta}} \tag{4}$$

received forward ant f at node w from node m **do**
  if destination(w) **then**
    $b.cost \leftarrow 0$; $b.list \leftarrow f.list$;
    send b to $b.list[1]$;
  else
    $f.list \leftarrow [w, f.list]$; % insert w at the head of the list
    $N^* \leftarrow N \setminus f.list$;
    if $N^* = \emptyset$ **then** $N^* \leftarrow N$; **end**
    choose $n' \in N^*$ according to $\frac{p_{n'}}{\Sigma_{n \in N^*} p_n}$
    if $n' \in f.list$ **then**
      $i \leftarrow find(f.list, n')$; $L \leftarrow |f.list|$;
      if $L < 2i$ **then return**; **end**
      $f.list \leftarrow f.list[i + 1 : L]$; % remove the loop
    end
    send f to n';
  end
end

received backward ant b at w **do**
  $b.cost \leftarrow b.cost + c_w$;
  $\mu \leftarrow \mu + \eta(b.cost - \mu)$; $\sigma^2 \leftarrow \sigma^2 + \eta((b.cost - \mu)^2 - \sigma^2)$;
  $W \leftarrow [b.cost, W]$;
  if $|W| > M$ **then** $W \leftarrow W[1 : M]$; **end** % keep only M costs
  calculate r according to (3);
  if source(w) **then** $I \leftarrow e^{r-0.5} I$;
  else
    $r \leftarrow \alpha r$;
    calculate p according to (2);
    $L \leftarrow |b.list|$; $n' \leftarrow b.list[1]$; $b.list \leftarrow b.list[2 : L]$;
    send b to n';
  end
end

received data ant d **do**
  choose $n' \in N^*$ according to $\frac{p_{n'}^\beta}{\Sigma_{n \in N^*} p_n^\beta}$;
  send b to n';
end

timeout release forward ant f at node w **do**
  if source(w) **then**
    $f.list \leftarrow [w]$;
    choose n' according to p_n;
    send f to n';
  end
  set next timeout after I;
end

Fig. 1. Basic ant-routing algorithm

The exponential term in this formular makes the probability distribution differ more, thus even more favoring the good links.

There are two ways in which Q_n can be obtained. First, for the feature-based destination specification, one may estimate the "distance" from the current node to the destination. For example, for geographic routing, one may estimate the node's cost by $k\sqrt{(x_d - x)^2 + (y_d - y)^2}$, where (x, y) is the location of the current node and (x_d, y_d) is the location of the destination. During initialization, each "hello" packet also includes the cost of a node. After initialization, each node has the costs of its neighbors. Alternatively, if the destination is known initially, which is the case for many sensor network applications, one can get the cost by flooding from the destination. The pseudo code for cost estimation and ant initialization is shown in Figure 2.

```
received initialization ant i from u do
    Q_u ← i.cost;
    i.cost ← min_{n∈N} (c_n + Q_n);
    broadcast i;
end

initialization at node w do
    if destination(w) then
        i.cost ← 0;
        broadcast i;
    end
end

ant-start at node w do
    p_n ← e^{(C-Q_n)^β} / Σ_{n∈N} e^{(C-Q_n)^β} ;
    if source(w) then
        release forward ant;
    end
end
```

Fig. 2. Initialization for cost estimation

3.2   Flooded Forward Ant Routing (FF)

If the destination is unknown initially or the cost estimation cannot be derived from the destination specification (e.g., address-based destination), SC reduces to the basic ant routing, and the problem of wandering around to find the destination still exists. A solution to that problem is to exploit the broadcast channel of wireless sensor networks. The idea is to flood forward ants to the destination; as before, successful forward ants will create backward ants to traverse back to the source. Multiple paths are updated by one flooding phase. Probabilities are updated in the same way as in the basic ant routing. The flooding can be stopped if the probability distribution is good enough for the data ants to the destination. In our case, we reduce the rate for releasing the flooding ants when a shorter path is traversed.

Two strategies are used to control the forward flooding. First, a neighbor node will broadcast a forward ant to join the forward search only if it is "closer" to the destination than the node that broadcasted at an earlier time. Link probabilities are used for the estimation, i.e., a forward ant is to broadcast only if $p_n < 1/|N|$, where n is the neighbor the ant is coming from and N is the set of neighbors. If initially there is no hint, i.e., $p_n = 1/|N|$ for all n, each node will broadcast once. Secondly, delayed transmission is used in that a random delay is added to each transmission, and if a node hears the same ant from other nodes, it will stop broadcasting. The pseudo code for the flooded forward ant routing is shown in Figure 3. The problem with FF is collisions. The flooded forward ants create a significant amount of traffic, which interferes with the data ants and the backward ants. Controlling the frequency of the forward flooding becomes the key for this method to succeed.

```
received forward ant f at node w from node m do
  if destination(w) then
    b.cost ← 0; b.list ← f.list;
    send b to b.list[1];
  elseif new(f) and p_m < 1/|N| then
    f.list ← [w, f.list];
    broadcast f after random delay t if not heard f within t;
  end
end

timeout release forward ant f at node w do
  if source(w) then
    f.list ← [w];
    broadcast f;
  end
  set next timeout after I;
end
```

Fig. 3. Flooded forward ant routing

3.3   Flooded Piggybacked Ant Routing (FP)

The flooding mechanism in wireless networks is very robust and works extremely well when the network is highly dynamic. In FP, we combine forward ants with data ants, with the data ants carrying the forward list. The same strategy to control the flooded forward ants as in FF is used to control the flooded data ants. In this case, the data ants not only pass the data to the destination, but also remember the paths which can be used by the backward ants to reinforce the probability on these links. The probablity distribution constrains the flooding towards the destination for future data ants. This method has a very high success rate with a relatively high energy consumption. Figure 4 shows the fragment of the code.

received data ant d at node w from node m **do**
  if destination(w) **then**
    $b.cost \leftarrow 0$; $b.list \leftarrow f.list$;
    send b to $b.list[1]$;
  elseif $new(d)$ and $p_m < 1/|N|$ **then**
    $d.list \leftarrow [w, d.list]$;
    broadcast d **after** random delay t if not heard d within t;
  end
end

Fig. 4. Flooded piggybacked ant routing

4 Performance Evaluations

We have simulated these routing strategies using Prowler [11], a probabilistic wireless network simulator designed for the Berkeley Motes.

4.1 Network Model

Prowler [11], written in Matlab, is an event-driven simulator that can be set to operate in either deterministic mode (to produce replicable results while testing an algorithm) or in probabilistic mode (to simulate the nondeterministic nature of the communication channel). Prowler consists of a *radio propagation model* and a *MAC-layer model*.

The radio propagation model determines the strength of a transmitted signal at a particular point of the space for all transmitters in the system. Based on this information the signal reception conditions for the receivers can be evaluated and collisions can be detected. The transmission model is given by:

$$P_{rec,ideal}(d) \leftarrow P_{transmit} \frac{1}{1 + d^\gamma}, \quad where \ 2 \leq \gamma \leq 4 \tag{5}$$

$$P_{rec}(i,j) \leftarrow P_{rec,ideal}(d_{i,j})(1 + \alpha(i,j))(1 + \beta(t)) \tag{6}$$

where $P_{transmit}$ is the signal strength at the transmission and $P_{rec,ideal}(d)$ is the *ideal* received signal strength at distance d, α and β are random variables with normal distributions $N(0, \sigma_\alpha)$ and $N(0, \sigma_\beta)$, respectively. A network is asymmetric if $\sigma_\alpha > 0$ or $\sigma_\beta > 0$. In (6), α is static depending on locations i and j only, but β changes over time. Furthermore, an additional parameter p_{error} models the probability of a transmission error by any unmodeled effects. The MAC-layer simulates the Berkeley motes' CSMA protocol, including random waiting and back-offs.

4.2 Performance Metrics

Various performance metrics are used for comparing different routing strategies in sensor networks. We have used the following:

- *latency* – the time delay of a packet from the source to the destination;
- *success rate* – the total number of packets received at the destinations vs. the total number of packets sent from the source;
- *energy consumption* – assuming each transmission consumes an energy unit, the total energy consumption is equivalent to the total number of packets sent in the network;
- *energy efficiency* – the ratio between the number of packets received at the destination vs. the total energy consumption in the network;

Some of these metrics are correlated and some conflict with each other. For different applications, different metrics may be relevant.

4.3   Example: Pursuer Evader Game

We use a real application to test the performance of these ant-routing algorithms. The Pursuer Evader Game (PEG) has been a standard testbed for sensor networks. In this testbed, sensors are deployed in a regular grid with random offsets. An evader is a tank or a car whose movement can be detected by the sensor nodes. A pursuer is another vehicle that is going to track the evader down. The communication problem in this task is to route packets sent out by sensor nodes who detected the evader to the mobile pursuer.

In our test, the network is a 7×7 sensor grid with small random offsets. For Berkeley motes, the maximum packet size is 960 bits and the radio transmission rate is 4Kbps. The default settings of Prowler are used, i.e., $\sigma_\alpha = 0.45, \sigma_\beta = 0.02$, and $p_{error} = 0.05$. The transmit signal strength is set to 1, and the maximum radio range is about $3d$ (a dense network), where d is the standard distance between two neighbor nodes in the grid. Figure 5 (a) shows an instance of the connectivity of such a network. Figures 5 (b) (c) and (d) show snapshots of ant traces during the first 100 seconds assuming both the evader and the pursuer are stationary. Figure 5 shows that the forward ants easily get lost in the basic ant routing, whereas SC can find destinations easily without wandering too much, and FF has shorter paths overall.

In the performance simulation, the source is changing from node to node, following the movement of the evader, and the destination is mobile. The average speed of both the evader and the pursuer is set to $0.01d/s$. The source rate is about 1 packet per second. For the basic ant routing, SC and FF, the ratio between data ants and forward ants is set to be 2 initially. The forward ant interval is changing according to the quality of the discovered path, but the data rate is fixed. There is an initialization phase of 3 seconds for establishing the connectivity of the network. In the basic ant routing, 3 "hello" packets are sent out from each node; for SC, FF, and FP, 3 backward floodings are used to establish the potential field and the initial probability distribution of links. Both initializations use about the same amount of energy, i.e., $3N$, where N is the number of nodes in the network. A total of 200 seconds are simulated for the four ant-routing algorithms, with 10 random runs for each.

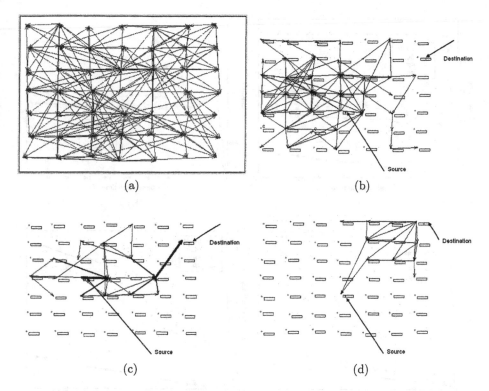

Fig. 5. (a) Instance of radio connectivity (b) Traces of the forward ants in the basic ant routing (c) Traces of forward ants in SC (line thickness corresponding to the probability of the link) (c) Traces of backward ants in FF

Performances are compared among the four ant-routing algorithms for this application. Figure 6 shows the average success rate, latency, total energy consumption, and energy efficiency. The basic ant routing has extremely poor success rates. FP has the highest success rates but is not very energy efficient. FF has in general the shortest delays, and SC is most energy efficient. We have developed other routing strategies for sensor networks [12]. Comparisons with these and other routing strategies will be discussed in future papers. The biggest problem with these ant-routing algorithms is the use of forward lists, which introduces limitations for small packet sizes in large-scale networks. Furthermore, even though ant-routing algorithms work for asymmetric networks, they all reinforce the best symmetric links. If a network does not have a symmetric path, these algorithms do not work as well.

5   Conclusions

In this paper, we have shown that the basic ant-routing algorithm does not perform well for dense, asymmetric, and dynamic networks. We then developed three

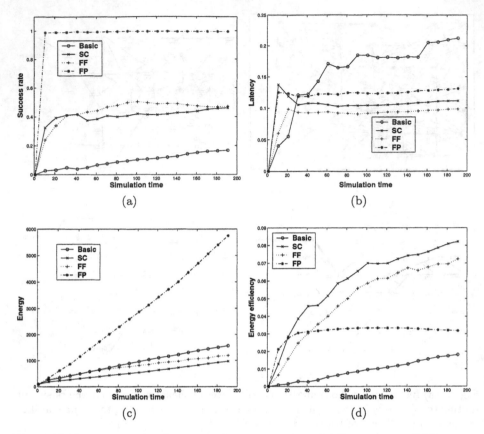

Fig. 6. Performance evaluation among four ant routings: (a) Success rates (b) Latency (c) Energy consumption (c) Energy efficiency

new ant-routing algorithms for sensor networks. The new ant algorithms work quite well overall, with each having some advantages in terms of performance: SC is energy efficient, FF has shorter delays, and FP has the highest success rates. In the near future, we will compare ant routing with other reinforcement learning based algorithms and with traditional ad-hoc routing algorithms such as AODV [8].

Acknowledgments

This work is funded in part by Defense Advanced Research Project Agency contract # F33615-01-C-1904.

References

1. G. Di Caro and M. Dorigo. Antnet: Distributed stigmergetic control for communication networks. *Journal of Artificial Research*, 9:317 – 365, 1998.

2. A. Cerpa and D. Estrin. Ascent: Adaptive self-configuring sensor networks topologies. In *Proc. IEEE InfoComm*, New York, NY, June 2002.
3. M. Chu, H. Haussecker, and F. Zhao. Scalable information-driven sensor querying and routing for ad hoc heterogeneous sensor networks. *Int. Journal on High Performance Computing Applications*, June 2002.
4. J. C. D. Subramanian, P. Druschel. Ants and reinforcement learning: A case study in routing in dynamic networks. Technical report tr96-259, Rice University, July 1998.
5. M. Dorigo, V. Maniezzo, and A. Colorni. The ant system: Optimization by a colony of cooperating agents. *IEEE Transactions on Systems, Man, and Cybernetics - Part B*, 26(1):1–13, 1996.
6. D. Estrin, R. Govindan, J. Heidemann, and S. Kumar. Next century challenges: Scale coordinations in sensor networks. In *Proc. ACM MobiComm*, pages 263–270, Seattle, WA, August 17-19 1999.
7. B. Karp and H. T. Kung. GPSR: Greedy perimeter stateless routing for wireless networks. In *Proc. 6th Int'l Conf. on Mobile Computing and Networks (ACM Mobicom)*, Boston, MA, 2000.
8. C. E. Perkins and E. M. Royer. Ad hoc on-demand distance vector routing. In *Proc. 2nd IEEE Workshop on Mobile Computing Systems and Applications*, pages 90–100, February 1999.
9. E. Royer and C. Toh. A review of current routing protocols for ad hoc mobile wireless networks. *IEEE Personal Communications*, April 1999.
10. R. Schoonderwoerd, O. Holland, J. Bruten, and L. Rothkrantz. Ant-based load balancing in telecommunications networks. *Adaptive Behavior*, 5(2):169 – 207, 1997.
11. G. Simon. Probabilistic wireless network simulator.
    http://www.isis.vanderbilt.edu/projects/nest/prowler/.
12. Y. Zhang and M. Fromherz. Message-initiated constraint-based routing for wireless ad-hoc sensor networks. In *Proc. IEEE Consumer Communication and Networking Conference*, 2004.

Integrating ACO and Constraint Propagation

Bernd Meyer[1,2] and Andreas Ernst[2]

[1] Monash University
[2] CSIRO, Clayton, Australia
bernd.meyer@acm.org, andreas.ernst@csiro.au

Abstract. Ant Colony Optimisation algorithms perform competitively with other meta-heuristics for many types of optimisation problems, but unfortunately their performance does not always degrade gracefully when the problem contains hard constraints. Many industrially relevant problems, such as fleet routing, rostering and timetabling, are typically subject to hard constraints. A complementary technique for solving combinatorial optimisation problems is Constraint Programming (CP). CP techniques are specialized for solving hard constraints, but they may be inefficient as an optimisation method if the feasible space is very large. A hybrid approach combining both techniques therefore holds the hope to combine these complementary advantages. The paper explores how such an integration can be achieved and presents a hybrid search method CPACS derived by embedding CP into ACS. We have tested CPACS on job scheduling problems. Initial benchmark results are encouraging and suggest that CPACS has the biggest advantage over the individual methods for problems of medium tightness, where the constraints cause a highly fragmented but still very large search space.

1 Introduction

While Ant Colony Optimisation (ACO) performs competitively to other meta-heuristics for many problems [8,10], its performance does not always degrade gracefully when hard constraints are present. As for other meta-heuristics, many different approaches for handling hard constraints in ACO have been tried. The basic types of methods that can be explored are the same as for other stochastic meta-heuristics and Evolutionary Algorithms: (1) relaxation and penalty-based techniques, (2) multi-phase methods, (3) repair techniques [7]. Relaxation techniques allow the search to use infeasible solutions (i.e. solutions violating hard constraints) and simply modify the objective function by a penalty based on the amount of constraint violation. Multi-phase methods attempt to split the search into phases that try to find feasible subspaces and phases that try to optimize feasible solutions (Section 5 will discuss these in more detail). Repair techniques attempt to explicitly modify infeasible solutions to make them feasible.

An alternative to these widely-used ways of handling hard constraints in constructive stochastic meta-heuristics is to incorporate some form of lookahead into the construction phase so that (almost) only feasible solutions are generated. A recent survey notes that as yet only very simple lookahead procedure have been

M. Dorigo et al. (Eds.): ANTS 2004, LNCS 3172, pp. 166–177, 2004.
© Springer-Verlag Berlin Heidelberg 2004

investigated with ACO [10]. Construction methods using lookahead, of course, need to be specific to the problem constraints. A generalized construction method based on a declarative mathematical model of the problem would be desirable. The hybridization of ACO with Constraint Programming (CP) gives us a way to achieve this goal. CP techniques are specialized for solving hard constraints, but on their own they are not always effective optimisation methods if the space of feasible solutions is very large. With some over-simplification, it could be said that meta-heuristics like ACO are a good approach for finding high quality solutions among an overwhelming number of feasible solutions, whereas CP is better where finding any feasible solutions is hard. A hybrid approach therefore holds the hope to combine their complementary advantages. Recent surveys have emphasized the importance of research into such hybridizations [4, 12].

The present paper explores a CP-ACO hybrid. To the best of our knowledge, such an integration has not been attempted before. We investigate the application of the hybrid algorithm to machine scheduling with sequence-dependent setup times, a problem that in our experience is difficult to solve with penalty-based and multi-phase ACO techniques.

Machine scheduling is an important application area that arises in a variety of contexts [13] and has been the subject of a significant amount of research by both the OR and AI communities. Formally this problem can be defined as follows. Let J be a set of n jobs to be scheduled on a single machine. Each job has a job identification $i \in \{1, \ldots, n\}$, a processing time $\overline{duration}_i$, a release time $\overline{release}_i$ and a due date $\overline{due}_i$. In addition between any pair of jobs i, $j \in J$ there is a changeover time $\overline{setup}_{i,j}$. A feasible schedule assigns each job j a start time $start_j$ and finish time end_j such that $start_j \geq \overline{release}_j$, $end_j = start_j + \overline{duration}_j \leq \overline{due}_j$ and for successive jobs i & j in the schedule $end_i + \overline{setup}_{i,j} \leq start_j$. The objective is to minimise the makespan $C_{\max} = \max_{j \in J} end_j$.

Previous work on job scheduling includes pure CP approaches [16] and an attempt by the same authors to include CP in a tabu-search approach [15]. Applications of ACO to job scheduling without hard constraints include [9, 3]. In practice due dates are however often hard requirements, dictated for example by shipment dates for exports. A good example of ACO handling hard constraints by a local search repair method in a different context (timetabling) is [17].

2   Constraint Programming

A brief introduction to CP may be in place. For a comprehensive introduction the interested reader is referred to [14]. The core idea of CP is to give a programmer support for maintaining and handling relations between (constraint) variables. A constraint solver allows variable values to be assigned or further constraints to be added ("posted" in CP terminology). There are many different forms of CP and the particular form used in this paper is finite domain constraint programming. This name refers to the fact that the variable domains (i.e. the set of possible values that the variable can take) have to be finite. We will only use integer variables.

```
Algorithm CP-basic
    setup domains for x_1, ..., x_n
    post intial constraints
    label([x_1, ..., x_n])
end.

procedure label(list xs)
    if xs=nil then return true
    else let x = first(xs) in
        if not bind(x) then return false
        else begin
            if label(rest(xs)) return true
            else begin
                unbind(x)
                return label(xs)
            end
        end
    end
end.

procedure bind(varname x)
    let d = domain(x) in
        if empty(d) then return false
        else begin
            v := first(d)
            success := post(x = v)
            if success return true
            else begin
                post(x ≠ v)
                return bind(x)
            end
        end
end.

procedure unbind(varname x)
    let v = current_value(x) in
        remove (x = v) from
            constraint store
        post(x ≠ v)
    end
end.
```

Fig. 1. Basic CP Search Algorithm

The constraint solver analyses the restrictions on variables automatically "behind the scenes" and provides at least two services to the program: (1) It analyses whether a new restriction is compatible with the already existing ones and signals this as success or failure when a new constraint is posted. (2) It automatically reduces the domains of constraint variables according to the explicit and implicit restrictions. The program can query the solver for the domain of a variable and obtain the set of values that have not been ruled out for this variable. An example will clarify this: assume that X is an integer variable and that the constraint $0 \leq X < 3$ has been posted. We have $domain(X) = \{0, 1, 2\}$. Later we post $X < Y$ and $Y \leq 2$, which results in $domain(X) = \{0, 1\}$. If we now post $Y = 0$, the solver will signal that this is not consistent with the previous ones as it does not leave any possible value for X. If, instead, we post $Y = 1$, the solver will automatically infer that the only remaining possible value for X is 0 and bind it to this value. It is important to note that $v \in domain(X)$ does not guarantee that there is a value assignment for all the remaining constraint variables that satisfies all the problem constraints if $X = v$. This is because constraint solvers necessarily have to be incomplete as they are trying to solve NP-hard problems.

To solve a problem with finite domain CP we have to find a mathematical model of the problem that captures its structure using finite domain variables. After setting up the decision variables and their domains and posting any further problem constraints the solver will reduce the domains of all decision variables as much as possible. The second phase of a constraint program is the so-called labeling phase, which is essentially a search through the remaining space of domain values (see Figure 1). In its simplest form, this search proceeds in individual labeling steps, each of which attempts to assign a particular decision variable a concrete value from its domain. A crucial question is, of course, in which order the variables are bound (the *variable ordering*) and in which or-

(1) $\forall i,j : \tau_{i,j} := \tau_0$ /* initialize pheromone matrix

(2) $\forall i,j : \eta_{i,j} := setup\_time(i,j)^{-1}$ /* initialize heuristic function

(3) $job_0 := 0$

(4) $\forall j : \tau_{0,j} := \tau_0 \wedge \eta_{0,j} := 1$ /* virtual start job 0

(5) $l_{gb} := +\infty;\ T^{gb} := nil;$ /* initialize global best

(6)   for $t := 1$ to max_iterations do

(7)   for $k := 1$ to number_of_ants do

(8) $T^k = nil$ /* intialize tour of ant k as empty

(9)   mark all jobs as unscheduled by ant k

(10)   for $n := 1$ to number_of_jobs

(11) $C :=$ set of jobs not yet scheduled by ant k

(12) $i := job_{n-1}$

(13)   if $random() > p$ then choose next job $j \in C$ to be scheduled by ant k with

$$probability\ p_j := \frac{\tau_{i,j}^{\alpha} \cdot \eta_{i,j}^{\beta}}{\sum_{j \in C} \tau_{i,j}^{\alpha} \cdot \eta_{i,j}^{\beta}}$$

(14)   else choose $j = argmax_{j \in C}(\tau_{i,j}^{\alpha} \cdot \eta_{i,j}^{\beta})$

(15) $T^k := append(T^k, (i,j))$

(16) $\tau_{i,j} := (1-\rho) \cdot \tau_{i,j} + \rho \cdot \tau_0$ /* local evaporation

(17)   mark job j as scheduled by ant k

(18) $job_n = j;$

(19) $start_j := max(release_j, end_i + setup\_time(i,j))$

(20)   end

(21)   if $feasible(T^k)$ then $l_k := length(T^k)$

(22)   else begin $l_k := \infty;\ T^k = nil$ end

(23)   end

(24) $ib := argmin_k(l_k)$ /* best tour index

(25)   if $l_{ib} < l_{gb}$ then begin $T^{gb} := T^{ib};\ l_{gb} := l_{ib}$ end

(26) $\forall(i,j) \in T^{gb} : \tau_{i,j} := (1-\rho)\tau_{i,j} + Q \cdot l_{gb}^{-1}$ /* evaporate and reward

(27)   end.

Fig. 2. Basic Ant Colony System (ACS) for JSP

der the corresponding domain values are tried (the *value ordering*). These two orderings significantly influence the efficiency of the search.

3 Integrating Constraint Propagation into ACO

A coupling of ACO and CP can be approached from two opposite directions: We can either take ACO or CP as the base algorithm and try to embed the respective other method into it. We will discuss both directions in turn and show that ultimately both approaches lead to the same method.

We use the basic Ant Colony System (ACS [11, 10]) as the point of departure. Consider ACS for Single Machine Job Scheduling as described in Figure 2. There are at least two possible interfaces through which ACS could interact with a constraint solver running in the background: (1) The candidate selection in line 11 and (2) the heuristic values $\eta_{i,j}$. Arguably the most straightforward form to integrate CP into ACO is to let it reduce the possible candidates by domain reduction of C. This effectively results in lookahead as to which candidate selections will not be able to be completed into a feasible solution. A more complex way to integrate CP would be to re-order the preferences for the candidate selection through the heuristic bias factor $\eta_{i,j}$. Additional information from the constraint solver, such as a bound on the objective, can be used

```
(0)     initialize solver; post initial constraints; s₀ := solver_state();
(1)     ∀i, j : τ_{i,j} := τ₀                      /* initialize pheromone matrix
(2)     ∀i, j : η_{i,j} := setup_time(i, j)^{-1}    /* initialize heuristic function
(3)     job₀ := 0
(4)     ∀j : τ_{0,j} := τ₀ ∧ η_{0,j} := 1          /* virtual start job 0
(5)     l_{gb} := +∞; T^{gb} := nil;     /* initialize global best
(6)     for t := 1 to max_iterations do
(7)         restore initial solver state s₀
(8)         for k := 1 to number_of_ants do
(9)             T^k = nil            /* intialize tour of ant k as empty
(10)            mark all jobs as unscheduled by ant k
(11)            n := 0; feasible := true;
(12)            while n < number_of_jobs and feasible do begin
(13)                n := n + 1
(14)                i := job^k_{n-1}
(15)                do
(16)                    C := domain(job^k_n)
(17)                    if random() > p then choose next job j ∈ C to be scheduled by ant k with
```

$$\text{probability } p_j := \frac{\tau^\alpha_{i,j} \cdot \eta^\beta_{i,j}}{\sum_{j \in C} \tau^\alpha_{i,j} \cdot \eta^\beta_{i,j}}$$

```
(18)                    else choose j = argmax_{j∈C}(τ^α_{i,j} · η^β_{i,j})
(19)                    feasible := post(job^k_n = j)∧
(20)                        post((start_j := max(release_j, end_i + setup_time(i, j))))
(21)                    if not(feasible) then post(job^k_n ≠ j)
(22)                until feasible ∨ C = ∅
(23)                if feasible then begin
(24)                    T^k := append(T^k, (i, j))
(25)                    τ_{i,j} := (1 − ρ) · τ_{i,j} + ρ · τ₀    /* local evaporation
(26)                    mark job j as scheduled by ant k
(27)                end
(28)            end
(29)            if feasible(T^k) then l_k := length(T^k)
(30)            else begin l_k := ∞; T^k = nil end
(31)        end
(32)        ib := argmin_k(l_k)        /* best tour index
(33)        if l_{ib} < l_{gb} then begin T^{gb} := T^{ib}; l_{gb} := l_{ib} end
(34)        ∀(i, j) ∈ T^{gb} : τ_{i,j} := (1 − ρ)τ_{i,j} + Q · l^{-1}_{gb}  /* evaporate and reward
(35)     end.
```

Fig. 3. Hybrid Constraint Propagation + Ant Colony System (CPACS) for JSP

to dynamically adjust $\eta_{i,j}$. However, a potential problem with this approach is that a dynamic η distorts the selection bias $\frac{\tau^\alpha \cdot \eta^\beta}{\Sigma \tau^\alpha \cdot \eta^\beta}$ as the relative weighting of competing pheromones changes. This can render the reinforcement learning ineffective. We will therefore focus on the first approach and modify the candidate list only. The basic idea is that C becomes a constraint variable and that the list of candidates is simply $domain(C)$. This requires a separate C^k_i for each decision step i and each ant k. When a candidate j is chosen, $C^k_i = j$ is posted to the solver, which can then propagate to reduce the domains of the remaining C^k_is.

However, depending on the model and the solver this propagation may not be strong enough. With the model and the propagation algorithms detailed below, the fact that a particular assignment $C^k_i = j$ cannot lead to a feasible solution is sometimes only discovered once $C^k_i = j$ is posted. To achieve a stronger propagation, we extend the coupling by using *single level backtracking*. The ant makes the usual probabilistic selection which job j to assign to C^k_i based on the reduced domain of C_i^k and posts the constraint $C_i^k = j$. If the solver signals failure the ant backtracks one step and posts $C_i^k \neq j$ instead. This removes j from the domain of C_i^k and potentially triggers other propagation steps. The ant then tries to bind C_i^k to another value from its further reduced domain. This process continues until either one of the assignments is accepted by the propagation solver or until $domain(C_i^k) = \emptyset$. In the first case the tour construction proceeds, while it fails in the second case. The complete algorithm for this coupling and the constraint model detailed in the next section is given in Figure 3. The generate&test step emulated by single level backtracking could to the same effect be embedded into the constraint solver instead of into CPACS.

A different approach would be to embed ACO within CP. The obvious point at which ACO can interact with CP is during the labeling phase (Figure 1). Two choices have to be made: the variable ordering and the value ordering. To simplify a first approach let us assume the variable ordering as fixed, say, lexicographically (we label job_i^k for each ant k in the order of increasing i). We use ACO to learn a value ordering that is more likely to produce good solutions.

Like in the CPACS model, pheromone values $\tau_{i,j}$ are used to learn a successor relation. Instead of using a fixed value ordering, we make a probabilistic decision based on $\tau_{i,j}$ using the same probabilistic choice function as in the CPACS algorithm. The complete Algorithm is given in Figure 4. Each ant independently attempts to construct a complete feasible solution. After all ants have completed their tours, the pheromones corresponding to the globally best solution are modified using the same global evaporation and reinforcement used in ACS. Local evaporation is implemented by reducing the corresponding pheromone value every time a variable is labelled. It is evident that this schema of integration is working in virtually the same way as CPACS. The only difference is that each ant uses full backtracking to construct a solution, i.e. in contrast to CPACS each ant will always construct a feasible solution (provided one exists).

4   The Constraint Model

The performance of the algorithm is to a large degree determined by the efficiency of the constraint propagation and therefore by how the task is modelled as a constraint problem. With ACO in mind it seems natural to have decision variables associated with sequence positions (the alternative of using successor variables is not explored here but used in e.g. [16]). We use variables job_n^k where $job_n^k = i$ if in the tour constructed by ant k the task with id number i is executed as the n-th job in sequence. We constrain:

$$\forall k, i : job_i^k \in \{1, \ldots, n\} \quad \text{and} \quad \forall k : all\_different(job_i^k)$$

Three further constraint variables per task model the temporal aspects: $start_n^k$ is the scheduled start time, end_n^k the end time, and $setup_n^k$ the setup time for the n-th task in the sequence of ant k.

For technical reasons (it is not straight forward to handle indexed data arrays in CP) we also use a set of auxiliary variables: $duration_n^k, release_n^k, due_n^k, setup_n^k$

```
Algorithm CP-with-ACO
    for each ant begin
        setup domains for x₁,...,xₙ
        post intial constraints
        if label([x₁,...,xₙ], nil)
        then update global best solution
    end
    evaporate and reward globally best solution
end.

procedure label(list xs, value last)
    if xs=nil then return true
    else let x = first(xs) in
        if not bind(x, last)  then return false
        else begin
            if label(rest(xs), value(x)) return true
            else begin
                unbind(x)
                return label(xs, last)
            end
        end
    end
end
end.
```

```
procedure bind(varname x, value last)
    let d = fd_domain_list(x) in
    if empty(d) then return false
    else begin
        choose v ∈ d probabilistically
        success := post(x = v)
        evaporate locally τₗₐₛₜ,ᵥ
        if success return true
        else begin
            post(x ≠ v)
            return bind(x, last)
        end
    end
end.
```

Fig. 4. CP Search with ACO labeling

to capture the respective times for the n-th task in the sequence of ant k. (Note that this is different from $\overline{duration}_j$ etc. which give the respective times for the task with id j). These are coupled to the data and each other via:

$$\forall k, n : end_n^k = start_n^k + duration_n^k$$

$$\forall k, n : start_n^k \geq release_n^k \wedge end_n^k \leq due_n^k$$

$$\forall n > 1, k : start_n^k \geq end_{n-1}^k + setup_n^k$$

$$\forall n > 1, k : setup_n \in \{\overline{setup}_{i,j} \mid i,j \in \{1\ldots n\}\}$$

$$\forall k, n : duration_n^k \in \{\overline{duration}_j \mid j = 1\ldots n\}$$

$$\forall k, n : release_n^k \in \{\overline{release}_j \mid j = 1\ldots n\}$$

$$\forall k, n : due_n^k \in \{\overline{due}_j \mid j = 1\ldots n\}$$

As we always assign the earliest possible start time to a scheduled task, it is clear that a valuation of job_n^k together with the data completely determines all other variables. Evidently this CP model naturally corresponds to a typical ACO approach if we label job_n^k in the order of increasing n.

We use reified constraints [14] to bind the auxiliary variables to the data as soon as the id of the n-th job in sequence is known. Note that these can only propagate once the pre-condition is fulfilled:

$$\forall i > 1, l, m : job_{i-1}^k = l \wedge job_i^k = m \Rightarrow setup_i = \overline{setup}_{l,m}.$$

$$\forall k, n, j : job_n^k = j \Rightarrow duration_n^k = \overline{duration}_j \wedge release_n^k = \overline{release}_j \wedge$$
$$due_n^k = \overline{due}_j \wedge start_n^k = min(\overline{release}_j, end_{n-1}^k + setup_n).$$

$$\forall i, j, k : end_i^k > \overline{due}_j \Rightarrow job_i^k \neq j.$$

Powerful high-level scheduling constraints with specialized propagation mechanisms are available in typical CP systems like Sicstus Prolog CLP(FD) [6] which we use to implement our algorithms. To make use of such constraints we need an additional second model, as we must capture the problem using a different set of decision variables. We use $\widetilde{start}_i^k$ and $\widetilde{end}_i^k$ to represent the start and end times of the task with id i for ant k. This is different from the decision variables $start_n^k$ and end_n^k used above which specify the start and end time for the n-th task in the k-th ant's sequence. These variables obviously introduces some redundancy and increase the problem size. The motivation for using them is that the gain in propagation effectiveness makes up for increased model size and improves the overall efficiency. We use a built-in global constraint *serialized* [6], which implements scheduling constraints directly and from our perspective has to be treated as a black-box with $\widetilde{start}_i^k$ and $\widetilde{end}_i^k$ as its interface. It is initialized with a complete problem specification containing release dates, due dates, durations and setup times. Finally, the two models have to be coupled for cross-propagation:

$$\forall k, n, j : job_n^k = j \Rightarrow start_n^k = \widetilde{start}_j^k \wedge end_n^k = \widetilde{end}_j^k$$

$$\forall k, n, j : \widetilde{start}_j^k < start_n^k \vee \widetilde{start}_j^k > start_n^k \Rightarrow job_n^k \neq j$$

5 Benchmarking

We have tested CPACS for various problem sizes and within those groups with problems of varying tightness (Table 1). The test data sets were drawn from two different sources. In all cases the initial letters indicate the problem type. This is followed by the number of jobs and further letters or numbers identifying the particular data set. The first source of data sets is based on one of the authors' experience with an application in wine bottling in the Australian wine industry. Setup times are taken from actual change-over times at a bottling plant. There are no release times as the wine is always available in the tanks. Due dates are selected randomly from one of seven possible spreads over the planning period (representing for example the end of each day over a week). Processing times are generated in a slightly more complicated manner to create problems with various degrees of tightness. Let μ be the average processing time available per job (calculated by removing n times the average setup time from the last due date). The duration for each job is chosen randomly in the interval $[\frac{1}{2\beta}\mu, \frac{3}{2\beta}\mu]$ for $\beta = 1.2, 2$ or 3 for problem class 1, 2 or 3. Hence problem 'w8.1' is the tightest problem involving eight jobs, while 'w8.3' is the loosest problem.

The remaining data sets are taken from the ATSP-TW literature [1] based on an application involving scheduling a stacker crane in a warehouse. In these data sets only some of the jobs have time window constraints (both release and due dates). The number of jobs with non-trivial time windows is indicated by the last set of digits in the problem name. Thus in RBG27.a.27 each of the 27 jobs has a time window, while RBG27.a.3 is identical except that the time window constraints for all but three of the jobs has been removed.

Table 1 shows averages and best over 10 runs, ACS/CPACS with 10 ants, $Q, \alpha, \beta = 1, \rho = 0.05, \tau_0 = Q/(20 \cdot N \cdot \rho \cdot \hat{l})$, where N is the number of jobs

Table 1. Benchmark Results

Problem	CP			ACS				CPACS			
	Best	Step	Finished	Best	Step	Avg	Failed	Best	Step	Avg	Failed
W8.1	8321	870	13,355	8321	93,151	8321	0 %	8321	8,040	8321	0 %
W8.2	5818	66,667	15,707	5818	32,894	5818	0 %	5818	9,153	5818	0 %
W8.3	4245	36,237	66,975	4245	13,037	4245	0 %	4245	1,899	4245	0 %
W20.1	8779	15,239	n/a	**8564**	474,080	8611	0 %	**8564**	95,386	**8604**	0 %
W20.2	5747	16,110	n/a	5102	454,360	5136	0 %	**5062**	38,361	**5117**	0 %
W20.3	4842	433,794	n/a	4372	449,054	4408	0 %	**4352**	46,693	**4385**	0 %
W30.1	8817	201,693	n/a	8232	1,094,843	8310	60 %	**8142**	122,367	**8182**	80 %
W30.2	5457	999,877	n/a	4787	892,083	4865	0 %	**4695**	94,486	**4743**	0 %
W30.3	5129	547,046	n/a	**4283**	1,175,283	4341	0 %	4288	125,814	**4337**	0 %
RBG10.a	3840	10	9,191	3840	958	3840	10 %	3840	13	3840	0 %
RBG16.a	2596	16	64,748	2596	198,270	2596	40 %	2596	1,134	2596	0 %
RBG16.b	2120	8,807	n/a	**2094**	316,657	2115	30 %	**2094**	87,353	**2094**	0 %
RBG21.9	6524	346,879	n/a	4546	721,818	4565	0 %	**4481**	677,044	**4489**	0 %
RBG27.a.3	1984	147,616	n/a	1715	397,299	1733	0 %	**927**	205,117	**940**	0 %
RBG27.a.15	1569	16,878	n/a	1507	482,355	n/a	90 %	**1068**	32,895	**1068**	0 %
RBG27.a.27	**1076**	27	113	n/a	n/a	n/a	100 %	**1076**	77	**1076**	0 %
BR17.a.3	1663	552,611	n/a	1527	393,808	1549	0 %	**1003**	40,072	**1003**	0 %
BR17.a.10	**1031**	650,120	n/a	1400	281,870	1431	30 %	**1031**	18,995	**1031**	0 %
BR17.a.17	1057	17	168	1057	10,759	1057	70 %	1057	35	1057	0 %

and $\hat{l}$ is an estimate on the average makespan, i.e. the sum of processing times plus $(N-1)$ times the average setup time. Labeling steps count the number of attempts to bind one of the decision variables to a value (i.e. execution of line 18 in ACS, line 19 in CPACS and calls to *bind* in CP). The runtime has been limited to 1,000,000 labeling steps for CP and to 5000 iterations for ACS and 500 iterations for CPACS. These low numbers of iterations are justified, because they result in a comparable number of labeling steps. The column "Steps" indicates the average number of labeling steps until the best solution was found. For CP "Finished" gives the number of labeling steps after which the complete CP search has exhausted the search space (if less than 1,000,000). "Failed" gives the percentage of runs for ACS and CPACS that did not produce any feasible solutions. Note that we use a multi-phase version of ACS. The reason is that the basic ACS failed to find feasible solutions for most of the medium to highly constrained problems for sizes $n \geq 16$ when reinforcement was purely based on the objective measure (makespan) and shortest setup time (SST) was used to define the heuristics $\eta_{i,j}$. SST is effective for minimizing the makespan, but is not a good guide for finding feasible solutions. Here, earliest due date (EDD) is much more effective. Therefore, the ACS version used here works in two phases. EDD ($\eta_{i,j} = due_j^{-1}$) is used and in the first phase the reward is inverse proportional to the degree of constraint violation (tardiness: $\Sigma_i max(0, end_i - due_i)$). This forces the search to first construct feasible solutions regardless of their quality. Once feasible solutions are found, we switch to the second phase which only reinforces feasible solutions with an amount of reinforcement proportional to the inverse of the makespan. Like CPACS, this ACS version uses single level backtracking.

We note that ACS and CPACS might perform better with other pheromone models [5]. We would, however, expect their relative performance to remain stable as such a change should impact on both algorithms in similar ways.

CP uses the same constraint model as CPACS and labels the decision variables job_i. To achieve a fairer comparison CP uses a variable ordering corresponding to increasing $\eta_{i,j}$ instead of an arbitrary ordering. Also, each time an improving solution with makespan m is found, CP posts a bound on the objective ($end_N < m$) as an additional constraint to further prune the search.

For most problem instances CPACS produces the best results. An exception are very small problem instances (w8) where all three algorithms perform equally. ACS produce a marginally better optimum in one problem instance (w30.3), but CPACS performs more reliable for this instance as can been seen from the average. CPACS produces the best average in all cases and generally needs less labeling steps to produce these results than ACS. This is not entirely surprising as the solver does additional work behind the scenes. For more tightly constrained problems CP needs less steps than CPACS, but in cases where the feasible space is larger CPACS tends to need fewer labeling steps. For medium constrained problems CP typically finds good solutions quickly, but often does not reach the optimum, because the number of solutions that have to be tested is exceedingly large. ACS often finds better solutions if the number of feasible solutions is large, but fails to find good solutions (or even fails to find any feasible solutions altogether) if the search space is highly fragmented. CPACS can exploit the combined advantages and converges more quickly towards good solutions. Exceptions to this comparison are only found for very small problems, which all three algorithms solve quickly to optimality, and very tight problems, where CP and CPACS both find the optimum solution after few labeling steps, whereas ACS struggles to find any feasible solution.

These very positive initial results should be taken with a grain of salt. Most importantly, CPACS and ACS are only heuristics and cannot guarantee an optimal solution whereas CP is a *complete* method and will eventually prove optimality (though it may take an exceedingly large time). Hence if an exact optimal solution is required CP is the right choice. However, we might still be interested in generating good solutions earlier during a complete search (for example, to produce a tight bound on the objective early). In this case CPACS-like integration of learning a good labeling order may still be useful.

The CP model used is not optimal and CP may be able to achieve a somewhat better performance with a different model. However, this should not change the our conclusions. Effectively, a model enabling better propagation would necessarily also improve the performance of CPAS. Therefore, the only change expected with a better model is that the performance comparison for a given problem shifts slightly in the direction of a more tightly constrained problem.

6   Conclusions and Future Directions

Clearly, this study is only a first step in the direction of hybridizing ACO algorithms with methods that are particularly targeted for solving hard-constrained problems. Apart from CP, typical OR techniques, such as integer programming [1], come to mind. However, even just for the combination of ACO and CP many interesting questions remain to be studied.

It appears that the ACO+CP hybrid delivers the greatest benefit for problems of intermediate tightness, where the space of feasible solutions is too large for complete search but already sufficiently fragmented to cause difficulties for ACO. We intend to investigate the correlation between problem tightness and relative performance more closely in subsequent studies.

A note that has to be made is that we have not given a comparison in terms of runtime. This study has only compared the performance of the algorithms in regards to the number of labeling steps. However, the cost of a labeling step can vary greatly. Obviously, in ACS a labeling step is simply a value assignment and therefore cheap. In CP much more is involved in a single labeling step, as it triggers propagation, i.e. a potentially complex deduction process in the background. The same holds for CPACS. Therefore a metric based on labeling steps does not give us a good feel for the CPU time tradeoff and this will have to be studied explicitly based on optimized implementations in subsequent studies. Such studies would also allow us to better investigate the impact of the fact that CP solvers are often optimized for backtracking via trailing [14], a benefit of which the CPACS integration described here cannot make effective use.

Probably the most challenging question is how the different approaches to modelling the domain of a variable in CP and in ACO can be unified. ACO requires an individual choice probability for each domain value, which can be problematic with very large domains. In contrast, a typical finite domain solver handles domains via unions of regions. For example, if we have $x \in \{1...10\} \land x \neq 5$, the domain will be represented as $domain(x) = \{1...4\} \lor \{6...10\}$. The difficulties in using union domains in an ACO+CP hybrid are obvious, but the prospect of being able to handle larger domains more effectively makes this a worthwhile area of study.

Here we have only investigated the learning of a value ordering. In future studies we are planning to address the integration with variable ordering, which could potentially also deliver substantial performance improvements.

Many, if not most, real-world applications are subject to hard constraints. This makes the integration of hard-constraint handling with meta-heuristics an important area of research. While more traditional methods, such as multiphase or relaxation techniques should not be forgotten, the hybridization of meta-heuristics with techniques for hard constraints, such as CP, is an important alternative branch of research. CP integrations with other meta heuristics, such as GA, have been explored before [2]. However, ACO holds a special position as it allows a comparatively straight-forward integration with CP. This initial study demonstrates the potential of such an integration. We hope that the benefits of this schema might show even more clearly in the context of complex real-world applications that are notoriously hard to solve for either approach on its own.

References

1. N. Ascheuer, M. Fischetti, and M. Grötschel. Solving the asymmetric travelling salesman problem with time windows by branch-and-cut. *Mathematical Programming*, 90(3):475–506, 2001.

2. N. Barnier and P. Brisset. Combine & conquer: Genetic algorithm and CP for optimization. In *Principles and Practice of Constraint Programming*, Pisa, October 1998.
3. A. Bauer, B. Bullnheimer, R.F. Hartl, and C. Strauss. An ant colony optimization approach for the single machine total tardiness problem. In *Proceedings of the Congress on Evolutionary Computation*, Washington/DC, July 1999.
4. C. Blum and A. Roli. Metaheuristics in combinatorial optimization: Overview and conceptual comparison. *ACM Computing Surveys*, 35(3):268–308, 2003.
5. C. Blum and M. Sampels. When model bias is stronger than selection pressure. In *Parallel Problem Solving From Nature (PPSN-VII)*, Granada, September 2002.
6. M. Carlsson, G. Ottosson, and B. Carlson. An open-ended finite domain constraint solver. In *Proc. PLILP'97 Programming Languages: Implementations, Logics, and Programs*, Southampton, September 1997.
7. C.A. Coello. Theoretical and numerical constraint-handling techniques used with evolutionary algorithms: a survey of the state of the art. *Computer Methods in Applied Mechanics and Engineering*, 191(11-12):1245–1287, 2002.
8. O. Cordon, F. Herrera, and T. Stützle. A review on the ant colony optimization metaheuristic: Basis, models and new trends. *Mathware and Soft Computing*, 9(2–3):141—175, 2002.
9. M. den Besten, T. Stützle, and M. Dorigo. Ant colony optimization for the total weighted tardiness problem. In *Parallel Problem Solving from Nature - PPSN VI*, Paris, France, September 2000.
10. M. Dorigo, G. Di Caro, and L.M. Gambardella. Ant algorithms for discrete optimization. *Artificial Life*, 5:137–172, 1999.
11. M. Dorigo and L.M. Gambardella. Ant colony system: A cooperative learning approach to the traveling salesman problem. *IEEE Transactions on Evolutionary Computation*, 1(1):53–66, 1997.
12. F. Focacci, F. Laburthe, and A. Lodi. Local search and constraint programming. In F. Glover and G. Kochenberger, editors, *Handbook of metaheuristics*. Kluwer, Boston/MA, 2003.
13. E.L. Lawler, J.K. Lenstra, A.H.G Rinnooy Kan, and D.B. Shmoys. Sequencing and scheduling: algorithms and complexity. In S.C. Graves, A.H.G. Rinnooy Kan, and P.H. Zipkin, editors, *Logistics of Production and Inventory*, pages 445–522. North Holland, Amsterdam, Netherlands, 1993.
14. K. Marriott and P. Stuckey. *Programming With Constraints*. MIT Press, Cambridge/MA, 1998.
15. G. Pesant and M. Gendreau. A constraint programming framework for local search methods. *Journal of Heuristics*, 5(3):255–279, 1999.
16. G. Pesant, M. Gendreau, J.-Y. Potvinand, and J.-M. Rousseau. An exact constraint logic programming algorithm for the traveling salesman problem with time windows. *Transportation Science*, 32(1):12–29, 1998.
17. K. Socha. MAX-MIN ant system for international timetabling competition. Technical report, Universite Libre de Bruxelles, September 2003. TR/IRIDIA/2003-30.

Logistic Constraints on 3D Termite Construction

Dan Ladley and Seth Bullock

School of Computing, University of Leeds, UK
{danl,seth}@comp.leeds.ac.uk

Abstract. The building behaviour of termites has previously been modelled mathematically in two dimensions. However, physical and logistic constraints were not taken into account in these models. Here, we develop and test a three-dimensional agent-based model of this process that places realistic constraints on the diffusion of pheromones, the movement of termites, and the integrity of the architecture that they construct. The following scenarios are modelled: the use of a pheromone template in the construction of a simple royal chamber, the effect of wind on this process, and the construction of covered pathways. We consider the role of the third dimension and the effect of logistic constraints on termite behaviour and, reciprocally, the structures that they create. For instance, when agents find it difficult to reach some elevated or exterior areas of the growing structure, building proceeds at a reduced rate in these areas, ultimately influencing the range of termite-buildable architectures.

1   Introduction

1.1   Termites

Termites create some of the most impressive structures seen in nature, building features such as air conditioning, fungus farms and royal chambers into their huge mounds. Recent research has started to explain how these complicated architectures are constructed, but we have yet to fully understand how communities of simple organisms can collaborate successfully on such grand constructions. What seems clear is that the simple termite cannot be relying on centralised control in the form of either a genetic blueprint or guidance from a single executive such as the termite queen. Rather, we now believe that much of the construction process is co-ordinated through *stigmergy*, a form of indirect communication via environmental cues that are typically produced as a side-effect of the very activity that requires co-ordination [8].

In the case of termite construction work, as building material is assembled, the local environment of the builders is altered in such a way as to encourage appropriate behaviours. Partially built structures "communicate" with workers in such a way as to facilitate their own completion. Over time, as building work continues, the "messages" conveyed by the environment change in subtle ways that allow complicated, heterogeneous architectures to be constructed.

In addition to the physical presence or absence of building material, the environment influences termite building behaviour through supporting various

M. Dorigo et al. (Eds.): ANTS 2004, LNCS 3172, pp. 178–189, 2004.
© Springer-Verlag Berlin Heidelberg 2004

diffusive processes. For instance, recently deposited material is a source of a particular kind of "cement" pheromone. This pheromone attracts termites, and thereby encourages the deposition of more building material at or near its source. The consequent positive feedback concentrates building effort in hot-spots, efficiently amplifying what are initially randomly-placed pieces of building material into well-spaced pillars, and subsequently walled enclosures, which may eventually be covered to form chambers. A second pheromone exuded by the queen termite encourages building activity when encountered at a particular concentration level. The pheromone gradient "template" created by a stationary queen encourages material to be deposited at a characteristic distance from her location, eventually leading to the construction of a "royal chamber" found at the centre of many termite mounds. Finally, a third "trail" pheromone, deposited by moving termites, also guides building activity, co-ordinating the formation of galleries and covered walkways. Even wind may affect the structures built by termites, who are thought to be sensitive to air currents and able to make decisions based on direct interaction with the wind. In addition, any wind will disturb pheromone diffusion and influence the structures being built as a result [6, 3].

While developing and exploring models of stigmergic, decentralised construction has direct application in insect biology, there also exists the possibility that improving our understanding of these processes might lead to powerful new engineering and design methodologies. Self-assembling robots, automatic repair of space craft or nuclear reactors, and the construction of nanoscale structures are all examples of engineering challenges that might benefit from or even require a termite-inspired approach due to the extreme scales at which the activity must take place, the inaccessibility of the building site, or simply the savings in time and money that could potentially be achieved.

1.2   Previous Models

Aspects of the termite mound-construction process have been modelled quite extensively. Deneubourg's mathematical model of pillar formation [4] explained the regular spacing observed in nature as resulting from a positive feedback cycle involving the cement pheromone emitted from recently placed building material. A set of differential equations tracked the movement of termites, diffusion of cement pheromone and location of "active" (recently placed) building material over a two dimensional world. From an initially random distribution of active material "pillars" could form through (i) the amplification of any initial variability and (ii) nearby pillars "competing" for termite attention.

This model was later expanded by Bonabeau et al. [3] to include equations representing factors such as wind, a pheromone template emitted by a stationary queen, and an artificially imposed net flow of ants across the world. The queen pheromone template allowed the structure of royal chambers to be modelled. A net flow of termites led to the construction of walkways and, where these walkways intersected, the formation of what was interpreted as a small chamber.

Strong winds could prevent any structure from forming, whereas light winds could explain the formation of galleries.

In addition to establishing that complicated structures reminiscent of natural termite mounds can result from very simple rules, these models provide a strong demonstration of the importance of environmental processes in the formation of such mounds. They show that spatio-temporal pheromone properties critically influence construction, and that additional influences such as wind or a net flow of termites can affect building behaviour in interesting ways.

However, these models, and others like them, are limited in certain important respects. First, since they do not explicitly represent the third spatial dimension, they cannot directly represent hollow termite-built structures such as arches, domes, and tunnels. Moreover, they neglect any climbing that termite structures might demand. More importantly, such treatments do not model the *logistic constraints* on termite movement and pheromone diffusion imposed by the physicality of building material.

In both models described above, the presence or absence of built material has no direct effect on either termite movement or pheromone diffusion. This is important, as it prevents the physical consequences of termite construction behaviour from directly impacting on subsequent termite behaviour, i.e., an important source of stigmergic effects is neglected. For instance, termites may continue to place building material in locations that are physically inaccessible to them, while pheromone may diffuse through solid structures influencing termites that in reality would be ignorant of the pheromone source. Here we label the constraints that these physical realities place on termite behaviour "logistic" since they concern the ability of termites to travel from one particular place to another. While some models of decentralised insect construction have incorporated these constraints, they have not included the role of pheromone diffusion [2, 10, 1]. In the next section, inspired by a combination of the models mentioned above, we describe a 3-d agent-based model developed in order to explore the impact of logistic constraints on termite construction. Subsequently, we present results from simulations of royal-chamber construction, walkway formation, and the effects of wind. We conclude with a discussion of the role of logistic constraints in decentralised termite construction.

2   Method

The simulation comprised a three-dimensional, rectangular lattice ($100 \times 100 \times 100$) with each location containing diffusing pheromones, and either a simple virtual termite, or solid building material, and, sometimes, wind. The model was updated synchronously in fixed discrete time steps. The lattice is initialised as empty of pheromone and building material and open in all directions, save that the lowest horizontal cross-section contains a single layer of inert building material that represents the ground.

2.1   Pheromone Behaviour

The distribution of each type of pheromone changes over time due to the action of three processes: emission, diffusion, and evaporation.

Emission. Three kinds of pheromone originate from three different kinds of point source. The queen produces a fixed volume of *queen pheromone* at each time step. Likewise, a trail-following termite produces a constant volume of *trail pheromone* at each time step. By contrast, the amount of pheromone emitted by each piece of building material is neither constant over time nor unlimited. Rather, each newly placed piece of building material contains an initial finite volume of *cement pheromone*, a proportion, $0 < r \leq 1$, of which is lost to the atmosphere at each time step.

Diffusion. Diffusion between lattice locations that share a common face was modelled as proportional to the pheromone gradient between them using a standard finite volume approach [9]. If the volume of pheromone at a particular location is x and the volume at one of its diffusion neighbours is y then the change of pheromone may be expressed as $\frac{\partial x}{\partial t} = -\alpha(x - y)$, where $0 \leq \alpha \leq \frac{1}{6}$ in order to prevent the creation of pheromone during diffusion. In all results reported here $\alpha = \frac{1}{7}$, which ensures that, at minimum, a pheromone concentration of $\frac{x}{7}$ remains after diffusion, while a maximum of $\frac{x}{7}$ can diffuse to any one diffusion neighbour.

Physical constraints on pheromones are modelled by returning any pheromone that diffuses into a location occupied by building material to its source location. Note that this scheme ensures that a piece of building material that neighbours other pieces on many sides will emit pheromone at a slower rate than a lone piece of building material. Boundary conditions assume that locations beyond the 3-d grid are always empty of pheromone.

Evaporation. At each time step, evaporation was modelled at a rate proportional to the concentration of pheromone at each location by multiplying each concentration by an evaporation constant, $0 < v < 1$.

2.2 Termites

The world contains a fixed number, n, of builder termites that remains constant over the course of a simulation. A termite can sense the levels of pheromones at its current location and the pheromone gradients in each direction.

Movement. Termite movement is physically constrained. Of the twenty-six possible adjacent locations available to a termite, it may only move to one that is unoccupied by building material. Additionally, termites may only move to locations that neighbour locations occupied by building material. The first constraint prevents termites moving through walls, while the second constraint forces the termite to move across surfaces, preventing it from flying around the world. If a termite leaves the lattice, which it may do freely, or cannot move to any adjacent location, it is discarded and a new replacement termite is introduced to the lattice.

Builder termites prefer to follow cement pheromone gradients, choosing a new location from the legal alternatives using a roulette-wheel constructed to reflect relative gradient strengths. However, with probability inversely proportional to the gradient strength of the selected direction, a random legal move will be made instead. This ensures that, on average, weak gradients exert less influence on termites than strong gradients.

A termite moves m times in this fashion every time step. Since termites move relatively fast by comparison with pheromone diffusion, for all results reported here, $m = 5$. Termites are not prevented from entering locations already occupied by other termites. In reality termites are much larger than the pieces of material that they deposit, however, in order to make our model tractable it was necessary to represent termites as single locations. This location was considered to be the head of the termite, i.e. the place it would perform its building and sensing activities. It seems plausible that two termites could be collecting sensory information from the same location at the same time.

Block Placement. For a piece of building material to be placed, the level of queen/trail pheromone at the site must lie within a predefined range ($[0.1, 0.5]$ for all results presented here) and the site must meet *at least one* of the following three conditions, which are intended to impose a crude physics on the world:

1. Either the location immediately underneath or immediately above the site must contain material.
2. The site must share a face with a horizontally adjacent location that contains material and satisfies (1)
3. One face of the site must neighbour three horizontally adjacent locations that each contain material.

The first constraint allows vertical stacks to be built. The second allows these stacks to be extended horizontally to a limited degree, while the third allows the gradual construction of elevated horizontal surfaces if sufficient support is present.

If these conditions are met, a block will be placed at the termites current location with probability, p, the termite will be removed from the lattice, and a replacement introduced. This is intended to represent the constraint that termites must forage for building material. Notice that once placed, building material cannot be removed or relocated.

Path-Following. In addition to the queen, and builder-termites, a third class was modelled. These termites lay and follow trail pheromone, but are not involved in any building activity. They represent termites who are leaving the nest to forage or returning to the nest having foraged. The model does not allow for foraging-termites or building-termites to change role during the simulation. A group of adjacent, ground-level entry points are specified at the edge of the lattice. At each time step, and with probability, c, each of a fixed number of termites may enter the lattice at a randomly chosen entry point. Each trail termite moves across the lattice in a direction roughly perpendicular to the lattice edge at which they enter. However, trail termites are not restricted to move in a straight line. We extend an approach employed by Deneubourg [5] to three dimensions in allowing a trail termite to move to any down-stream location that shares a face, edge or corner with its current location and is physically accessible. Trail termites are attracted to trail pheromone in the same way that builder termites are attracted to cement pheromone. If a termite moves off the lattice it is removed.

2.3 Wind

Wind was modelled in a deliberately simple manner, originating at a constant fixed strength, $0 < s < 1$, from one vertical face of the lattice, travelling perpendicular to this face, horizontally across the world, flowing around built structures, and influencing pheromone diffusion. In general, therefore, wind flows from one up-wind vertical cross-section of lattice, a_i, to the adjacent, down-wind, vertical cross-section, a_{i+1}. Wind strength in a_{i+1} can be calculated in the following manner. For every location, u in a_i, calculate the number of locations in a_{i+1} that share an edge, corner or face with u and are empty of building material. Divide the strength of wind at u equally among these locations in a_{i+1}, or (unrealistically) discard the value should no down-stream locations be empty of building material.

In areas lacking building material, this approach will result in wind strength remaining constant as the strength at each location will be distributed evenly across down-wind locations. In areas with obstructions wind strength will increase near exposed surfaces, but be reduced in "sheltered" areas, as the wind flows around the obstruction.

At each location, wind transfers a proportion of pheromone equal to its strength, s, to the location immediately down-wind (unless that location is occupied by building material). While wind can thus only carry pheromone in one direction, in combination with diffusion, it is capable of transporting pheromone around structures effectively.

3 Results

Simulations were carried out to examine the effects of (i) a pheromone gradient established by a stationary queen, (ii) wind, (iii) a net flow of termites in one direction, and (iv) a net flow of termites in two orthogonal directions.

3.1 Royal Chamber

The queen, represented by one or more contiguous blocks in the form of a rough half-cylinder, is placed at the centre of the world in contact with the ground. Each block is a fixed-rate source of queen pheromone. Throughout the simulation the queen remains stationary and is somewhat equivalent to building material in that she obstructs the flow of pheromone, and impedes movement of termites, allowing them to climb over her though not to build on her.

Fig. 1. A royal chamber being constructed. Parameters: $r = 0.5, \alpha = \frac{1}{7}, v = 0.1, p = 0.1, n = 300, m = 5, s = 0.0$

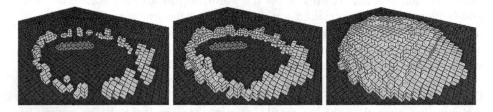

Fig. 2. A royal chamber being constructed under mildly windy conditions (wind emanates from the upper-left lattice edge). Parameters as figure 1, except: $s = 0.15$

The simulation is initially run for 1000 time steps without the presence of builder termites to allow the queen's pheromone template to become established. At this point termites are randomly allocated around the outside of the world in contact with the ground, and the model proceeds as described above. Unless otherwise noted the simulations were run with 300 building termites (n), a probability of block placement (p) of 0.1, a pheromone evaporation rate (v) of 0.1 and a pheromone output rate (r) of 0.5.

Figures 1 and 2 depict the construction of a royal chamber without wind, and under mildly windy conditions, respectively. First, pillar-like structures are formed at roughly regular spatial intervals, and at a specific distance from the queen. Subsequently, these pillars merge to form a wall that encircles the queen. Finally, the termites achieve a complete dome encompassing the pheromone source. In the absence of any disturbance, termites achieve a roughly hemispherical structure centred on the queen, that may echo the queen's physical shape if she is large enough in relation to the dome (see figure 1). However, in figure 2 the shape of the chamber has been influenced by wind blowing from the upper-left edge of the lattice. As a result, the chamber is not centred on the queen, and is distorted in both it's horizontal and vertical profile, e.g., exhibiting a steeply rising exposed face and a more gradually descending, sheltered, down-stream slope. At higher strengths, wind can prevent a chamber from being completed successfully, or even prevent any construction from becoming established at all.

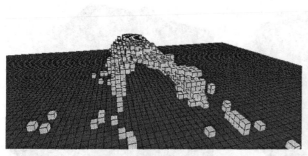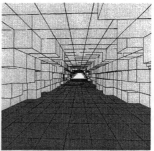

Fig. 3. A covered walkway is constructed. Parameters as figure 1, save that no wind has been modelled, and a flow of trail termites has been introduced: $t = 10, c = 0.5$. The tunnel's interior is clear of obstructions, and the cross-section is quite regular

3.2 Covered Walkway

A narrow flow of termites across the world is introduced by allowing trail termites to enter at ground level from central locations along one lattice edge. Builder termites enter the lattice at the same locations after 1000 time steps. Figure 3 shows the resulting tunnel, partially formed, and an internal view. Construction of the tunnel walls obviously precedes completion of the roof. At the end of the simulation the tunnel is straight, quite regular in cross-section, and clear of obstructions. Notice that, unlike previous models [3] the trail and distribution of trail pheromone are not explicitly defined, but arise from the ongoing activity of trail termites.

3.3 Crossing Paths

Two perpendicular flows of termites are introduced, crossing at the centre of the ground plane. Figure 4 depicts the cross-road structures that arise. Despite the potential for interference between trail pheromones from each path, and between the building work at each tunnel (especially given the stochastic rate at which both trails and tunnels are established), each tunnel remains unobstructed internally, and a working covered cross-road junction is achieved.

Figure 4 *left* depicts a scenario in which builder termites always enter the world from a random location along the ground-level edge of the lattice. The covered walkways that result reliably exhibit a tendency to "mushroom" at their down-stream ends. Constraining the building termites to enter at the same locations employed by the trail termites extinguishes this tendency (figure 4 *right*). In both cases, however, the interior of the tunnel remains clear of obstructions.

4 Discussion

The primary result exhibited by the model is that, pleasingly, the introduction of physical limitations on pheromone diffusion and termite movement have not

Fig. 4. Two examples of cross-road formation. Parameters again as figure 1, save that no wind is modelled, and two perpendicular streams of trail termites are introduced: $t_1 = 10, t_2 = 10, c = 0.5$. At each time step, between zero and 10 builder termites enter the lattice, with probability 0.5 per termite. *Left* – Builder termites enter from random locations along any edge of the ground plane. Some "mushrooming" of the down-stream sections of tunnel tends to occur. *Right* – Builders enter from the same entry points employed by trail termites. Mushrooming is suppressed

prevented the building of structures suggested by previous models that neglected such logistic constraints [4, 3].

As reported by Deneubourg [4], pillar formation can be driven by simple positive feedback involving a cement pheromone. His results from one- and two-dimensional models built on differential equations can be supported by our 3-d agent-based model. For instance, in our simulations, pillars form at regular spatial intervals in the early stages of the construction of the royal chamber (see figure 1).

Our model also agrees with the predictions of Bonabeau et al. [3], who showed, in an extension of [4], that the introduction of a source of diffusing queen pheromone can first guide the construction of pillars and subsequently an encircling wall. Our results show that this process can result in a fully-formed 3-d royal chamber. The manner in which our results are influenced by wind also agrees with previous findings, which reported a distortion of the horizontal cross-section of the royal chamber. However, out treatment of the third dimension also allows us to explore distortion to the elevation of the chamber.

Finally, the Bonabeau model suggested that an artificially imposed flow of termites emitting trail pheromone could result in the construction of walkways. Our simulations extend these results by showing that these structures tend to become covered walkways over time, and remain clear of building material (see figure 3). Previous speculation in [3] that two intersecting trails could give rise to the formation of a cross-road structure are confirmed by our model, which additionally demonstrates that such an intersection will not tend to become blocked by building activity (see figure 4).

In addition to confirming work from previous models, our approach has generated new insights. First, and most generally, it is perhaps remarkable that the introduction of physical constraints on termite motion and pheromone behaviour have not had more influence on the structures achieved. It was possible, for instance, that as a tunnel was formed and enclosed the trail within it, it would become difficult for trail pheromone to escape, distorting the pheromone distribution in such a way as to interfere catastrophically with new building activity. Moreover, when a piece of building material is placed it cannot be removed in this model meaning that tunnel blockages, should they arise cannot be rectified, making tunnel completion impossible.

In order to implement constrained movement and diffusion we introduced an explicit third dimension. This alone could have introduced dramatic differences between the behaviour observed in our system and in previous models. Climbing, for instance, plays no part in 2-d models, but is important in the current work, since agents must be able to reach a location in order to build there. Despite this, all of the structures suggested by the two relevant previous models were achieved. However, the fact that, in our model, it is harder to reach some locations than others causes agents to build more slowly in hard to reach places like roofs of structures, or the tops of pillars, which require time consuming assents compared to the walls which are at ground level.

Predictable, as the strength of simulated wind increased the building process became slower until a point was reached where no building was possible. While the effect of wind may simply be to disturb a pheromone template to the extent that builders are not encouraged to lay down building material at any location, it also has a more subtle influence on the ability of newly-formed structures to recruit builder termites. Without wind, cement pheromone is free to diffuse in every unobstructed direction, attracting termites to the emitting structure from a wide area. In the presence of wind, however, the cement pheromone cannot diffuse as effectively, hampering the recruitment of up-wind builder termites. This ensures that mildly windy conditions result in differential construction activity.

Rather than being severely limited by the physical obstruction imposed by the results of their own building behaviour, it appears that in some cases these effects can increase the efficiency of the building work and prevent some "problematic" kinds of construction activity such as the tendency for tunnels to "mushroom" observed in figure 4. This "overbuilding" stems from a combination of two factors, (i) a slight "fanning" of the trail laid by the termites due to their stochasticity and amplified by confusion at the intersection of two perpendicular pheromone trails, and (ii) a tendency for builder termites to approach the trail from all sides. These conditions lead to building behaviour near the outer edge of the trail "fan", which is amplified and ultimately results in wide tunnel walls. By contrast, where builder termites enter the lattice at the same points as the trail laying termites, they tend to begin building at a regular distance from the trail mid-line. As a tunnel is formed in this way, it limits the movement of builders, directing them to the tunnel end and preventing them from reaching the outer edge of the "fanned" distribution of trail pheromone. In this way, despite the inability of the trail termites to form a narrow, fixed-width pheromone template, building work is able to produce a narrow, fixed-width tunnel.

4.1 Future Work

Currently, several aspects of the model are under-developed, or unexplored. The three kinds of pheromone employed have not been explored in concert. In the results reported, only two types of pheromone were ever employed together. For instance, it would be interesting to explore the role that trail pheromone deposited by builder termites might play in directing them to sites of recent building activity. How would cement and trail pheromone interact in this situation? For example, it is possible that such an interaction might account for the maintenance of an entrance to the royal chamber, by inhibiting building work across routes employed by many builders. While several entrances might be maintained initially, over time one might expect a single location to dominate and the others to be closed.

Both the crude physics of construction, and the implementation of wind, which currently cannot support swirling eddies or any kind of back-flow, could be improved. The former might benefit from an approach similar to that taken by Funes & Pollack [7] in their work on simulating Lego structures. Currently, it is possible for unrealistic or impossible structures to be built. A more sophisticated physics might encourage termites to adopt more complicated strategies in order to achieve large-scale structures.

Several logistic factors have yet to be incorporated into our model. Currently, termites do not obstruct one another's movement, and, unrealistically, are assumed to occupy roughly the same amount of space as a piece of building material. For reasons of simplicity, we do not yet explicitly model the tasks involved in discovering raw materials or transporting them to the construction site. The significance of these factors remains to be seen.

More generally, closer comparison between the current model and the detailed behaviour of previous mathematical models is required, and a more extensive characterisation of the our simulation's dependencies on parameter values and initial conditions must be undertaken.

5   Conclusion

In conclusion, we have shown that physical and logistic constraints imposed on termite construction by the results of their own building behaviour do not prevent the formation of several structures reminiscent of natural termite architecture. Indeed, it appears that in some cases these constraints can have a positive effect on the ability of termites to achieve efficient, effective constructions.

References

1. E. Bonabeau, S. Guérin, D. Snyers, P. Kuntz, and G. Theraulaz. Three-dimensional architectures grown by simple 'stigmergic' agents. *BioSystems*, 56:13–32, 2000.
2. E. Bonabeau, G. Theraulaz, E. Arpin, and E. Sardet. The building behavior of lattice swarms. In R. A. Brooks and P. Maes, editors, *Artificial Life IV*, pages 307–312. MIT Press, Cambridge, MA, 1994.

3. E. Bonabeau, G. Theraulaz, J.-L. Deneubourg, N. R. Franks, O. Rafelsberger, J.-L. Joly, and S. Blanco. A model for the emergence of pillars, walls and royal chambers in termite nests. *Philosophical transactions of the Royal Society of London, Series B*, 353:1561–1576, 1997.

4. J.-L. Deneubourg. Application de l'ordre par fluctuations à la description de certaines étapes de la construction du nid chez les termites. *Insectes Sociaux*, 24:117–130, 1977.

5. J.-L. Deneubourg, S. Goss, N. Franks, and J.-M. Pasteels. The blind leading the blind: Modelling chemically mediated army ant raid patterns. *Journal of Insect Behaviour*, 2:719–725, 1989.

6. J.-L. Deneubourg, G. Theraulaz, and R. Beckers. Swarm-made architectures. In F. J. Varela and P. Bourgine, editors, *First European Conference on Artificial Life*, pages 123–133. MIT Press, Cambridge, MA, 1992.

7. P. Funes and J. Pollack. Computer evolution of buildable objects. In P. Husbands and I. Harvey, editors, *Fourth European Conference on Artificial Life*, pages 358–367. MIT press, Cambridge, MA, 1997.

8. P.-. P. Grassé. *Termitologia, Tome II – Foundations des societés – construction*. Masson, Paris, 1984.

9. C. Hirsch. *Numerical Computation of Internal and External Flows, Volume 1: Fundamentals of Numerical Discretization*. Wiley, Chichester, 1988.

10. G. Theraulaz and E. Bonabeau. Modelling the collective building of complex architectures in social insects with lattice swarms. *Journal of Theoretical Biology*, 177:381–400, 1995.

Modeling Ant Behavior
Under a Variable Environment

Karla Vittori[1], Jacques Gautrais[2], Aluizio F.R. Araújo[1],
Vincent Fourcassié[2], and Guy Theraulaz[2]

[1] Department of Electrical Engineering, University of São Paulo
Av. Trabalhador Sãocarlense, 400 - Centro - 13566-590, São Carlos, SP, Brazil
Tel: (055) 16.273.9365, Fax: (055) 16.273.9372
karlav@sel.eesc.usp.br
[2] Centre de Recherches sur la Cognition Animale, Université Paul Sabatier
118 route de Narbonne, 31062 Toulouse cedex 4
Tel: (033) 5.61.55.67.31, Fax: (033) 5.61.55. 61.54
theraula@cict.fr

Abstract. This paper studies the behavior of ants when moving in an
artificial network composed of several interconnected paths linking their
nest to a food source. The ant responses when temporarily blocking the
access to some branches of the maze were observed in order to study
which factors influenced their local decisions about the paths to follow.
We present a mathematical model based on experimental observations
that simulates the motion of ants through the network. In this model,
ants communicate through the deposition of a trail pheromone that attracts other ants. In addition to the trail laying/following process, several
other aspects of ant behavior were modeled. The paths selected by ants
in the simulations were compared to those selected by ants in the experiments. The results of the model were encouraging, indicating that
the same behavioral rules can lead ants to find the shortest paths under
different environmental conditions.

1   Introduction

Ant colonies are very interesting entities because of their capacities to collectively
achieve complex decisions and patterns through self-organization processes based
on simple behavioral rules and the use of local information and indirect communication [6,7]. The decisions of the ants are controlled by the laying/following
of trails of a chemical substance, called pheromone [22]. When given the choice
among several alternative paths, ants choose a path in a probabilistic way, based
on the pheromone concentration over the possible paths. This mechanism allows
the selection of the shortest path among several ones [1,10,11]. Shorter paths
are completed earlier than longer ones. Hence, the pheromone concentration on
those paths increase more rapidly and they attract more ants. This process of
indirect communication relies on a positive feedback mechanism [12,15] and depends on the environment characteristics, e.g. colony size [13], food type [18,19],

M. Dorigo et al. (Eds.): ANTS 2004, LNCS 3172, pp. 190–201, 2004.

number of food sources [16,17], and the nature of the substrate on which the ants are moving [9].

In this paper, we investigate the ant foraging behavior in a relatively complex environment, composed of an artificial network of paths where several interconnected routes can lead the insects to a food source. Our aim was to study the responses of ants when sudden changes occur in the structure of the network. The ant decisions were analyzed at two different levels: (1) at the collective level, we analyzed the paths chosen by ants to reach the food source and return to the nest; (2) at the individual level, we quantified several behavioral parameters of the ants' actions. Finally, we show through a mathematical model that the same behavioral rules can lead the ants to find the shortest paths under the different network conditions studied.

This paper is organized as follows. In Section 2 we describe the experiments and the results we obtained. In Section 3 we present the model we developed and compare its output with the responses of the ant colonies. Finally, in Section 4 we discuss the performance of the model and comment on future research.

2   Methods and Results

In this section, we describe the experimental set-up, the insects used, the experiments performed and the results obtained, both at the collective as at the individual level.

2.1   Experimental Set-Up and Procedures

The experimental set-up consisted in a maze carved in a plastic slab (31.5 x 18.0cm) covered by a glass plate. The maze was composed of four identical lozenges assembled together (fig. 1). The branches of each lozenge formed an angle of 60 [3] and the maze galleries had a 0.5 x 0.5 cm section. Five circular chambers (ϕ 2.0 cm) were built at the extremity of some of the branches of the maze.

We used the Argentine ant, Linepithema humile [14], to perform our experiments. Foragers of this species lay a pheromone trail both in their outbound and nestbound movements [2,3,8]. The ants were collected near Narbonne (France) and were distributed in ten experimental colonies, each containing around 2,000 individuals. All individuals were collected outside the nest and could thus be considered as foragers. The ants were fed with a solid food composed of carbohydrate and vitamin [5]. Each colony was housed in a plastic container (20.0 x 20.0 x 7.0 cm), with the walls coated with Fluon®to prevent escape. Ants nested in test tubes, partially filled with water behind a cotton plug, placed into each container. No interactions between nests were possible.

The colonies were starved for four days before the day of the experiments. At the beginning of each experiment, the nest was connected to a glass box (35.0cm^2) by a plastic bridge (40.0cm^2). Ants had then access from this box to the maze and to the food source (1 M sucrose solution, the food was spread on

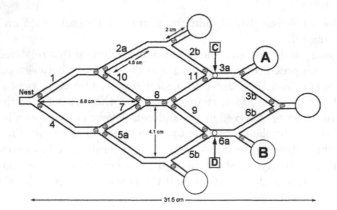

Fig. 1. Experimental set-up. The maze was carved in a plastic slab and the galleries had a 0.5 x 0.5 cm section. During the experiments, the ants had access to a food source (1M sucrose solution) located either in chamber A or B. The access to branch 3a and 6a can be blocked in C and D respectively

a surface large enough to avoid crowding effects). In half of the experiments, the food source was located in chamber A and in the other half, in chamber B (fig. 1). The access to the other chambers of the maze was denied during the experiments. After each experiment, the network was cleaned with alcohol to remove any residual chemical cues.

An experiment began when the first ant entered the network and the duration of each experiment was set to 60 min. The ant behavior was continuously recorded with a high definition camera (Sony CDR-VX 2000 E) placed above the set-up. The traffic of ants in the network was sampled during 20s every 3 min. The network offers twelve possible paths to connect the nest to the food source. These paths can be classified in four categories, according to their length: short, medium, long and very long (Table 1). The ten experimental colonies were tested under three conditions:

1. Control situation: the ants had access to the food source and no change was made on the network.
2. Block-after situation: in the first 30 minutes ants had access to all parts of the network, as in the control situation. After 30 minutes, the access to a branch next to the food source was denied (the branch 3a when the food was in chamber A, and the branch 6a when the food was in chamber B).
3. Block-before situation: the access to some branches was denied during the first phase of the experiments (0-30min.), forcing ants to use a long path. When the food was in chamber A, the ants could use only the route 1-10-7-5-9-11, and when the source was in chamber B, they could use the route 4-7-10-2-11-9-6a. After 30 min., the access to all the branches was allowed.

Whether a branch was selected or not at a bifurcation was tested by applying a binomial test [21] on the cumulated flow of ants on each branch. This allowed

Table 1. Possible paths to the food source and their lengths.s = short path, m = medium path, l = long path and vl = very long path.

Paths(source in A)	Paths(source in B)	Length
1-2a-2b-3a	4-5a-5b-6a	s - 21,5 cm
1-10-8-11-3a	1-10-8-9-6a	
4-7-8-11-3a	4-7-8-9-6a	
4-5a-5b-9-11-3a	1-2a-2b-11-9-6a	m - 30,5 cm
4-7-10-2a-2b-3a	1-10-7-5a-5b-6a	
4-5-6a-6b-3b	1-2-3a-3b-6b	
4-7-8-9-6a-6b-3b	1-10-8-11-3a-3b-6b	
1-10-8-9-6a-6b-3b	4-7-8-11-3a-3b-6b	
1-2-11-9-6a-6b-3b	4-5-9-11-3a-3b-6b	l - 39,5 cm
1-10-7-5-6a-6b-3b	4-7-10-2-3a-3b-6b	
1-10-7-5-9-11-3a	4-7-10-2-11-9-6a	
4-7-10-2-11-9-6a-6b-3b	1-10-7-5-9-11-3a-3b-6b	vl - 48,5 cm

us to determine the overall path selected by the ants through the network. Note that several paths could be selected in the same experiment when no significant choice was made at one or several bifurcations.

2.2   Experimental Results

In the control situation, ants selected one of the shortest paths of the network, in most experiments when going to the food source, and in all experiments when returning to the nest (fig. 2a).

In the first phase of the block-after situation, when ants had access to all parts of the network, the results were comparable to those obtained in the control situation: in the majority of experiments one of the shortest paths was selected (fig. 2b). In the second phase of the block-after situation, when the most direct access to the food was blocked, some colonies appeared to be unable to find a path, and instead performed loops in the network (fig. 2c). However, in most experiments ants chose a medium path, which was the shortest one towards the source in this situation. Some long paths were also selected. When returning to the nest on the other hand, all colonies selected one or two of the shortest paths.

In the second phase of the block-before situation, when the access to all the branches of the network was allowed again, the majority of the colonies found one of the shortest paths towards the food source (fig. 2d).

However, some colonies continued to use the long path they had selected in the first phase of the experiment, and one colony used both a short and a long path to reach the food. In the opposite direction, towards the nest, most of the colonies selected a short path. Only one colony chose a medium path.

We conclude that in most experiments ants found one of the shortest path in the network, both to reach the food source and to return to their nest.

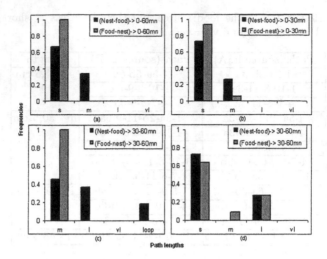

Fig. 2. Distribution of the outcome of the experiments.(a) in the control situation, when ants had access to all branches, (b)-(c) in the block-after situation, when one branch was blocked after 30 min.(d) in the block-before situation when several branches were blocked during the first 30 min of the experiment. s = short path, m = medium path, l = long path and vl = very long path. N= 10 experiments for each situation

2.3 Individual Behaviors

We measured three behavioral characteristics at the individual level in the control situation: (1) the ant speed (2) the probability to make a U-turn and (3) the time spent at the food source.

The speed of the ants was computed from the measure of the time spent by individuals to travel long and short paths towards the food source. We found a mean value of 1.10 cm/s(± SD 0.48, $N = 123$).

We measured the frequency of U-turns made by ants both in the outbound and nestbound direction over a long path. A total of 521 ants were considered. We found a frequency of U-turn per ant of 0.15 ($N = 521$ ants) and 0.10 ($N = 458$ ants) for the outbound and nestbound direction respectively.

The average time spent by ants at the food source was 185.30 s (± SD 167.62, $N = 43$). The maximum number of ants in the network was equal to 100 individuals on average.

3 Model Description

In this section, we describe the parameters of our model and compare its output to the experimental results we obtained in the three experimental situations we tested.

3.1   Structure of the Model

Each step of the model was equivalent to 1 s. As the maximum number of ants moving through the network at any moment in the experiments was assessed to 100 individuals on average, we decided to use this limit in our model.

In addition to the parameters measured at the individual level, we estimated other parameters from previous studies, and introduced them in the model. The values of all the parameters we used in our model are given at the end of this section.

As it is known that the ant flux is related to the size of the colony, we estimated the probability for an ant i to enter in the environment at each simulation step t as:

$$P_{ent_i}(t) \approx U(\phi) \tag{1}$$

where: ϕ = ant flux; $\phi = 1/$maximum number of ants, and U represents a uniform probability distribution.

The distance traveled by an ant i at each step t was estimated as the ants speed V_i incremented by a random error ε :

$$D_i(t) = (V_i + \varepsilon) \tag{2}$$

where $\varepsilon \approx N(0, V_i/10)$, N represents a Normal distribution and

$$V_i = \eta + \zeta \tag{3}$$

η= constant measured in the experiments; $\zeta \approx U[0; \iota]$.

When an ant i moves through the network, it lays a quantity f_l in each branch l it travels. The model considered that ants deposit trail pheromone more frequently on their way back to the nest than when going to the food source [8]. The quantity of pheromone Q_l gradually decreases over time in a proportion fixed by a parameter δ:

$$Q_l(t) = \delta Q_l(t-1) \tag{4}$$

where δ is a constant empirically obtained.

When reaching a bifurcation, the choice of a specific branch l depends on two characteristics: the pheromone concentration on the branches and their orientation relative to the ant direction of movement. As the ants have the capacity to locate the direction of the nest and the direction of the food source once they have found it [3], we considered that the orientation of a branch can be important in determining the course of an ant through the network.

The probability $P_c(j, k, t)$ for an ant of choosing branch j when it is at the bifurcation j-k at step t was thus estimated by using two probabilities: (i) $P_{pher}(j, k, t)$, the probability of choosing branch j as a function of the pheromone concentration on the branches j and k at the bifurcation j-k and (ii) $P_{dir}(j, k, t)$, the probability of choosing branch j as a function of its orientation.

$$P_c(j,k,t) \approx U\left(\frac{\rho(P_{pher}(j,k,t) + \tau P_{dir}(j,k,t))}{\rho + \tau}\right) \tag{5}$$

where:

$$P_{pher}(j,k,t) = \frac{(\sigma + Q_j)^\beta}{(\sigma + Q_j)^\beta + (\sigma + Q_k)^\beta} \tag{6}$$

$$P_{dir}(j,k,t) = \frac{(cos\theta)^\alpha}{(cos\theta)^\alpha + (cos\omega)^\alpha} \tag{7}$$

θ and ω are the angles formed respectively by the branch j and the branch k with the longitudinal axis of the set-up; ρ, τ, σ,α,β = constants empirically obtained. Equation (6) was proposed in earlier experiments with Argentine ants [4].

An ant i can change its direction of movement when traveling along a branch l by performing a U-turn. The estimation of the probability $P_{u_l}(t)$ for an ant i to make a U-turn at each step t over a short branch l, (labeled 3a, 6a and 8 in fig. 1), was based on previous results [4]:

$$P_{u_l}(t) \approx U\left(\frac{\varphi}{1,0 + \Pi Q_l(t)}\right) \tag{8}$$

where φ represents the maximum number of U-turns that can be performed on a short branch and Π is a constant empirically obtained.

Ants can also make a U-turn when the selected branch is not available, represented by Ω. Based on previous evidences [4], we estimated that the probability for an ant i to make a U-turn over a long branch l at each time t follows a sigmoidal law:

$$P_{u_l}(t) \approx U\left(\frac{\upsilon}{1.0 + exp(-\psi Q_l(t))}\right) \tag{9}$$

where υ represents the maximum number of U-turns that can be performed on a long branch and ψ is a constant empirically obtained.

Finally, ants can also make a U-turn when the selected branch is blocked. In this case the probability to make a U-turn is fixed:

$$P_{u_l}(t) \approx U\left(\Omega\right) \tag{10}$$

The time $T_{s_i}(t)$ spent by an ant i at the food source from instant t was also considered:

$$T_{s_i}(t) = (\kappa + \chi) \tag{11}$$

where: $\chi \approx N(0, \kappa/\mu)$ is a random error and κ, μ are two constants empirically obtained.

The values set to the parameters were: ϕ=0.01, η=1.0, ζ=0.3, f_l=1.0 in the outbound direction, f_l=5.0 in the nestbound direction, δ=0.99, ρ=1.0, τ=0.4, σ=5.0, β=3.5, α=3.0, φ=0.01, Π=1.0, υ=0.07, ψ=0.003, κ=60.0s, μ=5.0 and Ω=0.005. The simulations were run for 36,000 time steps and we performed 200 simulations for each situation.

When the parameter values could not be estimated from the experiments they were empirically adjusted in order to increase the fit between the results of the model and that of the experiments.

3.2   Comparison of the Model Output
with the Experimental Results

In the control situation, ants found 252 paths to reach the food source. Most of them corresponded to the shortest paths (237 out of 252)(fig. 3). There was a good agreement between the solutions found by ants in the model and in the experiments. In the nestbound direction, the ants found 204 paths, all of them corresponding to the shortest paths. Only in four simulations did ants use simultaneously two short paths to reach the nest.

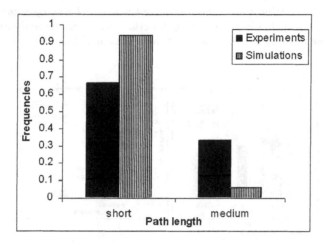

Fig. 3. Frequency distributions of the lengths of the paths chosen by the ants in the model and in the experiments for the outbound direction in the control situation (0-60 min.). s = short path and m = medium path

In the first phase of the block-after situation, ants selected 464 paths to reach the food source. Most of them corresponded to the shortest paths (248 out of 464), followed by medium paths (182 out of 464)(fig. 4a), in the same order as in the experiments. The model also generated some simulations where the long and very long paths were selected (7.33%). As this outcome was relatively rare, we considered that there was generally a good fit between the model and the experiments. In the nestbound direction, ants selected 539 paths in the simulations. Most of them corresponded to the short paths (275 out of 539) and medium paths (209 out of 539) (fig. 4b). Once again, the model generated some simulations where long and very long paths were selected (10.20%). However, the low level of occurrence of this outcome suggested a good agreement between the simulations and the experiments.

In the second phase of the block-after situation, the ants in the simulations selected 247 paths to reach the food source. The shortest paths (medium paths in this case) were selected in the majority of the simulations (164 out of 247) (fig. 5a). As in the experiments, the occurrence of loops was also observed (76

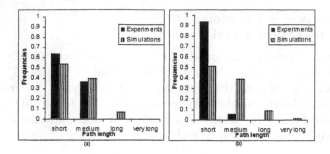

Fig. 4. Frequency distributions of the lengths of the paths chosen by ants in the model and in the experiments in the (a) outbound direction and (b) nestbound direction in the first phase of the block-after situation, when the access to all branches was allowed(0-30 min.)

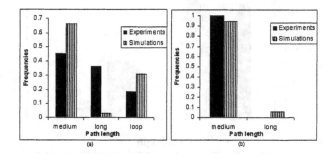

Fig. 5. Frequency distributions of the lengths of the paths chosen by ants in the model and in the experiments in the (a) outbound direction and (b) nestbound direction in the second phase of the block-after situation, when the access to some branches was denied(30-60 min.)

out of 247 paths). The long paths were also selected, representing a minor part of the paths selected (7 out of 247). In the nestbound direction, ants selected 227 paths; almost all of them corresponded to medium paths (214 out of 227)(fig.5b). This performance was similar to that observed in the experiments.

In the second phase of the block-before experiments, when all branches could be used, the ants selected 331 paths to reach the food (fig. 6a). These paths were distributed as followed: short paths (269 out of 331), medium paths (26 out of 331), long paths (35 out of 331) and loop (1 out of 331). As the occurrence of loops was reduced in this direction, we considered that the model performed satisfactorily. We noted nonetheless that the model performed rather badly in this situation. This was explained by the fact that, in the experiments, the high concentration of pheromone deposited on the branches used during the first 30 minutes attracted a lot of ants. These ants persisted therefore to use long paths.

When returning to the nest, ants selected 212 paths in the simulations (fig. 6b), most of them corresponding to the short paths (206 out of 212). Some

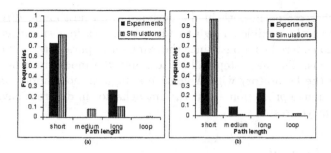

Fig. 6. Frequency distributions of the lengths of the paths chosen by ants in the model and experiments in the (a) outbound direction and (b) nestbound direction in the second phase of the block-before experiments, when the access to all branches was allowed again.(30-60 min.)

paths corresponded to loops (4 out of 212) and medium paths (2 out of 212). Otherwise, the ants in the simulations can be considered as more efficient than those in the experiments because they selected only short paths.

4   Conclusion

The aim of this work was to study how ants adapt their foraging behavior when environmental conditions are suddenly changed, e.g. when a path towards a food source is obstructed or when new and shorter routes are offered to the ants to reach a food source. Such changes are likely to arise in the natural environment and it is important to understand how ants react to them behaviorally and on which time-scale they are able to adapt.

In this study we introduced Argentine ants in a complex network, in which several paths of different lengths can lead to a food source. We changed the organization of the network, allowing or denying the access to some branches. The ant behavior was analyzed both at the collective and individual levels, and several aspects of their behavior and movement through the network were quantified. The paths selected by the ants, both when moving towards the food and the nest, were noted when the access to all branches was allowed and when one or several branches were blocked. The results show that ants were able to find the shortest paths in all situations, specially when returning to their nest. This result has been observed in previous studies [2,20], indicating the ability of the ants to locate the nest and return straight to it.

After the experimental study we developed an individual-based model whose parameters were based on the different decisions made by real ants when moving in the network when the access to all branches was allowed. The model was applied to the same situations tested in the laboratory with real ants, and the output of the model was compared to the results of the experiments. There was in general a good agreement between the results of the simulations and those of the experiments. In most simulations generated by the model the ants selected one of the shortest path, both in the outbound and nestbound direction. The model proved to be flexible enough to adapt to the various situations tested. It can also be considered as robust since the same set of parameter values was used in all situations. The next step will be to apply the model to other situations we tested in the laboratory with Argentine ants, and most importantly, to use the model to make prediction about the ant behavior in different environmental conditions not yet tested.

Acknowledgments

This work was supported by a doctoral grant from FAPESP to Karla Vittori.

References

1. Aron, S., Deneubourg, J.-L., Goss, S., Pasteels, J. M.: Functional self-organisation illustrated by inter-nest traffic in the argentine ant Iridomyrmex humilis. In Biological Motion, Eds W.Alt and G. Hoffman. Lecture Notes in Biomathematics, Springer-Verlag, Berlin, 533-547 (1990)
2. Aron, S., Beckers, R., Deneubourg, J.-L., Pasteels, J.M.: Memory and chemical communication in the orientation of two mass-recruiting ant species. Ins. Soc., 40, 369-380 (1993)
3. Beckers, R.: L'auto organisation - une reponse alternative à la complexité individuelle ? Le cas de la récolte alimentaire chez Lasius niger (L.) (Hymenoptera : Formicidae). Thèse, Université Paris-Nord, France (1992).
4. Beckers, R., Deneubourg, J.-L., Goss, S.: Trails and U-turns in the selection of a path by the ant Lasius niger. J. theor. Biol., 159, 397-415 (1992)
5. Bhatkar, A.W., Whitcomb W.: Artificial diet for rearing various species of ants. Fla. Entomol., 53, 229-232 (1970).
6. Bonabeau, E., Theraulaz, G., Deneubourg, J. L., Aron, S. and Camazine, S.: Self-organization in social insects. Trends Ecol. Evol., 12, 188-193 (1997)
7. Camazine, S., Deneubourg, J.-L., Franks, N., Sneyd, S., Bonabeau, E., Theraulaz, G.: Self-organisation in Biological Systems. Princeton University Press (2001)
8. Deneubourg, J. L., Aron, S., Goss, S., Pasteels, J. M.: The self-organizing exploratory pattern of the Argentine ant. J. Insect Behav., 3, 159-168 (1990)
9. Detrain, C., Natan, C., Deneubourg, J. L.: The influence of the physical environment on the self-organized foraging patterns of ants. Naturwissenschaften, 88, 4, 171-174 (2001)
10. Goss, S., Aron, S., Deneubourg, J.-L., Pasteels, J. M. Self-organized shortcuts in the argentine ant. Naturwissenschaften, 76, 579-581 (1989)
11. Goss, S., Beckers, R., Deneubourg, J.-L., Aron, S., Pasteels, J. M.: How trail laying and trail following can solve foraging problems for ant colonies. In R. N. Hughes, (Eds.), Behavioral mechanisms of food selection. NATO ASI Series 20, Berlin: Springer-Verlag, pp. 661-678 (1990)
12. Haken, H.: Synergetics, Springer-Verlag (1977)
13. Mailleux, A. C., Deneubourg, J.-L., Detrain, C.: How does colony growth influence communication in ants? Ins. Soc., 50, 24-31 (2003).
14. Newell, W.: The life history of the Argentine ant. J. Econ. Entomol., 2, 174-192(1909)

15. Nicolis, G., Prigogine, I.: Self-organization in non-equilibrium systems. Wiley (1977)
16. Nicolis, S. C., Deneubourg, J.-L.: Emerging patterns and food recruitment in ants: an analytical study. J. theor. Biol., 198, 575-592 (1999)
17. Nicolis, S. C., Detrain, C., Demolin, D., Deneubourg, J.-L.: Optimality of collective choices: a stochastic approach. Bull. math. Biol., 65, 795-808 (2003)
18. Portha, S., Deneubourg, J.-L., Detrain, C.: Self-organized asymmetries in ant foraging: a functional response to food type and colony needs. Behav. Ecol., 13, 776-791 (2002)
19. Portha, S., Deneubourg, J.-L., Detrain, C.: How food type and brood influence foraging decisions of Lasius niger scouts. Anim. Behav., in press (2004)
20. Shen, J.X., Xu, Z.M., Hankes, E.: Direct homing behaviour in the ant Tetramorium caespitum (Formicidae, Myrmicinae). Anim. Behav., 55, 1443-1450 (1998)
21. Siegel, S., Castellan, N.J.: Non parametric statistics for the behavioural sciences. McGraw-Hill, New-York (1988)
22. Wilson, E.O.: The insect societies. Harvard University Press. Cambridge, Massachussets (1971)

Multi-type Ant Colony:
The Edge Disjoint Paths Problem

Ann Nowé, Katja Verbeeck, and Peter Vrancx

Vrije Universiteit Brussel, COMO, Pleinlaan 2 1050 Brussel, Belgium
asnowe@info.vub.ac.be, {kaverbee,pvrancx}@vub.ac.be
http://como.vub.ac.be

Abstract. In this paper we propose the Multi-type Ant Colony system,
which is an extension of the well known Ant System. Unlike the Ant
System the ants are of a predefined type. In the Multi-type Ant Colony
System ants have the same goal as their fellow types ants, however are in
competition with the ants of different types. The collaborative behavior
between identical type ants is modeled the same way as in ant systems,
i.e. ants are attracted to pheromone of ants of the same type. The competition between different types is obtained because ants are repulsed by
the pheromone of ants of other types. This paradigm is interesting for
applications where collaboration as well as competition is needed in order
to obtain good solutions. In this paper we illustrate the algorithm on the
problem of finding disjoint paths in graphs. A first experiment shows on
a simple graph two ants types that find successfully two completely disjoint paths. A second experiment shows a more complex graph where the
number of required disjoint paths exceeds the number of possible disjoint
paths. The results show that the paths found by the ants are distributed
proportionally to the cost of the paths. A last experiment shows the influence of the exploration parameter on a complex graph. Good results
are obtained if the exploration parameter is gradually decreased.

1 Introduction

Colonies of social insects are capable of achieving very complex tasks. These
tasks are achieved without a governing intelligence. No single insect can complete the task alone or coordinate the actions of its colleagues, but from these
distributed, concurrent behaviors a very complex collective result can emerge.
Self-organization relies on interactions between agents. As such the communication system used by agents is an important part of a swarm intelligence system,
[6]. In ant based systems communication often takes the form of *stigmergy*.

Stigmergy is a term used to indicate interactions through the environment.
This form of communication requires no direct contact between the individuals.
Interaction occurs when one individual alters its environment in some way, and
other individuals later on respond to this change. In the Ant Colony System, [3]
on which our approach is based, the ants communicate through the environment
by leaving a pheromone trail. If all ants have the same type of pheromone, this

M. Dorigo et al. (Eds.): ANTS 2004, LNCS 3172, pp. 202–213, 2004.

results in a pure cooperative system. In this paper we propose an extension of the Ant Colony algorithm to the Multi-type case. Ants are attracted by pheromones coming from fellow type ants and repulsed by pheromones of non-fellow ant types. As such we get a system in which there is pure collaboration between fellow ants, but competition among the different ants types. This property has the potential to solve a whole array of interesting problems, such as production planning, traffic engineering and routing in telecommunication networks. In the last example for instance, routers have to collaborate in order to deliver the packets to the desired destination, while routers are also in competition for resources, i.e. bandwidth. Off course the routing problem has already been studied using the Antnet Algorithm, [2]. However our approach differs from Antnet because competition is modeled explicitly here.

The idea of using multi-type ants is not completely new. Parunak and Brueckner report in [4] on a multi-type ant approach and use the well-known Missionaries and Cannibals problem as a test case. In their approach 2 types with different pheromones are used to differentiate between actions or decisions that should be taken by the missionaries and by the cannibals. As such the problem is translated into a deterministic sequential decision problem. The approach clearly differs from our approach since there is no explicit competition between the ant types. In [5] the use of different pheromone types coming from different entity types is referred to as *Pheromones with different semantics*. In the air combat example described in [5], the pheromones are used in a collaborative way, and combined as a weighted sum into an overall pheromone. In the same paper, the authors introduce the concept of *Pheromones with different dynamics*. Pheromones with high propagation give a good long-range guidance but a poor short-term guidance, while low propagation gives a good short-range guidance but poor long-range guidance. Both dynamics are combined in a collaborative way, and as such are very different of our set up of multi-type ants. Blesa and Blum, also propose an ACO approach to the Maximum Edge-Disjoint Paths Problem. The approach described in [1], is an incremental multi start approach rather than a genuine parallel approach as we present in this paper. As the approach of [1] is only very recently published, we did not have the opportunity yet to compare our approaches in terms of quality of the solution and computation time.

In the following section, we give a brief introduction to the original Ant Colony System. Next we propose our extension in the Multi-type case. We illustrate the new approach on the problem of finding disjoint paths in graphs in section 4. Different settings are considered; a first experiment on a simple graph shows that two ants types find successfully two completely disjoint paths. A second experiment shows a more complex graph where the number of required disjoint paths exceeds the number of possible disjoint paths. The results show that the paths found by the ants are distributed proportionally to the cost of the paths. A last experiment shows the influence of the exploration parameter on a complex graph. Good results are obtained if the exploration parameter is gradually decreased. Finally we conclude with a discussion and reflection on possible future work.

2   Ant Colony System

The multi-type approach we present in this paper is based on the Ant Colony System, [3]. Before discussing the multi-type extension, we briefly describe the original Ant Colony System.

Ant Colony System (ACS) was originally developed by Marco Dorigo, [3]. It can be used to find good solutions for hard optimization problems such as the Traveling Salesman Problem. The algorithm uses a number of relatively simple agents that model the pheromone laying and following behavior of ants. Each ant in the system builds its own tour. It does this by starting at a random node and selecting the next neighboring node to go to, based on the distance from the current node and the amount of pheromone on the connecting edge. Starting from node r, ant k chooses the next node j from neighboring nodes that have not been visited yet. To do this it first generates a fixed length candidate list J_r^k, containing the nearest neighbors of the current node r. From the candidate list a node is chosen using the following probabilistic formula:

$$
j = \begin{cases} argMax_{u \in J_r^k} [\tau(r,u)].[\eta(r,u)]^\beta & \text{if } q \leq q_0 \\ J & \text{if } q > q_0 \end{cases} \tag{1}
$$

$\tau(r,u)$ is the amount of pheromone on edge (r,u) and $\eta(r,u)$ is a heuristic rating of edge (r,u). In the case of the TSP the heuristic value of an edge is taken to be the inverse of the length the edge. β is an algorithm parameter that determines the influence of the heuristic. q is randomly generated with a uniform distribution on the interval $[0,1]$. If q is larger than the algorithm parameter $q0$, formula (1) is used to determine the next node, else the next node J is chosen randomly from the unvisited neighboring nodes using the following formula :

$$
p^k(r,u) = \frac{[\tau(r,u)].[\eta(r,u)]^\beta}{\sum_{l \in J_r^k} [\tau(r,l)].[\eta(r,l)]^\beta} \tag{2}
$$

After ant k traverses an edge (i,j), it performs a local pheromone update that changes the amount of pheromone on the edge.

$$
\tau(i,j) = (1-\rho).\tau(i,j) + \rho.\tau_0 \tag{3}
$$

In this formula τ_0 is the initial amount of pheromone present on all edges. $\rho \in [0,1]$ simulates the decay of the pheromone trail that occurs in natural systems. Pheromone decay ensures that the pheromone intensity on a random edge does not increase without bound and cause stagnation.

On top of the local pheromone update an extra global pheromone update is performed once all ants have finished their tour. During the global update all edges of the best tour T^+ with length L^+ receive a pheromone update. The new trail value for each edge $(i,j) \in T^+$ is calculated as follows:

$$
\tau(i,j) = (1-\rho).\tau(i,j) + \rho.(1/L^+) \tag{4}
$$

After each iteration the solutions produced by each ant are checked and the best tour T^+ and its length L^+ are updated if needed. A complete description of the algorithm can be found in [3].

3   The Multi-type Ant Colony System

The traditional ant algorithms depend heavily on cooperation between the ants. A single ant on it's own cannot find a good solution; it takes the work of a whole colony of agents in order to obtain a useful solution. The goal of the extension to this algorithm is to introduce competition on top of the cooperation between agents.

The Multi-type ant algorithm we propose in this paper uses several different types of ants. Each ant helps to solve a problem by cooperating with other ants of the same type, in exactly the same way as in the original algorithm. The different ant types are in direct competition with each other. As there is no direct communication between ant agents, this competition is reflected by the pheromone communication system. An ant that encounters a pheromone trail left by an ant of the same type will still have a high probability of following it. Pheromone trails of other ant types however, will repel the ant and reduce the probability of it choosing a certain edge. The intention of this competition is to have the different types of ants find disjoint solutions.

The first part of the extension allows multiple ant types to leave pheromone trails on an edge, and to distinguish different types of pheromone. To achieve this each edge in the graph was associated with a set of separated values rather than a single value. Each value represents the pheromone trail left by a particular ant type. We also extended the ants with a type identifier to allow them to identify foreign and own pheromone trails in the graph.

To implement the ant's repulsion by foreign pheromone the ant's decision process had to be appropriately extended. In the appendix the adapted Ant Colony algorithm for finding disjoint paths is described. The most notable change to the algorithm is the extension of the probabilistic decision formula from ACS. This formula calculates the probability of an ant choosing a certain edge when building a solution. We extended formulas (1) and (2) to formula (5) and (6) respectively.

$$j = \begin{cases} argMax_{u \in J_r^k} [\tau(r,u)].[\eta(r,u)]^{\beta}.[1/\phi_s(r,u)]^{\gamma} & \text{if } q \leq q_0 \\ J & \text{if } q > q_0 \end{cases} \qquad (5)$$

$$p_s(r,u) = \frac{[\tau_s(r,u)]^{\alpha}.[\eta(r,u)]^{\beta}.[1/\phi_s(r,u)]^{\gamma}}{\sum_{l \in J_i^k} [\tau_s(r,l)]^{\alpha}.[\eta(r,l)]^{\beta}.[1/\phi_s(r,l)]^{\gamma}} \qquad (6)$$

In these formulas $\phi_s(r,u)$ represents the amount of pheromone trail not belonging to type s on the edge (r,u). We call this the amount of foreign pheromone. It is the sum of all pheromone trails left by all other ant types. The power γ indicates an ant's sensitivity to the presence of foreign trail. If γ is set to zero the ant will calculate the probability based on the problem heuristic and the pheromone of its own type, just like in the original algorithm, and thus the foreign pheromone trails will be ignored. If γ is increased, the probability of choosing an edge with a large amount of foreign trail will become increasingly smaller.

We also adapted the definition of the best solution used by the algorithm. Each ant type tries to find its own solution. Instead of trying to find the best

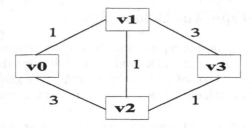

Fig. 1. Test graph 1, contains 2 disjoint paths from v1 to v3, neither of them coincides with the shortest path.

solution to the problem instance, the ants now prefer good problem solutions that are also disjoint to the solutions of other ant types. To ensure disjoint solutions for the disjoint path problem, the ant types try to find paths that minimize the following sum:

$$\sum_{e \in sharedEdges(P)} cost(e) * fa(e) \tag{7}$$

In this sum $fa(e)$ represents the number of other ant types using edge e, and $cost(e)$ represents the cost (in this case the length) of edge e. The sum is calculated over all edges from the path P that are shared with other ant types. Ideally we would like a path without shared edges, reducing the entire sum to zero and providing a completely disjoint solution for the current ant type. If two alternative paths result in the same value for equation 7, the shortest path is preferred. Using this definition the ant types all converge toward their own path, rather than all trying to converge to one single path. This is necessary for obtaining disjoint solutions. Usually it means one ant type converges on the best path found, while the next converges toward the second best path. There are situations where the shortest path is not in the solution set, because it would result in a suboptimal *set of paths*. The test graph in figure (1) above demonstrates such an instance. Assume we are trying to find two disjoint paths from $v0$ to $v3$. If an ant type takes the shortest path $v0 - v1 - v2 - v3$, it is not possible to find another path that is disjoint to this one. When this instance is presented to the Multi-type Ant Algorithm, one ant type will converge on the route $v0 - v1 - v3$, while the other will take $v0 - v2 - v3$. Neither solution found by the ant types is the shortest path, but the combined result is the optimal (and only) disjoint solution. This shows that our multi-type ant algorithm works in a parallel way, rather than an incremental way.

4   Multi-type Ants and the Disjoint Path Problem

The algorithm was tested on three different graphs, shown in figures (1), (2) and (3). These graphs were chosen to demonstrate the general behavior of the algorithm and the influence of its parameters. During testing the traditional ACS parameters were kept at the following constant values: $\beta = 2$, $\tau_0 = 0.05$,

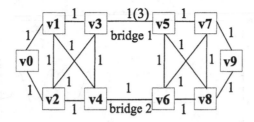

Fig. 2. Test graph 2: the bridges between both parts have to be shared when using more than 2 ant types. In the first experiment on this graph we use a weight equal to 1 for the edge v3-v5. In the second experiment we increase this weight to 3.

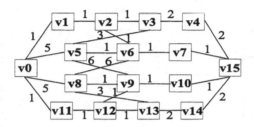

Fig. 3. Test graph 3: the graph allows for a maximum of 4 completely disjoint paths between v0 and v15.

$\rho = 0.1$ and candidate list length is $= 5$. These values are based on values used by Marco Dorigo [6] for experimenting with the TSP, and were found to be robust. The number of iterations, value of $q0$, γ and ants per type are included in the description of the experiments.

4.1 Experiment 1: Finding Disjoint Solutions

The first graph is the small 4 node graph from figure (1). It contains exactly 1 solution when finding 2 disjoint paths between the nodes $v0$ and $v3$. This solution consists of the paths $v0 - v1 - v3$ and $v0 - v2 - v3$ and has a total length of 8. It is interesting to note that the solution does not contain the single shortest path $v0 - v1 - v2 - v3$. This path excludes all other paths and cannot be part of a disjoint solution.

Table 1 shows the speed of convergence on graph 1 for different values of γ and $q0$. The result shows the percentage of runs that has reached the optimal solution after 20 iterations. On this simple problem this will normally reach 100 percent for all settings, given enough iterations. The table shows the influence that $q0$ and γ have on the speed with which the algorithm reaches the optimal solution. Remember that γ determines the ant's repulsion by foreign pheromones, while $q0$ is a parameter from the original Ant Colony System and determines the amount of exploration the ants do. We see that unlike the ACS traveling salesman algorithm, the disjoint paths algorithm benefits more from a low $q0$ and a high

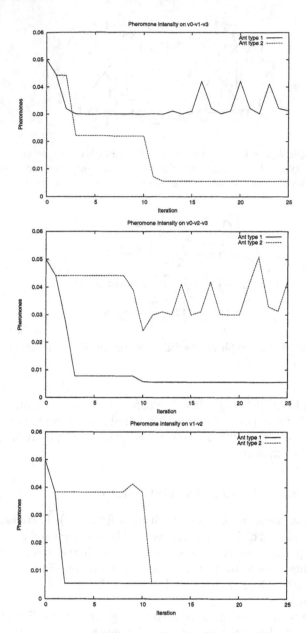

Fig. 4. Evolution of pheromones on the different paths in graph1 during experiment1. The top figure shows the evolution of pheromones on the first disjoint path $v0 - v1 - v3$. The middle figure shows the evolution of pheromones on path $v0 - v2 - v3$. The bottom figure shows the evolution on edge $v1 - v2$. This edge is taken to represent the pheromone intensity on the shortest path $v0 - v1 - v2 - v3$. The edges $v0 - v1$ and $v2 - 3$ are not included as they are also part of the disjoint solution, and as such have a higher pheromone value.

Table 1. Percentage of optimal solutions found on graph 1 for different values of γ and $q0$. The algorithm was run using 20 iterations and $m = 5$ ants for both ant types. Results are averaged over 100 runs.

gamma / q0	0	1	2	3	4	5
0	86	97	97	88	80	69
0.1	89	94	91	85	81	76
0.2	80	93	91	79	72	70
0.3	81	84	90	85	70	61
0.4	73	80	85	78	67	55
0.5	68	67	77	72	49	40
0.6	60	62	64	58	44	38
0.7	44	51	50	40	30	27
0.8	28	30	34	26	16	17
0.9	13	15	13	11	6	3

amount of exploration. Figure 4 shows the evolution of pheromone intensity on the different paths in graph 1. We see that each ant type starts preferring one of the disjoint paths relatively soon, and the amount of pheromones on the edge $v1 - v2$ belonging only to the non-disjoint shortest path drops sharply.

4.2   Experiment 2: Sharing Paths

We now examine the behavior of the algorithm, when the number of ant types is larger than the number of possible disjoint paths. In this case the different ant types compete for good paths between the source and destination nodes, while trying to minimize the number of shared edges. Ideally we would like the ant types to distribute themselves maximally among the possible paths. The path P an ant type chooses when there are no completely disjoint solutions available, depends on the measure it uses to compare paths. We compare 2 approaches in our experiments.

The first approach is to minimize the sum from equation 6. This approach tries to minimize the total cost by minimizing the number of shared edges and preferring to share lower cost edges. The result is that the number of ant types sharing a possible path will be proportional to the cost of that path. If all paths have the same cost, the ant types will distribute themselves equally among the possible paths.

The second approach is to minimize the average number of ant types on shared edges:

$$\frac{1}{\|sharedEdges(P)\|} \sum_{e \in sharedEdges(P)} fa(e)$$

This approach does not take into account the number of edges that is shared or their cost. It tries to minimize the number of ant types that use the same solution. The result is a more equally distributed solution, even when the paths differ in cost.

Table 2. Percentage of ant types choosing each bridge in graph 2. The algorithm was run using 1000 iterations, $\gamma = 2$, $q0 = 1$ and $m = 12$ ants for each ant type. Results are averaged over 100 runs.

	equal weights		different weights	
types	bridge1	bridge2	bridge1	bridge2
2	50.0	50.0	50.0	50.0
3	50.67	49.33	33.33	66.67
4	50.0	50.0	25.0	75.0
5	52.2	47.8	20.2	79.8
6	50.0	50.0	33.33	66.67

Table 3. The same experiment as in table 2, but using the "average number of ant types" measure.

	equal weights		different weights	
types	bridge1	bridge2	bridge1	bridge2
2	50.0	50.0	50.0	50.0
3	51.67	48.33	33.33	66.67
4	50.0	50.0	50.0	50.0
5	49.6	50.4	40.0	60.0
6	50.0	50.0	49.14	50.86

Tables 2 and 3 show the results for using both measures on the graph of figure (2). When trying to find paths between $v0$ and $v9$, the ants have to share one of the bridges connecting both parts of the graph. We examined the amount of times each bridge was selected for the case when both bridges have an equal cost, and for the case where the bottom bridge has a cost of 1 and the top bridge a cost of 3. We see that the use of both bridges is about equal for both measures when the bridges have equal weights. Using the first measure results in a lower use of bridge 1 when the cost of this bridge is increased. With the second measure, the ant types distribute themselves more equally even when bridge 1 has a higher cost.

4.3   Experiment 3: Multiple Disjoint Solutions

The third graph in figure (3) is used to demonstrate the algorithm behavior when more than 2 ant types are present, and there exists a possible disjoint path for every type. The graph allows for up to 4 disjoint paths between $v0$ and $v15$. The algorithm was tested using 2, 3 and 4 ant types. Table 4 shows the results for different values of the exploration parameter $q0$. We also experimented with a variable $q0$ value that was increased during the run of the algorithm, meaning that the exploration drops gradually with the number iterations. The value of $q0$ was increased from 0 to 0.9 in steps of 0.1. The best results are obtained with a constant, low value for $q0$, resulting in a high amount of exploration. Good results are also obtained using the variable $q0$, with a high amount of exploration only during the early phases of the algorithm.

Table 4. Comparison of algorithm results for different numbers of ant types finding paths in graph2. The disjoint columns show the percentage of disjoint solutions found. The optimal column shows the percentage of disjoint solutions found that are also optimal. The variance column shows the coefficient of variation of the solutions with respect to the optimal solution. The final row shows the results using a varying value for q0. This value is increased from 0 to 0.9 during the run of the algorithm. The algorithm was run using 1000 iterations with gamma = 2 and m = 12 ants for each ant type. Results are averaged over 100 runs.

	2 types			3 types			4 types		
q0	%Disj.	%Opt.	VarCoef	%Disj.	%Opt.	VarCoef	%Disj.	%Opt.	VarCoef
0	100	100	0	100	51	0.099	99	44	0.152
0.1	100	100	0	100	61	0.079	100	36	0.167
0.2	100	100	0	100	50	0.101	97	40.2	0.147
0.3	100	100	0	100	48	0.145	92	38	0.131
0.4	100	100	0	99	51.5	0.176	85	42.3	0.158
0.5	100	100	0	97	49.5	0.142	78	56,4	0.069
0.6	100	100	0	88	54.5	0.225	61	54.1	0.089
0.7	100	100	0	66	69.7	0.121	53	75.5	0.05
0.8	100	100	0	46	71.7	0.31	47	89.4	0.035
0.9	100	100	0	33	96.9	0.126	39	100	0
var	100	100	0	98	55.1	0.09	86	48.8	0.107

5   Conclusion

In this paper we have proposed an extension of the single type Ant-Colony algorithm to the multi-type case. Introducing multiple types allows to model competition as well as cooperation explicitly.

The working of the algorithm has been illustrated on the problem of finding a given number of disjoint paths in a graph such that the total cost of the paths is minimized. On the one hand, ants of different types are in competition for the use of links. On the other hand ants of the same type have to collaborate in order to find a path from source to destination that is a short as possible, however disjoint to the paths of ants of different types. Our Multi-type Ant Colony algorithm solves the problem in a non-incremental way, i.e. the different types of ants search for their best path simultaneously. An incremental approach to the disjoint path problem usually does not yield satisfactory results.

Currently the exploration parameter is set globally and is fixed. Except in the third experiment where the degree of exploration is gradually reduced. We believe that for complex graphs with a high demand of disjoint paths this is important in order to get disjoint paths that yield a good global solution, i.e. their total sum is minimized. We also think that setting the exploration rate locally high in areas with severe competition and low in other areas, is an interesting idea to explore in the future. Finally we are convinced that the multi-type approach has a broad spectrum of possible applications such as production planning, traffic engineering and routing in telecommunication networks.

6   Appendix: Pseudo Code

Parameters:
n: number of ant types, m: number of ants per type ,ρ: pheromone decay, β: sensitivity to heuristic, γ: sensitivity to foreign pheromone, q0: exploration rate, cl: candidate list length, t_{max}: maximum number of iterations
For every edge ij **do** {
    For s = 1 to n **do** {
        $\tau_s(i,j) = \tau_0$ }}
For s = 1 to n **do** {
    For k = 1 to m **do** {
        Place ant_s^k on source node. } }
For s = 1 to n **do** {
    Let P_s^+ be the path with length L_s^+ chosen by ant type s }
/* Main Loop */
For t = 1 to t_{max} **do** {
    For k = 1 to m **do** {
        For s = 1 to n **do** {
            While path $P_s^k(t)$ is not complete **do** {:
                If exists at least 1 node j $\in$ candidate list **then**
                Choose next node j $\in J_i^k$ among cl nodes in candidate list:

$$j = \begin{cases} argMax_{u \in J_i^k}[\tau_s(i,u)].[\eta(i,u)]^\beta[1/\phi_s(i,u)]^\gamma & if\ q \le q_0 \\ J & if\ q > q_0 \end{cases}$$

where J $\in J_i^k$ is chosen according to the probability

$$p_s^k(i,j) = \frac{[\tau_s(i,j)].[\eta(i,j)]^\beta.[1/\phi_s(i,j)]^\gamma}{\sum_{l \in J_i^k}[\tau_s(i,l)].[\eta(i,l)]^\beta.[1/\phi_s(i,j)]^\gamma}$$

where i is the current node
                Else
                choose random j $\in J_i^k$
                End if
                After each transition ant_s^k applies the local pheromone update:
                $\tau_s(i,j) = (1-\rho)\tau_s(i,j) + \rho\tau_0$ }}}
    For s = 1 to n **do** {
        For k = 1 to m **do** {
            Compute length $L_s^k(t)$ of path $P_s^k(t)$ by ant_s^k
            If better solution is found **then**:
                update P_s^+ and L_s^+
            End if }}
    For s = 1 to n **do** {
        For every edge ij **do** {
            Update pheromone trails with:
            $\tau_s(i,j) = (1-\rho).\tau_s(i,j) + \Delta\tau_s$, where $\Delta\tau_s = 1/L_s^+$ }}}
Print T^+ and L^+
Stop

References

1. Maria Blesa and Christian Blum. Ant Colony Optimization for the Maximum Edge-Disjoint Paths Problem. *EvoWorkshops 2004, Lecture Notes in Computer Science 3005, 160-169*, 2004.
2. Gianni Di Caro and Marco Dorigo. Antnet: Stigmergetic control for communications network. *Journal in Artificial Intelligence Research 9, 317 - 365*, 1998.
3. M. Dorigo and L.M. Gambardella. Ant colony system: A cooperative learning approach to the traveling salesman problem. *IEEE Transactions on Evolutionary Computation*, 1997.
4. H. Van Dycke Parunak and Sven Brueckner Ant-Like Missionaries and Cannibals: Synthetic Pheromones for distributed motion control. *Proceedings of the Fourth International Conference on Autonomous Agents (Agents'2000), Barcelona, Spain*, 2000.
5. H. Van Dycke Parunak, Sven Brueckner and John Sauter, Synthetic Pheromone Mechanisms for Coordination of Unmanned Vehicles. *Proceedings of the First International Joint Conference on Autonomous Agents and Multi-Agent Systems*, 2002.
6. Guy Theraulaz, Eric Bonabeau and Marco Dorigo. *Swarm Intelligence, From Natural to Artificial Systems*. Santa Fe Institute studies in the sciences of complexity. Oxford University Press, 1999.

On the Design of ACO for the Biobjective Quadratic Assignment Problem

Manuel López-Ibáñez, Luís Paquete, and Thomas Stützle

Darmstadt University of Technology, Computer Science Department
Intellectics Group, Hochschulstr. 10, 64283 Darmstadt, Germany

Abstract. Few applications of ACO algorithms to multiobjective problems have been presented so far and it is not clear how to design an effective ACO algorithms for such problems. In this article, we study the performance of several ACO variants for the biobjective Quadratic Assignment Problem that are based on two fundamentally different search strategies. The first strategy is based on dominance criteria, while the second one exploits different scalarizations of the objective function vector. Further variants differ in the use of multiple colonies, the use of local search, and the pheromone update strategy. The experimental results indicate that the use of local search procedures and the correlation between objectives play an essential role in the performance of the variants studied in this paper.

1 Introduction

Almost all the applications of Ant Colony Optimization (ACO) tackle problems for which solutions are evaluated according to only one objective [6]. However, many real-world problems involve multiple, often conflicting objectives and it is therefore highly desirable to extend the known best ACO techniques to tackle such problems. These multiobjective combinatorial optimization problems (MCOPs) replace the *scalar* value in single objective problems by an objective *vector*, where each component of the vector measures the quality of a candidate solution for one objective.

We consider MCOPs defined in terms of Pareto optimality. Therefore, we first have to introduce a dominance relation between objective vectors. Given two objective vectors u and v, $u \neq v$, we say that u *dominates* v if u is not worse than v for each objective and better for at least one objective. When neither u dominates v nor vice versa, we say that the two objective vectors are nondominated. (We use the same terminology among solutions as for the objective vectors.) The main goal then is to obtain a Pareto global optimum set, that is, a set of feasible solutions such that none of these solutions is dominated by any other feasible solution. As often done in the literature, we call *Pareto front* any set of nondominated objective vectors.

So far, few approaches of ACO algorithms to MCOPs defined in terms of Pareto optimality have been proposed [10, 8, 5]. (For a concise overview of ACO approaches to MCOPs we refer to [6].) In this article, we examine various of the

M. Dorigo et al. (Eds.): ANTS 2004, LNCS 3172, pp. 214–225, 2004.
© Springer-Verlag Berlin Heidelberg 2004

possible implementation choices for ACO algorithms when applied to MCOPs defined in terms of Pareto optimality using the biobjective Quadratic Assignment Problem (bQAP) as an example application. We also consider instances of MCOPs for which we systematically modified the correlation between the two objectives from high positive to high negative correlation. On the algorithmic side, we examine the influence of two essentially different search strategies of how to tackle MCOPs. A first strategy uses scalarizations of the objective function vector into a single value, while the second uses dominance criteria between solutions. In addition, we test several ACO specific parameters like the number of colonies, the pheromone update strategy, and the influence of the usage of local search. The algorithms are evaluated in terms of the binary and unary ϵ-measures [19] and median attainment surfaces [3]. The main results are that (i) for the bQAP local search is essential to achieve high performance, and (ii) the best choice of parameters and methods for the algorithmic components strongly depend on the correlation between the objectives.

The article is organized as follows. Section 2 introduces the bQAP. Section 3 gives an overview of the ACO components we considered for the algorithm configurations that are described in Section 4. Finally, Section 5 gives details on the experimental results and we conclude in Section 6.

2   The Multiobjective QAP

The quadratic assignment problem (QAP) is a well-known $\mathcal{NP}$-hard problem [15], which can intuitively be described as the problem of assigning a set of facilities to a set of locations with given distances between each pair of locations and given flows between each pair of facilities. The goal is to place the facilities on locations such that the sum of the products between flows and distances is minimal [2].

The multiobjective QAP (mQAP) proposed by Knowles and Corne [12] uses different flow matrices, and keeps the same distance matrix. Given n facilities and n locations, a $n \times n$ matrix A where a_{ij} is the distance between locations i and j, and Q $n \times n$ matrices B^q, $q = 1, ..., Q$, where b^q_{rs} is the qth flow between facilities r and s, the mQAP can be stated as follows:

$$\min_{\phi \in \Phi} \begin{cases} \sum_{i=1}^{n} \sum_{j=1}^{n} a_{ij} b^1_{\phi_i \phi_j} \\ \quad \vdots \\ \sum_{i=1}^{n} \sum_{j=1}^{n} a_{ij} b^Q_{\phi_i \phi_j} \end{cases} \tag{1}$$

where min refers to the notion of Pareto optimality. This problem arises in facilities layout of hospitals [12] and social institutions [9].

Local search algorithms for the biobjective QAP (bQAP) were presented in [14], where it was found that the best search strategy to be followed by the local search algorithms depended strongly on the correlation between the flow matrices (and, hence, objectives), which we denote as ξ. Based on these findings we also study the behavior of the ACO algorithms in dependence of ξ.

3   ACO Algorithms for MCOPs

In the following, we discuss some of the main implementation alternatives for tackling MCOPs in terms of Pareto optimality with ACO that are additional to those present for single objective problems. Some of the concepts were already examined in the literature [10, 8, 5].

Multiple Colonies. In a multiple colony approach, the total number of ants is divided into disjoint sets, each of these sets being called a colony. Multiple colonies have previously been used for the parallelization of ACO algorithms. In MCOPs, the main usage of multiple colonies of ants is to allow each colony to specialize on a particular region of the Pareto front [10]. In this case, each colony has its own pheromone information and an ant of a particular colony constructs solutions guided only by the pheromone information of its own colony. We will consider only a *cooperative* case where solutions can be exchanged among colonies, so that the decision of which solution updates the pheromone of a colony is affected by the solutions generated by the other colonies.

Pheromone Information. There exist two main alternatives to define the pheromone information. First, a single pheromone matrix can represent the desirability of the solution components with regard to all objectives [5]. In this case, the stochastic decision at each construction step is done as in single objective problems. The second option uses multiple pheromone matrices, where each of them represents the desirability of the solution components with respect to one objective. Stochastic decisions then may be made at each step according to only one objective, which may be randomly or deterministically chosen according to some criterion, or be based on aggregations of the different pheromone matrices using weights [10]. In the first case, the usual stochastic decisions of ACO algorithms can be made. The second case for the bQAP would be as follows. Given two pheromone matrices $[\tau_{ij}]$ and $[\tau'_{ij}]$, the probability of assigning a facility j to location i is

$$p_{ij}^k = \frac{\left[\tau_{ij}^{(1-\lambda)} \cdot \tau_{ij}'^{\lambda}\right]}{\sum\limits_{l \in \mathcal{N}_i^k} \left[\tau_{il}^{(1-\lambda)} \cdot \tau_{il}'^{\lambda}\right]} \qquad \text{if } j \in \mathcal{N}_i^k \tag{2}$$

where $\mathcal{N}_i^k$ is the feasible neighborhood of ant k, that is, those locations which are still free, and $\lambda \in [0, 1]$ weighs the relative importance of the two objectives. (In this formula we do not consider any heuristic information, which is anyway not used in state-of-the-art ACO approaches to the QAP [16]).

Weight Setting Strategies. Whenever we use multiple pheromone matrices that should be aggregated, *weights* need to be defined to regulate the influence of the individual terms. The usage of different strategies for setting the weights can then result in different search behaviors.

Iredi *et al.* [10] proposed that each ant uses a different weight (λ in Eq. 2), such that all weights are maximally dispersed in the interval $[0, 1]$. When multiple colonies are used, they proposed that each colony considers either the whole interval, disjoint subintervals, or overlapping subintervals. This weight setting strategy can be characterized as one of 'moving towards the Pareto front' in several directions at once [11]. A different strategy would be to assign all ants the same weight at each iteration and vary the weight between iterations. In this method, weights could be changed in such a way that the ants 'move along the Pareto front', trying to find in each iteration good solutions for a specific aggregation of objectives and then changing slightly the direction for the next iteration. These two strategies were also examined for local search procedures [14].

Pheromone Update Strategy. In single objective problems, the best performing ACO algorithms often use only the best solutions of each iteration (iteration-best strategy) or since the start of the algorithm (best-so-far strategy) for updating the pheromones [6]. One might expect that for MCOPs, similar pheromone update strategies may lead to very good performance. In the multiobjective case one can mimic *iteration-best* and *best-so-far* strategies by selecting among the set of solutions generated in the current iteration or since the start of the algorithm; however, it is more difficult to determine which are the best solutions to be chosen for the pheromone update.

Two different ways of implementing the pheromone update are possible. In *selection by dominance*, only nondominated solutions are allowed to deposit pheromone. An iteration-best strategy would consider the nondominated solutions among those generated in the current iteration; a best-so-far scheme would be obtained by choosing only solutions of an *archive* of the nondominated solutions found since the start of the algorithm. In the *selection by scalarization*, only the best solutions with respect to the scalarization chosen are used, either among the solutions found in the current iteration or since the start of the algorithm.

These possibilities can be applied, at least in principle, when using only one pheromone matrix or several pheromone matrices for each objective. In this paper, we restrict ourselves to a simple form of selection by scalarization that selects to update the pheromones of those solutions with the best value with respect to each of the two objectives. Moreover, when one pheromone matrix is used for each of the objectives, each pheromone matrix is updated by the solution from the candidate set with the best value for the respective objective. Therefore, in this case, only one ant per pheromone matrix will be allowed to deposit pheromone, as done in the best performing ACO algorithms.

A further, important factor to be considered in the pheromone update is whether one or multiple colonies are considered. While in the first case any of the above strategies may be applied, many additional possibilities arise in the latter case. For convenience, in this latter case we focus on the method called *update by region* [10], where first the nondominated ants are sorted according to one objective and then they are partitioned as equally as possible into a number of subsets equal to the number of colonies. Then, all solutions assigned to subset i are assigned to colony i.

Finally, one has to decide on the amount of pheromone to be deposited. In MCOPs, the solution cost cannot be used to define this amount. Therefore, all ants may deposit a same, constant amount of pheromone.

Local Search. For a large number of problems, ACO algorithms obtain the best performance when improving the solutions constructed by the ants through the use of local search procedures [6]. Additionally, it is frequently observed that the best parameter settings for ACO algorithms as well as the configuration of ACO algorithms (ie. how algorithmic features are applied and which features are useful at all) depends strongly on the interaction with the local search. Hence, if there exists the intuition that local search is important for the final performance, then local search should be considered from the first design phase.

We can roughly distinguish two classes of local search methods for MCOPs, one based on dominance criteria and the other based on scalarizations of the objective functions. For the bQAP, we implemented two local search algorithms. Pareto Local Search (PLS) [14, 13] is a local search that iteratively adds nondominated solutions to an archive whereas dominated solutions are discarded. The second local search is a single objective iterative improvement algorithm for a weighted sum scalarization of the objective function vector (W-LS). Both, PLS and W-LS use the same underlying 2-exchange neighborhood, where two solutions are neighbored if they differ in the location of exactly two facilities, and a best-improvement pivoting rule.

4   Configurations and Performance Assessment

Many different configurations of ACO algorithms can be designed mixing the concepts mentioned above. To limit the available choices for the underlying ACO algorithm, we decided to follow the rules of $\mathcal{MAX\text{-}MIN}$ Ant System ($\mathcal{MMAS}$) [16] like the management of the pheromones (evaporation, pheromone trail limits, etc.), since $\mathcal{MMAS}$ is known to be a state-of-the-art algorithm for the QAP. In our experimental evaluation, we focus on rather simple, straightforward ACO configurations that reflect the different types of search strategies available for MCOPs. The configurations we test are designed in such a way that they share some common principles and concepts to allow an understanding of what happens if some features of the configurations are varied. In this sense, our approach is very much as in experimental design and the goal is to systematically examine the influence of specific algorithm features on the final performance. In particular, we study the following features.

1. use of best-so-far vs. iteration-best pheromone update
2. use of one vs. multiple colonies
3. use of one pheromone matrix and a search strategy based on the dominance criterion (class D) vs. the use of multiple pheromone matrices and selection by scalarization (class S).

For algorithms of class D, the nondominated solutions generated in the current iteration (iteration-best) or since the start of the algorithm (best-so-far) are distributed among the colonies using update by region, and are then allowed to update the pheromone matrix of the corresponding colony. For algorithms of class S, after distributing the nondominated solutions among colonies, only the best solutions with respect to each objective is allowed to deposit pheromone in the respective pheromone matrix of each colony. (The same policy applies, independent of using only one colony or multiple colonies.)

For the class S algorithms we tested two ways of defining the weights for aggregating the pheromone matrices in the solution construction. The first is that every ant has a different weight and uses the same value in every iteration, denoted as S(all); the second is that all ants use the same weight in the current iteration but the weights are modified between the iterations, denoted as S(one). Finally, it must be mentioned that the algorithms tested were run with local search. In particular, for class D we use PLS, while in the case of class S we use W-LS for the local search. (Note that the local searches were chosen to match the main search strategy of either class D or class S.)

The different approaches are evaluated in terms of the unary ϵ-measure using a lower bound for the bQAP and the binary ϵ-measure [19]. The *unary ϵ-measure* gives the factor by which a Pareto front is worse than the Pareto global optimum set with respect to all objectives. Formally, given a Pareto front A and the Pareto global optimum set P, the unary ϵ-measure for Q objectives is defined as follows:

$$I_\epsilon(A, P) = \max_{p \in P} \min_{a \in A} \max_{q \in Q} \left(\frac{a^q}{p^q} \right) \qquad (3)$$

For the QAP, typically P is not known and a lower bound is needed. We use a lower bound to P extending the Gilmore-Lawler lower bound [7] to the biobjective case. In the single objective case, this bound is given by the optimal objective value of an associated Linear Assignment Problem (LAP) (see [1] for more details). We defined a biobjective LAP (bLAP) where each objective in the bLAP is associated to an objective in the bQAP. Then we solve several scalarizations of the bLAP to obtain a set of solutions that are not dominated by the Pareto global optimum set. In order to have a set that dominates the Pareto global optimum set, we added points as follows. First, we sort the objective vectors lexicographically and for each successive pair of objective vectors u, v we added a point $w = (\min\{u_1, v_1\}, \min\{u_2, v_2\})$. For the instances considered here we used 5000 weight vectors maximally dispersed in $[0, 1]$.

The *binary ϵ-measure* is the pairwise version of the unary ϵ-measure. Given two Pareto fronts A and B, it considers the values of $I_\epsilon(A, B)$ and $I_\epsilon(B, A)$. If $I_\epsilon(A, B) \leq 1$ and $I_\epsilon(B, A) > 1$ then A is better than B, that is, at least one solution in A is not worse in all objectives than any solution in B and $A \neq B$.

We use the binary ϵ-measure to detect if the Pareto front returned by an ACO algorithm is better than the other. If no conclusion can be drawn from the binary ϵ-measure, we use the unary ϵ-measure with the lower bound as defined above. As a next step, we perform a ANOVA analysis [4] on the values returned by the unary ϵ-measure in order to detect which of the components contribute to the overall performance of ACO. Finally, in order to visualize the differences of performance in the objective space, we also plot the median attainment surfaces [3] of the Pareto fronts obtained by some configurations.

5 Experimental Setup and Results

The experimental setup considers three components: (i) the search strategy that can be of class D, S(all) or S(one); (ii) iteration-best versus best-so-far phero-mone update; and (iii) one or multiple colonies. In addition, each algorithm was run with and without local search. The total number of ants (m) was equal to the instance size and we used 5 colonies in the multiple colony approach, where each colony had $m/5$ ants. For the management of the pheromones, we followed the rules of $\mathcal{MMAS}$ with $\rho = 0.9$ for the pheromone evaporation; $p_{best} = 0.05$ to derive the factor between the lower and upper pheromone trail limits; and τ_{max} was set to the theoretically largest value [16].

The algorithms were tested on six symmetric bQAP instances, three *un-structured* and three *structured* instances. The three unstructured instances were taken from an earlier experimental study [14] and are available at http://www.in tellektik.informatik.tu-darmstadt.de/~lpaquete/QAP. All instances were gen-erated using the instance generator of Knowles & Corne [12] with size $n = 50$ and $\xi \in \{0.75, 0.0, -0.75\}$, where ξ is a parameter that influences the correlation between the flow matrices. The QAP specific parameter settings for generating the unstructured instances were set the same as used for generating instances of class Taixxa [17]; parameter settings for the structured instances where analo-gous to those for generating Taixxb [17] instances.

Note that the correlations of the flow matrices also result in different corre-lations between the objective value vectors especially for the unstructured in-stances. For unstructured instances, the resulting correlations between the two objectives are of $0.90, -0.01$, and -0.90 for the three values of ξ; in the structured case, there is not anymore a clear correlation betwen the objectives (the corre-lations between the objectives are $0.23, 0.03$, and -0.08, respectively), which is probably due to the many zero entries in the flow matrices. The empirical cor-relations between objectives were determined through samples of 1,000 random solutions generated for each instance.

Each of the algorithms was run 20 times on each instance for a maximum of 300 CPU-seconds. The algorithms were coded in C and the experiments were run on a Pentium III 933 MHz CPU and 512 MB RAM under Debian Linux.

As a first step in the analysis of the experimental results, every algorithm was compared against each other using the binary ϵ-indicator. This analysis clearly showed that for unstructured as well as structured instances any algorithm that uses local search outperforms all algorithms that do not use local search. Interest-ingly, the second phase of our analysis, which is described below, also suggested that the best ACO configurations that do not use local search are very different from the best configurations that do make use of local search procedures. This indicates that (i) studying ACO algorithms for MCOPs without local search and then simply adding local search leads to suboptimal performance and (ii) that local search may be, as it is true for the single objective case, also an essential component of ACO algorithms for MCOPs. In the following analysis, we only consider ACO configurations that use local search.

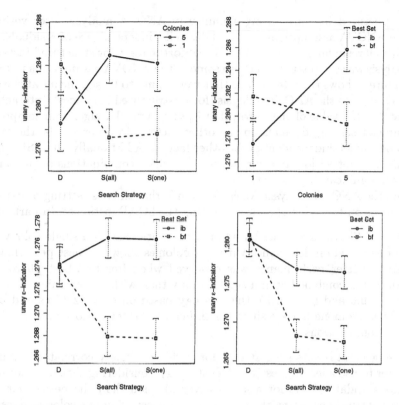

Fig. 1. Interaction plots for three unstructured instances of size 50 and with $\xi = \{\, 0.75$ (top), 0.0 (bottom left), -0.75 (bottom right)$\}$. The search strategy can be based on dominance with PLS (D) or on scalarizations with W-LS, either in several directions $(S(all))$ or in one direction at each iteration $(S(one))$. The pheromone update strategy can consider the best solutions found in the current iteration $(ib,$ iteration-best) or in all the previous iterations $(bf,$ best-so-far). The number of colonies can be one or five

The results of the first phase of the analysis also showed that for structured instances the configuration of class D with PLS performs worse than the one based on class S with W-LS. The main reason is that PLS becomes unfeasible, because of an enormous number of nondominated solutions, which is much larger than for the unstructured instances, resulting in very high computation times for PLS. In addition, this first phase showed that for unstructured instances with $\xi = 0.75$, the algorithms of class S using five colonies and iteration-best pheromone update were outperformed by the other variants and that the variant of class D using one colony and best-so-far pheromone update was slightly worse than the other variants. For the other unstructured instances the outcomes of the algorithms were mainly incomparable in terms of dominance. Therefore, we proceeded to the second phase of our analysis based on ANOVA.

In the second phase, we evaluated each run of all the algorithms using the unary ϵ-indicator based on the lower bounds and we analyzed the results for the unstructured instances using a multiway ANOVA analysis after verifying that the ANOVA assumptions (homoscedasticity, independence of residuals, and normality of residuals) were satisfied [4]. Unfortunately, for structured instances this analysis was not possible, because some of the ANOVA assumptions were not met. Figure 1 shows the interactions that were found to be significant at the 0.05 level according to the ANOVA analysis (for unstructured instances). Interactions indicate that the value of one factor, that is, one particular algorithm component or parameter setting, depends on the other factors. In other words, the factor cannot be studied independent of the other factors. Additionally, the HSD Tukey intervals are plotted [4], allowing us to infer which combinations of parameter values are significantly different.

From the ANOVA analysis, we can conclude that the best setting of parameters depends on the value of ξ for generating the bQAP instances. In particular,

- for $\xi = 0.75$, the strategies based on scalarizations with a single colony and the one based on dominance with five colonies show the best performance. Also, a better performance was observed when iteration-best pheromone update was combined with a single colony than with five colonies;
- for $\xi = 0.0$ and $\xi = -0.75$, the strategy based on scalarizations and best-so-far pheromone update show a significantly better performance than the other combinations.

These general results suggest that for highly positively correlated instances it is better to use a less aggressive ACO strategy, considering that iteration-best pheromone update allows for a larger diversification than the best-so-far one. The same is true considering the number of colonies: using less colonies does not allow a too strong focus on specific regions of the Pareto front, hence leading to a higher exploration. On the weakly or negatively correlated instances this is not anymore true; here a much stronger exploitation of the search experience appears to be necessary.

Figure 2 shows the median attainment surfaces [3] for unstructured (left side) and structured (right side) instances using search strategy S with W-LS. In addition, we plotted reference solutions (given as points in Fig. 2) that were obtained by a scalarized version of the Robust Tabu Search algorithm (W-RoTS) [14], which gives a high quality approximation to the Pareto global optimum set. For W-RoTS we run as many scalarizations as possible in the given 300 CPU seconds; each run of the underlying tabu search algorithm (RoTS) was stopped after $100 \cdot n$ iterations (see [18] for details on RoTS). In the given CPU-time approximately 136 scalarizations could be run for W-RoTS. All solutions generated in such a way were then filtered to yield the given nondominated points. The result of this comparison is that on the unstructured instances, W-RoTS yields typically better performance than the ACO algorithms, while on the structured instances, the median attainment surfaces of the ACO algorithms appear to dominate the result of W-RoTS. This result is somehow analogous to the single objective case, where $\mathcal{MMAS}$ using a simple 2-opt local search performs worse than RoTS for unstructured instances, whereas the same algorithm outperforms RoTS for structured instances that occur frequently in applications [16].

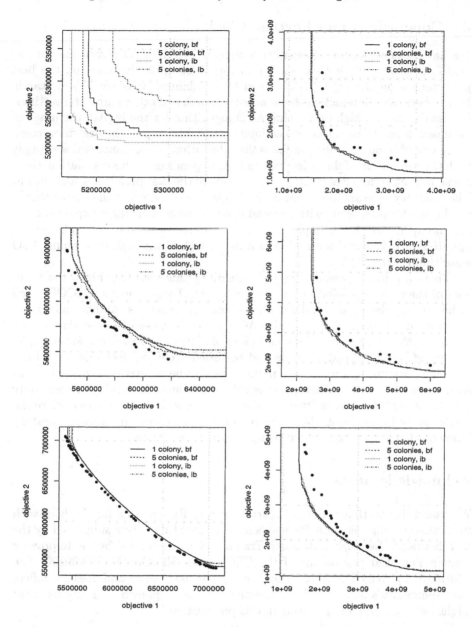

Fig. 2. Median attainment surfaces obtained for unstructured (left column) and structured (right column) instances of size 50 and ξ of 0.75 (top), 0.0 (center) and -0.75 (bottom) when using search strategy based on scalarizations with W-LS. In addition, the objective value vectors obtained by W-RoTS are plotted (points). (For details see text.)

6   Conclusions and Further Work

We have studied several alternative configurations of ACO algorithms for the bQAP. A first result is that, similar to the single objective case [6], the best performance on the bQAP is obtained by combining ACO with local search. Hence, because of interactions between local search and other parameter settings, for the design of a high performing ACO algorithm for the bQAP, the usage of local search has to be taken into account as the first step in the design process.

A second important observation is that algorithm performance varies strongly with the correlation of the objectives. In fact, depending on this correlation, there are significant differences in the configurations of the best performing algorithms. The tendency appears to be that for high positive correlations the search should not be too focused, while with decreasing correlation, a stronger exploitation of search experience, for example, through the usage of a best-so-far pheromone update strategy, becomes an important ingredient of any high performing ACO algorithm for the bQAP.

There are many possible ways of extending this research. First, the influence of the various possibilities of configuring ACO algorithms for MCOPs need to be studied also on other, differently structured problems. Additionally, more components useful for multiobjective ACO algorithms need to be explored. Furthermore, the exploration of fast and more effective local searches, for example, the use of RoTS instead of W-LS and bounded archiving techniques for PLS, is a highly promising direction to improve performance. Another interesting direction would be to use heterogeneous colonies at least for weakly or negatively correlated bQAP instances: two colonies based on scalarizations can explore the "tails" of the Pareto front while using one or more colonies with a search strategy based on dominance explore the center of the Pareto front.

Acknowledgments

We would like to thank Marco Chiarandini and Rubén Ruiz for the help with the statistical analysis. Luis Paquete and Thomas Stützle were supported by the Metaheuristics Network, a Research Training Network funded by the Improving Human Potential programme of the CEC, grant HPRN-CT-1999-00106. The information provided is the sole responsibility of the authors and does not reflect the Community's opinion. The Community is not responsible for any use that might be made of data appearing in this publication.

References

1. R. E. Burkard, E. Çela, P. M. Pardalos, and L. S. Pitsoulis. The quadratic assignment problem. In P. M. Pardalos and D.-Z. Du, editors, *Handbook of Combinatorial Optimization*, pages 241–338. Kluwer Academic Publishers, 1998.
2. E. Çela. *The Quadratic Assignment Problem: Theory and Algorithms*. Kluwer Academic Publishers, 1998.

3. V. Grunert da Fonseca, C. M. Fonseca, and A. O. Hall. Inferential performance assessment of stochastic optimisers and the attainment function. In Eckart Zitzler et al., editors, *Proc. of EMO'01*, LNCS 1993, pages 213–225. Springer Verlag, 2001.
4. A. Dean and D. Voss. *Design and Analysis of Experiments*. Springer, 1999.
5. K. Doerner, R. F. Hartl, and M. Reimann. Cooperative ant colonies for optimizing resource allocation in transportation. In E. J. W. Boers et al., editors, *Applications of Evolutionary Computing: Proc. of EvoWorkshops 2001*, volume 2037 of *LNCS*, pages 70–79. Springer Verlag, 2001.
6. M. Dorigo and T. Stützle. *Ant Colony Optimization*. MIT Press, Cambridge, MA, 2004.
7. P. C. Gilmore. Optimal and suboptimal algorithms for the quadratic assignment problem. *Journal of the SIAM*, 10:305–313, 1962.
8. M. G. Guntsch and M. Middendorf. Solving multi-criteria optimization problems with population-based ACO. In C. M. Fonseca et al., editors, *Evolutionary Multi-Criterion Optimization*, volume 2632 of *LNCS*, pages 464–478. Springer Verlag, 2003.
9. H. Hamacher, S. Nickel, and D. Tenfelde-Podehl. Facilities layout for social institutions. In *Operation Research Proceedings 2001*, pages 229–236. Springer Verlag, 2001.
10. S. Iredi, D. Merkle, and M. Middendorf. Bi-criterion optimization with multi colony ant algorithms. In E. Zitzler et al., editors, *Proc. of EMO'01*, volume 1993 of *Lecture Notes in Computer Science*, pages 359–372. Springer Verlag, 2001.
11. J. Knowles and D. Corne. Towards landscape analysis to inform the design of hybrid local search for the multiobjective quadratic assignment problem. In A. Abraham et al., editors, *Soft Computing Systems: Design, Management and Applications*, pages 271–279. IOS Press, 2002.
12. J. Knowles and D. Corne. Instance generators and test suites for the multiobjective quadratic assignment problem. In C. M. Fonseca et al., editors, *Proc. of EMO'03*, LNCS 2632, pages 295–310. Springer Verlag, 2003.
13. L. Paquete, M. Chiarandini, and T. Stützle. Pareto local optimum sets in the biobjective traveling salesman problem: An experimental study. In X. Gandibleux et al., editors, *Metaheuristics for Multiobjective Optimisation*, LNEMS 535. Springer Verlag, 2004.
14. L. Paquete and T. Stützle. A study of local search algorithms for the biobjective QAP with correlated flow matrices. *European Journal of Operational Research*, 2004. To appear.
15. S. Sahni and T. Gonzalez. P-complete approximation problems. *Journal of the ACM*, 23:555–565, 1976.
16. Thomas Stützle and Holger H. Hoos. $\mathcal{MAX}$-$\mathcal{MIN}$ Ant System. *Future Generation Computer Systems*, 16:889–914, 2000.
17. É. D. Taillard. A comparison of iterative searches for the quadratic assignment problem. *Location Science*, 3:87–105, 1995.
18. Éric D. Taillard. Robust Taboo Search for the Quadratic Assingnment Problem. *Parallel Computing*, 17:443–455, 1991.
19. E. Zitzler, L. Thiele, M. Laumanns, C. M. Fonseca, and V. Grunert da Fonseca. Performance assessment of multiobjective optimizers: an analysis and review. *IEEE Transactions on Evolutionary Computation*, 7:117–132, April 2003.

Reasons of ACO's Success in TSP

Osvaldo Gómez and Benjamín Barán

Centro Nacional de Computación
Universidad Nacional de Asunción
Paraguay
{ogomez,bbaran}@cnc.una.py
http://www.cnc.una.py

Abstract. Ant Colony Optimization (ACO) is a metaheuristic inspired by the foraging behavior of ant colonies that has empirically shown its effectiveness in the resolution of hard combinatorial optimization problems like the Traveling Salesman Problem (TSP). Still, very little theory is available to explain the reasons underlying ACO's success. An ACO alternative called Omicron ACO (OA), first designed as an analytical tool, is presented. This OA is used to explain the reasons of elitist ACO's success in the TSP, given a globally convex structure of its solution space.

1 Introduction

Ant Colony Optimization (ACO) is a metaheuristic proposed by Dorigo et al. [3] that has been inspired by the foraging behavior of ant colonies. In the last years ACO has empirically shown its effectiveness in the resolution of several different NP-hard combinatorial optimization problems [3]; however, still little theory is available to explain the reasons underlying ACO's success. Birattari et al. [1] developed a formal framework of ant programming with the goal of gaining deeper understanding of ACO, while Meuleau and Dorigo [10] studied the relationship between ACO and the Stochastic Gradient Descent technique. Gutjahr [7] presented a convergence proof for a particular ACO algorithm called Graph-based Ant System (GBAS) that has an unknown empirical performance. He proved that GBAS converges, with a probability that could be made arbitrarily close to 1, to the optimal solution of a given problem instance. Later, Gutjahr demonstrated for a time-dependent modification of the GBAS that its current solutions converge to an optimal solution with a probability exactly equal to 1 [8]. Stützle and Dorigo presented a short convergence proof for a class of ACO algorithms called $ACO_{gb,\tau_{min}}$ [11], where gb indicates that the global best pheromone update is used, while τ_{min} indicates that a lower limit on the range of the feasible pheromone trail is forced. They proved that the probability of finding the optimal solution could be made arbitrarily close to 1 if the algorithm is run for a sufficiently large number of iterations. Stützle and Hoos [12] calculated a positive correlation between the quality of a solution and its distance to a global optimum for the TSP, studying search space characteristics. Hence, it seems reasonable to assume that the concentration of the search around the best solutions

M. Dorigo et al. (Eds.): ANTS 2004, LNCS 3172, pp. 226–237, 2004.

found so far is a key aspect that led to the improved performance shown by ACO algorithms. However, there is no clear understanding of the real reasons of ACO's success, as recognized by Dorigo and Stützle [4, 11]. They stated that although it has been experimentally shown to be highly effective, only limited knowledge is available to explain why ACO metaheuristic is so successful [4].

Considering that elitist versions of ACO outperform non-elitist ones [12], this paper concentrates only on elitists. In search of a new ACO analytical tool to study their success, a simple algorithm preserving certain characteristics of elitist ACO was developed for this work. This is how the Omicron ACO (OA) was conceived. This name comes from the main parameter used (Section 3.2), which is *Omicron* (*O*). The OA simplicity facilitates the study of the main characteristics of an ACO in the TSP context, as explained in the following Sections.

The TSP is summarized in Section 2, while the standard ACO approach and the OA are presented in Section 3. The behavior of the OA for the problems berlin52, extracted from TSPLIB[1], and for a small randomly chosen TSP are shown in Section 4. In Section 5, the core of this paper is presented, analyzing the reasons of ACO's success in the TSP. Finally, the conclusions and future work are given in Section 6.

2   Test Problem

In this paper the symmetric Traveling Salesman Problem (TSP) is used as a test problem to study the OA, given the recognized ACO success in solving it [3, 12].

The TSP can be represented by a complete graph $G = (N, A)$ with N being the set of nodes, also called cities, and A being the set of arcs fully connecting the nodes. Each arc (i, j) is assigned a value $d(i, j)$ which represents the distance between cities i and j. The TSP is stated as the problem of finding a shortest closed tour r^* visiting each of the $n = |N|$ nodes of G exactly once.

Suppose that r_x and r_y are TSP tours or solutions over the same set of n cities. For this work, $l(r_x)$ denotes the length of tour r_x. The distance $\delta(r_x, r_y)$ between r_x and r_y is defined as n minus the number of edges contained in both r_x and r_y.

3   Ant Colony Optimization

Ant Colony Optimization (ACO) is a metaheuristic inspired by the behavior of ant colonies [3]. In the last years, elitist ACO has received increased attention by the scientific community as can be seen by the growing number of publications and its different fields of application [12]. Even though there exist several ACO variants, what can be considered a standard approach is next presented [5].

---

[1] Accessible at http://www.iwr.uni-heidelberg.de/iwr/comopt/software/TSPLIB95/

3.1 Standard Approach

ACO uses a pheromone matrix $\tau = \{\tau_{ij}\}$ for the construction of potential good solutions. It also exploits heuristic information using $\eta_{ij} = \frac{1}{d(i,j)}$. Parameters α and β define the relative influence between the heuristic information and the pheromone levels. While visiting city i, $\mathcal{N}_i$ represents the set of cities not yet visited and the probability of choosing a city j at city i is defined as

$$
\mathcal{P}_{ij} = \begin{cases} \dfrac{\tau_{ij}^{\alpha} \cdot \eta_{ij}^{\beta}}{\sum_{\forall g \in \mathcal{N}_i} \tau_{ig}^{\alpha} \cdot \eta_{ig}^{\beta}} & \text{if } j \in \mathcal{N}_i \\[2ex] 0 & \text{otherwise} \end{cases}
\tag{1}
$$

At every generation of the algorithm, each ant of a colony constructs a complete tour using (1). Pheromone evaporation is applied for all (i, j) according to $\tau_{ij} = (1 - \rho) \cdot \tau_{ij}$, where parameter $\rho \in (0, 1]$ determines the evaporation rate. Considering an elitist strategy, the best solution found so far r_{best} updates τ according to $\tau_{ij} = \tau_{ij} + \Delta\tau$, where $\Delta\tau = 1/l(r_{best})$ if $(i, j) \in r_{best}$ and $\Delta\tau = 0$ if $(i, j) \notin r_{best}$. For one of the best performing ACO algorithms, the $\mathcal{MAX}$-$\mathcal{MIN}$ Ant System ($\mathcal{MMAS}$) [12], minimum and maximum values are imposed to τ (τ_{min} and τ_{max}).

3.2 Omicron ACO

OA is inspired by $\mathcal{MMAS}$, an elitist ACO currently considered among the best performing algorithms for the TSP [12]. It is based on the hypothesis that it is convenient to search nearby good solutions [2, 12].

The main difference between $\mathcal{MMAS}$ and OA is the way the algorithms update the pheromone matrix. In OA, a constant pheromone matrix τ^0 with $\tau_{ij}^0 = 1$, $\forall i, j$ is defined. OA maintains a population $P = \{P_x\}$ of m individuals or solutions, the best unique ones found so far. The best individual of P at any moment is called P^*, while the worst individual P_{worst}.

In OA the first population is chosen using τ^0. At every iteration a new individual P_{new} is generated, replacing $P_{worst} \in P$ if P_{new} is better than P_{worst} and different from any other $P_x \in P$. After K iterations, τ is recalculated. First, $\tau = \tau^0$; then, $\frac{O}{m}$ is added to each element τ_{ij} for each time an arc (i, j) appears in any of the m individuals present in P. The above process is repeated every K iterations until the end condition is reached (see pseudocode for details). Note that $1 \leq \tau_{ij} \leq (1 + O)$, where $\tau_{ij} = 1$ if arc (i, j) is not present in any P_x, while $\tau_{ij} = (1 + O)$ if arc (i, j) is in every P_x.

Similar population based ACO algorithms (P-ACO) [5, 6] were designed by Guntsch and Middendorf for dynamic combinatorial optimization problems. The main difference between the OA and the *Quality* Strategy of P-ACO [6] is that OA does not allow identical individuals in its population. Also, OA updates τ every K iterations, while P-ACO updates τ every iteration. Notice that any elitist ACO can be considered somehow as a population based ACO with a population that increases at each iteration and where older individuals have less influence on τ because of the evaporation.

Pseudocode of the main *Omicron ACO*

> Input parameters: n, matrix $D = \{d_{ij}\}$, O, K, m, α, β
> Output parameter: P (m best found solutions)

$P =$ *Initialize population* (τ^0)
REPEAT UNTIL end condition
   $\tau =$ *Calculate pheromone matrix* (P)
   REPEAT K TIMES
      Construct a solution P_{new} using equation (1)
      IF $l(P_{new}) < l(P_{worst})$ AND $P_{new} \notin P$
       $P =$ *Update population* (P_{new}, P)

Pseudocode of the function *Initialize population* (τ^0)

Initialize set P as empty
WHILE $|P| < m$
   Construct a solution P_x using equation (1)
   IF $P_x \notin P$ THEN include P_x in P
Sort P from worst to best considering $l(P_x)$
$P_{worst} = P_0$

Pseudocode of the function *Calculate pheromone matrix* (P)

$\tau = \tau^0$
REPEAT for every P_x of P
   REPEAT for every arc (i, j) of P_x
     $\tau_{ij} = \tau_{ij} + \frac{O}{m}$

Pseudocode of the function *Update population* (P_{new}, P)

$P_0 = P_{new}$
Sort P efficiently from worst to best considering $l(P_x)$
$P_{worst} = P_0$

4 Behavior of Omicron ACO

Definition 1. *Mean distance from a tour r to P. $\delta(P, r) = \frac{1}{m} \sum_{i=1}^{m} \delta(P_i, r)$. If* $r = r^*$ it gives a notion of how close a population is to the optimal solution r^*.

Definition 2. *Mean distance of P. $\delta(P) = \frac{2}{m(m-1)} \sum_{i=1}^{m-1} \sum_{j=i+1}^{m} \delta(P_i, P_j)$. It* gives an idea of the convergence degree of a population.

Boese studied in [2] the space of solutions of the TSP att532 of 532 cities. 2,500 runs of different local search heuristic were made. For each heuristic, h different tours were stored in a set $H = \{H_i\}$. Each stored tour H_i has a length $l(H_i)$, a distance to r^* denoted as $\delta(H_i, r^*)$ and a mean distance to the other solutions of H called $\delta(H, H_i)$. Boese calculated a positive correlation between all

Table 1. Correlation of the OA behavior variables studied for the problem berlin52

	$\delta(P,r^*)_M$	$l(P^*)_M$	$l(P)_M$	$\zeta(P)_M$
$\delta(P)_M$	0.990	0.957	0.977	-0.972
$\delta(P,r^*)_M$		0.928	0.957	-0.995
$l(P^*)_M$			0.996	-0.900
$l(P)_M$				-0.934

these 3 variables. Given that the set $\{l(H_i)\}$ has a positive correlation with the set $\{\delta(H, H_i)\}$, Boese suggested a globally convex structure of the TSP solution space. In other words, the more central the position of a solution H_i within the set of solutions H is, the smaller its mean distance to the other solutions; therefore, the smaller is its expected length $l(H_i)$, i.e. the better the solution is. Global convexity is not convexity in the strict sense [9]. Boese suggested the analogy with a big valley structure, in which viewed from afar may appear to have a single minimum, but which up close has many local minima [2, Fig. 1]. Boese found similar results for two random geometric instances with 100 and 500 cities. At the same time, the authors of the present work are studying TSPLIB problems with identical conclusions. Also Stützle and Hoos calculated a positive correlation between the quality of a solution and its distance to a global optimum for the problems rat783 and fl1577 [12]. All these experimental results support the conjecture of a globally convex structure of the TSP's search space.

Based on the studies on local search heuristics mentioned above, the present work uses the globally convex structure of the TSP solution space concept as the main idea to explain the reasons of ACO's success.

It is also interesting to observe the length of P^*, $l(P^*)$; the mean length of a population, $l(P) = \frac{1}{m}\sum_{i=1}^{m}l(P_i)$ and the number of individuals $\zeta(P)$ that entered a population. Their mean values for several runs of the OA are denoted as $l(P^*)_M$, $l(P)_M$ and $\zeta(P)_M$ respectively. Accordingly, $\delta(P,r^*)_M$ and $\delta(P)_M$ represent the mean value of $\delta(P,r^*)$ and $\delta(P)$.

To maintain the number of possible tours to a manageable value, a random TSP called omi1 was designed with 8 cities of coordinates (58,12), (2,73), (14,71), (29,8), (54,50), (0,7), (2,91) and (44,53). Fig. 1 shows the evolution of the mean variables above defined as a function of the number of iterations in 10 runs of the OA. The left side of Fig. 1 presents the graphics for the TSP berlin52 (using the parameters $O = 600$, $m = 25$, $K = 1,000$, $\alpha = 1$ and $\beta = 2$), while the right side presents the graphics for the TSP omi1 (using the parameters $O = 30$, $m = 8$, $K = 10$, $\alpha = 1$ and $\beta = 2$).

The typical behaviors for both problems are similar to the mean behaviors shown in Fig. 1 respectively. The correlation values between $\delta(P,r^*)_M$, $\delta(P)_M$, $l(P^*)_M$, $l(P)_M$ and $\zeta(P)_M$ for the problem berlin52 are summarized in Table 1.

It can be observed in Fig. 1 that $\delta(P,r^*)_M$, $\delta(P)_M$, $l(P^*)_M$ and $l(P)_M$ decrease in the initial phase, while $\zeta(P)_M$ increases. In other words, new individuals with shorter length enter P at the beginning of a run; these individuals get closer to each other and at the same time they get closer to r^*. In the final phase the

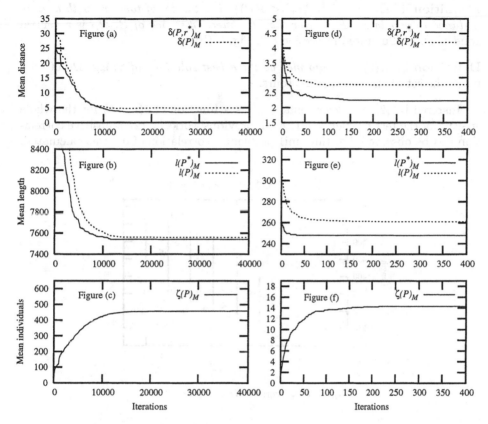

Fig. 1. Evolution as a function of iterations of $\delta(P, r^*)_M$, $\delta(P)_M$, $l(P^*)_M$, $l(P)_M$ and $\zeta(P)_M$ for 10 runs. Left, for the problem berlin52. Right, for the problem omi1

variables remain almost constant. It can be said that almost no individuals enter P and that $\delta(P, r^*)_M$ results smaller than $\delta(P)_M$, which means that the individuals finish closer to r^* than to the other individuals of P. These results motivate the analysis of the reasons of ACO's success in the next section.

5   Reasons of ACO's Success in TSP

The following exhaustive study is presented using the problem omi1 with 8 cities considering the space restrictions of this publication. The same exhaustive study was made using other randomly chosen problems with 7 and 9 cities and the results were very similar, making unnecessary any repetition in this paper.

5.1   Geometry of the Problem

Definition 3. $S = \{r_x\}$, *i.e. the whole discrete search space of a TSP. Ω will denote a subspace of S, i.e. $\Omega \subseteq S$.*

Definition 4. $\Omega_P = \{r_x|\ \delta(P, r_x) < \delta(P)\}$, *i.e. the set of tours r_x with a mean distance to population P shorter than the mean distance of P. Ω_P is a central zone of P, as illustrated in Section 5.3, Fig. 5 (b).*

Definition 5. $\Omega(e)$. *Ω conformed by the e best solutions of S; e.g. $\Omega(100)$ denotes the set of the 100 shortest tours.*

Inspired by [2], Fig. 2 presents $l(r_x)$ as a function of $\delta(r_x, r^*)$ for the whole space S of the test problem omi1. As in previous works [2, 12], a positive correlation can be observed. For this omi1 problem a correlation of 0.7 was calculated.

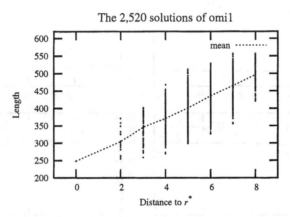

Fig. 2. $l(r_x)$ versus $\delta(r_x, r^*)$ $\forall r_x \in S$

Fig. 3 shows $l(r_x)$ as a function of $\delta(\Omega(e), r_x)$ $\forall r_x \in S$. For $e = 2,520$ ($\Omega(2,520) = S$), Fig. 3 (a) clearly shows that the correlation of the variables is 0 since the mean distance from any solution to all the others is the same. Fig. 3 (b) shows the same graph for $e = 2,519$, i.e. eliminating the worst solution from S. For this case the correlation increases to 0.521. Finally, Fig. 3 (c) draws the graph for $e = 1,260$ (best half solutions) and the correlation between the variables is 0.997. These results are consistent with the suggestion of a globally convex structure of the TSP solution space, since the smaller the distance of a solution r_x to a set of good solutions (and thus, more central its position in $\Omega(e) \subset S$), the smaller its expected tour length is.

Definition 6. $Q(e) = \{Q(e)_x\}$ *is defined as a set of randomly chosen elements of $\Omega(e)$ with cardinality $|Q(e)|$.*

Given the interesting geometrical characteristic of $\Omega(e)$, a good question is if this globally convex property is maintained for $Q(e)$. To understand the importance of this question, it should be noticed that a population P of an OA may be considered as $Q(e)$. Fig. 4 shows $l(r_x)$ as a function of $\delta(Q(e), r_x)$ $\forall r_x \in S$. Randomly selected $Q(e)$ with $|Q(e)| = 25$ for different values of e are presented in figures 4 (a) to (d). The figure shows individuals of $Q(e)$, denoted

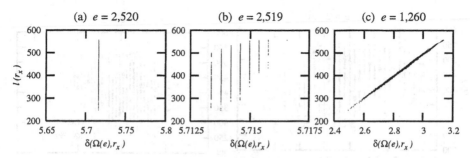

Fig. 3. $l(r_x)$ versus $\delta(\Omega(e), r_x)$ $\forall r_x \in S$ for different values of e

as $Q(e)_x$ and the elements of Ω_P for $P = Q(e)$, denoted as $\Omega_{Q(e)}$. As can be seen in Fig. 4 (b) to (d) the best solutions are in $\Omega_{Q(e)}$; therefore, it seems convenient to explore $\Omega_{Q(e)}$.

To interpret Fig. 4 better, Table 2 presents the correlation ϱ between $l(r_x)$ and $\delta(Q(e), r_x)$ $\forall r_x \in S$ for the four different $Q(e)$ of Fig. 4. To compare the experimental results of Fig. 4 with average values for 1,000 randomly chosen $Q(e)$, Table 2 also presents the calculated average ϱ_M for the same parameters e, showing that Fig. 4 represents pretty well an average case.

Table 2. Correlation between $l(r_x)$ and $\delta(Q(e), r_x)$ $\forall r_x \in S$ for different values of e

	$e = 2,520$	$e = 1,890$	$e = 1,260$	$e = 630$
Correlation ϱ (for Fig. 4)	-0.201	0.387	0.641	0.862
Experimental mean correlation ϱ_M	0.001	0.425	0.683	0.836

As seen in Table 2 there is no meaningful correlation ϱ_M in $Q(2,520)$, as happened with $\Omega_{2,520} = S$. When decreasing the value of e, ϱ_M increases as happened with the correlation calculated for $\Omega(e)$. Thus, with good probability, $Q(e)$ is also globally convex (this probability increases with $|Q(e)|$). Considering the global convexity property, it can be stated that given a population $P = Q(e)$ of good solutions, it is a good idea to search in a central zone of P and specially in Ω_P, which contains the solutions with the shortest distance to P, given the positive correlation between the quality of a solution and its distance to P.

5.2   OA in a Globally Convex Geometry

OA concentrates an important proportion of its search of new solutions in Ω_P. This can be understood because in the construction of a new solution P_{new}, a larger probability is assigned to the arcs of each individual of P. This can be seen as a search made close to each individual of P. As a consequence, P_{new}

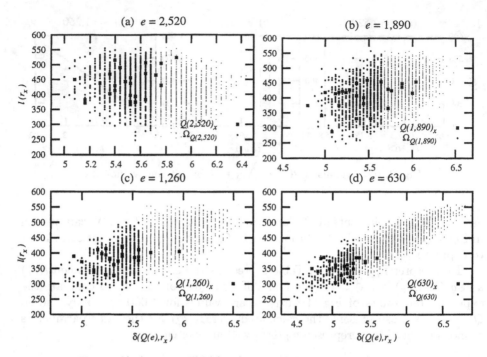

Fig. 4. $l(r_x)$ versus $\delta(Q(e), r_x)$ $\forall r_x \in S$ for different values of e

Table 3. Mean values p, w y $\frac{p}{w}$ for different values e

	$e = 2,520$	$e = 1,890$	$e = 1,260$	$e = 630$
p	0.758	0.718	0.636	0.516
w	1,282.06	1,098.42	761.83	372.06
$\frac{p}{w}$	5.91e-4	6.53e-4	8.34e-4	13.86e-4

is expected to have several arcs of $P_x \in P$, which means that the expected $\delta(P, P_{new})$ should be smaller than $\delta(P)$, i.e. P_{new} is expected to be in Ω_P.

Experimental results ratify this theory. 1,000 populations $Q(e)$ were taken randomly and 1,000 new solutions $Q(e)_{new}$ were generated with each population, using equation (1) with $O = 600$, $|Q(e)| = 25$, $\alpha = 1$, $\beta = 2$ for different values of e. Table 3 shows the proportion p of $Q(e)_{new}$ which lies inside $\Omega_{Q(e)}$, the mean cardinality of $\Omega_{Q(e)}$ (denoted w) and the relation $\frac{p}{w}$ (that may be understood as the proportion of $\Omega_{Q(e)}$ explored in average when generating each $Q(e)_{new}$).

At the very beginning of an OA computation, e is very large and there is a good probability of generating a solution in Ω_P (see p in Table 3 for $e = 2,520$). After progressing in its calculation, e decreases and so does $|\Omega_P|$; therefore, it becomes harder to find a new solution in Ω_P as shown in Table 3 (see how w and p decreases with e). Even though p decreases, it should be noticed that $\frac{p}{w}$, which is the proportion of Ω_P explored with each new individual, increases, i.e. OA searches more efficiently as computation continues.

5.3   Two-Dimension Interpretation of OA Exploration Space

For didactic reasons, an analogy between the n-dimensional TSP search space S and a two-dimension interpretation is presented. First, Fig. 5 (a) shows that the search nearby 3 points implies a concentration of the exploration in their central area. The intersection of the areas close to each point is their central area, i.e. the search nearby every $P_x \in P$, done by OA, implies an exploration in the central zone of P, a recommended search region according to Section 5.1.

Experimentally, the area composed by the points where its geometrical distance to a randomly chosen population of 25 points is smaller than the mean distance of the population is shown in Fig. 5 (b). This is the two-dimension interpretation of Ω_P and it is a central area in the population. As a consequence, OA's ability to search mainly in the central zone of P, means that it searches with a good probability in Ω_P.

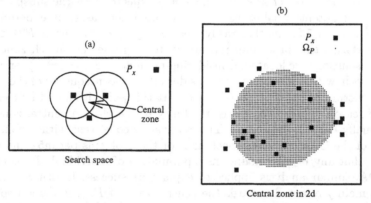

Fig. 5. (a) Simplified view of the search zone nearby all solutions of P (Ω_P) (b) Geometrical central zone of a population of 25 randomly chosen points

5.4   Reasons Underlying ACO's Success

In Fig. 6 the n-dimensional TSP search space is simplified to two dimensions for a geometrical view of the OA behavior. To understand the typical behavior of OA after the initial phase, a population $P1 = \{P1_x\} = Q(e)$ for $Q(e) \subset S$ of good solutions uniformly distributed is assumed in Fig. 6. As seen before, OA concentrates the search of new solutions in Ω_{P1} and replaces the worst solution of $P1$ ($P1_{worst}$) by a new solution P_{new} of smaller length $l(P_{new})$. A new population $P2$ is created including P_{new}. This is shown in Fig. 6 with a dotted line arrow. As a consequence, it is expected that $\delta(P2, r^*) < \delta(P1, r^*)$ because there is a positive correlation between $l(r_x)$ and $\delta(r_x, r^*)$. Similarly, $\delta(P, P_{new}) < \delta(P, P_{worst})$ because there is a positive correlation between $l(r_x)$

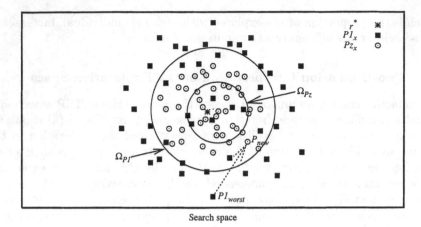

Search space

Fig. 6. Simplified view of OA behavior

and $\delta(P, r_x)$, therefore $\delta(P2) < \delta(P1)$, i.e. it is expected that the subspace where the search of potential solutions is concentrated decreases, as experimentally verified in Section 5.2. Another easily measured property is that $l(P2) < l(P1)$.

OA performs this procedure repeatedly to decrease the search zone where promising solutions are located. Considering population $Pz = \{Pz_x\}$ for $z >> 2$, Fig. 6 shows how Ω_{Pz} has decreased considerably as a consequence of the globally convex structure of the TSP search space. At this point it should be clear that the main reason of OA's success is its ability to search in a central zone of P, where usually better solutions lie. This analysis is consistent with the empirical behavior of the OA observed in Section 4 for the problems berlin52 and omi1.

Given that any P-ACO maintains a population of individuals P, as the presented OA, similar analysis applies to explain its success. In other words, the search is mainly oriented towards the central zone of P, where good solutions are usually found. Finally, as already mentioned in Section 3.2, any elitist ACO may be considered as a P-ACO and therefore the same explanation applies.

6   Conclusions and Future Work

OA concentrates the search in a central zone Ω_P of its population P. In globally convex problems, good solutions are usually found in this region; therefore, OA concentrates its search in a promising subspace. Every time a good solution is found, it enters the population reducing the promising search zone iteratively. Thus, this work explains the main reasons of OA's success, and any elitist ACO in general (e.g. $\mathcal{MMAS}$).

OA does not use positive feedback. Hence, elitist ACO does not necessarily share the same reasons of success with real ants, even though ACO was *inspired* by real ants behavior. This suggests not to limit the study of useful ACO properties to real ants behavior.

This work was limited to the analysis of elitist ACO in globally convex structures. Based on the presented framework, a future work will study local convergence, as well as other evolutionary algorithms and different problems. The authors are also studying all TSPLIB problems with known optimal solutions to experimentally confirm the globally convex property of the TSP search space.

References

1. M. Birattari, G. Di Caro, and M. Dorigo. For a Formal Foundation of the Ant Programming Approach to Combinatorial Optimization. Part 1: The problem, the representation, and the general solution strategy. Technical Report TR-H-301, ATR-Human Information Processing Labs, Kyoto, Japan, 2000.
2. K. D. Boese. Cost Versus Distance in the Traveling Salesman Problem. Technical Report 950018, Univ. of California, Los Angeles, Computer Science, May 19, 1995.
3. M. Dorigo and G. Di Caro. The Ant Colony Optimization Meta-Heuristic. In D. Corne, M. Dorigo, and F. Glover, editors, *New Ideas in Optimization*, pages 11–32. McGraw-Hill, London, 1999.
4. M. Dorigo and T. Stützle. An Experimental Study of the Simple Ant Colony Optimization Algorithm. In *2001 WSES International Conference on Evolutionary Computation (EC'01)*. WSES-Press International, 2001.
5. M. Guntsch and M. Middendorf. A Population Based Approach for ACO. In S. Cagnoni, J. Gottlieb, E. Hart, M. Middendorf, and G. Raidl, editors, *Applications of Evolutionary Computing, Proceedings of EvoWorkshops2002: EvoCOP, EvoIASP, EvoSTim*, volume 2279, pages 71–80, Kinsale, Ireland, 3-4 2002. Springer-Verlag.
6. M. Guntsch and M. Middendorf. Applying Population Based ACO to Dynamic Optimization Problems. In *Ant Algorithms, Proceedings of Third International Workshop ANTS 2002*, volume 2463 of *LNCS*, pages 111–122, 2002.
7. W. J. Gutjahr. A graph-based Ant System and its convergence. *Future Generation Computer Systems*, 16(8):873–888, June 2000.
8. W. J. Gutjahr. ACO Algorithms with Guaranteed Convergence to the Optimal Solution. *Information Processing Letters*, 82(3):145–153, May 2002.
9. T. C. Hu, V. Klee, and D. Larman. Optimization of globally convex functions. *SIAM Journal on Control and Optimization*, 27(5):1026–1047, Sept. 1989.
10. N. Meuleau and M. Dorigo. Ant colony optimization and stochastic gradient descent. *Artificial Life*, 8(2):103–121, 2002.
11. T. Stützle and M. Dorigo. A Short Convergence Proof for a Class of Ant Colony Optimization Algorithms. *IEEE Transactions on Evolutionary Computation*, 6:358–365, Aug. 2002.
12. T. Stützle and H. H. Hoos. $\mathcal{MAX}$-$\mathcal{MIN}$ Ant System. *Future Generation Computer Systems*, 16(8):889–914, June 2000.

S-ACO: An Ant-Based Approach to Combinatorial Optimization Under Uncertainty

Walter J. Gutjahr

Dept. of Statistics and Decision Support Systems, University of Vienna
walter.gutjahr@univie.ac.at
http://mailbox.univie.ac.at/walter.gutjahr/

Abstract. A general-purpose, simulation-based algorithm S-ACO for solving stochastic combinatorial optimization problems by means of the ant colony optimization (ACO) paradigm is investigated. Whereas in a prior publication, theoretical convergence of S-ACO to the globally optimal solution has been demonstrated, the present article is concerned with an experimental study of S-ACO on two stochastic problems of fixed-routes type: First, a pre-test is carried out on the probabilistic traveling salesman problem. Then, more comprehensive tests are performed for a traveling salesman problem with time windows (TSPTW) in the case of stochastic service times. As a yardstick, a stochastic simulated annealing (SSA) algorithm has been implemented for comparison. Both approaches are tested at randomly generated problem instances of different size. It turns out that S-ACO outperforms the SSA approach on the considered test instances. Some conclusions for fine-tuning S-ACO are drawn.

1 Introduction

The application of exact or heuristic methods of combinatorial optimization to real-world problems often faces the difficulty that for a particular problem solution considered, there is uncertainty on the objective function value achieved by it. A traditional way to represent uncertainty is by using a stochastic model. Based on such a model, the objective function becomes dependent not only on the solution, but also on a random influence, i.e., it becomes a random variable. Most frequently, the practical aim is then to optimize the *expected value* of this random variable.

When the expected value of the objective function can be represented as an explicit mathematical expression or at least be easily computed numerically, the solution of the stochastic combinatorial optimization problem needs not to be essentially different from that of a deterministic problem: The stochastic structure is then encapsulated in the representation of the expected objective function, and (possibly heuristic) techniques of deterministic optimization can be used. Very often, however, it is only possible to determine *estimates* of the expected objective function by means of sampling or simulation. This is the starting point for the area of *simulation optimization*, which has been a topic of intense research for several decades, but seems to undergo an interesting and

M. Dorigo et al. (Eds.): ANTS 2004, LNCS 3172, pp. 238–249, 2004.

challenging shift at present, the key features of which have recently been outlined by Michael C. Fu [8]. In Fu's opinion, there is a gap between research in stochastic optimization concentrated on sophisticated specialized algorithms on the one hand, and the current boom of integration of optimization routines into commercial simulation software packages, mainly based on metaheuristics such as Genetic Algorithms or Neural Nets, on the other hand. Fu argues that traditional stochastic optimization algorithms are often not (or not easily) adaptable to complex real-world simulation applications, whereas the mentioned metaheuristic approaches, in their stochastic variants, frequently lack methodological rigor and are not provably convergent.

As a promising candidate of a metaheuristic that holds the potential for an extension to simulation optimization purposes, Fu explicitly names the Ant Colony Optimization (ACO) metaheuristic approach as introduced by Dorigo and Di Caro [6]. To make ACO satisfy these expectations, one would have to fill the gap mentioned above by developing ACO algorithms for combinatorial optimization under uncertainty that are both broadly applicable and theoretically well-founded. The present article aims at a step in this direction by presenting first experimental results for S-ACO, an ACO-based general-purpose stochastic combinatorial optimization algorithm for which the convergence to the optimal solution has been shown in [11].

Before presenting the S-ACO approach, let us briefly refer to some alternative techniques. Traditional methods for optimization under uncertainty such as Stochastic Approximation or the Response Surface Method are not well-suited for an application in the context of *combinatorial* optimization. Problems that are both stochastic and combinatorial can, however, be treated by Sample Average Approximation [16], Variable-Sample Random Search Methods [14], the Stochastic Branch-and-Bound Method [18], the Stochastic Ruler Method [2], or the Nested Partition Method [19]. As approaches drawing from metaheuristic algorithms ideas, we mention Stochastic Simulated Annealing ([12], [1]) or Genetic Algorithm for Noisy Functions [9].

In the field of ACO, an early paper on a stochastic combinatorial optimization problem has been published by Bianchi, Gambardella and Dorigo [4], it investigates the solution of the Probabilistic Travelling Salesman Problem (PTSP). For the PTSP, an explicit formula for the expectation of the objective function value is known, so the chosen solution technique cannot be generalized to problems where sampling is necessary to obtain estimates of this expectation. We shall use, however, the PTSP as a benchmark for pre-tests with our general-purpose algorithm.

2   Problem Description and Cost Function Estimation

In the S-ACO approach, stochastic combinatorial optimization problems of the following very general form are considered:

$$\text{Minimize } F(x) = \mathrm{E}\left(f(x, \omega)\right) \quad \text{subject to} \quad x \in S. \tag{1}$$

Therein, x is the decision variable, f is the cost function, ω denotes the influence of randomness, E denotes the mathematical expectation, and S is a finite set of feasible decisions.

It is not necessary that $\mathrm{E}\,(f(x,\omega))$ is numerically computable, since it can be estimated by sampling: For this purpose, draw N random *scenarios* $\omega_1, \ldots, \omega_N$ independently from each other. A *sample estimate* is given by

$$\mathcal{E}F(x) \;=\; \frac{1}{N}\sum_{\nu=1}^{N} f(x,\omega_\nu) \;\approx\; \mathrm{E}\,(f(x,\omega)). \tag{2}$$

Obviously, $\mathcal{E}F(x)$ is an unbiased estimator for $F(x)$.

It should be mentioned that, contrary to its deterministic counterpart, problem (1) is typically nontrivial already for a very small number $|S|$ of feasible solutions: Even for $|S| = 2$, except when $F(x)$ can be computed directly, a nontrivial statistical hypothesis testing problem is obtained (see [18]).

3   Algorithms

3.1   The S-ACO Algorithm

In [11], the algorithm S-ACO indicated in Fig. 1 has been proposed for solving problems of type (1). S-ACO works based on the encoding of a given problem instance as a *construction graph* $\mathcal{C}$, a directed graph with a distinguished start node. (For examples, see section 4.) The stepwise construction of a solution is represented by a random walk in $\mathcal{C}$, beginning in the start node. Following the definition of the construction graph encoding given in [10], the walk must satisfy the condition that each node is visited at most once; already visited nodes are infeasible. There may also be additional rules defining particular nodes as infeasible after a certain partial walk has been traversed. When there is no feasible unvisited successor node anymore, the walk stops and is decoded as a complete solution for the problem. The conceptual unit performing such a walk is called an *ant*.

The encoding must be chosen in such a way that to each feasible walk in the sense above, there corresponds exactly one feasible solution. (The converse property that to each feasible solution there corresponds exactly one feasible walk is *not* required.) Since, if the indicated condition is satisfied, the objective function value is uniquely determined by a feasible walk, we may denote a walk by the same symbol x as a solution and consider S as the set of feasible walks.

When constructing a walk in the algorithm, the probability p_{kl} to go from a node k to a feasible successor node l is chosen as proportional to $\tau_{kl} \cdot \eta_{kl}(u)$, where τ_{kl} is the so-called *pheromone value*, a memory value storing how good step (k,l) has been in previous runs, and $\eta_{kl}(u)$ is the so-called *visibility*, a pre-evaluation of how good step (k,l) will presumably be, based on some problem-specific heuristic. $\eta_{kl}(u)$ is allowed to depend on the given partial walk u up to now. For the biological metaphors behind the notions "pheromone" and "visibility", we refer the readers to the basic texts on ACO, e.g., [7] [6].

In our stochastic context, the computation of the visibility values $\eta_{kl}(u)$ needs some additional explanation. The difficulty may arise that certain variables possibly used by a problem-specific heuristic are not known with certainty, because they depend on the random influence ω. This difficulty can be solved either by taking the *expected values* (with respect to the distribution of ω) of the required variables as the base of the visibility computation (these expected values are often directly given as model parameters), or by taking those variable values that result from a random scenario ω drawn for the current round.

Feasibility of a continuation (k, l) of a partial walk u ending with node k is defined in accordance with the condition above that node l is not yet contained in u, and none of the (eventual) additional rules specifies l as infeasible after u has been traversed.

As in some frequently used ACO variants for deterministic problems, we determine in each round a *round-winner*. In the stochastic context, we do this by comparing all walks that have been performed in this round on a *single* random scenario ω, drawn specifically for this round. We also experimented with determining the round winner based on *several* random scenarios, but this only increased the runtime and did not improve the solution quality.

In an ACO implementation for a deterministic problem, it is always reasonable to store the best solution seen so far in a special variable. A crucial difference to the deterministic case is that in the stochastic context, it is not possible anymore to decide with certainty whether a current solution x is better than the solution currently considered as the best found, $\hat{x}$, or not. We can only "make a good guess" by sampling: After a current round-winner x has been determined, x is compared with the solution considered currently as the overall best solution, $\hat{x}$. This is done based on a sample of N_m randomly drawn scenarios used by both solutions. Also these scenarios are round-specific, i.e., in the next round, new scenarios will be drawn. The larger N_m, the more reliable is the decision. The winner of the comparison is stored as the new "global-best" $\hat{x}$.

Both the solution $\hat{x}$ considered so far as global-best and the round-winner are reinforced on each of their arcs by pheromone increments. The parameters $c_1 > 0$ and $c_2 > 0$ in the algorithm determine the amount of pheromone increment on global-best and round-best walks, respectively. Experiments showed that c_2 should be chosen small compared to c_1, but a small *positive* c_2 produced better results than setting $c_2 = 0$. The parameters c_1 and c_2 are allowed to depend on the walks $\hat{x}$ resp. x; in particular, it is reasonable to choose them inversely proportional to the lengths of $\hat{x}$ resp. x, such that the overall amount of pheromone increment is constant. (In the construction graphs used in this paper, all walks have the same lengths, so c_1 and c_2 are chosen as constants.)

In [11], it has been shown that a slight modification of the algorithm S-ACO of Fig. 1 with $c_2 = 0$ (only global-best reinforcement) converges, on certain mild conditions, with probability one to the globally optimal solution of (1). The essential condition is that the sample size N_m grows at least linearly with the round number m. The mentioned modification consists in an extension of the pheromone update rule above by a rule that additionally uses a lower pheromone

---

Procedure S-ACO

---

set $\tau_{kl} = 1$ for all (k, l);
for round $m = 1, 2, \ldots$ {
    for ant $\sigma = 1, \ldots, s$ {
        set k, the current position of the ant, equal to the start node of $\mathcal{C}$;
        set u, the current walk of the ant, equal to the empty list;
        while (a feasible continuation (k, l) of the walk u of the ant exists) {
            select successor node l with probability p_{kl}, where
$$p_{kl} = \begin{cases} 0, & \text{if } (k, l) \text{ is infeasible,} \\ \tau_{kl} \cdot \eta_{kl}(u) \,/\, \left(\sum_{(k,r)} \tau_{kr} \cdot \eta_{kr}(u) \right), & \text{else,} \end{cases}$$
            the sum being over all feasible (k, r);
            set $k = l$, and append l to u;
        }
        set $x_\sigma = u$;
    }
    based on one random scenario ω, select the best walk x out of the
    walks $x_1, \ldots, x_s$;
    if $(m = 1)$ set $\hat{x} = x$; // $\hat{x}$ is the candidate for the best solution
    else {
        based on N_m random scenarios ω_ν, compute a sample estimate
        $\mathcal{E}(F(x) - F(\hat{x})) = \frac{1}{N_m} \sum_{\nu=1}^{N_m} (f(x, \omega_\nu) - f(\hat{x}, \omega_\nu))$
        for the difference between the costs of x and $\hat{x}$,
        if $(\mathcal{E}(F(x) - F(\hat{x})) < 0)$ set $\hat{x} = x$;
    }
    evaporation: set $\tau_{kl} = (1 - \rho)\,\tau_{kl}$ for all (k, l);
    global-best reinforcement: set $\tau_{kl} := \tau_{kl} + c_1$ for all $(k, l) \in \hat{x}$;
    round-best reinforcement: set $\tau_{kl} := \tau_{kl} + c_2$ for all $(k, l) \in x$;
}

Fig. 1. Pseudocode S-ACO.

bound (cf. [20]) of a certain type. In the present paper, we did not apply lower pheromone bounds.

3.2 Stochastic Simulated Annealing

In order to be able to compare the performance of the S-ACO algorithm with that of an alternative approach, we also implemented the Stochastic Simulated Annealing (SSA) algorithm described in [12]. SSA follows the philosophy of a standard Simulated Annealing algorithm. The only difference is that each time an objective function evaluation is required, which is the case when a current solution x is to be compared with a neighbor solution y, the evaluation is based on a sample estimate with some sample size N. In our implementations, we tried to keep equal conditions for S-ACO and SSA. Therefore, in analogy to

S-ACO, the comparative evaluation of x and y in SSA is done by means of the sample estimate $\mathcal{E}(F(x) - F(y))$, which is used in SSA as the input Δ for the probabilistic acceptance rule. As S-ACO, also SSA requires growing sample sizes N to satisfy theoretical convergence conditions, so we applied comparable sample size growth schemata to both algorithms. For two reasons, we did not use advanced SA concepts as *reheating*: First, also our S-ACO implementation was kept very basic (in particular, neither visibility values nor pheromone bounds were used). Secondly, only little experience is available at present about how to apply more elaborate SA concepts to *stochastic* problems.

4   Experimental Results

4.1   Pre-test on the PTSP

In order to get a first impression of the performance of S-ACO and an idea about suitable parameter choices, it seemed convenient to test the algorithm on a problem where, for comparison, the exact value of $F(x)$ can be determined, because a closed formula for the computation of the expectation in (1) exists. A good candidate for this purpose is the *Probabilistic Traveling Salesman Problem* (PTSP) introduced by Jaillet [15]. The PTSP consists in finding a fixed closed tour containing each of n customer nodes exactly once, such that the *expected* tour length is minimized, where uncertainty comes from the fact that each customer i has only a given probability p_i of requiring a visit. A realization of a tour only contains the subset of customers who actually require a visit, but the sequence in which these customers are visited is taken from the fixed a-priori tour x. For the PTSP, it is possible to compute the expected costs $F(x)$ directly, although by a somewhat clumsy formula containing double sums over double products. In [4], the homogeneous version of the PTSP where all p_i are equal has been investigated within an ACO context. We experimented with the more general inhomogeneous version, where the p_i may differ from each other.

We tested four different problem instances of sizes $n = 9$, 14, 29 and 256, respectively. The distance matrices $D = (d_{ij})$ were taken from diverse TSP benchmarks available in the Internet. These data were extended by probabilities p_i drawn uniformly at random in an interval $[\lambda, 1]$ with $\lambda = 0.3, 0.4, 0.5$ and 0.5, respectively, to obtain complete test instances. This part of the experiments was performed on a PC Pentium 933 Mhz, 256 MB RAM.

The 12 parameter combinations resulting from the following values were investigated for each problem instance: (i) number of ants: $s = 50, 500$, (ii) evaporation factor: $\rho = 0.05, 0.01$, (iii) visibility values: $\eta_{kl}(u) = 1/(d_{kl})^{\beta}$ with $\beta = 2$, 3, 4. For the PTSP experiments, we only applied global-best reinforcement, i.e., we chose $c_2 = 0$. The increment c_1 was chosen as 4ρ.

As the natural construction graph $\mathcal{C}$ of the S-ACO implementation for this problem, we took the complete graph on node set $\{1, \ldots, n\}$, where the nodes represent the customers. Node 1 was taken as the start node.

A sample scenario ω is obtained by making for each customer i a random decision (with probability p_i) whether (s)he requires a visit or not ($i = 1, \ldots, n$).

Let us briefly outline the main findings. The case $n = 14$ was the largest for which we were still able to compute the *exact* solution of the problem by the explicit formula, combined with complete enumeration. Using S-ACO, best results were achieved in this case by the parameter combination $s = 50$, $\rho = 0.05$ and $\beta = 2$. Already after 1 sec computation time, relatively good solutions (only 1.9 % worse than the optimal solution in the average) were obtained, with only moderate improvements later.

For the largest instance ($n = 256$), the optimal solution value is unknown. Best results were obtained by letting both $\beta = 2$ and the number $s = 50$ of ants unchanged, but decreasing ρ to the value 0.01, and increasing the number of rounds considerably: steepest decrements of the expected cost function were observed between 5 and 20 minutes after the start of the computation.

An important goal of the pre-test was an answer to the question whether sample size functions N_m growing linearly in m, as prescribed by theory, would turn out as useful. Indeed, the scheme $N_m = 50 + (0.0001 \cdot n^2) \cdot m$ yielded best results out of several alternatives, such as omitting the intercept 50, choosing factors 0.001 or 0.00001, or working with a sample size independent of m. The proportionality to n^2 has been chosen to parallel the space complexity $O(n^2)$ of the algorithm. Future work should address the question whether schemes N_m that are only sub-linear in m suffice to give good results.

4.2   The TSPTW with Stochastic Service Times

Our main experimental results concern the Travelling Salesman Problem with Time Windows and Stochastic Service Times (TSPTW-SST). Let us briefly recapitulate this NP-hard problem (cf. [5] and [13]):

As in the ordinary TSP, a set of customers $\{1, \ldots, n\}$ and a distance matrix $D = (d_{ij})$ are given. Distances are interpreted as driving times. Let us imagine that the travelling person is a service engineer. To each customer i, a time window $[a_i, b_i]$ can be assigned, indicating that customer i requests a visit by the service engineer starting at time t_i with $a_i \leq t_i \leq b_i$. Not every customer needs to have a time window for the visit. The service at customer i takes some time Y_i, where Y_i is a random variable with known distribution. After finishing the service at customer i, the service engineer drives to the next customer on the list given by the chosen permutation x of customers. The aim is to minimize the total driving time. As in the case of the PTSP, we restrict ourselves to the *fixed-routes* variant of the problem: The sequence of customers must be fixed in advance and cannot be changed when information on actual service times gets available.

We study a variant of this problem where time-window violations are possible, but penalized by two cost terms (which are added to the driving-time component of the objective function): If the service engineer arrives at customer i at a time t_i before time a_i, (s)he must wait until time a_i. This is penalized by a (low) cost factor C_w, multiplied by the waiting time $a_i - t_i$. If, on the other hand, the service engineer arrives too late, i.e., at a time t_i after time b_i, a (high) penalty C_t, multiplied by the tardiness $t_i - b_i$, is incurred.

4.3   Test Instance Generation

To perform a comparative test of S-ACO and SSA, we generated 20 problem instances of different sizes and different degrees of difficulty at random. In the case of each problem instance, n customer points were selected uniformly at random from a square. Distances were computed as Euclidean distances between these points. It was assumed that traversing an edge of the square takes 10 time units. For each customer, a random decision was made on wether or not to assign a time window, using a fixed probability p_{TW} for the existence of a time window. If a time window was assigned, its length was selected uniformly at random between 6 and 60 time units, and its start time was selected uniformly at random between time 0 and the maximum time such that the whole time window was still contained in an interval of 120 time units. The service time distributions were chosen as uniform distributions on the interval between 2 and 6 time units.

We experimented with the following parameter values: For n, we investigated the cases $n = 10$ and $n = 20$. For p_{TW}, we considered the cases $p_{TW} = 0.25$ and $p_{TW} = 0.50$ if $n = 10$, and the cases $p_{TW} = 0.20$ and $p_{TW} = 0.40$ if $n = 20$. For each of these four combinations, five random test instances were generated. For each of the 20 resulting test instances, we performed 20 test runs with each of the considered algorithms and their variants described below. The penalties for waiting time and tardiness were set to the values $C_w = 1$ and $C_t = 20$, respectively. In other words, it was assumed that the cost of waiting one time unit is considered equal to that of having to drive an additional time unit, and the cost of being tardy by one time unit is considered as equal to that of having to drive 20 additional time units.

4.4   Implementation Details and Outcome Evaluation

We chose the construction graph shown in Fig. 2. The start node is the leftmost node. A visit of node v_{ij} means that the ith visited customer is customer j. The additional constraint imposed on the walks in this solution encoding is that after some node v_{ij} has been visited, all nodes v_{kj} get infeasible for the rest of the walk. (An equivalent solution encoding has been used by Bauer et al. [3] and by Merkle and Middendorf [17] for scheduling problems.)

We did not use visibility values, i.e., we chose $\eta_{kl} = 1$.

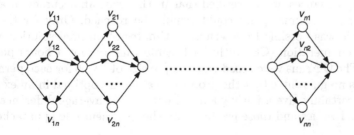

Fig. 2. Construction graph for the TSPTW-SST.

A sample scenario ω is obtained by drawing service times independently for each customer according to the described uniform distribution, and $f(x, \omega)$ is evaluated by simulating the arrival and departure times connected with the actual tour according to the drawn service times.

Besides the standard implementations of S-ACO and SSA as described in the previous section, we also experimented with variants of each approach obtained by the following possible modifications: First, inspired by ideas in [8] and [14], we also implemented variants where the sample size is not increased according to a fixed scheme, but computed in an adaptive way: If two solutions x and $\hat{x}$ in the case of S-ACO or two solutions x and y in the case of SSA are to be compared, the sample size is gradually increased until the absolute value of the sample estimate of the difference between the two objective function values is larger than the triple standard deviation of this sample estimate. In other words, we take the sample size in any case just sufficiently large to reject (at a significance level of 99.93 %) the null hypothesis that both objective function values are identical. This saves sampling time in cases where the objective function values have a large difference anyway, so that it can be decided soon that one of them is better. The resulting modifications of S-ACO and SSA will be denoted by S-ACOa and SSAa, respectively.

Second, since it is well-known that metaheuristic tour optimization approaches can gain from postoptimization by local search, we considered variants where after the run of the corresponding general-purpose algorithm, a 2-opt local optimization procedure is applied to the resulting tour, where each function evaluation is based on a random sample of size 5000. These variants are denoted by appending the suffix "+ls".

The following parameter values were used: For S-ACO, we executed 2000 resp. 4000 rounds for $n = 10$ resp. $n = 20$. We set $s = 50$, $c_1 = 2\rho$, $c_2 = 0.02\rho$, $N_m = 2m$. Furthermore, we set $\rho = 0.005$ resp. $\rho = 0.002$ for $n = 10$ resp. $n = 20$. For SSA, we chose the initial resp. final value of the temperature parameter as 1000 resp. 0.0001, reduced the temperature parameter by 5 % at each iteration, executed $2n$ neighbor selections and accept/reject decisions at each temperature level, and chose the sample size N_t at the tth temperature level as $20t$. The sample size increments for S-ACO and SSA were designed in such a way that, for both algorithms, a final sample size in the same order of magnitude resulted.

To evaluate the solutions produced by the different approaches, a final brute-force simulation run with sample size $N = 10^6$ was applied to each of them. By computing variances, it was verified that in this way, an accuracy of about one digit after the decimal point could usually be reached. Only for the $n = 10$ problems it was possible to determine estimates of optimal solution values by complete enumeration (CE) within a feasible time (about 4 hours per test instance). The CE runs were based on a sample size of 10000 for each permutation, so there is no guarantee that the produced values are optimal in an exact sense, but they certainly give a fairly good estimate of the average difference between the optimal values and those produced by the metaheuristic approaches.

4.5   Results

Table 1 summarizes the results for the four instance parameter combinations and the eight metaheuristic variants. Each entry contains: (i) the achieved average cost function value, averaged over the 100 runs in total (5 random instances, each with 20 test runs), (ii) the computation time (in seconds) for a run, averaged over the 100 runs, (iii) the superiority counts explained below. The TSPTW-SST computations were performed on a PC Pentium 2.4 Ghz, 1 GB RAM.

In general, averaging across different problem instances is problematic since they may have incompatible ranges of cost values (cf. [21]). Note, however, that in Table 1, averaging has only been performed across different random instances within the same application class, say, $n = 10$ and $p_{TW} = 0.25$. Thus, each reported cost value has a clear interpretation as an estimate of the expected cost produced by the considered algorithm if it is applied to a problem instance randomly selected — according to the given distribution — from the given application class. By *not* re-normalizing the results within an application class, we give "hard" random test cases (which are of a high practical importance in view of robustness) their natural weight in comparison with "easy" test cases. Nevertheless, to present also output data that do not depend on aggregation, we have indicated in Table 1 for how many random test instances of a given application class the S-ACO variant performed better than the corresponding SSA variant, and vice versa. E.g., the "superiority count" 4 in the entry for S-ACO, $n = 10$ and $p_{TW} = 0.25$ indicates that S-ACO outperformed SSA in 4 of the the 5 test instances of this application class. By the parameter-free *sign test*, based on the total of the 20 test instances, we judged which S-ACO variant outperformed the corresponding SSA variant *significantly* more often than vice versa.

In total, the S-ACO variants produced better results than the SSA variants, both with and without adaptive sample size, and both in the case of local postoptimization and that without it. Quantified by the number of test instances for which a variant of S-ACO outperforms the corresponding SSA variant, this result is statistically significant at the level 0.05 for the comparisons S-ACOa / SSAa and S-ACO+ls / SSA+ls, significant even at the level 0.01 for the comparison S-ACOa+ls / SSAa+ls, but only weakly significant (level 0.10) for the comparison S-ACO / SSA.

Postoptimation by local search improved all variants. However, for the test cases with $n = 10$, SSA profited more from this postoptimization than S-ACO (possibly because the S-ACO variants came already rather close to the optimal solutions in this case). For the test cases with $n = 20$, S-ACO profited more from postoptimization than SSA.

Adaptive sample size did not improve the results in *each* case. For S-ACO, it did not even always decrease the runtime, at least not for the smaller test instances. For the larger test instances, adaptive sample size *did* save time, and it achieved about the same level of solution quality. SSA always profited from adaptive sample size both in solution quality and in runtime, except in the most "difficult" parameter instance combination, $n = 20$ and $p_{TW} = 0.40$, where a slightly worse solution quality resulted.

Table 1. Average achieved cost values, runtimes (in sec), and superiority counts.

	$n = 10$ $p_{TW} = 0.25$			$n = 10$ $p_{TW} = 0.50$			$n = 20$ $p_{TW} = 0.20$			$n = 20$ $p_{TW} = 0.40$		
CE	52.8			57.0								
S-ACO	53.5	5	4	57.9	4	5	90.4	33	1	138.4	35	3
S-ACOa	53.6	8	4	57.9	10	5	89.2	19	2	139.8	21	2
S-ACO+ls	53.5	5	3	57.8	5	5	87.6	37	3	136.3	38	4
S-ACOa+ls	53.6	8	3	57.8	8	5	87.2	21	3	136.4	29	5
SSA	57.3	14	1	59.7	13	0	95.5	52	4	159.6	56	1
SSAa	57.1	7	0	59.7	5	0	91.3	15	2	160.6	14	3
SSA+ls	56.9	13	1	59.0	13	0	94.8	56	2	159.3	58	1
SSAa+ls	56.7	7	0	59.0	6	0	90.4	17	2	160.4	17	0

5 Conclusions

A general-purpose algorithm, S-ACO, for solving stochastic combinatorial optimization problems has been tested experimentally on two problems of fixed-routes type, a probabilistic travelling salesman problem (PTSP) and a travelling salesman problem with time windows and stochastic service times (TSPTW-SST). Encouraging results have been obtained. In particular, it has been shown that on the randomly generated test instances for the TSPTW-SST, the S-ACO algorithm outperformed a stochastic version of simulated annealing (SSA).

Future research should be directed to broader experiments with S-ACO on diverse other stochastic combinatorial problems in routing, scheduling, subset selection and many other areas, and to the experimental investigation of modifications of S-ACO as those outlined in [11], section 6. Furthermore, not only SSA, but also the other approaches indicated in the Introduction should be included into the comparative experiments. An especially challenging research topic for the future is the development of methods for analyzing stochastic combinatorial problems from a *multiobjective* point of view. Ongoing research investigates a possible combination of the S-ACO approach presented here with different ant-based multiobjective optimization techniques that have been developed in the ACO literature.

Acknowledgment

The author wants to express his thanks to Christian Grundner for his help in implementation and test of the PTSP part of the experiments.

References

1. Alrefaei, M.H., Andradóttir, S., "A simulated annealing algorithm with constant temperature for discrete stochastic optimization", Management Sci. 45 (1999), pp. 748–764.

2. Alrefaei, M.H., Andradóttir, S., "A modification of the stochastic ruler method for discrete stochastic optimization", *European J. of Operational Research* 133 (2001), pp. 160–182.
3. Bauer, A., Bullnheimer, B., Hartl, R.F., Strauss, C., "Minimizing Total Tardiness on a Single Machine Using Ant Colony Optimization", *Central European Journal of Operations Research* 8 (2000), pp. 125–141.
4. Bianchi, L., Gambardella. L.M., Dorigo, M., "Solving the homogeneous probabilistic travelling salesman problem by the ACO metaheuristic", *Proc. ANTS '02*, 3rd Int. Workshop on Ant Algorithms (2002), pp. 177–187.
5. Cordeau, J.-F., Desaulniers, G., Desrosiers, J., Solomon, M.M., Sounis, F., "VRP with Time Windows", in: *The Vehicle Routing Problem*, P. Toth and D. Vigo (eds.), SIAM Monographs: Philadelphia (2002), pp. 157–194.
6. Dorigo, M., Di Caro, G., "The Ant Colony Optimization metaheuristic", in: *New Ideas in Optimization*, D. Corne, M. Dorigo, F. Glover (eds.), pp. 11-32, McGraw–Hill (1999)
7. Dorigo, M., Maniezzo, V., Colorni, A., "The Ant System: Optimization by a colony of cooperating agents", *IEEE Trans. on Systems, Man, and Cybernetics* 26 (1996), pp. 1–13.
8. Fu, M.C., "Optimization for simulation: theory vs. practice", *INFORMS J. on Computing* 14 (2002), pp. 192–215.
9. Fitzpatrick, J.M., Grefenstette, J.J., "Genetic algorithms in noisy environments", *Machine Learning* 3 (1988), pp. 101–120.
10. Gutjahr, W.J., "A graph–based Ant System and its convergence", *Future Generation Computer Systems* 16 (2000), pp. 873–888.
11. Gutjahr, W.J., "A converging ACO algorithm for stochastic combinatorial optimization", *Proc. SAGA 2003* (Stochastic Algorithms: Foundations and Applications), eds.: A. Albrecht and K. Steinhöfl, Springer LNCS 2827 (2003), pp. 10–25.
12. Gutjahr, W.J., Pflug, G., "Simulated annealing for noisy cost functions", *J. of Global Optimization*, 8 (1996), pp. 1–13.
13. Hadjiconstantinou, E., Roberts, D., "Routing Under Uncertainty: An Application in the Scheduling of Field Service Engineers", in: *The Vehicle Routing Problem*, P. Toth and D. Vigo (eds.), SIAM Monographs: Philadelphia (2002), pp. 331–352.
14. Homem-de Mello, T., "Variable-sample methods for stochastic optimization", *ACM Trans. on Modeling and Computer Simulation* 13 (2003), pp. 108–133.
15. Jaillet, P., *Probabilistic Travelling Salesman Problems*, PhD thesis, MIT, Cambridge, MA (1985).
16. Kleywegt, A., Shapiro, A., Homem-de-Mello, T., "The sample average approximation method for stochastic discrete optimization", *SIAM J. Optim.* 12 (2001), pp. 479–502.
17. Merkle, D., Middendorf, M., "Modelling the Dynamics of Ant Colony Optimization Algorithms", *Evolutionary Computation* 10 (2002), pp. 235–262.
18. Norkin, V.I., Ermoliev, Y.M., Ruszczynski, A., "On optimal allocation of indivisibles under uncertainty", *Operations Research* 46 (1998), pp. 381–395.
19. Shi, L., Olafsson, S., "Nested partition method for global optimization", *Operations Reseacrh* 48, pp. 390–407.
20. Stützle, T., Hoos, H.H., "The MAX-MIN Ant system and local search for the travelling salesman problem", in: T. Baeck, Z. Michalewicz and X. Yao (eds.), *Proc. ICEC '97* (Int. Conf. on Evolutionary Computation) (1997), pp. 309–314.
21. Zlochin, M., Dorigo, M., "Model-based search for combinatorial optimization: a comparative study", *Proc. PPSN* (Parallel Problem Solving from Nature) '02 (2002), pp. 651–664.

Time-Scattered Heuristic for the Hardware Implementation of Population-Based ACO

Bernd Scheuermann[1], Michael Guntsch[1],
Martin Middendorf[2], and Hartmut Schmeck[1]

[1] Institute AIFB, University of Karlsruhe, Germany
{scheuermann,guntsch,schmeck}@aifb.uni-karlsruhe.de
http://www.aifb.uni-karlsruhe.de
[2] Institute of Computer Science, University of Leipzig, Germany
middendorf@informatik.uni-leipzig.de
http://pacosy.informatik.uni-leipzig.de

Abstract. We present a new kind of heuristic guidance as an extension to the Population-based Ant Colony Optimization (P-ACO) implemented in hardware on a Field Programmable Gate Array (FPGA). The heuristic information is obtained by transforming standard heuristic information into small time-scattered heuristic-vectors of favourable ant decisions. This approach is suited for heuristics which allow for an a priori calculation of the heuristics information. Using the proposed method, an ant can build-up a solution in quasi-linear time. Experimental studies measure the performance of the time-scattered heuristic. A comparison with the standard heuristic and candidate lists is also given.

1 Introduction

The Ant Colony Optimization (ACO) metaheuristic has been used to solve various hard optimization problems [3]. Usually, ACO algorithms are implemented in software on a sequential machine. However, if processing times are crucial, then there exist mainly two options to speed-up the execution. One option is to run ACO on parallel computers as proposed by several authors (see [9] for an overview). The other very promising approach is to directly map the ACO algorithm into hardware, thereby exploiting parallelism and pipelining features of the underlying architecture. A concept for an implementation of the ACO algorithm on reconfigurable processor arrays was presented in [8]. The first implementation of ACO in real hardware was presented by mapping an ACO variant onto Field Programmable Gate Arrays (FPGAs) [2, 12]. It was shown that Population-based ACO (P-ACO) [6] fits the architectural constraints of FPGAs better than the standard ACO approach and that the P-ACO hardware implementation leads to significant speed-ups in runtime compared to implementations in software on sequential machines. So far, the hardware P-ACO was implemented without considering heuristic guidance during ant decisions.

In this paper we introduce a new heuristic concept for the P-ACO algorithm that sustains the small runtimes of the hardware implementation. Standard heuristic information is divided into a sequence of small, time-scattered

M. Dorigo et al. (Eds.): ANTS 2004, LNCS 3172, pp. 250–261, 2004.

vectors of favourable ant decisions. Using this approach, ants can build solutions in quasi-linear time. We demonstrate the construction and the usage of the new heuristic concept. The proposed approach is then tested on various instances of the Traveling Salesperson Problem (TSP), where the goal is to find the shortest tour connecting n cities such that every city is visited only once. The performance of the time-scattered heuristic is compared with the standard heuristic approach and candidate lists.

2   Population-Based Ant Colony Optimization

In this section, we briefly explain the characteristics of the P-ACO algorithm introduced by Guntsch et al. [6]. Note that in this paper the description of the ACO/P-ACO metaheuristics is restricted to optimization problems with solutions represented by permutations of numbers $\{1, \ldots, n\}$. In ACO, an ant builds a solution by making a sequence of local decisions, i.e. successive selections of items. Every decision is made randomly with respect to a probability distribution over the so far unchosen items in selection set S:

$$\forall j \in S : \quad p_{ij} = \frac{\tau_{ij}^{\alpha} \cdot \eta_{ij}^{\beta}}{\sum_{z \in S} \tau_{iz}^{\alpha} \cdot \eta_{iz}^{\beta}} \tag{1}$$

where τ_{ij} denotes the pheromone information, η_{ij} describes the heuristic values. For example in TSP, the heuristic value of choosing to visit city j after the last chosen city i is considered to be inversely proportional to the distance d_{ij} separating them: $\eta_{ij} = 1/d_{ij}$. Parameters α and β are used to determine the relative influence of pheromone values and heuristic values. Note, that several variants of this rule have been proposed in the literature, see [1] for an overview.

An essential difference between standard ACO and P-ACO is how the knowledge gained in the optimization process is transferred from one iteration to the next. Instead of a pheromone matrix $[\tau_{ij}]$ as in the standard ACO, the P-ACO keeps and maintains a small population P of the best solutions of the preceding k iterations that is transferred to the next iteration. These solutions are stored in an $n \times k$ population matrix $Q = [q_{ij}]$, where each column contains one solution. The pheromone matrix $[\tau'_{ij}]$ which is used by the ants for solution construction can be derived from the population matrix Q as follows: Each element τ'_{ij} receives a minimum amount of pheromone $\tau_{init} > 0$ plus an additional pheromone amount Δ_P depending on the current solutions kept in the population according to:

$$\forall \, i, j \in \{1, \ldots, n\} : \tau'_{ij} = \tau_{init} + \zeta_{ij} \cdot \Delta_P, \tag{2}$$

with ζ_{ij} denoting the number of occurrences of item j in the i-th row of population matrix Q, i.e. $\zeta_{ij} = |\{h : q_{ih} = j\}|$.

Per iteration m solutions $\{\pi_1, \ldots, \pi_m\}$ are constructed. After every iteration, the population is updated by removing the oldest solution σ from the population and inserting the best solution π^* of the current iteration into the population. This means that the population is organized as a FIFO-queue of size k. Thus, a population update corresponds to an implicit update of the pheromone matrix $[\tau'_{ij}]$:

- A solution π^* entering the population corresponds to a positive update:
  $\forall\, i \in \{1,\ldots,n\} : \tau'_{i\pi^*(i)} = \tau'_{i\pi^*(i)} + \Delta_P$
- A solution σ leaving the population corresponds to a negative update:
  $\forall\, i \in \{1,\ldots,n\} : \tau'_{i\sigma(i)} = \tau'_{i\sigma(i)} - \Delta_P$

Empirical studies have shown that P-ACO is a competitive approach in comparison to other ACO algorithms and that for population size k small values $1 \leq k \leq 8$ are sufficient for all test instances regarded in [6]. Therefore, k can be considered to be a small constant.

3   FPGA Implementation of P-ACO

For brevity, the implementation of P-ACO on FPGA is explained on a rather functional level only. A detailed technical description is given in [2, 12].

The mapping of the P-ACO algorithm into the corresponding FPGA design (see Figure 1) consists of three main hardware modules: the Population Module, the Generator Module and the Evaluation Module (see Figure 1). The Population Module contains the population matrix Q. Here we assume that the population matrix also contains an elitist solution (i.e., the best solution found so far) in column $j = 1$, and the remaining columns $j \in \{2,...,k\}$ are organized as a FIFO-queue. Every entry $q_{ij} \in \{1,...,n\}$ is the number of the item located at place i of the j-th solution, e.g. for TSP, q_{ij} is the next city to be visited after city i in the j-th tour stored in the population. The Population Module is responsible for broadcasting the k items q_{ij} from the i-th row of the population matrix to the Generator Module. Furthermore, it receives the best solution of the current

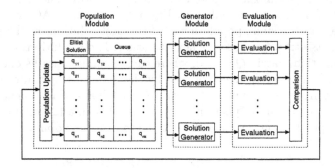

Fig. 1. P-ACO design with Population, Generator and Evaluation Modules

iteration from the Evaluation Module, which is then inserted into the queue. The Generator Module holds m Solution Generators working concurrently, one Solution Generator per ant. Every ant constructs its solution, e.g. a tour through

n cities for the TSP. These solutions are transferred to m parallel Evaluation blocks in the Evaluation Module. The evaluation results (e.g. tourlengths) of these m solutions are collected in a Comparison block, which determines the best solution of the current iteration and sends it to the Population Module. This best solution also becomes the new elitist solution if it is better than the current elitist solution.

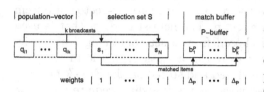

Fig. 2. Process of making a decision without heuristic guidance

The hardware implementation of P-ACO does not explicitly use any pheromone information. Instead ants make their decision solely based on the solutions stored inside the population as depicted in Figure 2. Consider a decision in row number i, then $(q_{i1}, \ldots, q_{ik})$ in matrix Q denotes the current population-vector and $\{s_1, \ldots, s_N\}$ the selection set of $N \leq n$ yet available items. All k items in the population-vector are broadcast to the selection set. If the respective item is still contained in S, then we call this a match. All matched items are stored into the population-buffer (P-buffer). After all broadcasts are finished, the P-buffer keeps $M_P \leq k$ matched items, each associated with the weight Δ_P. Now the next decision can be made according to the probability distribution p_{ij}:

$$\forall j \in S : p_{ij} = \frac{\tau'_{ij}}{\sum_{z \in S} \tau'_{iz}} = \frac{\tau_{init} + \zeta_{ij} \cdot \Delta_P}{\sum_{z \in S}(\tau_{init} + \zeta_{iz} \cdot \Delta_P)} = \frac{1 + \zeta_{ij} \cdot \Delta_P}{N + M_P \cdot \Delta_P} \quad (3)$$

Note that in the P-ACO hardware implementation we set $\tau_{init} := 1$. Heuristic information has so far been disregarded (but it is considered in this paper as outlined in Section 4). A decision is made by drawing an integer random number r from the range $r \in \{1, \ldots, N + M_P \cdot \Delta_P\}$. If $r \leq N$ then the r-th item s_r in set S is selected. Otherwise buffer item b_j^P with $j = \lceil \frac{r-N}{\Delta_P} \rceil$ is selected. Building up the probability distribution according to Equation (3) requires $\Theta(k)$ steps. Drawing a random number and selecting an item in hardware can be accomplished in $\Theta(1)$ (see [2, 12] for details). Hence, a complete ant decision can be computed in $\Theta(k)$ time, compared to $O(n)$ for the standard ACO algorithm.

4   Time-Scattered Heuristic

In general a characteristic of ACO is that heuristic information can often be easily integrated into the algorithm, and it is used successfully by most ACO algorithms for combinatorial optimization problems [4]. In this section, we describe a method of utilizing heuristic information in the FPGA implementation of the P-ACO algorithm without giving up the short runtime.

The integration of heuristic information into the P-ACO hardware algorithm poses two problems: heuristic values exist for all items of the set S, not just an

$O(k)$ size subset, and generally these are real numbers which are not readily supported by current FPGA technologies. Therefore, we propose the use of heuristic values only for a subset of the items and to change this subset regularly so that we have an approximation for the real heuristic values. More exactly, similarly to the P-buffer for pheromone, we use an additional H-buffer for heuristic values. This H-buffer is used analogously to the P-buffer for an ant's decision. The more often an item number is contained in the H-buffer, the higher the probability to be chosen. To fill the H-buffer, we use a so called heuristic-vector of c item numbers, for a small constant c with $c << n$. The heuristic-vector that is used to fill the H-buffer in an iteration is chosen from a set containing l heuristic-vectors. This set of heuristic-vectors is stored in a static heuristic matrix. We assume that the heuristic matrix is calculated and stored on the FPGA before a run of the ACO algorithm starts. Therefore, heuristics which require an online computation of the heuristic values cannot be handled by this method.

The general strategy to place item numbers into the heuristic matrix is such that the relative size of the heuristic value of an item should be approximated by its relative frequency in the matrix, and items with high heuristics value should not dominate a vector too much but instead should occur in many vectors. This can be achieved by the following algorithm (ALG-HM) calculating the heuristic matrix $[h_{uv}^i]_{c \times l}$ for row i from the heuristic values $\eta_i = (\eta_{i1}, \ldots, \eta_{in}) \in \mathbb{R}^n$:

Algorithm ALG-HM:
for all $j \in \{1, \ldots, n\}$
    calculate $\hat{\eta}_{ij} := \eta_{ij}^\beta$
calculate $\delta_i = \frac{1}{c \cdot l} \sum_{j=1}^n \hat{\eta}_{ij}$
for all $v \in \{1, \ldots, c\}$
    for all $u \in \{1, \ldots, l\}$ // column-wise traversal of heuristic-matrix
        determine j^* so that $\hat{\eta}_{ij^*} = \max_{j=1,\ldots,n} \hat{\eta}_{ij}$
        set $h_{u,v}^i := j^*$ // insert item j^* into heuristic-matrix
        update $\hat{\eta}_{ij^*} := \hat{\eta}_{ij^*} - \delta_i$

In accordance with standard ACO, $\beta > 0$ is a weight that determines the impact of the heuristic on an ant's decision. Clearly, the quality of the approximation attainable by this method depends on the values of η_{ij}, β, c and l. Note that, the order of the items within a row h_u^i of the heuristic-matrix is not important, where $h_u^i \in \{1, \ldots, n\}^c$ denotes the u-th heuristic-vector in the i-th matrix.

Since for $l > 1$ there is more than one heuristic-vector for the given row, only one of the heuristic-vectors $h_{u^*}^i$ is declared as active and the other $l-1$ are inactive. Only the active heuristic-vector affects the decision-process of the ant. After a specific number f_s of solution generations, the active heuristic-vector is shifted, i.e. it is replaced with some other h_u^i with $u \in \{1, \ldots, l\} \backslash \{u^*\}$. Parameter $f_s \geq 1$ is called the shift frequency. We consider two different policies of shifting the active heuristic-vector, either random (i.e. the next active vector u^* is chosen randomly from $\{1, \ldots, l\}$) or cyclic, i.e. $u^* := (u^* \bmod l) + 1$.

In the hardware P-ACO algorithm, the necessary calculations to transform row i in the population matrix Q into the respective values τ_{ij} are done according to Equation 2. Likewise it is possible to transform any heuristic-vector h^i_{u*} into the corresponding heuristic value η'_{ij} describing the heuristical impact of item j when making a decision in row i:

$$\forall\, j \in \{1, \ldots, n\} : \eta'_{ij} := \eta'_{init} + \gamma_{ij} \cdot \Delta_H, \tag{4}$$

where $\eta'_{init} > 0$ denotes a base heuristic value assigned to every item j and γ_{ij} describes the number of occurrences of item j in the active heuristic-vector, i.e. $\gamma_{ij} = |\{v : h^i_{u*v} = j, v = 1, ..., c\}|$ and $\Delta_H > 0$ is the weight assigned to items in the heuristic-vectors.

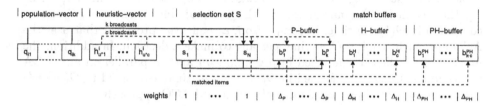

Fig. 3. Items in the current population row i and the respective heuristic-vector h^i_{u*} are broadcast. Match addresses are stored in match buffers. Weights for items in S and in the match buffers are indicated

The proposed method for heuristic guidance can be integrated into the P-ACO hardware implementation by extending the match buffer concept introduced in Section 3. Let population-vector $(q_{i1}, \ldots, q_{ik})$ be the current row in the population, and $h^i_{u*} = (h^i_{u*1}, \ldots, h^i_{u*c})$ the current heuristic-vector (cf. Figure 3). All $k + c$ items in both vectors are broadcast to the selection set S and matched items are then stored into three different types of Match Buffers. The P-Buffer stores $M_P \leq k$ items which are in the population-vector as well as in set S. The respective $M_H \leq c$ items which occur in the heuristic-vector and in selection set S are copied into a separate location called the H-Buffer. Furthermore, since the pheromone and heuristic values are multiplied in the ACO algorithm, we need an additional buffer – called PH-Buffer – storing the items which are in the heuristic-vector as well as in the population-vector and in set S. Therefore, items in the heuristic-vector are broadcast to selection set S and in parallel to the P-buffer. Let Δ_P be the weight associated with the population-vector and Δ_H the weight of the heuristic-vector derived from the heuristic information. Then $\Delta_{PH} = \Delta_P \cdot \Delta_H$ is the weight for an item j which is stored $\phi_{ij} = \zeta_{ij} \cdot \gamma_{ij}$ times in the PH-Buffer. After all broadcasts, the PH-Buffer contains $M_{PH} \leq k \cdot c$ items.

Using this extended buffer concept ants make random decisions according to the following probability distribution p_{ij}:

$$\forall j \in S: \ p_{ij} = \frac{\tau'_{ij} \cdot \eta'_{ij}}{\sum_{z \in S} \tau'_{iz} \cdot \eta'_{iz}} = \frac{1 + \zeta_{ij} \cdot \Delta_P + \gamma_{ij} \cdot \Delta_H + \phi_{ij} \cdot \Delta_{PH}}{N + M_P \cdot \Delta_P + M_H \cdot \Delta_H + M_{PH} \cdot \Delta_{PH}} \quad (5)$$

Note that the initial values are set to $\tau_{init} = \eta_{init} = 1$. Unlike standard ACO, exponentiations by β are calculated off-line prior to building the heuristic matrix. Pheromone and heuristic values are weighted by choosing the initial and Δ parameters accordingly. The required arithmetic operations better suit the resources provided by the FPGA architecture (cmp. [12]). To perform an ant decision, an integer random number $r > 0$ is drawn from the range limited by the denominator in Equation 5. This random number determines if an item in S or an item in one of the buffers was selected. An efficient parallel implementation of this new heuristic concept on an FPGA allows to make a decision in $\Theta(k + c)$ time, where k and c can be regarded as small constants. Also in the case of exploitation, according to the pseudo-random proportional action choice rule [5], the search for the maximum over the yet available items in set S can be computed in $\Theta(k + c)$ time. In comparison, a sequential processor that uses the ACO approach requires $O(n)$ time per ant to make a decision. In addition, the functional parallelism embodied in processing m ants in parallel on an FPGA allows m solutions to be formed in $\Theta((k + c) \cdot n)$ time, compared to $O(m \cdot n^2)$ time on a sequential processor using the standard ACO approach.

5   Experimental Results

We performed various experiments to study the performance of the proposed time-scattered heuristic approach. The experiments were conducted on Pentium 4 CPUs clocked at 1.5 GHz. The time-scattered heuristic was not implemented on an FPGA, but a comparison with the existing P-ACO hardware implementation and an estimate on the relative performance is given.

In the first experiment, the influence of parameters c and l, which determine the size of the heuristic-matrix, was studied (see Figure 4). The P-ACO algorithm with time-scattered heuristic (P-ACO-TSH) was applied to the symmetric TSP instance eil101 (101 cities) from the TSPLIB [10]. The results are shown in Figure 4. The lower part of the figure shows the deviation of the distribution of items in the heuristic matrix (obtained by applying the ALG-HM algorithm) from the distribution of the standard heuristic values η_{ij}^{β}. This deviation is determined by calculating the average quadratic distance between these two distributions. The results show that deviation decreases rapidly with the size of the heuristics matrix.

The upper part of Figure 4 shows the tourlength for different aspect ratios $r = c/l$ (average over 20 runs). The result for P-ACO with no heuristic guidance (P-ACO-NH) is also shown. Suitable parameter settings for P-ACO have been determined in a range of preliminary experiments. We chose $m = 8$ ants, a population size of $k = 3$ solutions plus 1 elitist solution, initial values $\tau_{init} = \eta_{init} = 1$, update values $\Delta_P = f_P \cdot n/(k+1)$ and $\Delta_H = f_H \cdot n/c$ with $f_P = f_H = 10$, weight $\beta = 4$, shift frequency $f_s = 8$ with random shift policy, and a runtime

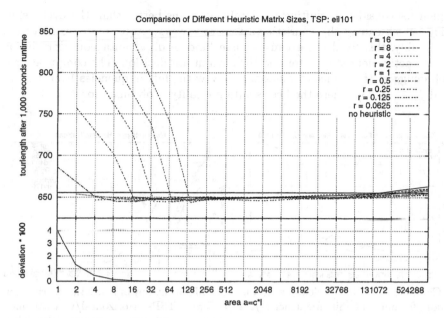

Fig. 4. Comparison of c and l combinations with different aspect ratios $r = c/l$ and heuristic matrix sizes $a = c \cdot l$

limit of 1000 seconds for every combination of c and l. For small values of a and aspect ratios $r \leq 1$ (i.e. $l \geq c$), P-ACO-TSH performs worse than P-ACO-NH. Obviously, in these cases the approximation of the original heuristic information is not good enough and misleads the decisions of the ants. However, as the mean quadratic deviation decreases, the algorithm benefits from the additional heuristic guidance. The solution qualities for $r \leq 1$ and $c \geq 3$ are better than those obtained for P-ACO-NH. Runs with high aspect ratios $r > 1$ perform better for small matrix sizes, since the ants are provided with more heuristic knowledge during individual decisions than in runs with low aspect ratios. For combinations with $r > 1$ solution qualities become worse with increasing matrix size a and are even worse than P-ACO-NH for $a > 2^{18}$. This behaviour is due to a higher execution time per ant decision which is proportional to c. The overall best solution quality was obtained for parameters $c = 3$ and $l = 48$, i.e. $a = 144$ and $r = 0.0625$.

In the next experiment, we examined the shift policy (either random or cyclic) and the shift frequency, which was scaled from $f_s = 1$ (shift after any solution generation) to $f_s = 2^{20}$ (no shift during the complete run). We set $c = 3$ and $l = 48$ and calculated the tourlength as the average over 20 repititions with different random seeds after 100000 iterations, all other parameter settings remained as in the previous experiment. Solution qualities for both policies worsen with increasing shift frequencies (see Figure 5a). Apparently, shifting the heuristic-vectors very often better approximates the original heuristic information and provides the ants with more accurate heuristic knowledge. In the majority of frequencies considered, shifting randomly performs better than the cyclic shift policy, which causes the heuristic-vectors to always rotate in the same order. Presumably, this fixed order causes some kind of disadvantageous correlation between active heuristic-vector and the current ant decision. The computational results suggest the choice of random policy with shift frequencies from the interval $f_s \in [4, 128]$, where the best solution quality is obtained for $f_s = 4$.

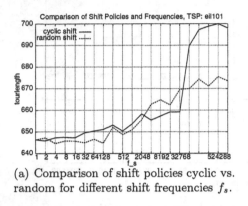

(a) Comparison of shift policies cyclic vs. random for different shift frequencies f_s.

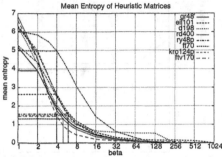

(b) Mean entropy of heuristic matrices for different TSPs. Horizontal/vertical lines indicate best mean entropies/β-weights resp.

Fig. 5. Determination of shift policy, shift frequency and weight β

Another experiment was performed to determine the influence of weight β for P-ACO-TSH. Therefore, β was scaled from 1 to 1024 (see Figure 5b), and heuristic-vectors were shifted after every $f_s = 8$ solution generations with random shift policy. All other parameter settings were chosen as in the previous experiment. For every β, we calculated the respective mean entropies of the heuristic matrices for a selection of TSP instances including four symmetric instances gr48 (48 cities), eil101 (101 cities), d198 (198 cities), rd400 (400 cities) and four asymmetric instances ry48p (48 cities), ft70 (70 cities), kro124p (100 cities) and ftv170 (171 cities). The entropy $H = -\sum_{i=1}^{i=n} p_i \cdot \log(p_i)$, with p_i denoting the relative frequency of item i, measures the uniformity of the distribution of items within the heuristic matrices (cmp. [13]). The maximum entropy value $H_{max} = \log(c \cdot l) = 7.17$ is reached for $\beta = 0$, i.e. all items in a matrix are different. For values $\beta \geq 1024$ the entropy converges to its minimum $H_{min} = 0$ for all considered TSP instances. This means all items in a matrix are identical. The vertical lines in Figure 5b represent the respective best β values found after executing P-ACO-TSH for 100000 iterations. The corresponding mean entropies are drawn as horizontal lines. All best mean entropies are located in the range $H^* \in [18.2\% \cdot H_{max}, 69.4\% \cdot H_{max}]$. Entropy maxima are associated with small TSP instances gr48 and ft70. Eliminating these instances constrains the best mean entropies to a smaller range $H^* \in [18.2\% \cdot H_{max}, 36.7\% \cdot H_{max}]$. Hence the P-ACO-TSH performs best, if heuristic values η_{ij} are weighted by $\beta > 1$ such that favourable items appear with a raised frequency. However, by choosing an appropriate value for β, the entropy of the heuristic matrices should be restricted to a small range around $27\% \cdot H_{max}$. This entropy range corresponds to a weight $\beta \approx 5$ which is in accordance with desirable β-parameters in the standard ACO algorithm.

In the final experiment, we implemented P-ACO with two kinds of heuristic guidance: the time-scattered and the standard heuristic. The latter was also tested in combination with candidate lists, which are a common technique to reduce the runtime for the Traveling Salesperson Problem [11]. The candidate list heuristic can be seen as an alternative to the time-scattered heuristic, although it is not suitable for a hardware realization. For every city i, the nearest neighbours are calculated and stored in a candidate list L_i of fixed size C_l. Using candidate lists, an ant located in city i can choose the next city j only from the candidate list. Only if none of these cities can be visited anymore ($S \cap L_i = \emptyset$), then j is chosen from S. Typically, the size of candidate lists is set to $C_l = 20$ (e.g. [7, 14]).

Table 1. Comparison of time-scattered heuristic, standard heuristic and standard heuristic with candidate lists. Columns show the TSP instance, the total computation time (TT), the average tourlength (AVG), the standard error (S-ERR) and the computation time per iteration (TI)

		P-ACO-TSH			P-ACO-SH			P-ACO-CL		
TSP	TT[s]	AVG	S-ERR	TI[ms]	AVG	S-ERR	TI[ms]	AVG	S-ERR	TI[ms]
gr48	562.0	5099.45	7.15	0.567	5075.85	4.26	5.620	**5075.45**	4.18	2.603
eil101	2432.4	**645.40**	0.87	1.258	649.99	1.06	24.324	647.37	0.89	5.576
d198	9215.4	15942.83	18.19	2.414	**15927.80**	15.38	92.154	15970.00	11.33	11.418
rd400	39483.4	16553.78	124.86	4.962	15557.17	16.77	394.834	**15483.90**	14.91	23.778
ry48p	562.0	14547.70	20.24	0.533	14554.95	28.26	5.620	**14518.45**	18.46	2.603
ft70	1176.1	**39071.65**	58.76	0.801	39150.45	51.58	11.761	39189.65	49.97	3.917
kro124p	2377.9	**36836.50**	88.51	1.134	37057.75	119.80	23.779	37227.30	102.17	5.564
ftv170	6884.3	**2836.05**	9.38	1.913	2858.10	14.54	68.843	2866.70	15.97	9.981

In our experiment, the parameter settings which are equal for all three heuristic approaches are: $k = 3$, one elitist ant, $m = 8$, $\tau_{init} = 1$, $f_P = 10$ and $\beta = 5$. For P-ACO-TSH, we set the following additional parameters: $q_0 = 0.9$, $\eta_{init} = 1$, $f_H = 10$, $c = 3$, $l = 48$, $f_s = 4$ and random shift policy. The standard heuristic approach (P-ACO-SH) was started with $q_0 = 0.3$ and $\alpha = 1$, the candidate list approach (P-ACO-CL) with $q_0 = 0.8$, $\alpha = 1$ and $C_l = 20$. Note that the q_0 parameters are different, preliminary experiments have shown that the P-ACO-TSH and the P-ACO-CL approach benefit from a high degree of exploitation. We studied the three approaches on a range of TSP instances shown in Table 1. In all cases, P-ACO-SH has the highest computation time per iteration. Therefore, we chose the maximum total computation time per instance (equal to all three approaches) such that P-ACO-SH could run for 100000 iterations. The results are shown as the average tourlength over 20 repititions with different random seeds.

Table 1 shows that in 50% of all TSP test instances P-ACO-TSH received the best average tourlengths. The time-scattered heuristic performs well compared to the standard heuristic and to the candidate list enhanced version in all cases but **rd400**. The reason for the relatively poor performance of P-ACO-TSH on this instance is probably due to the fact that c and l were chosen as constant over all instance sizes n, and while the values we used work very well for $n \approx 100$, we see that for smaller ($n \leq 48$) as well as larger ($n \geq 198$) instances, they lead to inferior performance. Preliminary tests show that tuning c and l to the instance size improves the solution quality significantly and results in a similar performance as the other two P-ACO variants.

In summary, Table 1 indicates that the three variants of the P-ACO algorithm implemented in software perform similarly over the specified amount of time. However, P-ACO-TSH is constructed specifically with a hardware implementation in mind, which, judging from earlier results obtained in [12,2], runs an order of magnitude faster than our software implementation on a sequential CPU. Therefore, P-ACO-TSH has the potential to outperform P-ACO-SH as well as P-ACO-CL, resulting in one of the most efficient implementations of the ACO metaheuristic available.

6   Conclusion

We proposed an extension to the P-ACO algorithm implemented in hardware on a Field Programmable Gate Array (FPGA). This extension allows to include heuristic knowledge into the ants' decisions without giving up the low asymptotic runtime complexity. However, the application of the proposed approach is restricted to such heuristics which allow for an a priori calculation of heuristic information. By enhancing the match buffer concept, the new heuristic approach can be integrated with only moderate modifications to the existing P-ACO hardware circuitry. In experimental studies comparing the software simulations of the new time-scattered heuristic with the standard heuristic and candidate lists, the proposed approach shows a competitive performance, which implies a superior performance with respect to the speed-ups attained by mapping P-ACO into hardware.

In future work, we will concentrate our efforts on implementing the modified P-ACO algorithm on an FPGA and verifying the expected speed-up by comparing it with the respective software implementation. Furthermore, we plan to investigate how the time-scattered heuristic works when applied to problem classes other than TSP.

Acknowledgments

This work is part of the projects "Optimization on Reconfigurable Architectures" and "Methods of Swarm Intelligence on Reconfigurable Architectures", both funded by the German Research Foundation (DFG).

References

1. E. Bonabeau, M. Dorigo, and G. Theraulaz. *Swarm Intelligence. From Natural to Artificial Systems.* Oxford University Press, 1999.
2. O. Diessel, H. ElGindy, M. Middendorf, M. Guntsch, B. Scheuermann, H. Schmeck, and K. So. Population based ant colony optimization on FPGA. In *IEEE International Conference on Field-Programmable Technology (FPT)*, pages 125–132, December 2002.
3. M. Dorigo and G. Di Caro. The ant colony optimization meta-heuristic. In D. Corne, M. Dorigo, and F. Glover, editors, *New Ideas in Optimization*, pages 11–32. McGraw-Hill, 1999.
4. M. Dorigo, G. Di Caro, and L. M. Gambardella. Ant algorithms for discrete optimization. *Artificial Life*, 5(2):137–172, 1999.
5. M. Dorigo and L. Gambardella. Ant colony system: A cooperative learning approach to the traveling salesman problem. *IEEE Transactions on Evolutionary Computation*, 1:53–66, 1997.
6. M. Guntsch and M. Middendorf. A population based approach for ACO. In S. Cagnoni et al., editor, *Applications of Evolutionary Computing - EvoWorkshops 2002: EvoCOP, EvoIASP, EvoSTIM/EvoPLAN*, number 2279 in Lecture Notes in Computer Science, pages 72–81. Springer Verlag, 2002.
7. D. S. Johnson and L. A. McGeoch. The traveling salesman problem: A case study in local optimization. In E. H. L. Aarts and J. K. Lenstra, editors, *Local Search in Combinatorial Optimization*. Wiley, 1995.
8. D. Merkle and M. Middendorf. Fast ant colony optimization on runtime reconfigurable processor arrays. *Genetic Programming and Evolvable Machines*, 3(4):345–361, 2002.
9. M. Randall and A. Lewis. A parallel implementation of ant colony optimization. *Journal of Parallel and Distributed Computing*, 62(9):1421–1432, 2002.
10. G. Reinelt. TSPLIB - a traveling salesman problem library. *ORSA Journal on Computing*, 3:376–384, 1991.
11. G. Reinelt. *The Traveling Salesman: Computational Solutions for TSP Applications*, volume 840 of *LNCS*. Springer-Verlag, 1994.
12. B. Scheuermann, K. So, M. Guntsch, M. Middendorf, O. Diessel, H. ElGindy, and H. Schmeck. FPGA implementation of population-based ant colony optimization. *Applied Soft Computing*, 4:303–322, 2004.
13. C. E. Shannon. A mathematical theory of communication. *Bell System Technical Journal*, 27:379–423 and 623–656, 1948.
14. T. Stützle and M. Dorigo. ACO algorithms for the quadratic assignment problem. In D. Corne, M. Dorigo, and F. Glover, editors, *New Ideas in Optimization*, pages 33–50. McGraw-Hill, 1999.

Ad Hoc Networking with Swarm Intelligence*

Chien-Chung Shen[1], Chaiporn Jaikaeo[2], Chavalit Srisathapornphat[3],
Zhuochuan Huang[1], and Sundaram Rajagopalan[1]

[1] Department of Computer and Information Sciences, University of Delaware, USA
[2] Department of Computer Engineering, Kasetsart University, Thailand
[3] Department of Computer Science, Kasetsart University, Thailand

Abstract. Ad hoc networks consist of mobile nodes that autonomously establish connectivity via multihop wireless communications. Swarm intelligence refers to complex behaviors that emerge from very simple individual behaviors and interactions. In this paper, we describe the ANSI project that adopts swarm intelligence as an adaptive distributed control mechanism to design multicast routing, topology control, and energy conservation protocols for mobile ad hoc networks.

1 Introduction

Mobile wireless ad hoc networks consist of mobile nodes that autonomously establish connectivity via multihop wireless communications. Without relying on any existing, pre-configured network infrastructure or centralized control, ad hoc networks are useful in many situations where impromptu communication facilities are required, such as battlefield communications and disaster relief missions.

Swarm intelligence [2] refers to complex behaviors that emerge from very simple individual behaviors and interactions. Most existing work on applying swarm intelligence to ad hoc networks concentrate on designing unicast routing protocols [3, 8, 7, 1]. In addition, there are work that use swarm intelligence to design algorithms for multicast routing [11, 4, 17], Steiner tree [6, 17], and minimum power broadcast [5] problems. However, these solutions either require global information and centralized computation, or are not adaptive to mobility, and hence can not be directly applied to mobile ad hoc networks.

In essence, swarm intelligence incorporates the following three components [2]: (1) **positive/negative feedback**, which searches for good solutions and stabilizes the result, (2) **amplification of fluctuation**, which discovers new or better solutions to adapt to changing environment, and (3) **multiple interaction**, which allows distributed entities to collaborate and self-organize. In this paper, we describe the ANSI (Ad hoc Networking with Swarm Intelligence) project that integrates these components to form a distributed, adaptive control mechanism to design multicast routing, topology control, and energy conservation protocols for mobile ad hoc networks. The ANSI project also designs a unicast ad hoc routing protocol [12].

---

* This work is supported in part by U.S. National Science Foundation under grant ANI-0240398.

M. Dorigo et al. (Eds.): ANTS 2004, LNCS 3172, pp. 262–269, 2004.

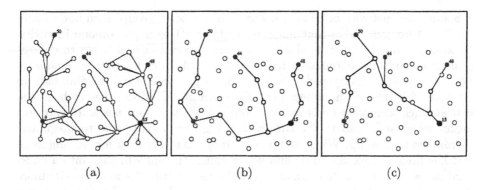

Fig. 1. Sample network snapshots illustrating the operations of MANSI with five group members (nodes 9, 15 44, 48 and 50) where node 15 is the core· (a) dissemination of an announcement from the core indicated by arrows, (b) initial forwarding set of twelve nodes (node 9 and the core are also forwarding nodes) as a result from the members sending requests back to the core; arrows indicate join-request directions, and (c) forwarding set of eight nodes evolved by ants later in time

2   Multicast Routing

We design the MANSI (Multicast for Ad hoc Networking with Swarm Intelligence) multicast routing protocol for mobile ad hoc networks [14]. MANSI require neither the knowledge of the global network topology, nor group memberships, and thus is suitable for ad hoc networks. For each multicast group, MANSI determines a set of forwarding nodes (forming a forwarding set) that connect all the group members together. Nodes belonging to the forwarding set are responsible for forwarding multicast data from one member to all the others by rebroadcasting non-duplicate data packets. By forwarding data in this manner, the group connectivity resembles a *mesh* structure rather than a tree, which is more robust to link breakage due to node mobility. In addition, MANSI is a reactive protocol in that no connectivity is maintained until some node has data to be sent to the group. The first member node who becomes an active source (i.e., starts sending data to the group) takes the role of the focal point for the other group members. This node, now known as the *core* of the group, then announces its existence to the others by flooding the network with an announcement packet. Each member node then relies on this announcement to send a join-request packet back to the core via the reverse path. Intermediate nodes who receive join-request packets addressed to themselves become forwarding nodes of the group. As a result, these forwarding nodes form an initial mesh structure that connects the group members together. To maintain connectivity and allow new members to join, the core periodically floods an announcement packet as long as there are more data to be sent. Since this process is performed only when there is an active source sending data to the group (i.e., on demand), valuable network bandwidth is not wasted to unnecessary maintenance of group connectivity in such dynamic environments. Figures 1(a) and 1(b) depict an announcement process and the formation of an initial forwarding set, respectively, in a simulated network of 50 nodes. For each multicast group, each node maintains a pheromone table that maps a neighbor ID to a pheromone intensity. An announcement packet flooded by the core also causes each node to deposit a small amount of pheromone on the reverse link from which the announcement was received. Every node, therefore, has a pheromone trail leading back to the core after the first announcement. Once an initial forwarding set has been formed, each member who is not the core periodically launches a forward ant packet which will head to the core with a high probability due to the initial pheromone trail setup. When an ant encounters the core or a forwarding node that has been join-requested by another member, the ant will turn into a backward ant and return to its originator. On the return trip, the ant deposits more pheromone on each node by an amount proportional to the distance (in terms of hop count) it has traveled back in order to attract more ants to explore the same path (i.e., positive feedback). Since a forward ant does not always follow a trail with the highest pheromone intensity, it sometimes chooses an alternate link to opportunistically explore different routes to the core (i.e., amplification of fluctuation). Negative feedback is implicitly carried out by pheromone evaporation, which means that a path on which a forward ant decides to travel but a backward ant never returns is considered a bad path and should be explored less often. Updating of pheromone can also trigger deformation of the forwarding set, as each member or forwarding node periodically sends a join-request to the neighbor whose pheromone intensity is the highest. If a member finds that a new neighbor has higher pheromone intensity than its currently requested neighbor, all subsequent join-requests will be sent toward that new neighbor instead. As a result, the forwarding set will keep evolving into new states that yield lower costs (expressed in terms of total cost of all the nodes in the forwarding set), as shown in Figure 1(c). Therefore, by depositing pheromone and following pheromone trails, multiple ants indirectly interact to cooperatively find a good solution, which is a low cost forwarding set in this case. This evolving (including exploring and learning) mechanism differentiates MANSI from all other existing ad hoc multicast routing protocols. Since a node's cost is abstract and may be defined to represent different metrics, MANSI can be applied to many variations of multicast routing problems such as load balancing and power-aware routing. MANSI also incorporates a mobility-adaptive mechanism that allows the protocol to remain effective as mobility increases. This is achieved by having members (or forwarders) located in highly dynamic areas send join-requests to the best two nodes (in terms of pheromone intensity) instead of one, resulting in more forwarding nodes and thus yielding more robust connectivity. Simulation results show that MANSI performs both effectively and efficiently in static or low-mobility environments, yet still effectively in highly dynamic environments.

3   Multicast Packet Delivery Improvement Service

To improve packet delivery of multicast routing protocols, we design a protocol-independent packet-delivery improvement service, termed PIDIS [15], for ad hoc multicast. PIDIS adopts swarm intelligence to adaptively search and quickly converge to good candidate routes through which lost packets could be recovered with the greatest probability, while adapting to changing topology.

PIDIS is a gossip-based scheme; however, unlike other gossip protocols, its mechanism for retrieving lost packets at a node is triggered exactly once, when the packet is considered lost. If the lost messages are not retrieved after a recovery process via a gossip (lost) request message (GREQ) is initiated, no further attempt is made to retrieve the lost messages again. Thereby, PIDIS is purely non-persistent and best-effort, and hence incurs low overhead.

The PIDIS mechanism works as follows: When packets are lost at a node, a packet-recovery service can either blindly gossip, without using any heuristics as to where a lost packet may be fetched from, or use available information about previously recovered packets to make a decision. PIDIS uses the information about previously recovered packets. When the first time a packet from source s is lost at a group member $i \in G = \{g_1, g_2, \ldots, g_m\}$ (the set of corresponding group members), i has no information about where to recover the lost packet from, and so broadcasts GREQ. This GREQ picks a random next hop at each hop visited, taking care to avoid loops, until it reaches a node $g \in \{s\} \cup G$, and $g \neq i$. If g has the packets which are lost at i, then g dispatches a GREP to i for each of the packets lost at i. The next hop j through which the GREPs are received at i is then positively reinforced (i.e., positive feedback) for recovering packets for the source-group pair s/G. In addition to finding an effective next gossip hop, from time to time, a gossip is broadcast instead of being unicast to facilitate amplification of fluctuations so as to find better routes to recover lost packets

4 Topology Control

Topology control in ad hoc networks aims to reduce interference, reduce energy consumption, and increase effective network capacity, while maintaining network connectivity. The primary method of topology control is to adjust the transmission powers of the mobile nodes. Topology control could be formulated as a combinatorial power assignment problem as follows. Let $V = \{v_1, v_2, \ldots, v_n\}$ represent the set of nodes in the network. Each node is equipped with the same, finite set P_s of power levels, where $P_s = \{p_1, p_2, \ldots, p_k\}$, $p_1 < p_2 < \ldots < p_k$, and $p_k \equiv P_{full}$ (the maximum available transmission power, termed *full power*). The network is assumed to be connected when every node is assigned P_{full}, which is the case when no topology control is in effect. The goal of topology control is to assign transmission power to each node (*i.e.*, to generate power assignment $A = \{P_{v_i} | \forall v_i \in V\}$, where P_{v_i} is the assigned power for node v_i) so as to achieve optimization objectives subject to the network connectivity constraints. Therefore, the size of the search space of the topology control problem is $O(k^n)$.

We use swarm intelligence to design a distributed topology control algorithm termed Ant-Based Topology Control (ABTC) [9, 16] for mobile ad hoc networks. ABTC adapts well to mobility, and its operations do not require any geographical location, angle-of-arrival, topology, or routing information. Further, ABTC is scalable, where each node determines its own power assignment based on local information collected from neighbor nodes located within its full power transmission range, In particular, ABTC attempts to achieve the following two objectives[1]: (1) minimizing the maximum power used by any node in the network, $P_{max} = \max_{i=1}^{n} P_{v_i}$, (MINMAX objective), and (2) minimizing the total power used by all of the nodes in the network, $P_{tot} = \sum_{i=1}^{n} P_{v_i}$, (MINTOTAL objective), subject to connectivity constraints.

ABTC works as follows. Every node v_i periodically executes a neighbor discovery protocol using the full power P_{full} to obtain its (current) neighbor set, $N(v_i)$ (*i.e.*, the set of nodes that are located within v_i's full power transmission range). In addition, the node v_i periodically broadcasts an ant packet using certain chosen transmission power, and the value of the transmission power is also carried inside the ant packet. This ant packet will then be relayed (via broadcast) by using the same transmission power by only nodes belonging to $N(v_i)$. In the meantime, upon receiving ant packets, the node v_i determines the minimal power $P_{i_{min}}$ such that it can receive ant packets originated from all, and relayed only by, the nodes in $N(v_i)$ using $P_{i_{min}}$, and assigns $P_{i_{min}}$ to P_{v_i}. It has been proved that when this condition is satisfied at every node, the resulting power assignment guarantees network connectivity.

When there is no mobility, every node v_i only needs to originate ant packets using each one of the power levels once, and in the meantime to collect ant packets from all of its neighbors to determine $P_{i_{min}}$. However, when there is mobility, the neighbor set of each node may keep changing. Therefore, a node may not have enough time to collect ant packets transmitted with all the different power levels originated from all its neighbors before its neighbor set changes, unless the broadcast period is short enough, which imposes undesirably control overhead. To address this issue, swarm intelligence is used as a heuristic search mechanism to discover a minimized power level that best meets the condition aforementioned as follows.

The (intensity of) pheromone in swarm intelligence is used to denote the *goodness* of power assignments. Initially, power is searched and assigned randomly. Gradually, better power assignments, those that better achieve the Min-Max or the MinTotal objective subject to connectivity constraints, are preferred in the search and their associated pheromone intensities increase, which in return further reinforce these power assignments. Through such a positive-feedback process, ABTC quickly converges to (locally) optimal power assignments. In contrast, negative feedback via pheromone *evaporation* stales old power assignments that are no longer effective. Furthermore, amplification of fluctuation explores and discovers new or better power assignments to accommodate changing topology due to mobility. Finally, nodes locally decide power assignments by collecting local information, via sending and receiving ant packets, without relying on any central entity (multiple interaction).

To demonstrate the capability of ABTC, an extensive simulation study has been conducted for different mobility speed, various density, and diverse node

---

[1] Since each node asynchronously collects local information from neighbor nodes to search for its appropriate transmission power, global optimality is not ensured.

distribution (uniform and non-uniform). In particular, comparison is performed, using several metrics, between the results of ABTC and (1) a centralized topology control algorithm based on CONNECT [13] that optimizes maximum power and approximates minimized total power, and (2) the Cone-Based Topology Control (CBTC) [10] algorithm that minimizes the node degree and power with the use of angle-of-arrival information. Simulation results show that ABTC exhibits a more balanced performance, as compared to CBTC and the CONNECT-based centralized algorithm, in terms of the following metrics: average power, average end-to-end hop count (length of shortest route from a source to a destination), average end-to-end total power (smallest total power to forward a packet from a source to a destination), and average node degree. In contrast, CONNECT and CBTC trade average end-to-end hop count for other performance metrics, for which CONNECT performs the best and CBTC performs the worst among the three.

5   Energy Conservation

We utilize swarm intelligence to design the Ant-Based Energy Conservation (ABEC) [18] protocol for mobile ad hoc networks. The idea of ABEC is to identify a small subset of active nodes which are connected and are reachable by other nodes in the network. These nodes serve as a forwarding infrastructure to maintain network forwarding performance. Energy conservation (or extended network operational lifetime) is achieved when other nodes turn to the *sleep* mode if they are idle from communications.

ABEC first identifies a set of nodes that have the highest energy within local proximity based on neighbor information collected from single-hop ant packet broadcast. Each local highest energy node is then responsible for searching a small number of neighbors to help maintaining network connectivity/performance. The node exploits amplification of fluctuation to discover the most suitable set of neighbors to serve this purpose. A certain number of neighbors will be selected (as a set) based on the size of the largest *gap* between them and their *affinity*. With an assumption that the direction of neighbors can be determined from the angle of arrival of incoming signals, the node is able to calculate the size of *gaps* between neighbors in the same set. *Affinity* of a neighbor indicates the likelihood that this neighbor will be reinforced by other nodes and depends on the amount of reinforcement the neighbor received in the past and it's remaining energy. Ant packets are directed to reinforce these sets of neighbors with different probability, which also enable ABEC to discover alternate set of active node when network topology or traffic condition changes. Negative feedback refers to the state of lacking reinforcement at the nodes and leads to the transition to the *sleep* mode if the nodes does not require any communication activity. ABEC also proposes a mechanism (through the pheromone model) for a node to adaptively adjust its sleep time based on the amount of time and forwarding effort the node has performed. Figure 2 depict ABEC operations using a small network scenario.

Simulation results shown that, compared to a network without energy conservation, ABEC extends the time before the first node exhausts its energy by

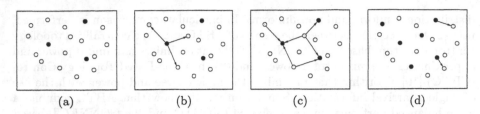

Fig. 2. A small network scenario illustrating ABEC mechanism: (a) an initial network with three local highest energy nodes displayed in black, (b) one local highest node starts to reinforce (using ant packets) a subset of neighbors shown in gray which are chosen through the amplification of fluctuation technique, (c) once the grey neighbors receive ant packets, they further reinforce other sets of their neighbors which are chosen in a similar fashion, (d) the gray nodes in (c) become active (black nodes) and continue the reinforcement process. Other white nodes, which are not reinforced by ants, may enter the *sleep* mode after a certain wait period

approximately 10-25%, and also extends the network operational lifetime to more than three times for high density static networks, with comparable packet delivery ratio and end-to-end delay. The slight reduction of forwarding performance is due to the increase of channel utilization from ABEC protocol overhead and the decrease of the number of active nodes.

6   Conclusion and Future Work

By integrating the positive/negative feedback, amplification of fluctuation, and multiple interaction components, the ANSI project exploits swarm intelligence as a distributed, adaptive control mechanism to design networking protocols for mobile ad hoc networks. In essence, the positive/negative feedback component is exploited to search for good solutions (such as forwarding sets of lower cost, lower power assignment, good candidate active nodes, etc.), the amplification of fluctuation component is exploited to discover new or better solutions to adapt to topology and/or traffic dynamics, and the multiple interaction component is exploited to design distributed protocols that enable mobile nodes to use local information to collaborate without any centralized coordination.

Research is in progress to explore *feature interactions* and *cross-layer* protocol design by deploying a common set of ants that are capable of performing multiple networking functions. Research is also in progress to adopt swarm intelligence to designing robust peer-to-peer systems over mobile ad hoc networks.

References

1. Baras, J.S., Mehta, H.: A Probabilistic Emergent Routing Algorithm for Mobile Ad hoc Networks. In: WiOpt'03: Modeling and Optimization in Mobile, Ad Hoc and Wireless Networks. (2003)
2. Bonabeau, E., Dorigo, M., Theraulaz, G.: Swarm Intelligence: From Natural to Artificial Systems. Oxford University Press, New York (1999)

3. Camara, D., Loureiro, A.A.F.: A GPS/Ant-Like Routing Algorithm for Ad Hoc Networks. In: IEEE Wireless Communications and Networking Conference (WCNC?00), Chicago, IL (2000)

4. Chu, C.H., Gu, J., Hou, X.D., Gu, Q.: A Heuristic Ant Algorithm for Solving QoS Multicast Routing Problem. In: 2002 Congress on Evolutionary Computation (CEC). (2002) 1630–1635

5. Das, A., II, R.J.M., El-Sharkawi, M., Arabshahi, P., Gray, A.: The Minimum Power Broadcast Problem in Wireless Networks: An Ant Colony System Approach. In: IEEE CAS Workshop, Pasadena, CA (2002)

6. Gosavi, S., Das, S., Vaze, S., Singh, G., Buehler, E.: Obtaining Subtrees from Graphs: An Ant Colony Approach. In: Swarm Intelligence Symposium, Indianapolis, Indiana (2003)

7. Gunes, M., Sorges, U., Bouazizi, I.: ARA–The Ant-Colony Based Routing Algorithm for MANETs. In: International Conference on Parallel Processing Workshops (ICPPW), Vancouver, B.C., Canada (2001)

8. Heissenbttel, M., Braun, T.: Ants-Based Routing in Large Scale Mobile Ad-Hoc Networks. Technical report, University of Bern (2003)

9. Huang, Z., Shen, C.-C.: Poster: Distributed Topology Control Mechanism for Mobile Ad Hoc Networks with Swarm Intelligence. In: ACM MobiHoc, Annapolis, Maryland (2003) (poster paper, also appeared in ACM Mobile Computing and Communications Review, Vol. 7, Number 3, July 2003).

10. Li, L., Halpern, J.Y., Bahl, P., Wang, Y.M., Wattenhofer, R.: Analysis of a Cone-Based Distributed Topology Control Algorithms for Wireless Multi-hop Networks. In: ACM Symposium on Principle of Distributed Computing (PODC), Newport, Rhode Island (2001)

11. Lu, G., Liu, Z., Zhou, Z.: Multicast Routing Based on Ant Algorithm for Delay-Bounded and Load Balancing Traffic. In: 25th Annual IEEE Conference on Local Computer Networks (LCN'00), Tampa, FL (2000)

12. Rajagopalan, S., Shen, C.-C.: A Routing Suite for Mobile Ad hoc Networks Using Swarm Intelligence. Technical Report UD CIS TR# 2004-014, University of Delaware, Newark, Delaware (2004)

13. Ramanathan, R., Rosales-Hain, R.: Topology Control of Multihop Wireless Networks using Transmit Power Adjustment. In: Infocom, Tel-Aviv, Israel (2000) 404–413

14. Shen, C.-C., Jaikaeo, C.: Ad hoc Multicast Routing Algorithm with Swarm Intelligence. ACM Mobile Networks and Applications (MONET) Journal, special issue on Algorithmic Solutions for Wireless, Mobile, Ad Hoc and Sensor Networks **10** (2005) (to appear).

15. Shen, C.-C., Rajagopalan, S.: Protocol-independent Packet Delivery Improvement Service for Mobile Ad hoc Networks. Technical Report UD CIS TR# 2004-05, University of Delaware, Newark, Delaware (2004)

16. Shen, C.-C., Huang, Z., Jaikaeo, C.: Ant-Based Distributed Topology Control Algorithms for Mobile Ad Hoc Networks. submitted to ACM Wireless Networks (WINET) Journal (2004) (revision under review).

17. Singh, G., Das, S., Gosavi, S., Pujar, S.: Ant Colony Algorithms for Steiner Trees: An Application to Routing in Sensor Networks. In de Castro, L.N., von Zuben, F.J., eds.: Recent Developments in Biologically Inspired Computing. Idea Group Inc. (2004)

18. Srisathapornphat, C., Shen, C.-C.: Ant-Based Energy Conservation for Ad hoc Networks. In: 12th International Conference on Computer Communications and Networks (ICCCN), Dallas, Texas (2003)

An Ant Colony Heuristic for the Design of Two-Edge Connected Flow Networks

Efstratios Rappos[1] and Eleni Hadjiconstantinou[2]

[1] Information and Analysis Directorate (Operational Research)
Department for Work and Pensions
The Adelphi, 1–11 John Adam Street
London, WC2N 6HT, United Kingdom
efstratios.rappos@dwp.gsi.gov.uk
[2] Tanaka Business School, Imperial College London
South Kensington Campus
London, SW7 2AZ, United Kingdom
e.hconstantinou@imperial.ac.uk

Abstract. We consider the problem of designing a reliable network defined as follows: Given an undirected graph, the objective of the problem is to find a minimum cost network configuration such that the flow balance equations are satisfied and the network is two-edge connected. The cost function for each edge consists of a fixed component and a variable component, which is proportional to the flow passing through the edge. We present a novel ant colony approach for the solution of the above problem. Computational experience is reported.

1 Introduction

Many successful developments in algorithms for hard optimization problems are inspired by processes occurring in Nature (genetic algorithms, simulated annealing and neural algorithms [2]). Ant Colony Optimization (ACO) is a distributed meta-heuristic approach for solving hard optimization problems and was inspired from the behavior of social insects [1], [6]. It has been successfully initially implemented, among others, for the Travelling Salesman Problem (TSP) by [4] and [5], for the transportation and the vehicle routing problem [7], the quadratic assignment problem [10], graph coloring [3] and network design [8].

This article describes an ACO heuristic to solve the following network design problem with reliability constraints: We consider an undirected network (V, E) without parallel edges or loops. The network has a single source, vertex 0, and the remaining vertices are demand (customer) vertices. We denote $V' = V \backslash \{0\}$ and $N = |V|$. Associated with each demand vertex is an integer d_i representing the amount of flow that this vertex requires (the demand). Attached to each edge e are two numbers, b_e and c_e, representing the fixed and variable edge costs respectively. The cost of using an edge is the sum of the fixed cost plus the variable cost multiplied with the flow that passes through the edge. The total cost of a network is the sum of the costs of all the edges it contains.

M. Dorigo et al. (Eds.): ANTS 2004, LNCS 3172, pp. 270–277, 2004.
© Springer-Verlag Berlin Heidelberg 2004

The flows that pass through the edges must be such that each customer receives exactly the amount of flow that it requires given that the flow is conserved. We further impose the following reliability requirement: *We seek a network that can satisfy the flow constraints if any single edge is removed from the network.*

The optimization objective is to minimize the cost of the solution network among all possibilities that satisfy the design requirements. There are no capacity constraints on the amount of flow that can pass through an edge.

This optimization problem shall be referred to as the *Reliable Network Design Problem (RNDP)*. The RNDP problem is in general $\mathcal{NP}$-hard even in the case without flows, since it contains both the two-edge connected network design problem and the fixed charge problem as special cases.

2   An Ant Colony Heuristic Algorithm for the RNDP

Like most ACO algorithms, the proposed algorithm is based on the original implementation for the TSP. There are however a few important differences that are worth mentioning: First of all, in the RNDP there are *two* costs per edge, one linear and one fixed, and the total cost of the solution depends on both these costs and also on the flows that pass through each edge. Secondly, we need to devise a way to incorporate reliability into the design of the network, i.e., to force an ant algorithm to produce a solution that satisfies the two-edge connectivity property. A new approach that include these issues in an ACO algorithm was developed and is discussed in the next sections.

2.1   The Ant Colony Setup

The first key question is how to use ants or agents to represent a solution to the RNDP. We can observe that, given the linear cost function with respect to the edge flow, any optimal solution will consist of a set of *flow* edges (which supply the customers) and form a tree, and a set of *reliability* edges which have zero flow and are included in the layout to guarantee the required reliability properties. A key observation is that any tree supplying $N - 1$ customers can be decomposed into $N - 1$ paths from the source to the customers, as shown in Fig. 1. Each path carries the demand that the customer at the end of the path requires. This observation enables us to represent each of these $N - 1$ paths with an ant trip.

The proposed ACO algorithm for the RNDP employs the following setup: We define an *ant group* as a set of artificial agents (ants) which move on the graph using predefined rules, and each ant group represents a feasible solution to the RNDP. In the implementation many ant groups can be present simultaneously.

Each edge of the graph has *two* trails attached to it, the *edge trail* and the *flow trail*, represented by two real numbers (the pheromone variables). The edge trails are related to the fixed cost of using the edge, whereas the flow trails reflect the additional cost incurred when the edge has some flow passing through it.

Each ant group consists of two types of agents: *flow ants* and *reliability ants*. When an ant group is first created, it contains the following agents: (i) $N - 1$ flow

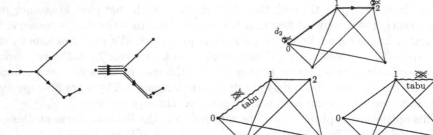

Fig. 1. A tree with N vertices is equivalent to $N - 1$ paths leaving the source

Fig. 2. Flow ants and reliability ants

ants which start from the source and have a customer vertex as their destination (each ant having a different customer as its destination); and (ii) $N - 1$ flow ants that perform the reverse trip: from the customer vertices to the source vertex.

Every flow ant also has a integer number associated with it, the *demand* of the flow ant, which is equal to the demand of the customer vertex it is going to or coming from. Every reliability ant has an edge associated with it, the *tabu edge*, which is an edge that it cannot use in its path as this represents the edge that has broken down in the reliability requirement of the RNDP.

Once an ant group has been created, the flow ants of the group start moving along the vertices of the graph using a probabilistic decision rule until they reach their destinations. When all the flow ants have reached their destinations, then reliability ants are added to the ant group. In particular, for every flow ant its path is tracked and a reliability ant is added for every edge visited by that ant. Each reliability ant has a particular edge (among the edges that were used by that flow ant) as *tabu* meaning that it cannot pass over it. The newly added ants also have the same source and destination vertices as the corresponding flow ant.

Consider the network shown in Fig. 2. A flow ant leaves vertex 0 (the source) and travels to vertex 2, carrying a flow of d_2. Assume that the ant followed the path $0 \rightarrow 1 \rightarrow 2$. For this particular ant, two reliability ants will be added since it used two edges in its path. Each of the two reliability ants will have the same source and destination as the original ant (i.e., vertices 0 and 2), and one of the two edges as their tabu edge, as shown in the last two diagrams of Fig. 2.

2.2   Trail Deposition and Updating

In this section we describe how the two types of pheromone (the flow pheromone variable T_f and the edge pheromone variable T_e) are deposited. We assume that: (i) *flow ants deposit and smell both types of pheromone*; whereas, (ii) *reliability ants deposit and smell the edge pheromone only*.

This is because the inclusion of a reliability edge into the layout will only incur the fixed cost, whereas a flow edge will incur both the fixed and the variable cost components. The contribution to the trails is inversely proportional to the cost incurred. In particular, an ant a, with a demand d_a, moving along edge (i, j)

will yield a contribution $\delta T_e(i,j)$ and $\delta T_f(i,j)$ to the edge and flow pheromone variables respectively, given by the following rules:

Reliability ants: $\delta T_e(i,j) := \delta T_e^a(i,j) = 1/b_{ij}, \qquad \delta T_f^a(i,j) = 0$ (1)

Flow ants: $\delta T_e(i,j) := \delta T_f^a(i,j) = 1/b_{ij}, \qquad \delta T_f^a(i,j) = 1/(c_{ij}d_a),$ (2)

The reason behind the above rules is that when an ant carrying a flow of d_a passes over edge (i,j), it will incur a fixed cost of b_{ij} and a variable cost of $c_{ij}d_a$. Reliability ants carry no flow and thus only the fixed cost is taken into account.

The pheromone variables are updated when all the ants in a group have finished. Note that some of the edges may be visited by more that one ant, and care should be taken in the calculation of the pheromone variables T_f. This is because in the RNDP solution, the fixed cost b_{ij} is incurred *only once*, no matter how many ants of the group used the edge. In order to maintain the balance between the fixed and linear trails, the edge trails must receive only one contribution for any edge visited. The trail update rules are:

Flow pheromone: $T_f(i,j) \leftarrow \rho T_f(i,j) + Q \sum_{a \in A_{ij}} \delta T_f^a(i,j)$ (3)

Edge pheromone: $T_e(i,j) \leftarrow \rho T_e(i,j) + Q \Xi_{ij}\, \delta T_e(i,j)$ (4)

where ρ is the evaporation coefficient, Q is a constant, A_{ij} is the set of ants that used edge (i,j) in their path and $\delta T_e(i,j)$, $\delta T_f^a(i,j)$ are given by (1) and (2) respectively. The variable Ξ_{ij} takes the value 1 if the edge (i,j) has been used by an ant of the group, and the value 0 otherwise.

2.3   Path Selection – Flow Ants

In this section we describe the strategy of selecting where an ant will move to at each step. This will be different for each of the two types of ant. Assume that a flow ant a is at vertex i. A random number q is selected from a uniform $(0,1)$ distribution and compared with a fixed parameter q_0 $(0 < q_0 < 1)$.

If $q \leq q_0$ the ant moves to vertex j, where

$$j = \arg \max_{k \in K} \left\{ [F(T_f(i,k), T_e(i,k))]^\alpha [H(i,k)]^\beta \right\}$$ (5)

Here $T_f(i,k)$ and $T_e(i,k)$ are the flow and edge pheromone variables for edge (i,k), and K is the set of allowed potential vertices, i.e.,

$$K = \{k \in V : (i,k) \in E, \text{ and vertex } k \text{ has not been yet visited}\}$$ (6)

The function H is a heuristic function and has been chosen to be equal to the inverse of the total cost of using the edge, i.e., $H(i,k) = 1/(c_{ik}d_a + b_{ik})$, where d_a is the demand carried by the ant. The two trails are combined using the function *trail function* F, which was chosen to be

$$F(T_f, T_e) = (1/T_f + 1/T_e)^{-1}.$$ (7)

The reason for the form (7) above is that it best represents the cost incurred by an ant when using an edge. Given that the two pheromones are the inverse of the fixed and variable costs, an equivalent pheromone combining the two costs will be of the form (7) because the total cost is the sum of the two cost components. Equation (5) specifies that the ant should move to the 'best' vertex, i.e., the one that has the highest combined value of 'strong trail' and 'small cost'. If $q > q_0$, then the ant may move to any allowed vertex using:

$$\mathbf{P}(\text{move to } j \in K) = \frac{[F\,(T_f(i,j), T_e(i,j))]^\alpha\,[H(i,j)]^\beta}{\sum_{k \in K}[F\,(T_f(i,k), T_e(i,k))]^\alpha\,[H(i,k)]^\beta}, \tag{8}$$

and, $\mathbf{P}(\text{move to } j \notin K) = 0$.

2.4   Path Selection – Reliability Ants

The reliability ants move in a similar way as the flow ants, but because they do not 'smell' the flow trails the corresponding functions are simpler. These ants, however, are not allowed to move along their tabu edge. Just as in the case for flow ants, if $q \leq q_0$ the ant moves from i to the 'best' overall vertex j, where

$$j = \arg\max_{k \in K}\left\{[T_e(i,k)]^\alpha\,[H(i,k)]^\beta\right\}, \tag{9}$$

whereas if, on the other hand, $q > q_0$ the ant moves to j with probability:

$$\mathbf{P}(\text{move to } j \in K) = \frac{[T_e(i,j)]^\alpha\,[H(i,j)]^\beta}{\sum_{k \in K}[T_e(i,k)]^\alpha\,[H(i,k)]^\beta}, \tag{10}$$

and $\mathbf{P}(\text{move to } j \notin K) = 0$. The notation used is the same as before, with the difference that the set K does not allow moves over tabu edges, i.e.,

$$K = \{k \in V : (i,k) \in E,\ (i,k) \neq \text{tabu}_a,\ \text{and } k \text{ has not been yet visited}\} \tag{11}$$

where tabu_a is the tabu edge of ant a.

2.5   Converting Ant Group Paths to a RNDP Solution

The following procedure is used to convert an ant group to a solution for the RNDP, after all the ants in the group have terminated:

1. Construct a network with the same vertex set as the input network containing the edges that were used in the motion of the flow ants.
2. Find the minimum cost tree solution to the flow problem on this network.
3. Add the edges that were used by the reliability ants.
4. Remove any redundant edges using the following procedure:
   (a) Scan all edges that carry no flow, in descending order of fixed cost (starting with the edge that has the largest fixed cost).
   (b) Test if that edge can be removed without violating the reliability requirements. If so, remove that edge and iterate the process.
   (c) If no more edges can be removed, stop.

3   Computational Implementation and Results

The algorithm runs until a specified number of ant groups have been generated, which shall be denoted by NG (number of groups). The output of the algorithm is the best solution found by the ant groups during the execution (recall that every ant group generates one RNDP solution).

When the algorithm starts, the trails are initialized as follows: The edge trails are all set to a small positive number, ϵ. For the flow trails we solve the flow problem on the input network with the variable costs only (no reliability) and we set the trails on the edges of the flow tree to a positive number, ϵ'. The remaining edges are initialized to ϵ, where $\epsilon' > \epsilon$. This is done in order to assist the flow ants to discover promising paths, since the flow tree above often contains 'good' edges that help the algorithm converge more quickly. In the implementation of the algorithm, these initial values were set to $\epsilon = 10$ and $\epsilon' = 20$.

3.1   Ant Group Generation

Note that the time taken for each ant group to terminate is not constant because in every group the ants follow different paths. It is useful to restrict the number of ant groups that are present at the same time on the network. This is beneficial for two reasons: it improves computational performance because the number of ants present could grow very large; and since each ant group represents a feasible solution, it is better to wait until a few of them have been produced before creating new ant groups in order to take better advantage of the cooperation between the groups. (Also note that the pheromone trail is updated after an ant group has finished and therefore having some groups start later allows the 'experience' of the earlier groups to be propagated.) The maximum number of groups that can be present at any time is denoted by NG_p. The algorithm starts by generating NG_p ant groups and thereafter, when an group has terminated, a new ant group is created until the total number of groups NG is achieved.

Moreover, we note that whenever a group discovers a solution with a smaller cost that the incumbent, we reward its pheromone contributions by multiplying them with the *trail enhancement factor h*. This is done to increase the pheromone levels there, as it is possible that better solutions may be found in that vicinity.

3.2   Preliminary Computational Results

The heuristic algorithm was implemented using C++ on a *Pentium III* PC with a clock speed of 600MHz. The algorithm was executed two times for each test instance and both the best and the average performance were recorded.

We experimented by modifying the values of all the parameters of the algorithm in an effort to find those that produce the best results. These are represented by NG, NG_p, Q, α, β, ρ, h, q_0, and were described in previous sections.

Recall that the parameters α and β represent the relative weight between the trail strength and small cost in the path selection part of the algorithm. For this reason one of the two can be fixed. Following [4], we choose to fix α to the value $\alpha = 1$ and vary β in the computational experimentation. The parameter Q can be similarly fixed because it represents a measure of the quantity of the trail deposited. By experimenting, the following values were observed to be good choices for the parameters: $\rho = 0.999$, $h = 1.001$, $q_0 = 0.7$, $\beta = 3$.

The test problems used were randomly generated problems with 12 to 50 vertices and an edge density of 75%. Following [9], the demands and the linear costs were chosen uniformly in the set $\{1, 2, \ldots, M\}$ and the fixed costs f were chosen in such a way that the quantity $\gamma = Ml/2f$ is equal to a given constant. Table 1 presents the average performance over the five instances of each group. The notation used is the following:

z_1^*, z_2^* The cost of the solution to the RNDP reported by the ACO algorithm in each of the two executions.

z_{opt} The cost of the optimal solution or the cost of the best feasible solution found at the end of the branch and cut algorithm.

$\%z_1^*, \%z_2^*$ Percentage ratios $z_1^*/z_{opt} - 1$ and $z_2^*/z_{opt} - 1$, respectively.

$\%z_{best}^*, \%z_{av}^*$ The smallest and average values among $\%z_1^*$ and $\%z_2^*$.

t_{av}, t_b The computational time (in seconds) taken to execute the ACO algorithm, averaged over the two runs, and the branch and cut.

Table 1. Computational results (averages)

Vert.	Edges	γ	$\%z_{best}^*$	$\%z_{av}^*$	t_{av}	t_b
12	94.4	0.1	6.24	7.00	38.4	2.1
12	94.2	1	0.33	0.44	35.4	1.7
12	96.4	2	0.83	1.07	31.9	0.8
Average			*4.98*	*5.40*	*36.8*	
15	144.6	0.1	3.31	4.20	143.5	4.0
15	144.8	1	1.43	1.76	173.4	3.6
15	146.8	2	0.55	0.67	126.9	3.0
Average			*4.31*	*4.70*	*156.1*	
20	255	0.1	7.16	7.16	348.1	25.0
20	254.6	1	0.85	0.91	347.3	24.5
20	259.4	2	0.86	0.98	325.6	11.8
Average			*7.54*	*7.67*	*339.7*	
25	397.4	0.1	10.41	11.23	470.8	271.5
25	398.4	1	1.11	1.26	526.9	294.0
25	400.6	2	1.88	1.93	551.5	59.6
Average			*11.14*	*11.80*	*541.6*	

Vert.	Edges	γ	$\%z_{best}^*$	$\%z_{av}^*$	t_{av}	t_b
30	569.8	0.1	16.71	16.86	950.3	5841.3
30	571	1	2.49	2.60	844.5	1087.6
30	569.2	2	1.87	1.88	859.5	1239.6
Average			*12.41*	*12.74*	*912.5*	
40	1020.6	0.1	21.27	21.77	3297.2	29797*
40	1016.4	1	2.67	2.84	3423.7	33741*
40	1022.4	2	2.54	2.56	2912.0	33286*
Average			*14.71*	*15.13*	*3217.5*	
50	1585.8	0.1	15.82	16.66	5166.4	36000*
50	1570.2	1	2.77	2.93	5876.1	36000*
50	1584	2	3.27	3.49	5095.2	36000*
Average			*13.00*	*13.69*	*5309.4*	

Vert.	NG	NG_p	Vert.	NG	NG_p	Vert.	NG	NG_p	Vert.	NG	NG_p
12	2000	15	20	1000	15	30	300	15	50	100	10
15	2000	15	25	500	15	40	200	10			

The optimal solutions were obtained by formulating the problem as a mixed integer program and solving it to optimality using a branch and cut algorithm and a commercial solver. A star means that the no optimal was reached for some instances in 10 hours. We note that the worst performance was obtained for the problem groups with small γ values, $\gamma = 0.1$ and $\gamma = 0.05$, where the largest percentage errors were produced. For the instances corresponding to $\gamma = 1$ and $\gamma = 2$ the algorithm performed extremely well even for the largest network sizes, with all percentage errors being less than 4%. The most difficult instances detected by the exact branch and cut algorithm also corresponded to the case of

$\gamma = 0.1$. For the smaller instances, the exact algorithm finds the optimum quickly and thus outperforms the ACO. The ACO however performs significantly better for the largest instances (e.g., the gaps for $|V| = 50$, $\gamma = 1$ and $\gamma = 2$ were between 7.8% and 8.6% after 10h using the exact algorithm).

4   Conclusions

In this article, a new ACO heuristic algorithm was developed for solving the two-edge connected network design problem with the presence of flows. The approach is based on the development of an ant group, where each member of the group exhibits an ant-like behaviour (depositing trails and preferentially selecting pheromone-rich paths) while the ant group represents a feasible solution. The idea that many ant groups are present simultaneously enhances the collaboration between the agents and yields a better solution. The technique could be extended to other reliable network flow problems with a tree-like solution, for instance the case with concave cost functions. Further experimentation is however necessary to evaluate the potential of ACO in the field, especially due to the complexity of combining flows and reliability.

References

1. E. Bonabeau, M. Dorigo, and G. Theraulaz. Inspiration for optimization from social insect behavior. *Nature*, 406:39–42, 2000.
2. A. Colorni, M. Dorigo, F. Maffioli, V. Maniezzo, G. Righini, and M. Trubian. Heuristics from nature for hard combinatorial optimization problems. *International Transactions in Operational Research*, 3(1):1–21, 1996.
3. D. Costa and A. Hertz. Ants can colour graphs. *Journal of the Operational Research Sociaty*, 48:295–305, 1997.
4. M. Dorigo and L. M. Gambardella. Ant colonies for the traveling salesman problem. Technical Report TR/IRIDIA/1996-3, Université Libre de Bruxelles, 1996.
5. M. Dorigo, V. Maniezzo, and A. Colorni. The ant system: optimization by a colony of cooperating agents. *IEEE Transactions on Systems, Man and Cybernetics (Part B)*, 26(1):29–41, 1996.
6. M. Dorigo and T. Stützle. The ant colony optimization metaheuristic: algorithms, applications, and advances. In F. Glover and G. A. Kochenberger, editors, *Handbook of Metaheuristics*, chapter 9, pages 251–285. Kluwer Academic, 2003.
7. L. M. Gambardella, É. Taillard, and G. Agazzi. MACS-VRPTW: A multiple ant colony system for vehicle routing problems with time windows. In D. Corne, M. Dorigo, and F. Glover, editors, *New ideas in optimization*, pages 63–76. McGraw-Hill, London, 1999.
8. J. Gomez, H. Khodr, P. De Oliveira, L. Ocque, J. Yusta, R. Villasana, and A. Urdaneta. Ant colony system algorithm for the planning of primary distribution circuits. *IEEE Transactions on Power Systems*, 19(2):996–1004, 2004.
9. D. S. Hochbaum and A. Segev. Analysis of a flow problem with fixed charges. *Networks*, 19:291–312, 1989.
10. E.-G. Talbi, O. Roux, C. Fonlupt, and D. Robillard. Parallel and colonies for the quadratic assignment problem. *Future Generation Computer Systems*, 17(4):441–449, 2001.

An Experimental Analysis of Loop-Free Algorithms for Scale-Free Networks
Focusing on a Degree of Each Node

Shigeo Doi and Masayuki Yamamura

Interdisciplinary Graduate School of Science and Engineering
Tokyo Institute of Technology
4259 Nagatsuta-cho, Midori-ku, Yokohama-shi, Kanagawa-ken, 226-8502 Japan
doi@es.dis.titech.ac.jp, my@dis.titech.ac.jp
http://www.es.dis.titech.ac.jp/en/index_en.html

Abstract. To use AntNet-FA globally, the ability of routing algorithms must be clear. The Internet has special topology and a hierarchy (AS and router). The topology have power-laws or scale-free property in other words. In this paper, we focused on the network topology and we applied AntNet algorithm to the network such as the Internet. We examined a node should use either a Loop-Free algorithm or a non-Loop-Free algorithm depending on its degree in heavy traffic condition. The Loop-Free feature means that when an ant decides to visit an adjacent node, then the ant selects the next node from its unvisited node. The non-Loop-Free algorithm is the same to the original AntNet. As a result, we found that network topology affects the ability of AntNet algorithms.

1 Introduction

AntNet algorithms [6] [7] [8] [9] showed better performance than the traditional best-effort routing algorithms, OSPF[14] and so on, through computer simulations. Now, The power-laws [5] [11], or scale-free [2] on the Internet topology are introduced by Faloutsos in 1999. In the power-laws, the nodes are categorized in two types, hub nodes and the others, in the context of the degree of each node. Moreover, the Internet is composed of Autonomous Systems(ASs) and routers. An AS is a set of routers. A router corresponds to a node in a graph representation. In other words, the Internet has a hierarchy. The average degree of routers is different from that of ASs through the analysis of recent real Internet topology [17] [18].

In this paper, we integrated these two results. We ran AntNet and its variations on a scale-free network under the hierarchy of the Internet and checked algorithms' ability. Also we inspected what criteria is appropriate whether each node should use the original AntNet or AntNet with Loop-Free feature to enhance the potential of AntNet-FA. The Loop-Free feature means that an ant selects a next node from its unvisited nodes when the ant decides to go an adjacent node. To use AntNet-FA on a wide area network (WAN) as a replacement of

M. Dorigo et al. (Eds.): ANTS 2004, LNCS 3172, pp. 278–285, 2004.

IGP (Interior Gateway Protocol) and BGP(Border Gateway Protocol), it is considered that AntNet should be examined on networks like the Internet, which is subject to the power-laws. We tested AntNet and AntNet with Loop-Free feature on two networks with 125 routers and 5 ASs as a borderless routing algorithm and other two networks with 100 ASs as an IGP routing algorithm. Loop-Free feature is effective on the network with 125 routers but causes side-effect with 100 ASs. In this paper, we show some results through computer simulations. As a conclusion, we propose either a Loop-Free algorithm or the original AntNet depending on the degree of each node or the diameter of the network to enhance the potential of AntNet.

The structure of this paper is as follows: In section 2, we review the network metrics on the recent Internet and explain the power-laws. In section 3, we add some features to AntNet and explain Loop-Free feature. In section 4, we explain our computer simulation environments and show results. Also we consider the results. In section 5, we discuss the future work.

2 Internet Topology and Power-Laws

The whole Internet topology can't be captured in real time. This is because the network nodes and links are changed dynamically in every moment [12]. However CAIDA (the Cooperative Association for Internet Data Analysis) provides the snapshot of the Internet topology.

In general, the Internet has a hierarchy, AS and router [5]. Each AS has about 25 routers on average in 2000 and 17 routers in 2001, respectively [17] [18]. The average degree of ASs is nearly equal to 4 or greater and that of routers is nearly equal to 3 or less [18] [17]. Faloutsos and et al. pointed out that the Internet topology has power-laws through analyzing the real network topology [5]. Power-Laws means that (1) the constant rank exponent d_v is proportional to the rank of the node r_v , to the power of constant R. That is, $d_v \propto r_v^R$. This power-law is evaluated by computing the out-degree for every node, sorting the out-degrees in descending order, and plotting in log-log space. (2) the frequency f_d of an out-degree d is proportional to the out-degree, to the power of a constant O. That is, $f_d \propto d^O$. (3) The total number of pairs of nodes within h hops $P(h)$ is proportional to the number of hops, to the power of a constant H. That is, $P(h) \propto h^H for h << \delta$, where δ is the diameter of the graph. (4) The eigenvalues of a graph λ_i are proportional to the order i, to the power of a constant E. That is, $\lambda_i \propto i^E$. [16] [5].

3 Combination of AntNet-FA and Loop-Free Routing

AntNet is a better algorithm for best-effort routing but we consider the ability of AntNet depends on the topology. In addition, the network topology can be easily changed. The word "Best-effort routing" implies that suddenly either a router or a link is possibly down. Therefore the routing algorithm must be robust for such the change. The network routing problem is like a dynamic traveling salesman problem [4]. In the network routing problem, there are many factors such as traffic density, traffic locality, we focus on network topology in this paper.

First, we explain the Lopp-Free feature below. When a node routes an ant to a neighbor node, the ant is routed only to the nodes which the ant hasn't visited yet. This feature is like the tabu-search heuristic in TSP. The reason why we implement the feature is that the algorithm ignores routing tables. This violation can reorganize the routing tables after network condition is changed. However the network condition is stable for a longtime, we consider the performance can be worse than the original AntNet-FA. It is inevitable for AntNet algorithms to generate a loop because nodes only learn an adjacent node corresponding to each destination [3]. The Loop-Free feature is like an anti-pheromone algorithm [13] from the view of exploring search space, thus the feature can be more robust than AntNet-FA. By the way, we consider the Loop-Free feature is appropriate for sparse networks. For example, a network has the node which has two links. If the Loop-Free feature is not applied to the node, an ant can make a loop. On the other hand, the Loop-Free algorithm is applied to the node, any ant isn't able to make a loop. Especially once network condition is changed, an ant without the Loop-Free feature tends to make a loop as we consider because it requires time to propagate the event that the condition was changed. Hence, at least the degree of the node is an important factor whether a node runs the Loop-Free algorithm or not.

Second, when an ant has a loop in its experience and its length of the loop is greater than half of its length, the ant is removed the loop from its experience and then it is forced to go back its source. If the current node is same to the source, the ant does nothing. This means that the intermediate nodes between the current node and the source node can make good use of the ant's experiences.

Finally, when focusing on properties of scale-free networks, we consider that the routing policy of hub nodes must be different from that of other nodes and the hub nodes routes ant without the Loop-Free feature. Therefore we set a threshold degree in a node whether the node routes ants with Loop-Free feature or not. The Loop-Free algorithm on stable condition because performance of the algorithms should be achieved same to AntNet-FA if the Loop-Free algorithm was robust.

3.1   Making Good Use of Ant's Experiences

In AntNet, when an ant has a loop in its stack and the length of the loop is greater than half of the total length, then the ant is killed. To use its experience effectively, if a loop is detected in an ant's experience, then the loop is removed and the ant is forced to go its source node. In other words, The ant is treated as if the ant had arrived its destination node. This is because both forward ants and backward ants have high priority in link queues in nodes all over the network, it is considered that the experience isn't obsolete after the loop is removed. This feature sometimes can cause that the time to transfer datagram packets becomes longer and longer. The procedure when detecting a loop in an ant's stack is as follows:

1. An ant arrives at a node.
2. Whenever the ant has a loop in its experience, then the loop is removed from the experience and the ant is forced to go back its source.
3. For every intermediate node on the ant's way, the node updates its routing tables using the ant's experience. It is same to AntNet-FA.

4   Experiments

4.1   Topology and Network Metrics

We used BRITE [10] to generate networks that we used in the subsequently described experiments. BRITE can generate the Internet-like network described above. We used Barabási's model [2] and Preferential Connection model in connection model in BRITE. In BRITE, We use "BA" model as the connection model between two routes and between two ASs and "Preferential Attachment" is enabled. This means that each AS has power-laws in router level. We generated 4 networks. The networks are categorized into two types: (1)100 ASs (AS100PC2, AS100PC3), and (2)5 ASs and 25 Routers (AS5RT25, AS5RT25-2). Table 1 shows some metrics of each network.

Table 1. Network Metrics of AS100PC2, AS100PC3, AS5RT25 and AS5RT25-2.

Parameters / Topology	AS100PC2	AS100PC3	AS5RT25	AS5RT25-2
Nodes	100	100	125	125
Diameter	5	4	8	9
Maximum Degree	24	29	16	15
Minimum Degree	2	3	3	2
Average Degree	3.94	5.88	5.68	3.92

4.2   Simulation Parameters

Table 2 shows the parameters that are used in our simulations. These parameters are almost same to the paper [15]. We assume each link has 1.5Mbps bandwidth and 1ms propagation delay. You think these two parameters are not up to date, but we consider that ants have to collect the trip time effectively such a poor network. If ants didn't work appropriately, the whole network throughput of datagram packets degraded and their transfer time got longer and longer.

We used the Poissonian Process to generate datagram packets and each node in a network generates one datagram packet with the average generation interval. We also considered IP packet fragmentation, which means the maximum length of each packet is limited to Maximum Transfer Unit (MTU). We set that MTU is equal to 1500 Bytes used in general LAN.

We selected the following eight competitors: the original AntNet-FA, AntNet-FAA, AntNet-FAA-Ln ($2 \leq n \leq 6$) , respectively. AntNet-FAA is an algorithm

Table 2. Simulation Parameters.

Regular interval	300ms	Average length of each packet	512 Bytes
Bandwidth of each link	1.5Mbit/s	Each input buffer size	∞
Each output buffer size	1Gbit	TTL for each datagram packet	15s
MTU	1500Bytes	Unit time of each simulation	0.1ms
Initial probability	uniform	Total time for each simulation	1000s
Learning time	500s		

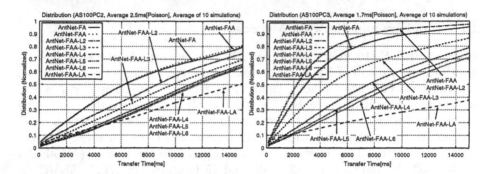

Fig. 1. Normalized distribution on AS100PC2 and AS100PC3.

that adds a "Making good use of ants' experience" in the section 3.1 to AntNet-FA. For AntNet-FAA-Ln, n means the degree whether the algorithm applies the Loop-Free feature to nodes in the network or not. For example, AntNet-FAA-L2 is the algorithm that applies Loop-Free feature to the nodes which its degree equals to or is less than 2. That is, when an algorithm that applies Loop-Free feature to the nodes which its degree equals to or is less than n, we call it AntNet-FAA-Ln. Last, AntNet-FAA-LA is an algorithm that applies Loop-Free feature to all the nodes.

4.3   Results

We got these results through 10 simulations for each network topology. We used the criteria of the experiments, packet transfer time distribution, which is also used in AntNet. In addition to the criteria, we use the criteria, average packet transfer time and packet loss rate (i.e. Loss Rate = (the number of all the packet sent) / (the number of all the packet received)) through 10 simulations.

The average packet generation interval of Poissonian process for each node is 2.5ms in AS100PC2, 1.7ms in AS100PC3 and 10ms in AS5RT25 and AS5RT25-2, respectively. The graphs about normalized packet distribution are shown in Figure 1 and Figure 2.

Figure 1 and Table 3 illustrate that the results of AntNet-FAA-L2 are identical with that of AntNet-FAA since the degree of all the nodes in AS100PC3 and AS5RT25 is greater than 2. Accordingly the behavior of AntNet-FAA-L2 is exactly alike that of AntNet-FAA. Futuremore, the results of AntNet-FAA

Table 3. Statistics: Average Transfer Time(ATT, milliseconds) and Loss Rate(LR, %). The best items are shown in bold face.

Topology	AS100 PC2	AS100 PC2	AS100 PC3	AS100 PC3	AS5 RT25	AS5 RT25	AS5 RT25-2	AS5 RT25-2
Algorithm	ATT	LR	ATT	LR	ATT	LR	ATT	LR
AntNet-FA	**5522.0**	19.6	3670.6	4.5	9801.1	35.8	9345.1	37.8
AntNet-FAA	5556.4	**18.3**	**2986.8**	**2.3**	9633.4	35.3	9395.7	37.9
AntNet-FAA-L2	6449.1	25.0	**2986.8**	**2.3**	9633.4	35.3	9553.0	36.5
AntNet-FAA-L3	6828.4	30.3	4934.4	13.0	9720.4	33.3	9357.9	31.8
AntNet-FAA-L4	7435.2	33.7	6237.0	20.9	8580.8	**31.4**	8296.3	29.1
AntNet-FAA-L5	7619.3	35.1	6742.7	25.3	8065.2	31.5	8434.0	28.6
AntNet-FAA-L6	7681.4	36.1	6929.5	28.5	**7641.3**	31.9	8405.4	**28.5**
AntNet-FAA-LA	7007.3	48.8	6095.5	61.5	8404.9	38.2	**8027.1**	31.2

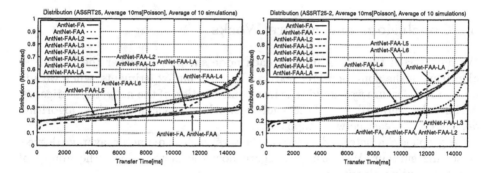

Fig. 2. Normalized distributions on AS5RT25 and AS5RT25-2.

showed the best performance of competitors. Consequently the Loop-Free feature is inappropriate for such the scale-free networks through these experiments.

Hence, when applying the Loop-Free feature, it requires to check the network metrics, at least the average degree of the nodes in the network. Indeed we consider that a Loop-Free feature violates the routing tables. Once a node learns a pathway to a certain destination, then it should modify the learned pathway gradually if network condition is changed.

In the Table 3, AntNet-FA, AntNet-FAA and AntNet-FAA-L2 showed lower performance than the other competitors in AS5RT25 and AS5RT25-2. This implies a Loop-Free feature can achieve higher performance on a sparse connected network. Consequently, the result of experiments on AS5RT25 and AS5RT25-2 indicates that the diameter or the maximum degree of the graph can affect the ability of routing algorithms. AS5RT25 includes the structure, AS and router. When we introduce the algorithm to the nodes which have a certain degree or less, the Loop-Free features showed better performance than AntNet without Loop-Free feature about packet transfer time and the statistics at last especially in a heavy traffic situation. In the other words, if the whole network is congested, the Loop-Free feature is possible to deliver datagram packets faster for their destinations.

On the other hand, In the light traffic situation, little difference between these algorithms can be observed. When comparing AS5RT25-2 with AS100PC2 and AS5RT25 with AS100PC3, the main differences are the diameter and the maximum degree. Hence, the Loop-Free features can be more adaptive on the sparse network. In addition, AntNet without the Loop-Free feature is a useful algorithm as a replacement of IGP or BGP.

These experiments resulted in getting the following knowledge: (1) A node-degree-based Loop-Free feature can get the same performance to the original AntNet but the parameter tuning is required. (2) The degree whether the Loop-Free feature or not depends on at least the topology. If the average degree all over the network could be measured, AntNet algorithm could exploit the network resource.

5   Conclusions and Future Works

We have tested the algorithm that each node should run either the Loop-Free algorithm or the non-Loop-Free feature. Through our experiments, when we add a Loop-Free feature we should take into consideration of the network metrics, especially the average degree of the whole network. We conclude the best algorithm depends on at least network topology.

A graph is characterized by various factors, such as diameter, average degree, clustering efficient. We must investigate how the factor affects the ability of adaptive routing algorithms.

We have to add ability, which ants collect the degree of intermediate nodes on their way. By collecting the degree of intermediate nodes which an ant visits, a node can determine whether a Loop-Free feature should be performed or not. It goes without saying that network condition is dynamically changing and then the factor requires to detect locally as possible.

Now we discuss the future work. In AntNet, the interval when a node launches an ant is constant. We have been developing algorithm that changes the interval corresponding to the load of links. When an ant is about to carry meaningless experience, this only consumes the bandwidth and degrades quality of services. The "Node Pheromone" [1] concept seems to be a good approach but it isn't able to adapt sudden changes of traffic input and network topology. We consider that if this case happened, nodes should launch ants as many as not to disturb to send datagram packets.

Forward ants and backward ants should be safe, not malicious, for the routers when using AntNet globally for a routing algorithm. Hence malicious mobile agents have to be removed by the routers and the security level of mobile agents has to be assured.

References

1. K. Amin and A. Mikler. Dynamic agent population in agent-based distance vector routing. *In Proc. of Intelligent Systems Design and Architecture (ISDA)*, 2002.
2. A. Barabási, R. Albert, and H. Jeong. Scale-free characteristics of random networks: The topology of the world wide web. *Physica A*, pages 69–77, 2000.
3. G. Canright. Ants and loops. *Proceedings of Ants2002*, pages 235–242, 2002.
4. C. J. Eykelhof and M. Snoek. Ant systems for a dynamic tsp - ants caught in a traffic jam. *Proceedings of Ants2002*, pages 88–99, 2002.
5. M. Faloutsos, P. Faloutsos, and C. Faloutsos. On power-law relationships of the internet topology. *In Proceedings of ACM SIGCOMM*, pages 251–262, 1999.
6. G. Di Caro and M. Dorigo. AntNet: A mobile agents approach to adaptive routing. *Technical Report IRIDIA 97-12*, 1997.
7. G. Di Caro and M. Dorigo. An adaptive multi-agent routing algorithm inspired by ants behaviour. *Proceedings of PART'98 - fifth Annual Australasian Conference on Parallel and Real-Time Systems*, 1998.
8. G. Di Caro and M. Dorigo. AntNet: Distributed stigmergetic control for communications networks. *Journal of Artificial Intelligence Research (JAIR)*, 9:317–365, 1998.
9. G. Di Caro and M. Dorigo. Two ant colony algorithms for best-effort routing in datagram networks. *Proceedings of PDCS'98 - 10th International Conference on Parallel and Distributed Computing and Systems*, 1998.
10. A. Medina, A. Lakhina, I. Matta, and J. Byers. BRITE: An approach to universal topology generation. *In Proceedings of the International Workshop on Modeling, Analysis and Simulation of Computer and Telecommunications Systems- MAS-COTS '01*, 8 2001.
11. A. Medina, I. Matta, and J. Byers. On the origin of power laws in internet topologies. *ACM Computer Communications Review*, 2000.
12. N. Minar, K. H. Kramer, and P. Maes. Cooperating mobile agents for mapping networks. In *Proceedings of the First Hungarian National Conference on Agent Based Computation*, 1998.
13. J. Montgomery and M. Randall. Anti-pheromone as a tool for better exploration of search space. *Proceedings of Ants2002*, pages 100–110, 2002.
14. J. Moy. *RFC 2178, OSPF Specification Version 2*. Network Working Group, July 1997.
15. K. Oida and A. Kataoka. Lock-free AntNets and their adaptability evaluations. *Electronics and Communications in Japan (Part I: Communications)*, 84:51–61, 2000.
16. C. R. Palmer and J. G. Steffan. Generating network topologies that obey power laws. *In Proceedings of the Global Internet Symposium*, 2000.
17. H. Tangmunarunkit, R. Govindan, S. Jamin, S. Shenker, and W. Willinger. Network topologies, power laws, and hierarchy. Technical report, Computer Science Department, University of Southern Calfornia, 2001.
18. H. Tangmunarunkit, R. Govindan, S. Jamin, S. Shenker, and W. Willinger. Network topology generators: Degree-based vs structural. *ACM SIGCOMM*, 2002.

An Experimental Study of the Ant Colony System for the Period Vehicle Routing Problem

Ana Cristina Matos[1] and Rui Carvalho Oliveira[2]

[1] Escola Superior de Tecnologia - Instituto Politécnico de Viseu
Campus Politécnico - Repeses, 3504-510 Viseu, Portugal
amatos@mat.estv.ipv.pt
[2] CESUR/DEC- Instituto Superior Técnico
Av. Rovisco Pais, 1049-001 Lisboa, Portugal
roliv@ist.utl.pt

Abstract. In this paper, a new Ant System approach to the Period Vehicle Routing Problem (PVRP) is presented. In PVRP, visit days have to be assigned to customers in order to find efficient routes over the period. We suggest a new technique for defining the initial solution and a novel strategy to update the pheromone trails that is especially suited for solving large scale problems. An illustrative example for a waste collection system involving 202 localities in the municipality of Viseu, Portugal, demonstrates the effectiveness of the model.

1 Introduction

This research addresses the study of a solid waste collection system involving 8087 containers within the Municipality of Viseu, Portugal. The company gathers the containers with a fleet of $15m^3$-capacity trucks operating during the 5 working days and Sundays. The trucks depart from a common garage each morning and return to the depot for the lunch break. A second tour is operated in the afternoon and each tour must take less than 3 hours.

The number of weekly visits to the different containers depends on their location since the filling rates are different: there are containers that are emptied twice a week and others which must be emptied daily.

This collection system has been managed by splitting the territory in geographical areas defined on a purely empirical basis. Each team is responsible for collecting all the containers in one area. The company has no efficient means for routing decision making and this generates an ineffective collection because it sometimes cannot assure the number of visits that matches weekly (collection) needs. This operation represents an instance of a periodic vehicle routing problem (PVRP).

In the classic Vehicle Routing Problem (VRP) m vehicles with a fixed capacity must deliver quantities of goods to n customers from a single depot. Given the distance between customers and the nonnegative service duration, the objective of the problem consists in designing m vehicle routes such that each route starts and ends at the depot; each customer belongs to exactly one route; the

M. Dorigo et al. (Eds.): ANTS 2004, LNCS 3172, pp. 286–293, 2004.

total demand of the route assign to each vehicle does not exceed its capacity; its total duration does not exceed the allowable maximum time; and the total cost of all routes is minimized. In classic VRPs, the planning period is typically a single day.

The Period Vehicle Routing Problem (PVRP) generalizes the classic VRP by extending the planning period to t days and each customer i specifies a service frequency e_i and a set C_i of allowable combinations of visit days. The problem then becomes simultaneously selecting a visit combination for each customer and establishing vehicle routes for each day of the planning horizon, according to VRP rules.The PVRP is thus a challenging NP-hard "multilevel combinatorial problem" (Chao et al.[4]).

Beltrami and Bodin [1] pioneered the PVRP concept. Heuristic methods are often used to solve the problem (e.g. see [13], [4]). More recently, metaheuristics were used to solve PVRP (e.g. [7], [14]). With the exception of Beltrami and Bodin [1] procedures, all PVRP approaches were considered as a multilevel combinatorial optimization problem, where in the first level an allowable day is assigned to each customer and, in the second level, we need to solve a classic VRP for each day of the period. In one of the approaches suggested by Beltrami and Bodin [1], in the first level routes are developed and then assigned to days of the week, but this analysis was limited to instances which require service either 3 or 6 times per week.

In this paper, a two-phase approach is presented where Beltrami and Bodin's approach is enhanced. In the first phase of our analysis, we construct an initial infeasible solution to PVRP where routes for the whole week are designed using a new Ant Colony Optimization (ACO) model. In this first phase the scheduling constraints are not considered. In the second phase, we resort to a graph colouring problem and to an interchange procedure to convert this good infeasible solution into a good feasible one.

The first ACO algorithm, *Ant System* (AS), was initially proposed by Colorni, Dorigo and Maniezzo (see [6],[9]), applied to the travelling salesman problem (TSP). Its application to VRPs was later presented in 1999 by Bullnheimer et al. [2] and followed namely by Doerner et al. [10] and Reimann et al. [12]. ACO is inspired by the behaviour of real ants that, when searching for food, mark the used paths with a certain amount of pheromone(s) depending on the quality of the food source. Other ants observe these pheromone trails and are enticed to follow them, thus reinforcing the paths. Gradually, paths leading to rich food sources will be used more frequently, while other paths, leading to remote food sources, will not be used anymore.

The remainder of this paper is organized as follows. Section 2 deals with the description of the problem and the presentation of our model. In section 3, the proposed ACO algorithm is presented. A numerical analysis and experimental results are shown in 4. Conclusions are presented in section 5.

2   Description of the Problem and Model Formulation

In order to provide an efficient solution to this solid waste collection problem, the its size must be consider . By examining the containers locations we observe that they are only spread out over 202 localities. Assuming that it is likely that a team collects all the containers in a locality that it visits, and all the containers in a locality have the same periodicity of visit, we can break down the number of nodes to 202.

The approach that we take is a two-phase procedure. In phase I, we initially create multiple replicas for localities where the containers require collection more than once a week. Creating multiple replicas expands the problem size from 202 to 610 nodes. The problem at this stage is to minimize total costs, define a set of routes that visits all the localities and their replicas, so that the replicas of one node always belong to distinct routes, not exceeding the time allowed per route nor the capacity of each vehicle. This is a large scale VRP with some additional restrictions. We resorted to ACO in order to solve this phase of the study efficiently, introducing a new strategy and managing the information contained in the pheromone in a novel way.

In the second phase of the algorithm, an automatic system of route assignments to different weekdays is constructed. These assignments have to comply with a set of additional restrictions which emerge from the fact that we are dealing with a PVRP: each locality cannot be visited more than once a day and the schedule must respect a certain time spacing between consecutive visits (collections). We resort to generating a colour-coded graph which, along with an interchange procedure, will lead to a feasible solution. Finally, a VRP is solved for each day of the week.

3   Heuristic Solution Procedures

This section describes the implementation of the ACO algorithm for solving the problem. The two basic ACO phases are briefly described, namely the construction of a solution for the VRP (in subsection 3.1) and the trail update scheme (in subsection 3.2). This is followed by a post-optimization procedure which aims at improving the first-phase solution (in subsection 3.3). Subsection 3.4 describes the second phase of the algorithm that converts the solution of the first phase into a feasible solution to the PVRP, i.e. one that verifies the restrictions related to the collection schedule.

3.1   Construction of a Solution for the VRP

The ACO applications for the VRP found in the literature are iterative processes where, in each iteration, artificial ants are created and assigned to nodes that serve as starting points to sequentially construct a set of routes. Each ant generates a solution to the problem. These solutions will be evaluated later and the information on the best solutions will be retained through differentiating the amount of pheromone leased.

The meta-heuristic we propose differs from the previous right from the start, i.e., we shall consider an iterative process where each iteration generates only one solution - which is equivalent to creating a sole ant. The information retained in memory depends on the quality of the solution generated. In the route construction we adopt the procedure proposed by Reimann et al. [12]: each ant constructs a solution based on the well known Savings algorithm (Clarke and Wright [5]). We apply the parallel Savings algorithm, but in order to be able to use it in the Ant System, we propose some modifications. Let us now describe how this is implemented.

For each pair of localities i and j, the savings measure s_{ij} represents the saving of combining two localities i and j in one tour as opposed to serving them with two different tours. In each iteration, an artificial ant is created and assigned to the locality with the greatest saving. From i, the node j is chosen among the set of all possible nodes with a connection from i. For the selection of (not yet visited) locations j, two aspects are taken into account: how good *was* the choice of that point, an information that is stored in the pheromone trails τ_{ij} associated with each arc (i, j), and how promising *is* the choice of that node . This latter measure of desirability, or visibility, denoted by η_{ij} takes into account the distance between two nodes d_{ij}, and the relative location of the two localities and the depot (measured through the saving s_{ij}) and it is expressed by $\eta_{ij} = (\frac{1}{d_{ij}})^{\beta_1} \times (s_{ij})^{\beta_2}$. A node j is selected to be visited after node i according with the probability distribution:

$$
p_{ij} = \begin{cases} \dfrac{[\tau_{ij}]^\alpha [\eta_{ij}]}{\sum_{h \in \Omega} [\tau_{ih}]^\alpha [\eta_{ih}]} & if \ j \in \Omega \\ \\ 0 & otherwise, \end{cases} \tag{1}
$$

where Ω denotes the set of all feasible nodes to be visited starting from i. This probability distribution depends on the α, β_1 and β_2 parameters that determine the relative influence of the trails and the visibility.

3.2   Trail Update

After an artificial ant has constructed a set of routes, the pheromone trails are laid depending on the value of the objective function for that solution. In early ant system approaches, all the ants contributed to the trail update [9]. Later on, only the best ant contributes to the pheromone trails [8]. In more recent papers, Bullnheimer et al. [3] use a set of so-called elitist ants to update the pheromone trails. In our procedure, we suggest updating the pheromone intensities according to the expression:

$$
\tau_{ij}(t + 1) = \rho \tau_{ij}(t) + \Delta\sigma_{ij} + \Delta\sigma_{ij}^* \tag{2}
$$

where ρ is the trail persistence (with $0 \leq \rho \leq 1$), thus the trail evaporations given by $(1 - \rho)$; $\Delta\sigma_{ij}$ refers to the emphasis that must be given to all of the links which belong to the current solution (with the value of the objective function equal to L_i). This emphasis is directly proportional to the quality of the solution obtained: the incentive given to links belonging to better solutions - lower cost - is greater than to poor solutions:

$$
\Delta\sigma_{ij} = \begin{cases} \omega^{\frac{(L_i - L^*)}{L^*} * 100} & if \ link \ (i, j) \ belongs \ to \ the \ last \ solution \\ \\ 0 & otherwise, \end{cases} \tag{3}
$$

where L^* is the objective value of the best solution found so far and ω a parameter such that $0 < \omega < 1$. The greater ω, the greater the intensity of the pheromone to be applied for one variation. Additionally, all arcs belonging to the best current solution (objective value L^*) are emphasized with an amount $\Delta\sigma_{ij}^* = \delta$, where $\delta \in \{1, 2, 3, 4, 5\}$.

3.3 Post-optimization

Bullnheimer et al. [2] applied the 2-opt algorithm to all vehicle routes built by the ants, before updating the pheromone information. Additionally, Reimann et al. [12] applied a local search based on swap moves to an ant solution, by exchanging two nodes from different tours. We apply the 3-opt algorithm to all routes before updating the pheromone information. In order to reduce the number of routes we also propose an algorithm that merges, if possible, some routes, in what we call *fusion*. Later, a mechanism of exchanges presented by Osman [11] called $\lambda - interchanges$ is carried out with insertion moves and exchange moves between two routes and a node. These procedures are only applied if, during the last iterations, no improvement in the solutions has been found.

3.4 Assignment Procedure

When constructing the routes, there is no concern regarding the admissibility of the solutions generated for a later assignment to the days of the week. If two routes have a common node, they cannot be assigned to the same day. Furthermore, the existence of 7 routes, all of which have a common node, generates a solution that is impossible for the problem of period 6. In order to identify similar situations and find a solution, we resort to a graph colouring problem and an exchange mechanism for the schedule patterns of the nodes to visit. A graph is constructed based on the first-phase solution, where the vertices represent the routes and the arcs the conflicts between pairs of routes. That is, two adjacent routes mean that they have at least one common node, therefore they cannot be operated at the same time. In graph theory, the colouring problem consists in attributing the lowest possible different colours to the vertices so that no pair of vertices have the same colour. Through colouration, we can perceive which routes must occur on different days (the different coloured ones) and which can be operated in the same days (the ones with the same colour). The problem may be formulated in Whole Linear Programming. In colouring the graph, it may happen that the lowest number of necessary colours (which in graph theory is known as the chromatic number) is higher than the number of periods. If this happens, it means that some of the arcs must be eliminated (we would suggest those whose existence is owing to a lower number of common nodes between routes). Then the relaxed problem can be coloured. The next step consists in verifying the admissibility of the patterns of each node. Inadmissibility may occur due to the relaxation of the graph. In this situation, nodes with these inadmissible patterns would be missing a visit. If the collection points have a periodicity 6, we schedule visits in all six days. If the collection points have a periodicity of 3 or 2, the existing two or one visits will be maintained. Moreover, it is necessary to identify the closest node with the same periodicity whose pattern is admissible. If this neighbour has a collection in the days of the node in question, the scheduling pattern will be the same. Otherwise, we will find a day in which the neighbour's container will be collected that does not coincide with the problematic node and which will satisfy the time intervals required between two successive collections. Problems of inadmissibility are solved in a similar way. After checking all the patterns, the classic VRP is solved for each day of the period under study using the ACO algorithm described in the first phase, but this time without creating node replicas.

4 Numerical Analysis

In this section some numerical studies are presented in order to assess the proposed heuristic. The impact of the parameters and of the pheromone information on the first-phase solution quality are shown.

The pheromone intensity updating procedure (expressions 2 and 3) depends on three parameters: ρ, ω and δ. Similarly to what has been proposed by several authors, the best results were obtained for $\rho = 0.9$ and for 0.8. The parameter ω does not act alone since the value of δ also influences the path to be followed on emphasizing the arcs belonging to the best current solution. It is therefore necessary to find the best combination of δ and ω. The values $\omega = 0.6, 0.7, 0.8$ and 0.9 were applied to various problems instances and we concluded that the most promising values were $\omega = 0.9$ and 0.6. The combination of these parameters, albeit important as the search can be pursued in different directions, is not very important because, generally speaking, any combination of the parameters values among the values suggested above leads to good results. Similarly to what is presented in the literature, the best choice of probability expression (1) parameters are $\alpha = \beta_1 = \beta_2 = 1$, though a later change in the value of α may be beneficial, namely if the solutions found initially are of poor quality. Thus, after post-optimization emphasizing, the importance of the pheromone (through $\alpha > 1$) may direct the search to better quality solutions (see Figure 1).

We confronted the proposed ACO algorithm with the procedure suggested by Bullnheimer et al. [3] where, in each iteration, m ants are created and the best

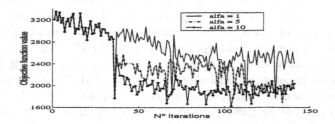

Fig. 1. The effect of alfa parameter after post-optimization

current solution, L^*, and the best solution for the current iteration contribute to updating the pheromone. Figure 2 compares the evolutions of both processes by observation of the objective function value for each iteration. We applied the algorithm to the 7 largest PVRP described in literature [7]. As illustrative examples, we show the results for our case study and for example number 30 described in [7]. In Bullnheimer's algorithm, $m = 10$ and $\sigma = 2$ were used. Post-optimization techniques refered above (3-opt after each solution and the fusion and λ-interchange), $\alpha = \beta_1 = \beta_2 = 1$ were applied. In both cases, as in the other

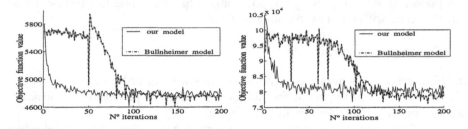

Fig. 2. The Convergence of our algorithm vs Bullnheimer's algorithm (our case study, on the left; instance number 30 from [7], on the right)

instances tested, despite the fact that there were not significant differences in the value of the best solution found, the computational effort of our methodology is much lower since each iteration only generates one solution, as opposed to what is described in the literature where m solutions are generated per iteration and only the value of the objective function of the best among the m solutions is recorded. This aspect is more relevant as the problem size increases, and therefore our approach is specially suited for large-scale problems. Furthermore after few iterations, the effects of the pheromone are already quite visible, whereas in the methodology proposed by Bullnheimer the pheromone reinforcement will depend on the magnitude of the objective function: it will be faster if the value of L^* is lower, or slower if L^* is large.

5   Conclusions

In this article, a two-phase algorithm to solve a real world periodic vehicle routing problem was presented. The experience obtained by applying our algorithm to this case study and to a number of instances available in literature clearly suggests that this is a promising approach to solve PVRPs. The proposed ACO system is particularly suited to solve large-scale problems since it involves a major reduction in computational effort when compared with other models presented in the literature, without compromising the quality of the final solutions. Further research will be directed towards the comparison of the formulation presented here with the one suggested by most authors, in which scheduling collection points in the first phase precedes the construction of the routes.

References

1. E. J. Beltrami and L.D.Bodin. Networks and vehicle routing for municipal waste collection. *Networks*, 4: 65-94, 1974.
2. B. Bullnheimer, R.F.Hartl, and C.Strauss. An improved ant system algorithm for the vehicle routing problem. *Annals of Operations Research*, 89: 319-328, 1999.
3. B. Bullnheimer, R.F.Hartl, and C.Strauss. A new rank based version of the ant system: a computational study. *Working Paper No.1,SFB Adaptative Information Systems and Modelling in Economics and Management Science*, Vienna, 1997.
4. I.M. Chao, B.L. Golden,and E.A. Wasil. An improved heuristic for the period vehicle routing problem. *Networks*, 26: 25-44, 1995.
5. G. Clarke, and J.W. Wright. Scheduling of vehicles from a central depot to a number of delivery points. *Operations Research*, 12: 568-581, 1964.
6. A. Colorni, M. Dorigo, and V. Maniezzo. Distributed optimization by ant colonies. F. Varela, P. Bourgine, eds. *Proc.Europ. Conf. Artificial Life*, Elsevier, Amsterdam, pages 134-142, 1992.
7. J. F.Cordeau, M. Gendreau, and G.Laporte. A tabu search heuristic for periodic and multi-depot vehicle routing problems. *Networks*, 30: 105-119, 1997.
8. M. Dorigo and L.M. Gambardella. Ant colony system: a cooperative learning approach to the traveling salesman problem. *IEEE Transactions on Evolutionary Computation* 1(1): 53-66, 1997.
9. M. Dorigo, V. Maniezzo and A. Colorni. The ant system: optimization by a colony of cooperating agents. *IEEE Transactions on Systems, Man and Cybernetics - Part B*, 26(1): 29-41, 1996.
10. K. F. Doerner, M. Gronalt, R. F. Hartl, M. Reimann. Optimizing pickup and delivery operations in a hub network with ant systems. *POM Working Paper 02/2000*, University of Vienna, Vienna, Austria, 2000.
11. I. H. Osman. Metastrategy simulated annealing and tabu search algorithms for vehicle routing problem. *Annals of Opertations Research* 41: 412-451, 1993.
12. M. Reimann, M. Stummer, K. Doerner. A savings based ant system for the vehicle routing problem. *Proc. of the Genetic and Evolutionary Computation Conference*, pages 1317-1362, 2002.
13. R. A.Russell and D. Gribbin. A multiphase approach to the period routing problem. *Networks* 21: 747-765, 1991.
14. D. Vianna, L.Ochi, L.Drummond. A parallel hybrid evolutionary metaheuristic for the period vehicle routing problem. *Proc. IPDPS*, 1999.

An Extension of Ant Colony System to Continuous Optimization Problems

Seid H. Pourtakdoust and Hadi Nobahari

Sharif University of Technology, P.O. Box: 11365-8639, Tehran, Iran
pourtak@sharif.edu, nobahari@mehr.sharif.edu

Abstract. A new method for global minimization of continuous functions has been proposed based on Ant Colony Optimization. In contrast with the previous researches on continuous ant-based methods, the proposed scheme is purely pheromone-based. The algorithm has been applied to several standard test functions and the results are compared with those of two other meta-heuristics. The overall results are compatible, in good agreement and in some cases even better than the two other methods. In addition the proposed algorithm is much simpler, which is mainly due to its simpler structure. Also it has fewer control parameters, which makes the parameter settings process easier than many other methods.

1 Introduction

Ant algorithms were first proposed by Marco Dorigo and colleagues [3, 2] as a multi-agent approach to solve difficult combinatorial optimization problems. The first algorithm inspired from the ant colony functioning, is Ant System (AS) [3, 7], which is considered the base of many other approaches such as Max-Min AS (MMAS) [15], Ant Colony System (ACS) [6] and Ant-Q [9].

Ant algorithms have been applied to many different discrete optimization problems such as the Traveling Salesman Problem, Quadratic Assignment Problem, Job-Shop scheduling, Vehicle Routing, Sequential Ordering, Graph Coloring, Routing in Communications Networks and so on [5, 4]. But there are few adaptations of such algorithms to continuous problems, whereas these problems are frequently met, especially in engineering. The first algorithm designed for continuous functions optimization is CACO (Continuous Ant Colony Optimization) [1, 17], which uses ant colony framework to perform local searches where as the global search is handled by a genetic algorithm.

The main idea utilized in all of the above algorithms has been adopted from the ants pheromone trails-laying behavior, which is an indirect form of communication mediated by modifications of the environment. But this behavior is also a part of the recruitment process, which is defined as a form of communication that leads some individuals to gather at the same place in order to perform a particular action. According to this definition, some other optimization methods such as API [13, 14] have been developed in a continuous framework. This method is inspired by a primitive ant's recruitment behavior. It performs a

M. Dorigo et al. (Eds.): ANTS 2004, LNCS 3172, pp. 294–301, 2004.

tandem-running which involves two ants and leads them to gather on a same hunting site. The authors use this particular recruitment technique to make the population proceed towards the optimum solution, by selecting the best point among those evaluated by the ants.

A recent research on modeling ants' behavior [8] has shown that it is possible to start a recruitment sequence even without taking pheromone trails into account. In this model, the stigmergic process is deliberately ignored and only inter-individuals relationships are considered. The model tries to reproduce the flow of ants exiting the nest after the entry of a scout who has discovered a new food source. To differentiate this process from the stigmergic recruitment, the authors have called it mobilization. Another utilized approach to solve continuous optimization problems is by converting them to discrete form. Then discrete versions of ant algorithms can be applied [11, 10].

In this paper a pure pheromone based method has been developed to find the global minimum of continuous functions. Our approach is to extend the well-known ACS [6] to continuous optimization problems. So we will call it *Continuous Ant Colony System* (CACS).

2   Continuous Ant Colony System

It is desired to find the absolute minimum of a positive non-zero function $f(x)$, within a given interval $[a, b]$, in which the minimum occurs at a point x_s. In general f can be a multi-variable function, defined on a subset of $\mathbb{R}^n$ delimited by n intervals $[a_i, b_i]$, $i = 1, ..., n$.

2.1   Ant Colony System Basic Features

Ant Colony System uses a graph representation. In addition to the cost measure, each edge has also a desirability measure, called *pheromone intensity*. To solve the problem, each ant generates a complete tour by choosing the nodes according to a so called *pseudo-random-proportional state transition rule,* which has two major features. Ants prefer to move to the nodes, which are connected by the edges with a high amount of pheromone, while in some instances, their selection may be completely random. The first feature is called *exploitation* and the second is a kind of *exploration*.

While constructing a tour, ants also modify the amount of pheromone on the visited edges by applying a *local updating rule*. It concurrently simulates the *evaporation* of the previous pheromone and the *accumulation* of the new pheromone deposited by the ants while they are building their solutions. Once all the ants have completed their tours, the amount of pheromone is modified again, by applying a *global updating rule*. Again a part of pheromone evaporates and all edges that belong to the global best tour, receive additional pheromone conversely proportional to their length.

2.2   Continuous Pheromone Model

The first step to develop a continuous version of ACS is to define a continuous pheromone model. Although pheromone distribution has been first modeled over discrete sets, like the edges of a traveling salesman problem, in the case of real ants, pheromone deposition occurs over a continuous space. In this regard, consider a food source surrounded by several ants. The ants aggregation around the food source causes the most pheromone intensity to occur at the food source position. Then increasing the distance of a sample point from the food source will decrease its pheromone intensity. To model this variation, a normal probability distribution function (pdf) is proposed as follows:

$$\tau(x) = e^{-\dfrac{(x - x_{\min})^2}{2\sigma^2}} \tag{1}$$

where $x_{\min}$ is the best point found within the interval $[a, b]$ from the beginning of the trial and σ is an index of the ants aggregation around the current minimum. To initialize the algorithm, $x_{\min}$ is randomly chosen within the interval $[a, b]$, using a uniform pdf and σ is taken at least 3 times greater than the length of the search interval, $(b - a)$, to uniformly locate the ants within it.

2.3   State Transition Rule

In ACS, ants choose their paths by both exploitation of the accumulated knowledge about the problem and exploration of new edges. In CACS, pheromone intensity is modeled using a normal pdf, the center of which is the last best global solution and its variance depends on the aggregation of the promising areas around the best one. So it contains exploitation behavior. In the other hand, a normal pdf permits all points of the search space to be chosen, either close to or far from the current solution. So it also contains exploration behavior. It means that ants can use a random generator with a normal pdf as the state transition rule to choose the next point to move to.

2.4   Pheromone Update

During each iteration, ants choose their destinations through the probabilistic strategy of equation (1). At the first iteration, there isn't any knowledge about the minimum point and the ants choose their destinations only by exploration. During each iteration pheromone distribution over the search space will be updated using the acquired knowledge of the evaluated points by the ants. This process gradually increases the exploitation behavior of the algorithm, while its exploration behavior will decrease. Pheromone updating can be stated as follows: The value of objective function is evaluated for the new selected points by the ants. Then, the best point found from the beginning of the trial is assigned to $x_{\min}$. Also the value of σ is updated based on the evaluated points during the last iteration and the aggregation of those points around $x_{\min}$. To satisfy simultaneously the fitness and aggregation criteria, a concept of weighted variance is defined as follows:

$$\sigma^2 = \frac{\sum\limits_{j=1}^{k} \frac{1}{f_j - f_{min}}(x_j - x_{min})^2}{\sum\limits_{j=1}^{k} \frac{1}{f_j - f_{min}}} \tag{2}$$

where k is the number of ants. This strategy means that the center of region discovered during the subsequent iteration is the last best point and the narrowness of its width is dependent on the aggregation of the other competitors around the best one. The closer the better solutions get (during the last iteration) to the best one, the smaller σ is assigned to the next iteration.

During each iteration, the height of pheromone distribution function increases with respect to the previous iteration and its narrowness decreases. So this strategy concurrently simulates pheromone accumulation over the promising regions and pheromone evaporation from the others, which are the two major characteristics of ACS pheromone updating rule.

3   Results and Discussion

For a comparative study of CACS with two other metaheuristic methods, namely API and a Genetic Algorithm [13, 14], similar test functions are chosen as proposed in [16]. These functions are tabulated in table 1.

Table 1. Utilized test functions (All have zero minimum value)

Function		Interval of x_i
f_1	$x_1^2 + x_2^2 + x_3^2$	$[-5.12, 5.12]$
f_2	$100(x_1^2 - x_2)^2 + (1 - x_1)^2$	$[-2.05, 2.05]$
f_3	$50 + \sum_{i=1}^{5}(x_i^2 - 10\cos(2\pi x_i))$	$[-5.12, 5.12]$
f_4	$1 + \sum_{i=1}^{2} \frac{x_i^2}{4000} - \prod_{i=1}^{2}\cos(\frac{x_i}{\sqrt{i}})$	$[-5.12, 5.12]$
f_5	$1 + \sum_{i=1}^{5} \frac{x_i^2}{4000} - \prod_{i=1}^{5}\cos(\frac{x_i}{\sqrt{i}})$	$[-5.12, 5.12]$
f_6	$0.5 + (\sin^2(x_1^2 + x_2^2)^{1/2} - 0.5)/(1 + 0.001(x_1^2 + x_2^2))$	$[-100, 100]$
f_7	$(x_1^2 + x_2^2)^{0.25}(1 + \sin^2 50(x_1^2 + x_2^2)^{0.1})$	$[-100, 100]$

3.1   Parameter Study

The proposed CACS has only one parameter to set which is the number of ants, k. In order to determine the optimal value of k, the minimum values obtained with different number of ants have been considered. To make the results comparable with those of the previous researches [13, 14], the maximum number of function evaluations was limited to 10000.

Table 2. The minimum values obtained with different number of ants (k)

k	$\min(f_1)$	$\min(f_2)$	$\min(f_3)$	$\min(f_4)$	$\min(f_5)$	$\min(f_6)$	$\min(f_7)$
1	$9.9e+000$	$6.0e+01$	$5.6e+01$	$3.2e-01$	$8.0e-01$	$3.7e-01$	$7.1e+00$
	$(8.0e+000)$	$(1.4e+02)$	$(1.8e+01)$	$(3.0e-01)$	$(2.0e-01)$	$(1.1e-01)$	$(2.6e+00)$
2	$2.0e+000$	$7.9e+00$	$3.2e+01$	$4.9e-02$	$4.7e-01$	$2.7e-01$	$4.5e+00$
	$(3.2e+000)$	$(3.6e+01)$	$(1.5e+01)$	$(1.2e-01)$	$(2.4e-01)$	$(1.5e-01)$	$(2.6e+00)$
5	$\mathbf{2.1e-242}$	$5.6e-05$	$1.5e+01$	$5.2e-03$	$7.5e-02$	$4.9e-02$	$9.1e-01$
	$(9.2e-243)$	$(1.5e-04)$	$(7.8e+00)$	$(3.4e-03)$	$(1.2e-01)$	$(7.0e-02)$	$(1.6e+00)$
10	$\mathbf{1.8e-228}$	$6.5e-13$	$1.1e+01$	$5.8e-03$	$3.1e-02$	$1.0e-02$	$2.9e-03$
	$(2.1e-228)$	$(2.7e-12)$	$(7.1e+00)$	$(3.1e-03)$	$(6.6e-02)$	$(1.4e-02)$	$(8.3e-03)$
20	$\mathbf{2.0e-135}$	$\mathbf{1.5e-19}$	$\mathbf{7.1e+00}$	$\mathbf{5.3e-03}$	$\mathbf{2.8e-02}$	$\mathbf{5.1e-03}$	$\mathbf{7.7e-16}$
	$(1.9e-134)$	$(1.4e-18)$	$(4.5e+00)$	$(3.3e-03)$	$(5.5e-02)$	$(2.1e-03)$	$(1.6e-15)$
50	$\mathbf{1.5e-067}$	$\mathbf{1.2e-31}$	$\mathbf{4.8e+00}$	$\mathbf{5.0e-03}$	$\mathbf{1.1e-02}$	$\mathbf{4.6e-03}$	$\mathbf{4.2e-06}$
	$(9.5e-067)$	$(6.8e-31)$	$(3.9e+00)$	$(3.5e-03)$	$(2.5e-02)$	$(1.1e-03)$	$(7.7e-06)$
100	$\mathbf{3.6e-037}$	$\mathbf{1.6e-33}$	$\mathbf{4.9e+00}$	$\mathbf{4.1e-03}$	$\mathbf{7.8e-03}$	$\mathbf{3.9e-03}$	$\mathbf{2.5e-03}$
	$(2.2e-036)$	$(8.5e-33)$	$(2.8e+00)$	$(3.7e-03)$	$(9.0e-03)$	$(1.8e-03)$	$(1.5e-03)$
200	$\mathbf{1.5e-020}$	$\mathbf{3.2e-22}$	$\mathbf{7.1e+00}$	$\mathbf{2.7e-03}$	$\mathbf{7.7e-03}$	$\mathbf{3.7e-03}$	$\mathbf{5.9e-02}$
	$(3.0e-020)$	$(2.0e-21)$	$(2.4e+00)$	$(3.6e-03)$	$(8.4e-03)$	$(1.8e-03)$	$(1.8e-02)$
500	$3.0e-009$	$1.7e-12$	$9.4e+00$	$1.1e-03$	$1.4e-02$	$4.2e-03$	$3.8e-01$
	$(3.2e-009)$	$(4.4e-12)$	$(2.3e+00)$	$(2.6e-03)$	$(1.5e-02)$	$(1.4e-03)$	$(1.2e-01)$
1000	$2.6e-005$	$1.0e-07$	$1.0e+01$	$1.6e-04$	$5.5e-02$	$5.2e-03$	$6.5e-01$
	$(2.4e-005)$	$(2.9e-07)$	$(2.9e+00)$	$(1.1e-03)$	$(2.3e-02)$	$(1.3e-03)$	$(1.8e-01)$
2000	$2.8e-003$	$2.3e-05$	$1.1e+01$	$1.6e-04$	$7.2e-02$	$7.1e-03$	$8.3e-01$
	$(2.0e-003)$	$(4.7e-05)$	$(2.6e+00)$	$(2.3e-04)$	$(3.0e-02)$	$(4.0e-03)$	$(2.4e-01)$
5000	$3.2e-002$	$1.2e-03$	$1.3e+01$	$6.1e-04$	$9.0e-02$	$9.5e-03$	$1.1e+00$
	$(2.2e-002)$	$(1.1e-03)$	$(3.4e+00)$	$(5.6e-04)$	$(3.7e-02)$	$(5.9e-03)$	$(3.3e-01)$
10000	$7.8e-002$	$4.5e-03$	$1.7e+01$	$1.2e-03$	$1.1e-01$	$1.3e-02$	$1.2e+00$
	$(4.9e-002)$	$(4.6e-03)$	$(3.8e+00)$	$(1.1e-03)$	$(4.0e-02)$	$(8.1e-03)$	$(3.3e-01)$

Table 2 gives the average (m), and the standard deviation (ρ) of minima found for each function. All presented results are averaged over 50 different runs. The columns report the average solutions where the numbers in the parentheses are the standard deviations. Since the maximum number of function evaluations is limited to 10000, k can vary from 1 to 10000. The first interesting observation is that the order of minima reached by this algorithm is widely dependent on the complexity of test function, similar to the other known optimization methods. This means that to determine the admissible range of k, each function must be considered separately. The admissible range of k for which the minimum values are relatively accurate with respect to the other results, are shown in bold face. This remark shows that choosing the value of k between 20 and 200 can guarantee a relatively good performance against these test functions.

To further study the effect of k on the algorithm performance, two performance indices are defined. Since all test functions have a minimum of zero, a normalized summation of the minima, has been proposed as a good index for performance evaluation [13]. This parameter is defined as follows:

$$S_1 = \sum_{i=1}^{7} \frac{m_i - \min_{j=1}^{7}\{m_j\}}{\max_{j=1}^{7}\{m_j\} - \min_{j=1}^{7}\{m_j\}} \tag{3}$$

where m_i is the average of the minima found for f_i. The second index can also be defined as follows:

$$S_2 = \sum_{i=1}^{7} \frac{\rho_i}{m_i} \tag{4}$$

which is the summation of the normalized standard deviations obtained for the given test functions. Since it is desirable to reach to the smallest possible values of S_1 and S_2, their variations with k are investigated (Figure 1). One interesting observation is that CACS is not very robust or responsive with small number of ants ($k < 10$). In this case the probability of falling in a local minimum increases because it is very probable that during one iteration all ants choose the points near to the current minimum. This in turn decreases the values of the weighted variances magnifying the aggregation of the later iteration points. When the value of the weighted variance becomes several times smaller than the distance to the nearest local minimum, it's no longer possible to jump to a better local minimum and the minimization process stops.

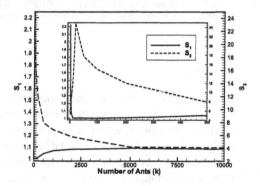

Fig. 1. Variation of S_1 and S_2 with the number of ants

Figure 1 shows that S_1 is relatively constant when the number of ants is between 10 and 200 and it begins to increase for greater values. This is consistent with the previous results about the admissible range of the number of ants. On the other hand, S_2 first increases with the number of ants up to $k = 20$ and then begins to decrease. However the reduction rate slows down as the number of ants increases. According to figure 1 and the results of Table 2, the range $[50, 100]$ seems to be an admissible range for the number of ants. The convergence rate will decrease for simpler functions like f_1 if the number of ants increases further.

3.2   Comparison with Other Methods

Table 3 shows the CACS results, under 10000 function evaluations with ($k = 50, 100$), which are compared with those of two other methods. The first one is API algorithm [13, 14], which is another continuous ant-based method and the second one is a generational genetic algorithm (GA) with binary tournament selection, 1 point crossover and with a real-coded representation [12]. Table 3 shows that CACS can find the global minima better than or at least as good as API for all functions. It can also find the global minima better than GA for all functions except for the case of f_3 which is a complex function containing many local minima.

Table 3. Comparision of CACS Scheme with API and GA

Method	$\min(f_1)$	$\min(f_2)$	$\min(f_3)$	$\min(f_4)$	$\min(f_5)$	$\min(f_6)$	$\min(f_7)$
CACS	$1.5e-067$	$1.2e-31$	$4.8e+00$	$5.0e-03$	$1.1e-02$	$4.6e-03$	$4.2e-06$
$N_a = 50$	$(9.5e-067)$	$(6.8e-31)$	$(3.9e+00)$	$(3.5e-03)$	$(2.5e-02)$	$(1.1e-03)$	$(7.7e-06)$
CACS	$3.6e-037$	$1.6e-33$	$4.9e+00$	$4.1e-03$	$7.8e-03$	$3.9e-03$	$2.5e-03$
$N_a = 100$	$(2.2e-036)$	$(8.5e-33)$	$(2.8e+00)$	$(3.7e-03)$	$(9.0e-03)$	$(1.8e-03)$	$(1.5e-03)$
API	0.00000	0.00000	7.47651	0.00413	0.25034	0.00659	0.09307
	(0.00000)	(0.00000)	(2.98922)	(0.00402)	(0.12254)	(0.00443)	(0.01886)
GA	0.00000	0.04029	2.12457	0.03095	0.13955	0.07376	0.13358
	(0.00000)	(0.06515)	(1.30328)	(0.03531)	(0.07620)	(0.06590)	(0.06271)

4   Conclusion

In this study a new continuous optimization method was proposed based on Ant Colony Optimization. The proposed scheme was tested over several standard test functions in order to set its control parameters and to compare its results with two other methods, namely API and a Genetic Algorithm. The overall results of CACS are compatible and in some cases even better than those of the two other methods. Also CACS possesses higher convergence rate for some functions due to its adaptive pheromone updating rule strategy. One of the best advantages of CACS is its simplicity, mainly due to its simpler structure and having only one control parameter. Obviously, this feature, makes the parameter setting process easier than many other methods. In addition CACS does not seem to be sensitive with respect to its control parameter. Numerical results show that choosing the number of ants within the range [20, 200] provides relatively good accuracy for the considered test functions.

References

1. Bilchev, G. and Parmee, I.C.: The ant colony metaphor for searching continuous design spaces. Lecture Notes in Computer Science **993** (1995) 25-39
2. Colorni A., Dorigo, M. and Maniezzo, V.: Distributed optimization by ant colonies. Proceedings of the First European Conference on Artificial Life, Elsevier Science Publisher (1992) 134-142
3. Dorigo, M.: Optimization, learning and natural algorithms. PhD thesis, Dipartimento di Elettronica, Politecnico di Milano, IT (1992)

4. Dorigo, M., Bonabeau, E. and Theraulaz, G.: Ant algorithms and stigmergy. Future Generation Computer Systems **16** (2000) 851-871
5. Dorigo, M., di Caro, G. and Gambardella, L. M.: Ant algorithms for discrete optimization. Artificial Life **5** (3) (1999) 137-172
6. Dorigo, M. and Gambardella, L. M.: Ant colony system: A cooperative learning approach to the traveling salesman problem. IEEE Transactions on Evolutionary Computation **1** (1) (1997) 53-66
7. Dorigo, M., Maniezzo, V. and Colorni, A.: The ant system: optimization by a colony of cooperating agents. IEEE Transactions on Systems, Man and Cybernetics-Part B **26** (1) (1996) 29-41
8. Dreo, J. and Siarry, P.: A new ant colony algorithm using the heterarchical concept aimed at optimization of multi-minima continuous functions. Lecture Notes in Computer Science **2463** (2002) 216-221
9. Gambardella, L. M. and Dorigo, M.: Ant-Q: A reinforcement learning approach to the traveling salesman problem. Proceedings of the 12th International Conference on Machine Learning, ML-95, Palo Alto (1995) 252-260
10. Jun L. Y. and Jun, W. T.: An adaptive ant colony system algorithm for continuous-space optimization problems. Journal of Zhejiang University Science **4** (1) (2003) 40-46
11. Ling, C., Jie, S., Ling, O. and Hongjian, C.: A method for solving optimization problems in continuous space using ant colony algorithm. Lecture Notes in Computer Science **2463** (2002) 288-289
12. Michalewicz, Z. Genetic Algorithms+Data Structures=Evolution Programs. Springer, Berlin, 3rd Edition (1996)
13. Monmarche, N., Venturini, G. and Slimane, M.: On how the ants Pachycondyla apicalis suggesting a new search algorithm. Internal Report No. 214, E3i, Downloadable from website http://www.antsearch.univ-tours.fr/webrtic (1999)
14. Monmarche, N., Venturini, G. and Slimane, M.: On how Pachycondyla apicalis ants suggest a new search algorithm. Future Generation Computer Systems **16** (2000) 937-946
15. Stützle, T. and Hoos, H.: MAX-MIN Ant System. Future Generation System **16** (8) (2000) 889-914
16. Whitley, D., Mathias, K., Rana S. and Dzubera, J. Building better test functions. Proceedings of the 6th International Conference on Genetic Algorithms, Morgan Kaufmann Publishers (1995) 239-246
17. Wodrich, M. and Bilchev, G.: Cooperative distributed search: the ant's way. Control and Cybernetics **26** (1997) 413-445

Ant Algorithms
for Urban Waste Collection Routing*

Joaquín Bautista and Jordi Pereira

Escola Politècnica Superior d'Edificació de Barcelona
Avda. Grergorio Marañón 44, 08028 Barcelona Spain
{joaquin.bautista,jorge.pereira}@upc.es

Abstract. Problems arising on Urban Waste Management are broad
and varied. This paper is focused on designing collection routes for urban
wastes, a problem existing in most European waste collection systems.
The relationship between the real world problem and the Arc Routing
literature is established, and the Capacitated Arc Routing Problem is
extended to comply with traffic rules. Afterwards, an Ant Algorithm is
designed to solve this problem, and its efficiency is tested using the instance sets from the CARP literature and a set of real life instances from
the Metropolitan Area of Barcelona. Finally, the integration between the
proposed algorithms and a Decision Support System for Urban Waste
Management is shown.

1   Introduction

During the last years, social concerns regarding the environment has acquired
great relevance. In the EU a growing body of community directives, the basis of each country legislation, obliges its members to recover and recycle many
products and components (e.g. containers, glass, paper, plastic, consumer goods,
automotive components, electronics, etc). From the Municipal Waste Management perspective, the problems related to the recovery and recycling of waste,
as well as the design, management and control of systems oriented towards the
return and treatment of disposable goods should be studied.

Waste generated in urban areas is collected by municipal organizations or its
collection is adjudicated to private companies. It is their responsibility to collect
the waste and transport it to its final destination.

Collection techniques have evolved in step with technological advances, and
the adaptation to the particularities of each city, trying to offer an optimal
service with minimum cost. The population of the EU is highly concentrated in
cities with a very high population density, as people usually live in apartment
buildings. As a result, the collection technique has evolved away from door-to-
door collection, very common in the USA, opting in the majority of cases for the
establishment of curbside collection points where citizens leave their refuse. Each
collection point is made up of one or more refuse bins (with dimensions ranging

---

* This research has been partially funded by BEC2003-03809 Grant.

M. Dorigo et al. (Eds.): ANTS 2004, LNCS 3172, pp. 302–309, 2004.

from 1 to 3.2 m^3). Different trucks collect each fraction of waste and transport it to their final destination (a refuse dump, an incinerator or a recycling plant).

One key factor for proper operation of collection systems is to design collection routes for fractions, which are separetely collected under some casesdue to technical design constraints and different treatment processes. The present research work focuses on the study of models and methods to design collection routes individually, in comparison to [10]. Section 2 is devoted to studying arc routing literature, a starting model for the problem, and describing a model to solve the Capacitated Arc Routing Problem (CARP) taking into account traffic constraints. Section 3 presents the literature devoted to solve the CARP, while section 4 shows the implementation of an Ant Colony Optimization algorithm, [7], to solve the problem. Section 5 is given over to a computational experience with the algorithms proposed, while section 6 deals with the integration of the proposed method in a Decision Support System and the conclusion of this work.

2   Modelling the Problem

In arc routing problems (ARPs) the aim is to determine a least-cost traversal of a specified arc subset of a graph, $G = (V, A \cup E)$, with or without additional constraints. Every arc and edge has associated a nonnegative cost, distance or length, l_{ij}, and a subset R of edges and arcs are said to be required, i.e., they must be serviced or covered by a vehicle. An up-to-date state of the art on arc routing problems can be found in [9].

When more than a single route is required, the problem is known as the Capacitated Arc Routing Problem (CARP). CARP instances associate a nonnegative weight or demand, d_{ij}, to each edge and arc of R and the problem consists of designing a set of feasible vehicle routes of least total cost, such that each required edge and arc appears in at least one route and is serviced by exactly one vehicle, with limited capacity D.

Most of the work on solutions procedures for CARP problems has been based on heuristics, due to the lack of success from exact procedures, see [15]. Solution procedures found in the first real applications of arc routing, see [3], were based on ad hoc heuristics to comply with real cases constraints. Christofides, see [6], is the first author to introduce a general purpose heuristic for the CARP. Local search procedures, see [14], and metaheuristic approaches [13] and [16], are also available.

If forbidden turns and traffic signals are not taken into account, the urban waste collection problem can be seen as the previously defined CARP problem. For the scope of this paper, the section of the city where the collection routes are being planned represents the graph G. The vertices of the graph are associated to street junctions and dead ends and the arcs and edges are associated to existing street sections between street junctions. Arcs represent single-way streets while edges represent two-way streets. Any arc or edge with a collection point will be considered to belong to the subset R of required arcs and edges of G, with an associated demand equal to the total waste generated by the citizens allocated to the collection point.

Even if CARP model shows most of the desired characteristics, routes generated by a CARP solution procedure would not be applicable to real life circumstances as forbidden turns and others traffic signals are not observed. Some approaches, based on post-processing illegal solutions have been proposed in the literature to take turns into account, see [5], but when the number of forbidden turns is high, as in real circumstances, it is better to take turns into account during the construction procedure.

To comply with the additional constraints imposed by turns, an auxiliary array of distance between midpoints of arcs and edges is used. Each arc and edge, required or not, has an associated set of midpoint composed by one or two points depending on the number of directions that the street section has.

Forbidden turns and other traffic signals are taken into account when the auxiliary distance array is initialised. The distance between a pair of midpoints is initialised to a prohibitive high value unless the arcs or edges of the midpoints share the vertex used to move from the initial midpoint to the final midpoint and no traffic signal or traffic direction bans direct movement between them (for example, between two arcs a turn prohibition exists). If direct movement is allowed, the distance associated is equal to half the sum of lengths of ech arc or edge associated. This array can be used to trace valid routes between any pair of midpoints using any conventional shortest path algorithm, and report the direction used to visit each required arc. A mathematical formulation of the problem can be found in [20] based on a previous model found in [17].

3   Ant Algorithms to Solve the Problem

The procedures found in the literature are based on properties of the original graph for constructing solutions. As they do not include information about additional constraints and penalties these procedures, if not modified, are not useful to solve real problems where additional constraints appear. This section is focused on the modification of some of the procedures from the literature and their adaptation to use the auxiliary cost matrix showed above, to grant the algorithms the possibility to solve the real life problem found in urban waste collection, as well as their hybridization with the Ant Colony Optimization metaheuristic. The proposed algorithms use the Ant System scheme (AS), [8], with several modifications: (1)The AS scheme uses as heuristic information the auxiliary distance array between midpoints, and ants will leave pheromone between pairs of required midpoints; (2) during the construction phase, the candidate list is composed by midpoints associated to required arcs and edges not present in the solution under construction; (3) optionally, a dynamic Restricted Candidate List, RCL, based on heuristic information only and composed by cl elements is used. The algorithms using this candidate list of limited size is denoted as AS-II, while the algorithm not using this scheme is denoted as AS-I. An additional approach is tested based on the populational ACO algorithms, [12] where only a subset of elite solutions leave trail. This algorithm is denoted as P-AS, and does not make use of a restricted candidate list.

The following subsections cover the proposed construction, local search and pheromone updating phases of the implementation.

3.1   Generation of Solutions

Two of the constructive heuristics present in the literature have been chosen for the generation phase of the ant algorithms.

The first algorithm is the path scanning heuristic [11]. The algorithm appends required midpoints to the solution choosing between candidates complying with capacity constraints. When no candidate midpoints exist, the route returns to the depot, and a new route is created unless all required arcs and edges are serviced. The heuristic rule is based on considering two cases: (1) the load of the current route is less than half the vehicle capacity, chooing midpoints maximizing the distance to return to the depot, and (2) the load of the current route is greater than or equal to half the vehicle capacity, chuusing midpoints minimizing the return distance.

This heuristic rule is used under a probabilistic rule, discriminating both cases and modifying visibility accordingly to the case. The proposed procedure works as follows: first a visibility criteria, η_{ij}, equal to the inverse of the distance between the last midpoint visited and the candidate midpoint is obtained, and is modified in case the midpoint does not fulfill the current case dividing the visibility by two. Pheromone is read as usually. Candidates are chosen from the candidate list, cl with probability as in formula 1.

$$p_{ij} = \frac{[\tau_{ij}]^\alpha \, [\eta_{ij}]^\beta}{\sum_{\forall h \in cl} [\tau_{ih}]^\alpha \, [\eta_{ih}]^\beta} \tag{1}$$

On the other hand, the Augment and Combine heuristic, see [19], has two different construction phases, each requiring a different treatment, and thus two different AS construction phases are used. Each phase has a different paper on the construction procedure and keeps different transition rules and pheromone.

During the first phase, a number of selected midpoints are chosen as a "seed" to build the routes, and required midpoints appearing in the shortest routes from depot to seed and from seed to depot are directly added to initial routes. First phase uses as visibility, denoted η_i, the distance between the depot and the arc, and equal to the distance between the depot and the midpoint, as it favors farther arcs to depot than nearer ones. Pheromone trail, denoted as τ_i, is deposited in midpoints used as seeds for routes. Candidates are chosen from the candidate list, cl with probability as in formula 2.

$$p_i = \frac{[\tau_i]^\alpha \, [\eta_i]^\beta}{\sum_{\forall h \in cl} [\tau_h]^\alpha \, [\eta_h]^\beta} \tag{2}$$

The second phase of the algorithm insert midpoints to routes trying to minimize the additional cost incurred for servicing the midpoint. In this case, heuristic information, η_{ij}, is related to the coveringcost of midpoint j, immediately after midpoint i and immediately before midpoint k. As the midpoint is inserted between two midpoints currently in the solution, two sources of heuristics information appear, one from the previous midpoint and the other from the following midpoint where the solution will be inserted, and substract the heuristic information between the current midpoints, see formula 3.

$$p_{ijk} = \frac{[\tau_{ij} + \tau_{jk} - \tau_{ik}]^{\alpha} [\eta_{ij} + \eta_{jk} - \eta_{ik}]^{\beta}}{\sum_{\forall h \in cl} [\tau_{ih} + \tau_{hk} - \tau_{ik}]^{\alpha} [\eta_{ih} + \eta_{hk} - \eta_{ik}]^{\beta}} \tag{3}$$

In addition to the previous procedures, a nearest neighbourhood heuristic as in the original AS algorithm has also been used, just taking into account midpoints from required arcs and edges, as well as capacity constraints.

3.2   Local Search

Two local search procedures have been tested. The first one is a *3-opt* local search using the auxiliary cost matrix. The other exchange is an *Or-Opt* exchange, where required edges are allowed to exchange position as well as direction.

3.3   Pheromone Updating

The last element required to describe the implementation of the metaheuristic is the usage of pheromone. This phase has a different behavior depending on the usage of the population scheme, P-AS, or the use of the original Ant System approach.

As in the original Ant System, the pheromone update is carried once a solution has been built and improved by the local search procedure. The P-AS scheme, keeps pheromone for a reduced set of elite solution, el, and updates the trail when a solution enters the elite set. When a new elite solution is found, the pheromone from a previous elite solution is evaporated and new pheromone is left by the new elite solution, following the same pheromone updating formula from the Ant System case.

4   Computational Experience

To ascertain the quality of the proposed algorithms and determine a proper set of parameters, the algorithms were implemented and compared to best known solutions found in the literature. The computational experience was carried on a 600 Mhz. Pentium III computer.

A previous computational experience without local search, see [20] for details, was conducted to test the quality of each construction procedure, metaheuristic variation and five different control parameters. The most promising combinations were chosen for a broader comptational experience, including local search, documented below.

The computational experience compares the solutions offered by each algorithm with the best found results reported in the literature, due to a Genetic

Algorithm [16] and a Tabu Search [13] for three reference datasets. A fourth dataset from real life urban waste management design problem from the municipality of Sant Boi del Llobregat, part of the metropolitan area of Barcelona is also tested. Under all circumstances, the algorithms were limited to 1000 iterations. Table 1 shows the characteristics of each dataset, number of optimal solutions found and deviation versus best known solutions, see [2].

Table 1. Number of instances, range of Required Edges and Vertices, number of best solutions found by proposed procedures and standard deviation from best known solution are given for each instance set. (NA stands for not applicable)

Dataset	Instances	RE	V	N. Best	Deviation
GDB [11]	23	11-55	7-27	18	0.5%
BCCM [4]	34	34-97	24-50	12	1.8%
Li-Eglese [18]	24	77-140	51-190	0	3.8%
Real Instances	5	20-50	15-55	NA	NA

On the other hand, table 2 shows the mean deviation of each of the five ant algorithms and proposed local searches for each instance set. Each algorithm is denoted by the metaheuristic approach, the construction procedure (n, stands for the nearest neighborhood heuristic, i for the Augment and Insert heuristic and ps for the Path Scanning heuristic), and the local search procedure in use, (0 stands for *Or-opt* and 1 for *3-opt*). Paramenters in use are $\alpha=1$,$\beta=5$ and $\rho=0.1$ for AS-I, $\alpha=0.2$,$\beta=1$, $\rho=0.1$,$RLC=5$ for AS-II and $\alpha=3$,$\beta=15$,$el=10$ for P-AS. The results show near optimal solutions within reduced running times. For the two first datasets, GDB and BCCM the results obtained by the best procedure are within less than 4% of the best known solution, while the gap grows for the Li-Eglese dataset up to 5.08%, but with a very limited running time. It is also important to note that running times for the best known solution are not reported. If we analyze the solutions for real life instances, as well as the solutions to the datasets from the literature, the AS algorithm behave better than the other proposed algorithms, even if their quality without the local search procedure was not the best. Additionally the running times are reduced, around a minute for 1000 iterations.

5   Integrating the Algorithms with a DSS

The applicability of the proposed procedures to real life circumstances is subject to their integration within decision support systems (DSS) to aid planners in their decisions. SIRUS, see [1], is a software application capable of assisting the decision-making, design and management steps related to urban waste collection

Table 2. Deviation from best known solution for 1000 iterations, and running times for Li-Eglese instances

Alg.	GDB %	BCCM %	Li-Eglese %	Real %	Time (s.)
AS-I,n,0	3.80	6.21	8.1	0.43	1.33
AS-I,n,1	2	3.5	5.08	0.19	60.02
P-AS,n,0	5.89	6.3	6.47	0.62	0.68
P-AS,n,1	4.03	4.66	5.47	0.49	26.43
AS-II,i,0	4.36	5.65	9.19	0.64	1.01
AS-II,i,1	2.42	3.92	5.94	0.41	35.15
AS-II,n,0	3.29	5.78	7.73	0.43	1.5
AS-II,n,1	1.69	3.87	5.4	0.23	35.98
AS-II,ps,0	4.81	7.3	8.64	0.45	1.13
AS-II,ps,1	2.15	4.7	6.45	0.32	27.58

in an urban area. The system simplifies the periodical tasks of locating collection points and Dumpsters as well as designing routes when conditions change.

SIRUS was designed for integration with a geographic information system, making possible to work with CAD-like geographic data and link this data with databases of the municipality's management systems, such as the census of inhabitants, taxation on economic activities and data relevant to traffic.

The decision maker starts the design process by selecting, using a modern user interface, the section of the city where the new collection system is established. The application obtains information about the streets, their sections, collection points and traffic regulations. The data is used to identify the vertices, and arcs of the graph, as well as the length and sense of arcs, forbidden turns and turn penalties. The results of the algorithm are graphically shown, and additional data is provided.

It is important to point out that high quality, short running time, algorithms should be used in combination with both a Geographical Information and Decision Support System to obtain an application allowing a methodological approach to the design of municipal waste management systems, while reducing operation costs. A tool with these characteristics facilitates the calculations performed in the decision-making step for collection routes, which might not be properly solved in absence of such a tool.

References

1. Bautista, J.: Proyecto integral de gestión de residuos urbanos en el municipio de Sant Boi de Llobregat CPDA Barcelona (2001)
2. Belenger, J.M., Benavent, E. A cutting plane algorithm for the capacitated arc routing problem Computers and Operations Research : **30** (2003) 705–728

3. Beltrami, E., Bodin, L.: Networks and vehicle routing for municipal waste collection Networks **4:1** (1974) 65–94
4. Benavent, E., Campos, A., Corberán, A., Mota, E.: The capacitated arc routing problem: Lower bounds Networks **22:4** (1992) 669–690
5. Corberán, A., Martí, R., Martínez, E., Soler, D.: The Rural Postman Problem on Mixed Graphs with turn penalties Computers and Operations Research **29** (2002) 887–903
6. Christofides, N.: The optimum traversal of a graph Omega **1:6** (1973) 719–732
7. Dorigo, M., Maniezzo, V., Colorni, A.: Positive feedback as a search strategy. Technical Report 91-016, Dip. Elettronica, Politecnico di Milano, Italy (1991)
8. Dorigo, M., Maniezzo, V., Colorni, A.: The Ant System: Optimization by a Colony of Cooperating Agents. IEEE Transactions on Systems, Man and Cybernetics - Part B **26:1** (1996) 29–41
9. Dror, M. (ed.): Arc Routing: Theory, Solutions and Applications Kluwer Academic Publishers (2000)
10. Fischer, M.; Meier, B.; Teich, T.; Vogel, A.: Inner city disposal of waste with ant colony optimization. In: Proceedings of ANTS'2000 - From Ant Colonies to Artificial Ants: Second International Workshop on Ant Algorithms. Eds.: Dorigo et al., Brussels, Belgium, September 7-9 (2000) pp. 51-58.
11. Golden, B.L., DeArmon, J.S., Baker, E.K.: Computational experiments with algorithms for a class of routing problems Computers and Operations Research **11:1** (1983) 49–66
12. Guntsch, M., Middendorf, M.: A population based approach for ACO. Lecture Notes in Computer Science : **2037** (2002) 72–81
13. Hertz, A., Laporte, G., Mittaz, M.: A Tabu Search heuristic for the capacitated arc routing problem Operations Research **48:1** (2000) 129–135
14. Hertz, A., Laporte, G., Nanchen, P.: Improvement procedures for the undirected rural postman problem INFORMS Journal of Computing **11** (1999) 53–62
15. Hirabayashi, R., Saruwatari, Y., Nishida, N.: Tour construction algorithm for the capacitated arc routing problems Asia Pacific Journal of Operations Research **9:2** (1992) 155–175
16. Lacomme, P., Prins, C., Ramdane-Chérif, W. A genetic algorithm for the capacitated arc routing problem and its extensions, in Applications of evolutionary computing Lecture Notes in Computer Science **2037** (2001)
17. Laporte, G., Mercure, H., Nobert, Y.: A branch and bound algorithm for a class of Assymmetrical Vehicle Routeing Problems Journal of the Operational Research Society **43:5** (1992) 469–481
18. Li, L.Y.O., Eglese, R.W.: An interactive Algorithm for Vehicle Routing for Winter-Gritting Journal of the Operational Research Society **47** (1996) 217–228
19. Pearn, W.L. Augment algorithms for the capacitated arc routing problem Computers and Operations Research **18:2** (1991) 189–198
20. Pereira, J.: Modelización y resolución de problemas de diseño de sistemas de recogida de residuos urbanos Unpublished Ph.D. Thesis, UPC (2004)

Ants Can Play Music

Christelle Guéret, Nicolas Monmarché, and Mohamed Slimane

Laboratoire d'Informatique de l'Université de Tours
École Polytechnique de l'Université de Tours, Département Informatique
64, Avenue Jean Portalis, 37200 Tours, France
{nicolas.monmarche,mohamed.slimane}@univ-tours.fr

Abstract. In this paper, we describe how we can generate music by simulating moves of artificial ants on a graph where vertices represent notes and edges represent possible transitions between notes. As ants can deposit pheromones on edges, they collectively build a melody which is a sequence of Midi events. Different parameter settings are tested to produce different styles of generated music with several instruments. We also introduce a mechanism that takes into account music files to initialize the pheromone matrix.

1 Introduction

Science and music have a long common history made of mutual interactions [19]. Many examples can be given to show how music can lead to new perceptions or theories of scientific realities or how scientific researchers can improve music performance or composition.

This paper investigates more precise relations between music and science with computer generated music and Artificial Intelligence (AI) from the swarm intelligence point of view. We will demonstrate how the collective behavior of ants can lead to a compositional system. As far as we know, artificial ants have never been used to generate music and this paper is the first attempt to address this type of method. Real ants are known to produce sounds like crickets do but we will be interested in this paper by their collective behavior such as their capacity to build paths between their nest and a food site in their environment.

After a little introduction to ant algorithms and AI methods that have been used to generate music, we will present the AntMusic project and give some details about the algorithm and its parameters. Finally, we will present results as far as we can do it on paper before investigating future directions.

1.1 Little Things About Artificial Ants

Ants have recently been used as a new kind of inspiration for computer science researchers and many successful works deal with combinatorial optimization (see a review in [4]). For most of the cases, a global memory is used to guide the search agents toward promising solutions as real ants spread volatile substances, known as pheromones, on their path leading to food sources. In this case, artificial pheromones are real values that are used with a positive feedback

M. Dorigo et al. (Eds.): ANTS 2004, LNCS 3172, pp. 310–317, 2004.
© Springer-Verlag Berlin Heidelberg 2004

mechanism which reinforce promising solutions. We can also notice that as not all of the ant species use pheromones, for instance, we have shown that their artificial translation can exploit other capabilities to solve optimization [16] or clustering problems [15]. All these collective problem solving methods belong to the emerging "swarm intelligence" research field.

We have already used artificial ants to produce another kind of computer generated art stuff: artificial paintings [1]. In this application, ants are used to spread colors on a painting and compete against each other to make their color dominant. A interactive genetic algorithm have been used to find the best parameters (i.e. colors and ants properties) from the user point of view.

1.2   Generating Music with AI

Artificial Intelligence (AI) has been used to conceive computer music systems since the beginning of computer science (see a review in [13]). Several models have been proposed to capture musical knowledge such as finite and infinite automata [11], neural networks [5, 8], analogical reasoning [13], constraint programming [10], multi agent systems [22] or Evolutionary Computation (EC) (see for instance [18] and [21, 2] for a larger review).

These systems can be classified in three fields: compositional, improvisation and performance systems. In the following, we will focus on the two first types since the last one deal with producing artificial realistic performances. In the former category, we can cite the pioneering work of Hiller and Isaacson's (in 1858) where notes are pseudo randomly generated by Markov chains and selected accordingly to classical harmony and counterpoint rules. In the EMI project [6], the style of various composers was emulated by searching patterns in several compositions of a given author. In the later category, neural networks are used for instance to learn how to improvise jazz solos [9] or with an interactive genetic algorithm [3]. Interactions with the user can also be integrated for instance as an agent by the way of a Midi instrument [22], by the evaluation of melodies in an interactive genetic algorithm [18] or by the body gestures [17].

In this paper, we will be interested by using the collective behavior of ants to produce music. Todd and Miranda [20] has proposed that there are three ways to generate music with an artificial life system: (1) music can be an expression of the movement of agents which are not aware of what they produce. In this case, music can be considered as a representation of the artificial world. (2) Each individual produces music and its survival depends on it. This approach belongs to evolutionary algorithms technics. (3) Agents produce music that has an influence on other agents' behavior. According to this classification, we will demonstrate that our system belongs to the third category: music is the result of multiple social interactions of ant like agents.

2   The AntMusic Project

2.1   Artificial Ants on the Graph of Notes

In the AntMusic project, we use artificial ants to build a melody according to transition probabilities and taking advantage of the collective behavior of marking paths with pheromones. Artificial ants are agents that are located on vertices in a graph and can move and deposit pheromones on edges. The more ants choose an edge, the more pheromones will be present and the more other ants will choose this edge. As mentioned in section 1.1, this general principle has been exploited for combinatorial optimization problems like the well known Traveling Salesman Problem (TSP). One of the first algorithm designed to deal with this problem was called Ant System (AS) [7] and we have adopted in this project the same principles. In our case, the vertices correspond to Midi events (note, duration,...) and a melody corresponds to a path between several vertices. Each edge (i,j), between vertices i and j, is weighted by a positive pheromone value τ_{ij} (see figure1).

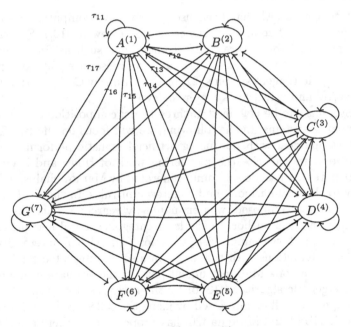

Fig. 1. Example of graph with 7 notes: $A^{(1)}, \ldots, G^{(7)}$ (numbers are indices) on which ants can move (pheromones have only been represented when moving from note A).

A transition rule is used to decide which of the vertices an ant will choose when located in vertex i: the probability of choosing vertex j is given by:

$$p_{ij} = \frac{\tau_{ij} \times \eta_{ij}^{\beta}}{\sum_{k \in N_i} \tau_{ik} \times \eta_{ik}^{\beta}} \tag{1}$$

where η_{ij} is a numerical value representing the desirability and which is set to $1/(d(i,j)+1)$ with $d(i,j)$ representing a distance (in our case it corresponds to the number of half tones between the two notes i and j but we could use another distance). This desirability encourages ants to choose closest notes from their current position to form a melody. The parameter β controls the influence of the style: high values increase the importance of the desirability and low values give more importance to the pheromone values for the transition choices between notes. Finally, N_i stands for the set of possible vertices that can be reached from i.

Each time an ant moves from a vertex i to a vertex j, it deposits a pheromone quantity τ_0:

$$\tau_{ij} \leftarrow \tau_{ij} + \tau_0 \tag{2}$$

Finally, as it occurs in natural systems, pheromones slowly evaporates:

$$\tau_{ij} \leftarrow (1 - \rho) \times \tau_{ij} \tag{3}$$

where ρ is a parameter called evaporation ($0 \le \rho \le 1$).

For one ant, the algorithm that simulates its movements on the graph and generates a melody of $T_{\max}$ notes is the following:

1. Initialization:
   (a) randomly choose a vertex for the initial position of each ant
   (b) initialize pheromone values: $\tau_{ij} \leftarrow \tau_0 \quad \forall i, j$
   (c) $t \leftarrow 0$
2. while $t < T_{\max}$ do
   (a) choose the next Midi event j according to the transition rule (formula 1)
   (b) deposit pheromones on chosen edge (formula 2)
   (c) move to vertex j and play the corresponding Midi event
   (d) $t \leftarrow t + 1$

The same algorithm can be used simultaneously for several ants in order to compose several voices for the selected instrument. We can notice that it is not necessary that all ants play the Midi event they encounter. This can be useful for instance to build quickly a path in the graph: a few ants play the instrument while a large number collectively build the melody and remain silent. Finally, the evaporation step is performed regularly and independently from the ants movements. Pheromones evaporation is useful in optimization methods because it allows the search not to be kept in local minima. For the compositional system we have conceived, it allows the generated melody to evolve during long runs.

The main underling principle that we emphasize in this system is that a musical melody is often identified by the human mind as a repetition of notes and rhythms in the same way which is presented in [14]. So, the reinforcement performed when an ant lays down pheromones can be considered as an encouragement to repeat cycles in the graph.

2.2 Additional Features

Several Instruments. As many levels of complexity can be added to the system, we have introduced several instruments that can be played at the same time. In this case, we only have to build one graph for each desired instrument and different ants can move independently on these graphs in the same way that has been described before.

Limiting the Graph Size. The graph that contains Midi events can be very large. In order to decrease its size, we can use any Midi file to initialize the graph: the file is scanned to build a graph that only contains the Midi events that are present in the file. With this method, it is possible to obtain a melody which sounds like the music contained in the Midi file since only these events can be played. Moreover, pheromones can be initialized according to the Midi event transitions that can be found in the file. In this case, ants will not only use the same events found in the file, they will also have a tendency to build a melody that sounds like the melody contained in the Midi file.

2.3   Additional Parameters

In order to let the user define his/her will, for each instrument, the following parameters can be independently set (default values are indicated within brackets):

- instrument: one of the 128 Midi instruments (1: piano)
- minimum and maximum values of notes (0–127)
- volume (0–127)
- length of the melody: T_{max} (25)
- number of ants: silent (0) and playing ones (1)
- style: parameter β (1.0)
- possible durations: 2, 1, 1/2, 1/4 (1)
- quantity of pheromones laid down by ants: τ_0 (0.1)
- evaporation parameter: ρ (0.01)

3   Results

As the generated music is saved in Midi files we present in this section the scores that correspond to the obtained files[1].

The score of the figure 2 has been generated with the following parameters: one instrument (piano), 3 playing and 10 silent ants, 50 and 70 as minimum and maximum notes, $\beta = 1.5$, $T_{max} = 15$, possible durations are 1-1/2-1/4, $\tau_0 = 0.1$ and $\rho = 0.01$. We can observe that 3 voices can be found (3 notes can be played at the same time) because we have used 3 playing ants.

The score of the figure 3 has been generated with the following parameters: two instruments (two pianos), for the first one: only 1 playing ant, 60 and 80 as minimum and maximum notes, $\beta = 2.0$, $T_{max} = 15$, possible durations are 1-1/2-1/4, $\tau_0 = 0.1$ and $\rho = 0.01$. For the second piano: 2 playing and 5 silent ants, 40 and 65 as minimum and maximum notes, $\beta = 1.5$, $T_{max} = 15$, possible durations are 2-1-1/2, $\tau_0 = 0.3$ and $\rho = 0.01$. This example shows how it is possible to use two instruments to obtain two independent voices for the same

---

[1] The corresponding Midi files can be listen on this url:
http://www.hant.li.univ-tours.fr/webhant/index.php?pageid=39

Fig. 2. Example of score obtained with 3 playing ants (see the text for other parameters).

Fig. 3. Example of score obtained with 2 pianos (see the text for other parameters).

Fig. 4. Example of score obtained with 3 instruments (piano and strings).

instrument: the first piano plays in a high range of notes (60-80) whereas the second one plays in a low range (40-65).

The last example in figure 4 shows a score obtained with three instruments: 2 pianos and 1 strings.

We have noticed that the parameters β (weight for desirability) and ρ (evaporation) have an evident impact on the generated music: the former allows us to build melodies without any wide gap between notes (the style of the music is smoother) and the latter allows us to control how often the sequences of notes are repeated. The influence of this last parameter also depends on the number of ants on the graph, either playing or silent ones.

4   Conclusion

The compositional system presented in this paper is inspired of the natural behavior of ants that build paths using pheromones. Their natural ability to build paths in their environment has been exploited to make paths between notes, that is building melodies.

The results can only be evaluated more regarding to the strategy presented in this paper than from earlier composing and elaboration technics.

Many directions remain to be explored. For instance, in its current version, the system can produce melodies for different instruments simultaneously but without any harmonization between them. This could be addressed by introducing harmonization rules that would reduce the sets N_i at the transition step. Moreover, the style variation can only be influenced by the β parameter and its influence only concerns the distance between notes. Several ways can be proposed: we can add a Midi entry which could be used by a a midi instrument musician or by a given midi file to influence the pheromone values either as an initialization step or as an interactive and real time system.

We have also worked on a system that generates paintings according to the movements of ants (in this case, pheromones correspond to colors) [1]. We plan to merge the two systems to build a compositional systems with a graphical aesthetic view of the process as this type of experiments can be found in [12].

Finally, one drawback of this compositional system is its large number of parameters and consequently the huge number of possible melodies. To improve this point, we need to help the user to find easily the parameters that will produce a music he/she appreciates. One possible way is to use an interactive genetic algorithm to perform an intuitive and very subjective exploration of this huge search space.

Acknowledgment

The authors would like to thanks the computer science department students R. Conchon and V. Bordeau for their contribution to this work.

References

1. S. Aupetit, V. Bordeau, N. Monmarché, M. Slimane, and G. Venturini. Interactive Evolution of Ant Paintings. In *IEEE Congress on Evolutionary Computation*, volume 2, pages 1376–1383, Canberra, 8-12 december 2003. IEEE Press.
2. P. Bentley and D. Corne, editors. *Creative Evolutionary Systems*. Morgan Kaufmann, 2001.
3. J. Biles. GEMJAM: a genetic algorithm for generating jazz solos. In *Proceedings of the International Computer Music Conference*, pages 131–137, San Francisco, 1994. International Computer Music Association.
4. E. Bonabeau, M. Dorigo, and G. Theraulaz. *Swarm Intelligence: From Natural to Artificial Systems*. Oxford University Press, New York, 1999.
5. C. Chen and R. Miikkulainen. Creating melodies with evolving recurrent neural networks. In *Proceedings of the 2001 International Joint Conference on Neural Networks (IJCNN-2001)*, pages 2241–2246, 2001.
6. D. Cope. Pattern matching as an engine for the computer simulation of musical style. In I. C. M. Association, editor, *Proceedings of the International Computer Music Conference*, pages 288–291, San Francisco, 1990.

7. M. Dorigo, V. Maniezzo, and A. Colorni. The Ant System: Optimization by a colony of cooperating agents. *IEEE Transactions on Systems, Man, and Cybernetics-Part B*, 26(1):29–41, 1996.

8. D. Eck and J. Schmidhuber. Finding temporal structure in music: Blues improvisation with LSTM recurrent networks. In H. Bourlard, editor, *Proc. of IEEE Workshop on Neural Networks for Signal Processing XII*, pages 747–756, 2002.

9. J. Franklin. Multi-phase learning for jazz improvisation and interaction. In *Proceedings of the eighth Biennal Symposium on Art and Technology*, New London, Connecticut, 1-3 March 2001.

10. M. Henz, S. Lauer, and D. Zimmermann. COMPOzE – intention-based music composition through constraint programming. In *Proceedings of the 8th IEEE International Conference on Tools with Artificial Intelligence*, pages 118–121, Toulouse, France, Nov.16–19 1996. IEEE Computer Society Press.

11. T. Johnson. *Self-Similar Melodies*. Edition 75, 1996.

12. V. Lesbros. From images to sounds, a dual representation. *Computer Music Journal*, 20(3):59–69, 1996.

13. R. Lopez de Mantaras and J. Arcos. AI and music, from composition to expressive performance. *AI magazine*, 23(3):43–57, 2002.

14. M. Minsky. Music, Mind and Meaning. In S. Schwanauer and D. Levitt, editors, *Machine Models of Music*, pages 327–354. MIT Press, 1993.

15. N. Monmarché, M. Slimane, and G. Venturini. On improving clustering in numerical databases with artificial ants. In D. Floreano, J. Nicoud, and F. Mondala, editors, *5th European Conference on Artificial Life (ECAL'99)*, volume 1674 of *Lecture Notes in Artificial Intelligence*, pages 626–635, Swiss Federal Institute of Technology, Lausanne, Switzerland, 13-17 September 1999. Springer-Verlag.

16. N. Monmarché, G. Venturini, and M. Slimane. On how *Pachycondyla apicalis* ants suggest a new search algorithm. *Future Generation Computer Systems*, 16(8):937–946, 2000.

17. R. Morales-Manzanares, E. Morales, R. Danenberg, and J. Berger. SICIB: An interactive music composition system using body movements. *New Music Research*, 25(2):25–36, 2001.

18. A. Moroni, J. Manzolli, F. Von Zuben, and R. Gudwin. Vox populi: An interactive evolutionary system for algorithmic music composition. *Leonardo Music Journal*, 10:49–54, 2000.

19. R. Root-Bernstein. Music, creativity and scientific thinking. *Leonardo*, 34(1):63–68, 2001.

20. P. Todd and E. Miranda. Putting some (artificial) life into models of musical creativity. In I. Deliege and G. Wiggins, editors, *Musical creativity: Current research in theory and practise*. Psychology Press, 2003.

21. P. Todd and G. Werner. Frankensteinian Methods for Evolutionary Music Composition. In N. Griffith and P. Todd, editors, *Musical Networks: Parallel distributed perception and performance*. MIT Press/Bradford Books, 1998.

22. R.D. Wulfhorst, L.V. Flores, L. Nakayama, C.D. Flores, L.O.C. Alvares, and R.M. Viccari. An open architecture for a musical multi-agent system. In *Proceedings of Brazilian symposium on computer music*, 2001.

Backtracking Ant System
for the Traveling Salesman Problem

Sameh Al-Shihabi

Industrial Engineering Department, University of Jordan, Amman 11942, Jordan
shihabi_sameh@hotmail.com

Abstract. In this work, we adopt the concept of backtracking from the Nested Partition (NP) algorithm and apply it to the Max-Min Ant System (MMAS) to solve the Traveling Salesman Problem (TSP). A new type of ants that is called backtracking ants (BA) is used to challenge a subset of the solution feasible space that is expected to have the global optimum solution. The size of this subset is decreased if the BAs find a better solution out of this subset or increased if the BAs fail in their challenge. The BAs don't have to generate full tours like previous ant systems, which leads to a considerable reduction in the computation effort. A computational experiment is conducted to check the validity of the proposed algorithm.

1 Introduction

The Traveling Salesman Problem (TSP) consists of a number of nodes that are fully connected by edges and it is required to visit all of these nodes, return to the starting city and avoid the formation of sub tours. An arc between city i and city j is penalized an amount d_{ij} and the objective of the TSP is to minimize the sum of these penalties[4].

Ant Colony Optimization (ACO) algorithms imitate the natural behavior of ants in finding the shortest distance between their nests and food sources. Ants exchange information about good routes through a chemical substance called pheromone that accumulates for short routes and evaporate for long routes. The first ACO algorithm is the Ant system (AS) [2] that did not show good results with respect to solving large TSPs.

Different modification to the original AS are introduced to improve the quality of the solution such as having local and global updating of the pheromone matrix as in the Ant Colony system (ACS) [3] or allowing certain ants only to update the pheromone matrix [2] . The Max-Min Ant System (MMAS)[8] modifies the AS by keeping the pheromone values within a range $[\tau_{max}, \tau_{min}]$ to ensure that there is an upper bound to the probability that a certain edge is selected [8].

This work adopts a number of concepts from the Nested Partition (NP) algorithm and adds it to the MMAS. The Nested partition (NP) [6,5] algorithm is based on adaptive sampling where the algorithm moves from a subregion to another in the feasible space rather than moving between solution points. At

M. Dorigo et al. (Eds.): ANTS 2004, LNCS 3172, pp. 318–325, 2004.
© Springer-Verlag Berlin Heidelberg 2004

each step of the NP method, the most promising subregion is divided into M subregions while the abandoned subregions are aggregated to form a surrounding subregion. These $M+1$ subregions are sampled and a promising index is used to choose which subregion is the most promising to be partitioned further, hence, more closely inspected. The algorithm backtracks to a larger subregion if the surrounding subregion is found to be the most promising. The algorithm stops once it reaches a subregion that cannot be partitioned further.

This work presents a new type of AS that called Backtracking Ant System (BAS). Similar to the NP algorithm, the BAS moves from a subset of the feasible region to another. The proposed algorithm uses a new type of ant called Backtracking Ant (BA) to find new routes, in a manner similar to the sampling of the surrounding subregion in the NP algorithm. If a BA finds a shorter tour than the incumbent one, the algorithm backtracks to a larger subregion containing the new sample. The BAS is tested in this work by solving the TSP.

A hybrid between the MMAS and NP is already implemented in [1]. The sampling scheme implemented in [1] uses ants for finding some samples then finds other samples by perturbing the ants' tours. This work is different from [1] since it does not implement the partitioning step and does not sample the subregions resulting from partitioning the most promising subregion. Additionally, all samples are found by ants in this work.

2   BAS Algorithm

Due to the large number of symbols used to describe the proposed algorithm, this section begins by summarizing the notation used. We try to keep the same MMAS notation used in [8] and some of the NP notation used in [6]. The BAS algorithm consists of a number of steps and these steps are explained using an assumed problem instance having 5 cities that are numbered as $0, 1, ..., 4$.

k	The iteration number.
N_c	The number of cities of the TSP instance.
d_{ij}	The distance between city i and city j.
$L_g(k)$	The value of the best solution found by iteration k.
$T_g(k)$	A vector showing the best tour found by iteration k.
$T_c(k)$	A vector of variable size that shows the subset of edges challenged by the backtracking ants.
$d(k)$	The depth of iteration k which represent the number of edges challenged by the backtracking ants.
m	Number of backtracking ants used to challenge $T_c(k)$.
τ_{ij}	The pheromone value between city i and j.
$\tau_{max}(k)$	The maximum value of the pheromone at iteration k.
$\tau_{min}(k)$	The minimum value of the pheromone at iteration k.
ρ	The evaporation rate of the pheromone.
η_{ij}	A heuristic information which is taken as $\frac{1.0}{d_{ij}}$ for the TSP.
α	A constant indicating the importance of the pheromone value.
β	A constant Indicating the importance of the heuristic information.

P_{best} The probability that the best solution is constructed again by ants.
f The pheromone updating frequency.
GT Number of tours generated.
ET Number of cities selected using Equation 1 divided by N_c.

The BAS algorithm can be explained through the following metaphoric description. Imagine that a group of sophisticated ants wants to visit N_c cities starting and ending with at nest in a manner similar to the TSP. Imagine also that these ants are divided into two groups; engineer ants and scout ants. The engineer ants are attaching a road sign to each city showing what is the next city to visit such that other ants could easily follow. These ants follow an up to date map about the shortest route. Waiting at the $d(k)^{th.}$ city to mark, a group of m scout ants "BAs" are sent back to the nest without a map and are asked to find other routes than the one shown in the map. These ants are instructed to follow as many as they want of the already placed $d(k) - 1$ road signs but then they need to find the rest of the route by their own.

The m scout ants need to report back to the ant group if they found a shorter route or not. If these m scout ants fail in finding such a route then the $d(k)^{th.}$ road sign is added and the group of engineer ants move to the $d(k) + 1$ city according to their original map. However, if one of the scout ants finds a better route by following the road signs till city q then finds the rest of the route by its own, the group of ants need to remove all the signs between city q and city $d(k) - 1$ and return back to city q. This group of ants update its map and again send a new group of m scout ants to discover a shorter route. The ants keep repeating this process until they mark a full feasible route. The above metaphoric description can be represented through the following steps:

1. **Initialization:** Starting with city 0, an initial solution is found using any heuristic algorithm such as the minimum distance heuristic. This solution is considered as the global best solution $L_g(0)$. No cities are fixed by this step, $d(0) = 0$ and the backtracking ants are not used to generate tours. For the initialization step, $T_c(0) = [0]$. The initial value of the pheromones is considered as τ_{max} where τ_{max} is calculated according to Equation 2.

2. **Challenging:** This is the main step of the BAS algorithm where m BAs try to find a better route than the one already found. As with other AS, each BA is equipped with a list where it saves the sequence of cities already visited such that it can only generate a feasible tour. Each BA would first generate an integer number $q \in (0, d(k) - 1)$ then copies to its list the sequence $T_c(k) = [0, x, ..., q]$ and choose the rest of cities to visit according to equation 1. The BAs are not allowed to choose the qth edge of $T_c(k)$.

$$p_{ij} = \frac{\tau_{ij}^\alpha \cdot \eta_{ij}^\beta}{\sum_{l \in \{cities \ not \ visited\}} \tau_{il}^\alpha \cdot \eta_{il}^\beta}, \tag{1}$$

Example 1. Assuming that at $k = 4$, the depth $d(4) = 2$ and $T_g(4) = [0, 4, 2, 3, 1, 0]$, then $T_c(4) = [0, 4, 2]$. A BA might have $q = 1$, then this ant copies the first edge only to its list and finds the rest of the cities using Equation 1. A feasible backtracking tour for this ant might be $[0, 4, 3, 2, 1, 0]$ while the tour $[0, 4, 2, 1, 3, 0]$ is not a feasible backtracking tour. Another BA might randomly generate $q = 0$, so a feasible BA tour for this ant might be $[0, x/\{4\}, ..., 0]$.

3. **Pheromone Updating:** As with other AS, the pheromone matrix is updated every f generate BA tours. The best of these f BA tours is used to update the pheromone matrix where its tour length is denoted by L_{bf}. For the MMAS, the trail's strengths are limited by a maximum $\tau_{max}(k)$ and minimum $\tau_{min}(k)$ values. Both of these values are changed based the value of the global solution $L_g(k)$. The value of $\tau_{max}(k)$ is calculated according to:

$$\tau_{max}(k) = \frac{1}{1 - \rho} \cdot \frac{1}{L_g(k)}, \tag{2}$$

while the value of $\tau_{min}(k)$ is found using:

$$\tau_{min}(k) = \frac{\tau_{max}(k) \cdot (1 - \sqrt[n]{p_{best}})}{(avg) \cdot \sqrt[n]{p_{best}}}. \tag{3}$$

The avg represents the average number of cities that an ant needs to visit to build a full tour. The value of avg is equivalent to the $N_c/2$. The pheromone matrix is updated as shown in Equation 4.

$$\tau_{ij} = \rho \cdot \tau_{ij} + \frac{1.0}{L_{bf}} \tag{4}$$

4. **Backtracking:** If the BAs find a better solution than $L_g(k)$ then the algorithm decreases the size of the challenged subset. The algorithm compares the BA list that has the new best solution to the old $T_g(k)$ starting from city 0. The algorithm backtracks to the first edge that shows a difference between the two solutions.

Example 2. Assuming that at $k = 6$, the depth $d(6) = 3$ and $T_g(6) = [0, 4, 2, 3, 1, 0]$, then $T_c(6) = [0, 4, 2, 3]$. Assuming also that a BA generates the tour $[0, 4, 3, 2, 1, 0]$ and this tour is penalized less than $T_g(6)$. The algorithm backtracks to $d(7) = 1$ and $T_c(7) = [0, 4]$.

5. **Covering:** If the BAs fail to find a better solution than the global best solution, the algorithm adds a new edge to $T_c(k)$. The edge selected is the edge connecting cities located at position $d(k)$ and $d(k)+1$ of $T_g(k)$. It needs to be noted that more than one edge can be covered. This might be beneficial for large TSP instances but this would affect also the value of q selected by the BAs.

Example 3. Using the same situation described in Example 2 but assuming that the BAs fail to find a better solution, then $T_c(7) = [0, 4, 2, 3, 1]$ and $d(7) = 4$.

6. **Stopping:** The algorithm stops once $d(k) = N_c - 1$.

Although we are not presenting any mathematical proof about the convergence of the algorithm to the global optimum solution, the algorithm is capable of recognizing this solution for a number of instances. We think that such a proof would follow the same steps used to prove the convergence of the NP method [7]. In other words, the algorithm forms a Markov chain and keeps moving between transient states till it get absorbed in the global optimum solution state. The MMAS will guarantee that every solution point in the feasible space has a positive probability of being selected which is needed to recognize such an absorbing state.

3   Numerical Experiment

A number of ad-hoc experiments that are not reported here were conducted to gain an insight about the performance of the algorithm with respect to the big number of parameters involved. It needs to be noted that the parameters used are not the optimum ones and more work is needed to understand the relations between these parameters and the solution quality. However, the results obtained using the chosen parameters are quite encouraging and deserve presenting.

In the experiments conducted, we used the same parameters of [8] such as having $\rho = .98$, $\alpha = 1.0$, $\beta = 2.0$ and $P_{best} = .05$. When forced to choose a city according to Equation 1, a BA would check first a candidate list of 20 cities and if all these 20 cities are already visited by the BA then it chooses deterministically the next city with the maximum value of $\tau^\alpha \cdot \eta^\beta$. The values of m and f used are changed based on the experiment conducted. These values are presented for each experiment.

The algorithm described is coded with C programming language and run on a PC equipped with Celeron/1.8Ghz processor. The CPU time presented is in seconds. The term Generated Tours (GT) used in presenting the results, represent the number of full tours generated while running the algorithm. As presented in the algorithm description, the BAs would copy to their lists part of the global optimum solution and generate the rest of their tours using Equation 1. The term Equivalent Tours (ET) used in reporting the results is calculated by counting the number of cities selected using Equation 1 during the sampling step divided by the instance's number of cities N_c. All of the problem instances are found at the TSPLIB website at: http://www.iwr.uni-heidelberg.de/groups/comopt/software/TSPLIB95/.

3.1   Symmetric TSP

A number of symmetric TSP instances are solved using the proposed algorithm. The results of each instance presented are generated by running the algorithm 15 times using different random seeds.

Table 1 clearly shows that the algorithm is capable of finding the global optimum solution for a number of instances and good quality results are obtained for the rest. The nonlinear increase in the computation time with respect to the

Table 1. The results of solving a number of symmetric TSP instances using BAS where 15 independent runs are used to obtain the results.

Instance	m	f	opt.	best	worst	avg. sol	$\sigma_{sol.}$	avg. ET	σ_{ET}	avg. GT	CPU time
eil51	300	10	426	426	432	427.8	1.35	19048	2124	23160	16.4
berlin52	300	10	7542	7542	7762	7639.2	74.4	20652	4485	25380	19.4
eil76	300	10	538	538	544	540.7	2.4	31569	6972	37900	50.2
eil101	300	10	629	636	648	642.7	3.4	54997	9578	66760	215.3
eil101	300	20	629	637	650	643.2	3.12	58238	9037	69260	218.5
krob150	300	20	26130	26257	28826	26527.0	174.3	116176	34928	138880	1547.2
d198	300	20	15780	15953	16129	16000.7	50.4	90661	21894	110573	2538.3

problem instance is due to the increase in the number of tours generated and as mentioned in [2], the larger the TSP gets the more probable that an ant needs to search the cities not included in its candidate list. The high CPU time can be reduced by using a fast local search heuristic like the 2-Opt and minimizing the number of tours generated by ants. Additionally, covering more than one edge would speed up the above algorithm specially for large TSP instances.

3.2 Asymmetric TSP

The algorithm is also tested against some asymmetric TSP instances. The deviation from the optimum solution obtained in the case of the asymmetric TSP is more than that obtained for the symmetric TSP. The result obtained for Kro124 problem instance is worse than most of the earlier versions of AS [8].

Table 2. The results of solving a number of asymmetric TSP instances using BAS where, 15 independent runs are used to obtain the results.

Instance	m	f	opt.	best	worst	avg. sol.	$\sigma_{sol.}$	avg. ET	σ_{ET}	avg. GT	CPU time
ry48p	400	20	14422	14507	15201	14734.1	224.0	24914	5050	30960	18.2
ftv55	400	20	1608	1612	1708	1645.7	26.7	34423	6322	42213	31.6
ftv64	400	20	1839	1842	1889	1864.2	15.2	34144	8276	42133	53.9
ft70	400	20	38673	39207	39500	39368.0	91.1	41468	6567	49520	86.8
Kro124p	400	20	36230	36956	38332	37539.6	393.6	60400	17392	72933	243.4
ftv170	400	20	2755	2808	2972	2888.7	46.7	95918	35866	119866	1295.8

3.3 Forcing Ants to Generate Full Tours

It is noticed while running the proposed algorithm that two types of backtracking take place. The first type, which we call full backtracking, happens when ants discover that avoiding the first edge would generate a better solution than the one already found. Partial backtracking is the second type of backtracking where ants backtrack to an edge other than the first edge.

It is also noticed that the two cities connected to city 0 compete in forming the first edge which results in the full backtracking. For example, in eil51 problem instance, city 0 is connected to city 31 and city 21, these two cities compete in forming the first edge. If the first edge is $(0, 31)$ it is noticed that the ants find the best end of tour sequence which ends with the edge $(21, 0)$. If the ants are forced to replace the first edge $(0, 31)$ then ants would choose the edge $(0, 21)$ and form a tour in the opposite direction. It is noticed that most improvements obtained in the value of $L_g(k)$ happens due to full backtracking.

Another experiment is conducted which guarantees that at least 5% of backtracking tours generated are full tours. Using this modification in addition to using a high computation budget of $m = 1000$ and $f = 50$, some instances are solved again as shown in Table 6. This table also shows some results quoted from [8] for the same instances averaged using 25 runs.

The table shows that the results obtained for the symmetric TSP instances are quite comparable to those obtained using MMAS without smoothing [8]. The MMAS shows better results for all the asymmetric TSP instances. The MMAS results presented are the average of 25 runs and the number of tours generated for each run is $10000N_c$ for the symmetric TSP and $20000N_c$ for the asymmetric TSP. The computation budget needed to obtain the BAS results is less than 15% of that used for MMAS in case of the symmetric TSP and less than 8% for the asymmetric instances. For the proposed parameter, the BAS performed better than ACS for the symetric TSP, worse than ACS in case of the Asymetric TSP and better than AS for the symetric and asymetric TSP.

Table 3. The results of solving a number of symmetric and asymmetric TSP instances using BAS where at least 5% of the ants generated full tours. Computation parameters of $m = 1000$ and $f = 50$ are used and the results are averaged over 15 independent runs. The table present first the optimum solution value for each problem instance followed by the solutions found by BAS, MMAS and the relative deviation of the BAS from that of MMAS. The following two colums present the results for the same problem instance using ACS and AS. The last two columns present the average ET and GT. The results of MMAS, ACS and AS presented are all quoted from [8].

Instance	opt.	BAS	MMAS	dev.	ACS	AS	avg. ET	avg. GT
eil51	426	427.6	427.6	0.0	428.1	437.3	60126	72600
kro100	21282	21321.3	21320.3	< 0.1	21420.0	22471.4	143346	172812
d198	15780	15979.7	15972.5	< 0.1	16054.0	16702.1	294385	358266
ry48p	14422	14572.8	14553.2	0.1	14565.4	15396.4	53156	65133
ft70	38673	39260.7	39040.2	0.6	39099.0	39596.3	96176	115800
kro124p	36230	37583.1	36773.5	2.2	36857.0	38733.1	138992	168466
ftv170	2755	2880.2	2828.8	1.8	2826.5	3154.5	207203	258533

4   Conclusion and Future Work

The proposed algorithm presents a new modification to the ACO techniques. The algorithm challenges a subset of the feasible solution space that is expected to include the optimum solution. Although not finely tuned, the results obtained are comparable to the MMAS in case of the symmetric TSP instances and shows slight deviation in case of the asymmetric TSP.

In addition to using a maximum and minimum values for the pheromone's trails to avoid stagnation, the algorithm obliged the ants to select new routes by randomly selecting an edge of the best solution and forcing the ants to avoid this edge and then update the pheromone matrix using a tour different than the best found. Unlike the MMAS, the algorithm did not need a smoothing step nor using different types of ants to update the pheromone matrix [8].

More work is still needed to fine tune the algorithm and understand the relations between the different parameters involved in the algorithm. Distributing the computation budget based on depth or stagnation condition can be tried. Adding a local heuristic search algorithm like the 2-Opt would improve the proposed algorithm and make it more suitable for large problem instances.

References

1. Al-Shihabi, S.: Ants for sampling in the nested partition algorithm. In: C. Blum, A. Roli, and M. Sampels, editors: Proceedings of the 1st International Workshop on Hybrid Metaheuristics (2004) 11–18.
2. Dorigo, M.,Maniezzo V., Colorni, A.: The ant system: Optimization by a colony of cooperating agents. IEEE Transactions on Systems, Man and Cybernetics-Part B, **26**(1996) 29–42.
3. Dorigo, M., Gambardella, L.: Ant colony system: A cooperative learning approach to the traveling salesman problem. IEEE Transactions on Evolutionary Computation**1**(1997) 53–66.
4. Lawler, E., Lenstra, A., Kan, R. , Shmoys, D.: The traveling salesman problem. John Wiley & Sons, 1985.
5. Shi, L., Ólafsson, S.: An integrated framework for determistic and stochastic optimization. Proceedings of the 1997 Winter Simulation Conference (1997) 358–365.
6. Shi, L., Ólafsson, S.: New parallel randomized algorithms for the traveling salesman problem. Computers & Operations Research **26** (1999) 371–394.
7. Shi, L., Ólafsson, S.: Nested partition method for global optimization. Operations Research **48**(2000) 390–407
8. Stützle, T., Hoos, H.: Max-Min ant system. Future Generation Computer Systems **18**(2000) 889–914

Colored Ants for Distributed Simulations

Cyrille Bertelle, Antoine Dutot, Frédéric Guinand, and Damien Olivier

LIH - 25 rue Philippe Lebon 76600 Le Havre
Antoine.Dutot@univ-lehavre.fr

Abstract. Complex system simulations can often be represented by an evolving graph which evolves with a one-to-one mapping between vertices and entities and between edges and communications. Performances depend directly on a good load balancing of the entities between available computing devices and on the minimization of the impact of the communications between them. We use competing colonies of numerical ants, each depositing distinctly colored pheromones, to find clusters of highly communicating entities. Ants are attracted by communications and their own colored pheromones, while repulsion interactions between colonies allow to preserve a good distribution.

1   Introduction

Complex system simulations are composed of interacting entities. We deal here with their implementation on distributed computing environment. The induced interaction graph and its dynamic discourages a static distribution before application execution. As the system evolves, communications between entities change. Communications and entities may appear or disappear, creating stable or unstable organizations. Therefore we need an anytime distribution method that advices the application on better locations for each entity preserving load-balancing between computing resources, but ensuring that entities communicating heavily are close together. In this paper, a method based on the Ant System [3] is described that recommends a location for entities according to the trade-off between load balancing and minimization of communications overhead.

2   Graph Model and General Algorithm

We model the application by a graph $G = (\mathcal{V}, \mathcal{E})$ where $\mathcal{V}$ is a set of vertices representing entities of the application and $\mathcal{E} = \mathcal{V} \times \mathcal{V}$ is a set of edges $e = (v_i, v_j)$ representing communications between entities associated to vertices v_i and v_j. Edges are labeled by weights which correspond to communication volumes (and possibly more attributes). Each vertex is assigned to an initial processing resource at start. No assumption is made about this initial mapping.

We distinguish two different kinds of communications: local ones between some computing resources, and *actual* ones between distinct computing resources. Our goal is to reduce actual communications (source of the overhead)

M. Dorigo et al. (Eds.): ANTS 2004, LNCS 3172, pp. 326–333, 2004.

by identifying sets of highly communicating entities in order to map all entities belonging to one set on the same computing resource. An extension of ant algorithm is used to detect clusters of highly communicating entities. Thus to solve load balancing problems we introduce *colored ants* and *colored pheromones* that correspond to available processing resources, ant colonies producing colored sets. To suit our algorithm we extend our graph definition:

A dynamic communication colored graph is a weighted undirected graph noted $G = (\mathcal{V}(t), \mathcal{E}(t), \mathcal{C}(t))$ such that:

- $\mathcal{C}(t)$ is a set of p colors where p is the number of processing resources of the distributed system at time t.
- $\mathcal{V}(t)$ is the set of vertices at time t. Each vertex has a color belonging to $\mathcal{C}(t)$.
- $\mathcal{E}(t)$ is the set of edges at time t. Each edge is labelled with a weight. A weight $w(t, u, v) \in \mathbb{N}^+$ associated with an edge $(u, v) \in \mathcal{V}(t) \times \mathcal{V}(t)$ corresponds to the importance of communications between the couple of entities at time t, corresponding to vertices u and v.

The figure 5 shows an example of a dynamic communication colored graph at several steps of its evolution. The proposed method changes the color of vertices if this change can improve communications or processing resource load. The algorithm tries to color vertices of highly communicating clusters with the same colors. Therefore a vertex may change color several times, depending on the variations of data exchange between entities.

The graph represents the application as it runs, it is therefore dynamic at several levels: weights change continuously; edges and vertices can appear and disappear at any time; processing resources can appear or disappear and change power at any time. These changes are one of the major motivation for using ant algorithms. So, the problem may be formulated as the on-line allocation, reallocation and migration of tasks to computing resources composing a distributed environment that may be subject to failure.

Many variants of this problem have been extensively studied, and two main classical approaches may be distinguished. The static one consists in computing at compile-time an allocation of the tasks using the knowledge about the application and about the computing environment. This problem is known as the mapping problem [2]. However this approach is not suitable as-is in our context since we don't have information at compile-time. The second approach is dynamic load-balancing. Many works have been dedicated to the special case of independent tasks [4, 9]. When communications are considered, the problem is often constrained by some information about the precedence relation between tasks. Ant-based clustering algorithms first introduced in [7] and then modified by [6] have been proposed in [8] for graph partitioning, able to detect "natural" clusters within a graph.

2.1   Color Ant Algorithm

Our algorithm is inspired by the Ant System[3]. We consider a dynamic communication colored graph G.

- Each processing resource is assigned to a color. Each vertex gets its initial color from the processing resource where it appears. For each processing resource, ants are allocated proportionally to its power.
- The process is iterative. Between steps $t-1$ and t, each ant crosses one edge and reaches a new vertex. During its move, it drops pheromone of its color on the crossed edge.

We define the following elements:

- The quantity of pheromone of color c dropped by one ant x on the edge (u, v), between the steps $t-1$ and t is noted $\Delta_x^{(t)}(u, v, c)$.
- The quantity of pheromone of color c dropped by the ants when they cross edge (u, v) between steps $t-1$ and t is noted:

$$\Delta^{(t)}(u, v, c) = \sum_{x \in \mathcal{F}(t)} \Delta_x^{(t)}(u, v, c) \text{ with } \mathcal{F}(t) \text{ ant population at } t \quad (1)$$

- The total quantity of pheromone of all colors dropped by ants on edge (u, v) between steps $t-1$ and t is noted:

$$\Delta^{(t)}(u, v) = \sum_{c \in \mathcal{C}(t)} \Delta^{(t)}(u, v, c) \quad (2)$$

- If $\Delta^{(t)}(u, v) \neq 0$, the rate of pheromone of color c on the edge (u, v) between the steps $t-1$ and t is noted

$$K_c^{(t)}(u, v) = \frac{\Delta^{(t)}(u, v, c)}{\Delta^{(t)}(u, v)} \text{ with } K_c^{(t)}(u, v) \in [0, 1] \quad (3)$$

- The current quantity of pheromone of color c present on the edge (u, v) at step t is denoted by $\tau^{(t)}(u, v, c)$. Its initial value (when $t = 0$) is 0 and then is computed following the recurrent equation:

$$\tau^{(t)}(u, v, c) = \rho \tau^{(t-1)}(u, v, c) + \Delta^{(t)}(u, v, c)$$

where $\rho \in \,]0, 1]$ represents the pheromone persistence.
- We need now to take into account the load balancing in this self-organization process. So we balance the reinforcement factor $\tau^{(t)}(u, v, c)$ with $K_c^{(t)}(u, v)$, the relative importance of considered color with regard to all other colors. This corrected reinforcement factor is computed as: $K_c^{(t)}(u, v)\tau^{(t)}(u, v, c)$. We use a delay-based relative importance of considered color with regard to all other colors to avoid unstable processes. For a time range $q \in \mathbb{N}^+$, The corrected reinforcement factor becomes:

$$\Omega^{(t)}(u, v, c) = \sum_{s=max(t-q,0)}^{t} K_c^{(s)}(u, v)\tau^{(t)}(u, v, c) \quad (4)$$

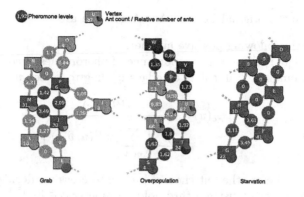

Fig. 1. Problems encountered by our algorithm

- Let $p(u, v_k, c)$ the probability for one arbitrary ant of color c, on the vertex u, to walk over the edge (u, v_k) whose weight is noted $w(t, u, v_k)$ at time t.

$$
\begin{cases}
p(u, v_k, c) = \dfrac{w(0, u, v_k)}{\sum\limits_{v \in \mathcal{V}_u} w(0, u, v)} & \text{if } t = 0 \\[6mm]
p(u, v_k, c) = \dfrac{(\Omega^{(t)}(u, v_k, c))^\alpha (w(t, u, v_k))^\beta}{\sum\limits_{v \in \mathcal{V}_u} (\Omega^{(t)}(u, v, c))^\alpha (w(t, u, v))^\beta} & \text{if } t \neq 0
\end{cases}
\tag{5}
$$

Where $\mathcal{V}_u$ is the set of vertices adjacent to u. The relative values of α and β give the relative importance between pheromone factor and weights.
- Vertex color $\xi(u)$ is obtained from the main color of its incident arcs:

$$
\xi(u) = \arg\max_{c \in C} \sum_{v \in \mathcal{V}_u} \tau^{(t)}(u, v, c)
\tag{6}
$$

3 Local Corrections on the General Algorithm

The original algorithm reveals several difficulties due to lack of control over the relations between the numerous parameters. In clusters of high communication, ants tend to follow privileged paths that form *loops*. This is due to the fact communications in such areas are mostly the same and pheromone take a too large importance in ant path choices. Such paths exclude some nodes that could be settled and leads to three problems: *grab*, *starvation* and *overpopulation* as shown in figure 1.

To solve these problems death and hatching mechanisms have been added. We perturb the ants repartition generating small stable clusters which are the result of local minima. Furthermore this procedure makes senses since our algorithm runs continuously not to find a static solution as the standard Ant System, but to provide anytime solutions to a continuously changing environment. However, to maintain a constant population we make one hatch for one death.

Additions to the original Colored Ant Algorithm are:

1. We define the following positive number:
   - $\varphi_c(u) \in [0,1]$ the relative importance of pheromones of color c compared to pheromones of all colors on edges leading to vertex u:

$$\varphi_c(u) = \frac{\tau^{(t)}(u,c)}{\displaystyle\sum_{c \in C}\left(\sum_{v \in \mathcal{V}_u} \tau^{(t)}(u,v,c)\right)} \tag{7}$$

2. At each step, before the ant chooses an arc to cross we decide whether the ant will die or not by using a threshold parameter $\phi \in [0,1]$ [1]. For an ant of color c on vertex u:
   - if $\varphi_c(u) < \phi$ we kill the ant and create a new one choosing a new location. We select randomly a set $\mathcal{V}_n$ of n vertices. Let $\mathcal{N}_v$ be the number of ants on vertex v, we select a vertex u in $\mathcal{V}_n$ using: $u = \arg\min_{v \in \mathcal{V}_n}(\mathcal{N}_v)$, and make the new ant hatch on it.

This procedure eliminates grab and starvation. Grabbed ants die, and hatch in starvation areas. However, it does not eliminate loops that sometimes tend to reappear. In order to break them, we introduce memory in ants defined by $\mathcal{W}_x$ the set of the last visited vertices by ant x with $|\mathcal{W}_x| < M$. We introduce in equation 5 a penalisation factor $\eta \in]0,1]$. The new probability formula for the specific ant x is:

$$p_x(u,v_k,c) = \frac{(\Omega^{(t)}(u,v_k,c))^\alpha (w(t,u,v_k))^\beta \eta_x(v_k)}{\displaystyle\sum_{v_q \in \mathcal{V}_u} (\Omega^{(t)}(u,v_q,c))^\alpha (w(t,u,v_q))^\beta \eta_x(v_q)} \tag{8}$$

$$\text{Where} \quad \eta_x(v) = \begin{cases} 1 \text{ if } v \notin \mathcal{W}_x \\ \eta \text{ if } v \in \mathcal{W}_x \end{cases} \tag{9}$$

To avoid vertices that already contain too many other ants, and better spread ants in the graph, we introduce a penalisation factor $\gamma(v)$ on vertices that have a population greater than a given threshold. This factor is introduced in the formula 8. Given $N(v)$ the ant count on the vertex v and $N^*(v)$ the threshold.

$$\gamma(u,v) = \begin{cases} 1 & \text{if } N(v) \le N^*(v) \\ \gamma \in]0,1] & \text{else} \end{cases} \tag{10}$$

The formula 8 becomes:

$$p_x(u,v_k,c) = \frac{(\Omega^{(t)}(u,v_k,c))^\alpha (w(t,u,v_k))^\beta \eta_x(v_k)\gamma(v_k)}{\displaystyle\sum_{v_q \in \mathcal{V}_u} (\Omega^{(t)}(u,v_q,c))^\alpha (w(t,u,v_q))^\beta \eta_x(v_q)\gamma(v_q)} \tag{11}$$

---

[1] Preferably small, under 0.1.

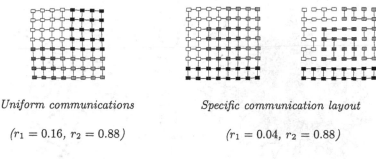

Uniform communications   Specific communication layout

$(r_1 = 0.16, r_2 = 0.88)$ $(r_1 = 0.04, r_2 = 0.88)$

Fig. 2. Grids $(\alpha = 1, \beta = 1, \rho = 0.86, \eta = 0.0001, N = 7)$

4   Experiments

Experiments have been made with static and dynamic graphs with different properties as detailed below. Some are representative of possible simulations, such as grids for regular applications with for example distributed domains or meshes, or such as scale-free graphs for interaction networks... But we also tested the algorithm on random graphs as well as hierarchical graphs.

It is necessary to measure the solution quality found on these graphs. There are two antagonist aspects to take into account. First, the global costs of communications, to which we map a criterion $r_1 = a/s$ where a is the sum of all actual communication costs (as previously defined) and s the total communication volume. Second the application load-balancing, to which we map a criterion r_2. For each color c, we have v_c the number of vertices of color c and p_c the power of processing resource affected to c. Then we have:

$$r_2 = \frac{min\mathcal{K}}{max\mathcal{K}} \quad \text{where} \quad \mathcal{K} = \left\{ \frac{v_c}{p_c}; c \in C \right\}$$

These criteria are used to compare different solutions obtained during the computation, essentially to verify if we improve the solution during the steps.

The figure 2 shows two 8×8 grids. In this kind of regular graph, communications play a major role. The first grid has uniform communications across while the another defines four clusters of high communication. In the second one, ants find clusters of high communications exactly as shown by creating a maximum spanning tree on the graph and cutting the less communicating edges.

The figure 3 shows two views of a random graph of 300 vertices. Such graphs[5] have a edge distribution that follows a normal law. Inside this graph, we have four large clusters of highly communicating entities and several other small clusters. The structure of this graph only appears in communications. The first representation shows the graph with highly communicating vertices highlighted, whereas the second shows the graph with low communication edges removed. Some errors are present due to the large number of low communication edges that still impact the algorithm.

The figure 4 shows a scale-free graph of 400 vertices. Scale-free graphs[1] have an edge distribution that follows a power law. Such graphs are often generated as a growing network using preferential attachment.

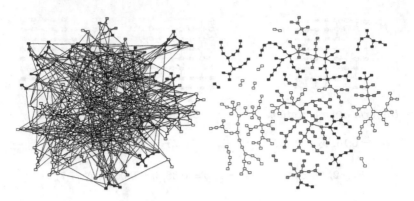

Fig. 3. A random graph with highly communicating edges highlighted ($t = 847$, $r_1 = 0.18$, $r_2 = 0.91$, $\alpha = 1$, $\beta = 1$, $\rho = 0.86$, $\eta = 0.0001$, $N = 8$)

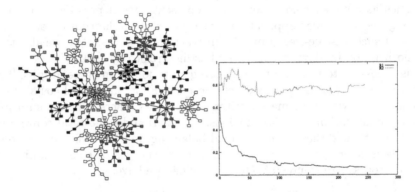

Fig. 4. A scale-free graph of 400 vertices colored with 4 colonies ($t = 249$, $r_1 = 0.06$, $r_2 = 0.8$, *colors* = 4, $\alpha = 1$, $\beta = 1$, $\rho = 0.86$, $\eta = 0.0001$, $N = 5$

Dynamic aspects have been tested, for that we used a program that simulates the application by creating and then applying events to it. Events are the appearance and disappearance of a vertex or processing resource, and weight modifications on edges.

The graph presented in figure 5 is a dynamic graph of 32 vertices that continuously switch between four configurations. At each step criteria r_1 stayed between 0.12 and 0.2 and r_2 at 0.77.

5   Conclusion

We have described a colored ant algorithm, observed its behaviour with dynamic graphs and provided methods to handle them. We have shown several experiments. We are currently developing an heuristic layer allowing to take into account the migration recommendations from ant algoritm and to handle some constraints tied to the application. This work takes place within the con-

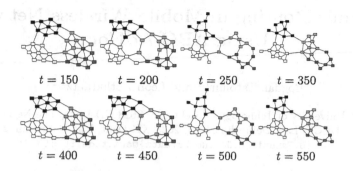

$t = 150$ $t = 200$ $t = 250$ $t = 350$

$t = 400$ $t = 450$ $t = 500$ $t = 550$

Fig. 5. A dynamic graph at eight steps of its evolution (left to right, top to bottom, $colors = 4$, $\alpha = 1$, $\beta = 4$, $\rho = 0.8$, $\eta = 0.0001$, $N = 10$)

text of aquatic ecosystem models, where we are faced to a very large number of heterogeneous auto-organizing entities, from fluid flows to living creatures presenting a peculiar behaviour.

References

1. R. Albert and A. Barabaśi. Statistical mechanics of complex networks. *Reviews of modern physics*, 74:47–97, 2002.
2. S.H. Bokhari. On the Mapping Problem. *IEEE Transactions on Computers*, 30:207–214, Mar. 1981.
3. M. Dorigo, V. Maniezzo, and A. Colorni. The ant system: optimization by a colony of cooperating agents. *IEEE Trans. Systems Man Cybernet.*, 26:29–41, 1996.
4. D.L. Eager, E.D. Lazowska, and J. Zahorjan. A comparison of receiver-initiated and sender-initiated adaptive load sharing. *Performance evaluation*, 6:53–68, 1986.
5. P. Erdös and A. Rényi. On random graphs. *Pubiones Mathematicaelicat*, 6:290–297, 1959.
6. B. Faieta and E. Lumer. Diversity and adaptation in populations of clustering ants. In *Simulation of Adaptive Behavior*, pages 501–508, Cambridge, MA:MIT Press, 1994.
7. J.-L. Deneubourg and S. Goss and N. Francks and C. Detrain and L. Chrétien. The dynamics of collective sorting: Robot-like ants and ant-like robots. In *Simulation of Adaptive Behavior*, pages 356–363, Cambridge, MA:MIT Press, 1991.
8. P. Kuntz and D. Snyers. Emergent colonization and graph partitionning. In *Simulation of Adaptive Behavior*, pages 494–500, Cambridge, MA:MIT Press, 1994.
9. F.C.H. Lin and R.M. Keller. The gradient model load balancing method. *IEEE TOSE*, 13:32–38, 1987.

Dynamic Routing in Mobile Wireless Networks Using ABC-AdHoc

Bogdan Tatomir[1,2] and Leon Rothkrantz[1,2]

[1] Delft University of Tehnology, Mekelweg 4, 2628 CD Delft, The Netherlands
[2] DECIS Lab, Delftechpark 24, 2628 XH Delft, The Netherlands
{B.Tatomir,L.J.M.Rothkrantz}@ewi.tudelft.nl

Abstract. In case of disaster people want to escape from the dangerous area (a building or an area of a city). Because of changing environment dynamic routing is requested. The main idea is to provide some actors involved in crisis situations with PDAs connected via a wireless network. The PDAs will form a mobile ad-hoc network (MANET). These kinds of networks do not require any existing infrastructure or centralized administration. The challenging tasks in such networks, of dynamically changing collection of wireless mobile nodes, are intelligent routing, bandwidth allocation, and power control techniques. In this paper we introduce a routing algorithm for MANET based on ideas from artificial life (swarm intelligence). We implemented the algorithm and compared its performance with the well-known AntNet.

1 Introduction

At DECIS Lab Delft there is a project running concerning Multi-agent based intelligent network decision support systems in a chaotic open world (COMBINED). A complex chaotic world is characterized by always changing environments and events caused by unexpected autonomous causes. To control such an environment and taking appropriate decisions is a far from trivial problem.

The problems of critical information not being available at the right time and at the right place, especially when time is precious, can occur in various situations and on different orders of magnitude. Real world data domains are often noisy, incomplete and inconsistent. Our approach to solve these problems is smart communication. We assume that actors in a Combined system communicate in different ways, using computer networks, GSM etc. To improve the communication and to solve the problems mentioned above, we add a new device, a handheld, connected by a wireless network. The handhelds are a collection of wireless mobile nodes, which dynamically form an ad-hoc network (MANET), without using any existing infrastructure or centralized administration. The mobile nodes in the network establish routing paths between themselves. In this paper we will focus on intelligent network routing, which takes into account that the network topology can change continuously.

M. Dorigo et al. (Eds.): ANTS 2004, LNCS 3172, pp. 334–341, 2004.

2   Related Work

A common feature of all the routing algorithms is the presence in every network node of data structure, called routing table, holding all the information used by the algorithm to make the local forwarding decisions. The routing table is both a local database and a local model of the global network status. The idea of emergent behavior of natural ants can be used in wireless networks to build the routing tables e.g. [1], [3], [4]. The networks cannot sense the pheromone trails, but they can use probabilities. Each node in the network has a probability table for every possible final destination. The tables have entries for each next neighbouring node. So for a node i, $P_i = (p_{dn})$ where p_{dn} is a probability value. This expresses the goodness of choosing node n as its next node from the current node i if the packet has to go to the destination node d. For any connected node i it holds that: $\sum_n P_{dn} = 1$.

2.1   AntNet

The AntNet adaptive agent-based routing algorithm e.g. [2], is one of the best-known routing algorithms for packetswitched communications networks, which is inspired from the ants life. Besides the probability tables a node keeps a second data-structure, which its main task is to follow the traffic fluctuations in the network. It is given by an array $M_i(\mu_d, \sigma_d{}^2, W_d)$ that represents a sample means μ_d and variance $\sigma_d{}^2$ computed over the packet's delay from node i to all the nodes d in the network, and an observation window W_d where the last packet's best trip time towards destination d are stored. Routing is determined through complex interactions of network exploration agents. These agents (ants) are divided into two classes, the forward ants and the backward ants. The idea behind this sub-division of agents is to allow the backward ants to utilize the useful information gathered by the forward ants on their trip from source to destination. Based on this principle, no node routing updates are performed by the forward ants, whose only purpose in life is to report network delay conditions to the backward ants. This information appears in the form of trip times between each network node. The backward ants inherit this raw data and use it to update the routing tables of the nodes.

2.2   ABC

Ant-based Control (ABC) is another successful swarm based algorithm designed for telephone networks e.g. [5]. This algorithm shares many of its key features with AntNet, but has also a few differences. The update philosophy of the routing table is a little different. There is only one class of ants, which travel along the network gathering information. They use by themselves this information, for updating the routing tables.

3   ABC–AdHoc

Lacking the backward ants, the ABC algorithm, might be more suitable for routing in an Ad-Hoc wireless network. In dynamic environment, because of the nodes mobility, the path discovered by the forward ant can happen not to be available for the backward ant. We compared a forward-backward version of the ABC, with the AntNet, in a static network. AntNet proved to be better and more adaptive. This is not only because of the extra local traffic statistics, which a node is maintaining, but also because of the more complex formulae used for maintaining the system.

For a wireless ad-hoc network we create a new algorithm (ABC-AdHoc) adding to the ABC some features from the AntNet. Moreover, a new structure was added to the local data system of a node. It is an array U_d, where for each possible destination d in the network, is stored the last time when the actual node received traffic from the node d. By the concept traffic we mean not only agents but also packets. If the time U_d becomes higher than a certain value(ex. twice the ant generation period multiplied by the number of nodes in a network), the destination d is considered unreachable and the actual node stops generating packets for it. Also the buffer structure is changed. Besides the two buffers for each neighbour (a high priority one for ants and a low priority for packets), a new waiting buffer is added. Here are stored for a while the packets for which the destination is seen as unreachable by the node.

The ABC-AdHoc algorithms works as follows:

1. The mobile agents $F_{s \longrightarrow d}$ are launched at regular time intervals from every network node s.
2. Each ant keeps a memory about its path (visited nodes). When an ant arrives in a node i coming from a node j, it memorises the identifier of the visited node (i) and the trip time necessary for ant to travel from node i to j. These data are pushed onto the memory stack $S_{s \longrightarrow d}(i)$. where:

$$T_{i \longrightarrow j}[s] = \frac{q_j + Size(F_{s \longrightarrow d})}{Bandwidth_{i \longrightarrow j}} + D_{i \longrightarrow j} \qquad (1)$$

   (a) $T_{i \longrightarrow j}[s]$ is the time (virtual delay) that a packet of its size would have to wait to move from node i to get to the previous node j;
   (b) q_j [bits] is the length of the packets buffer queue towards the link connecting node i and its neighbour j;
   (c) $Bandwidth_{i \longrightarrow j}$ is the bandwidth of the link between i and j in [bit/s];
   (d) $Size(F_{s \longrightarrow d})$ [bits] is the size of the forward ant $F_{s \longrightarrow d}$;
   (e) $D_{i \longrightarrow j}[s]$ is the propagation delay of the link $i \longrightarrow j$.
3. When an ant comes in the node i, it has to select a next node n to move to. The selection is done according with the probabilities P_d and the traffic load in the node i.

$$P'_{dn} = \frac{P_{dn} + \alpha \cdot l_n}{1 + \alpha \cdot (|N_i| - 1)} \qquad (2)$$

$l_n \in [0, 1]$ is a normalized value proportional to the amount x_n (in bits waiting to be sent) of the incoming traffic from the link connecting the node i with its neighbour n:

$$l_n = 1 - \frac{x_n}{\sum_{j=1}^{|N_i|} x_j} \tag{3}$$

This happens because we can have completely different traffic load between one way the ant is travelling and the other way where the probabilities are updated. We set $\alpha = 0.4 \in [0, 1]$.

4. If the ant reaches a node where its destination d is seen unreachable, it is killed.

5. In every node i, the forward ant $F_{s \longrightarrow d}$ updates the data structures of the node: the local traffic statistics and the routing table for the source node s, and also for the other nodes the $F_{s \longrightarrow d}$ visited.

First the value U_s is refreshed. If the trip time $T_{i \longrightarrow s} < I_{sup}(s)$ where:

$$I_{sup}(s) = \mu_s + z \frac{\sigma_s}{|W_s|} \tag{4}$$

$$z = \frac{1}{\sqrt{1 - \gamma}}, \quad \gamma = 0.8 \in [0, 1) \text{ gives a selected confidence level.} \tag{5}$$

the updating process goes on, following the steps:

(a) the local data structure is changed:

$$\mu_s = \mu_s + \eta(T_{i \longrightarrow s} - \mu_s) \tag{6}$$

$$\sigma_s{}^2 = \sigma_s{}^2 + \eta((T_{i \longrightarrow s} - \mu_s)^2 - \sigma_s{}^2) \tag{7}$$

where $\eta = 0.1$ is a value that makes the old measurements to defuse in time (like the pheromone).

(b) a reinforcement value r is computed as follows:

$$r = c_1 \frac{I_{inf}(s)}{T_{i \longrightarrow s}} + c_2 \frac{I_{sup}(s) - I_{inf}(s)}{(I_{sup}(s) - I_{inf}(s)) + (T_{i \longrightarrow s} - I_{inf}(s))} \tag{8}$$

$$r = \frac{1 + e^{\frac{a}{|N_i|}}}{1 + e^{\frac{a}{r|N_i|}}} \tag{9}$$

where for an entry s, $I_{inf}(s) = min\{W_s\}$, the best value in the observation window W_s. $c_1 = 0.7, c_2 = 0.3, a = 5$ were the values we used for these parameters.

(c) The probabilities are updated with the reinforcement value,

$$P'_{sk} = P_{sk} + r(1 - P_{sk}), k = j \text{ the previous node chosen by the ant} \tag{10}$$

$$P'_{sk} = P_{sk} - rP_{sk}, \quad \text{for } j \neq n \tag{11}$$

6. If a cycle is detected, that is, if an ant is forced to return to an already visited node the ant is killed. But this is not enough. In the new algorithm killing the ant just when the cycle is found is too late. The changes in the tables are already done. So a cycle should be avoided. This can be done also comparing, in every node i, on the way of $F_{s \longrightarrow d}$, for all the items k on the ant stack, if

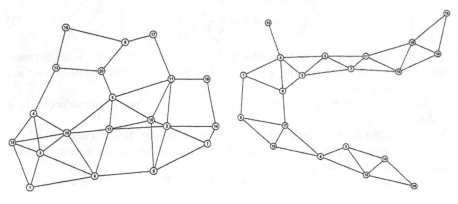

Fig. 1. Net1   **Fig. 2.** Net2

$T_{i \longrightarrow k} > I_{sup}(k)$. In this case the path the ant followed from k to i, is very possible not to be the best one, according with the local information in node i. This means that the ant started to get closer to the already visited node k, and might end in a cycle after some steps. We decided to kill the ant in this situation, unless the ant is very young (it visited less than 3 other nodes).

7. When the destination node d is reached, the agent $F_{s \longrightarrow d}$ is making the updates and dies.

The packets have a different behavior than the ants. In the nodes they don't share the same buffers with the ants, having a lower priority. They are routed to destination according with the probabilities. When a packet $PK_{s \longrightarrow d}$ arrives in a node i, the value U_s in this node is refreshed.

4 Simulation

We implemented and tested both algorithms the new ABC-AdHoc and AntNet. Two networks with different topologies were chosen for testing see figures 1 and 2. Both networks characteristics and the algorithms parameters were set to the same values for as for AntNet in [2]. During the simulations, there were measured 4 metrics for evaluation: the troughput, the bandwidth capacity usage, the arrived packets delay and the number of arrived and lost packets per time unit. First we tested the algorithms in a static environment, without moving the nodes, but with a high traffic load, so that packet loss occurred. Then, the algorithms behaviors were tested in a dynamic environment, where nodes were moved. Each test least 300 s with a training period (just ants were generated) of 20s.

5 Results

5.1 Static Environment – High Traffic Load

For crowding the networks we set the mean packet generation period(PGP) to 5 ms for both, and the mean packet size(MPS) to 8 Kbites for Net1 and 4 Kbites

Table 1. Results for high traffic load test

	Bandwidth usage [Mbits/s^{-1}]		Troughput [%]		Packet delay [ms]		Arrived packets/ Lost packets	
	Net1	Net2	Net1	Net2	Net1	Net2	Net1	Net2
ABC-AdHoc	83	68-80	2700	1400-1300	4300-5000	5400	370/70	370/70
AntNet	81	68-73	2700	1400-1200	5000-5500	6000	370/70	370-330/50-100

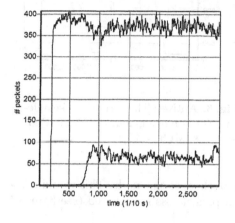

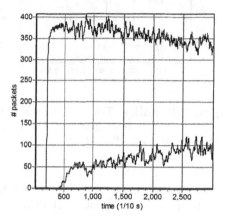

Fig. 3. ABC-AdHoc: The number of arrived/lost packets (Net2, PGP = 5 ms, MPS = 4 Kbits)

Fig. 4. AntNet: The number of arrived/lost packets (Net2, PGP = 5 ms, MPS = 4 Kbits)

for Net2. The results are displayed in table 1. Per total ABC-AdHoc obtains less delay and less lost packets than AntNet. In figures 3 and 4 are presented the packet arrived/lost graphs for Net2.

5.2   Dynamic Environment

The simulator we created, allows us, also to test the algorithms in a very dynamic environment. All the nodes can move periodically in random directions and with different speeds. It is supposed that each node has a limited range of the signal. So the connections between them appear and disappear according with the distances between the nodes. To obtain consistent data we moved just one node at a certain time.

So in case of Net2 we moved node 17 after 50 s to the right until was completely disconnected from the network. At second 120 we brought it back to its old place following the same path. Then we repeated the process moving to left node 9 at second 180 and by bringing it back at second 250.

Disconnecting one node produced an increase in the traffic load in the rest of the network. Some links are no longer available. So the bandwidth usage is growing. The node, which was moved would stop sending packets once it is disconnected from the network. Also the other nodes, after a while, detecting no

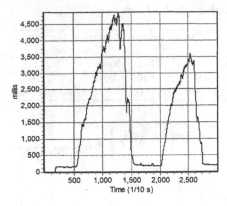 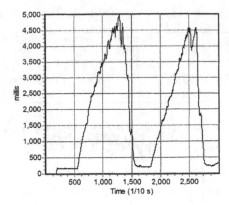

Fig. 5. ABC-AdHoc: The arrived packets delay (Net2 - moving nodes 17 and 9, PGP = 10 ms, MPS = 4 Kbits)

Fig. 6. AntNet: The arrived packets delay (Net2 - moving nodes 17 and 9, PGP = 10 ms, MPS = 4 Kbits)

traffic coming from it, would stop sending packets to this node. This would cause a drop in the throughput. High delay and packet loss is occurred (see figures 5 and 6).

When the node is reconnected to the network the ants will detect this and packets will be generated at this node again. Also the other nodes will start sending traffic to the new connected node. The packets which were kept in the waiting buffer are put in the traffic too. This generates the increase of throughput and bandwidth usage. Most of the packets stored in the waiting buffers are killed. After a while all the variables come close to the values they had at the beginning. Comparing the packet delay graphs the ABC-AdHoc algorithm performed better than AntNet in this situation. Also the number of lost packets was less in case of ABC-AdHoc (see figures 7 and 8).

6   Conclusions and Future Work

We took the most important features from the two well known routing algorithms, ABC and AntNet, and combined them. We obtained a new ABC algorithm which was designed to be used in wireless networks. Its performance was tested in a simulation environment. The AntNet was also adapted to run in a wireless ad-hoc network and used to compare with the new ABC-AdHoc. ABC-AdHoc performed better not only in a dynamic network, but also in a static one. In all the tests, it delivered faster the packets to destination decreasing also the number of the lost ones. At TUDelft we also have a city traffic environment which uses a version of AntNet for routing the cars. We assume that some cars are having a PDA and can form an ad-hoc network. We plan to connect the wireless network simulator two this application and move the nodes not in random way but according with the cars path on the streets network of a city.

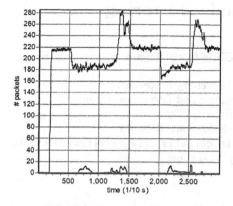
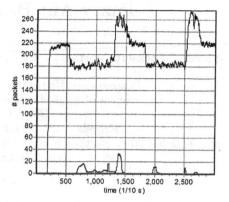

Fig. 7. ABC-AdHoc: The number of arrived/lost packets (Net2 - moving nodes 17 and 9, PGP = 10 ms, MPS = 4 Kbits)

Fig. 8. AntNet: The number of arrived/lost packets (Net2 - moving nodes 17 and 9, PGP = 10 ms, MPS = 4 Kbits)

References

1. P. Arabshahi, A. Gray, I. Kassabalidis, A. Das, S. Narayanan, M. El- Sharkawi, R.J. Marks II. Adaptive Routing in Wireless Communication Networks using Swarm Intelligence, (2001)
2. G. Di Caro and M. Dorigo. AntNet: distributed stigmergetic control for communication networks. Journal of Artificial Intelligence Research (JAIR), 9, (1998), 317-365
3. Mesut Gunes, Udo Sorges, Imed Bouazizi. ARA – The Ant-Colony Based Routing Algorithm for MANETs, (2002)
4. Mesut Gunes, Otto Spaniol. Routing Algorithms for Mobile Multi-Hop Ad-Hoc Networks, International Workshop NGNT, (2002)
5. R. Schoonderwoerd, O. Holland, J. Bruten, L.J.M. Rothkrantz. Load Balancing in Telecommunication Networks, Adaptive Behaviour, vol. 5 no. 2, (1997)

Fuzzy Ant Based Clustering

Steven Schockaert, Martine De Cock, Chris Cornelis, and Etienne E. Kerre

Fuzziness and Uncertainty Modelling Research Unit
Department of Applied Mathematics and Computer Science
Ghent University
Krijgslaan 281 (S9), B–9000 Gent, Belgium
{steven.schockaert,martine.decock,chris.cornelis,etienne.kerre}@UGent.be
http://fuzzy.UGent.be

Abstract. Various clustering methods based on the behaviour of real ants have been proposed. In this paper, we develop a new algorithm in which the behaviour of the artificial ants is governed by fuzzy IF–THEN rules. Our algorithm is conceptually simple, robust and easy to use due to observed dataset independence of the parameter values involved.

1 Introduction

While the behaviour of individual ants is very primitive, the resulting behaviour on the colony-level can be quite complex. A particularly interesting example is the clustering of dead nestmates, as observed with several ant species under laboratory conditions [3]. Without negotiating about where to gather the corpses, ants manage to cluster all corpses into 1 or 2 piles. The conceptual simplicity of this phenomenon, together with the lack of centralized control and a priori information, are the main motivations for designing a clustering algorithm inspired by this behaviour. Real ants are, because of their very limited brain capacity, often assumed to reason only by means of rules of thumb [5]. Inspired by this observation, we propose a clustering method in which the desired behaviour of artificial ants (and more precisely, their stimuli for picking up and dropping items) is expressed flexibly by fuzzy IF–THEN rules.

The paper is organized as follows: in Section 2, we review existing work in the same direction, in particular the algorithm of Monmarché which served as our main source of inspiration. Section 3 familiarizes the reader with important notions about fuzzy set theory and fuzzy IF–THEN rules, while in Section 4 we outline the structure of our clustering algorithm and motivate its key design principles. Some experimental results are presented in Section 5. Finally, Section 6 offers some concluding remarks.

2 Related Work

Deneubourg *et al.* [3] proposed an agent–based model to explain the clustering behaviour of real ants. In this model, artificial ants (or agents) are moving randomly on a square grid of cells on which some items are scattered. Each cell

M. Dorigo et al. (Eds.): ANTS 2004, LNCS 3172, pp. 342–349, 2004.

can only contain a single item and each ant can move the items on the grid by picking up and dropping these items with a certain probability which depends on an estimation of the density of items of the same type in the neighbourhood.

Lumer and Faieta [8] extended the model of Deneubourg *et al.*, using a dissimilarity–based evaluation of the local density, in order to make it suitable for data clustering. Unfortunately, the resulting number of clusters is often too high and convergence is slow. Therefore, a number of modifications were proposed, by Lumer and Faieta themselves as well as by others (e.g. [4, 12]).

Monmarché [10] proposed an algorithm in which several items are allowed to be on the same cell. Each cell with a non–zero number of items corresponds to a cluster. Each (artificial) ant a is endowed with a certain capacity $c(a)$. Instead of carrying one item at a time, an ant a can carry a heap of $c(a)$ items. Probabilities for picking up at most $c(a)$ items from a heap and for dropping the load onto a heap are based on characteristics of the heap, such as the average dissimilarity between items of the heap. Monmarché proposes to apply this algorithm twice. The first time, the capacity of all ants is 1, which results in a high number of tight clusters. Subsequently the algorithm is repeated with the clusters of the first pass as atomic objects and ants with infinite capacity, to obtain a smaller number of large clusters. After each pass k–means clustering is applied for handling small classification errors.

In a similar way, in [6] an ant–based clustering algorithm is combined with the fuzzy c–means algorithm. Although some work has been done on combining fuzzy rules with ant-based algorithms for optimization problems [7], to our knowledge until now fuzzy IF–THEN rules have not yet been used to control the behaviour of artificial ants in a clustering algorithm.

3   Fuzzy IF–THEN Rules

A major asset of humans is their flexibility in dealing with imprecise, granular information; i.e. their ability to abstract from superfluous details and to concentrate instead on more abstract *concepts* (represented by words from natural language). One way to allow a machine to mimic such behaviour, is to construct an explicit interface between the abstract symbolic level (i.e. linguistic terms like "high", "old", ...) and an underlying, numerical representation that allows for efficient processing; this strategy lies at the heart of fuzzy set theory [13], which since its introduction in the sixties has rapidly acquired an immense popularity as a formalism for the representation of vague, linguistic information, and which in this paper we exploit as a convenient vehicle for constructing commonsense rules that guide the behaviour of artificial ants in our clustering algorithm.

Let us recall some basic definitions. A fuzzy set A in a universe U is a mapping from U to the unit interval $[0, 1]$. For any u in U, the number $A(u)$ is called the membership degree of u to A; it expresses to what extent the element u exhibits the property A. A fuzzy set R in $U \times V$ is also called a fuzzy relation from U to V. Fuzzy relations embody the principle that elements may be related to each other to a certain extent only. When $U = V$, R is also called a binary fuzzy relation in U. Classical set theory is tightly linked to boolean logic, in a sense that e.g. the operations of set complement, intersection and union are defined by means of logical negation, conjunction and disjunction respectively. This link is also maintained under the generalization from $\{0, 1\}$ to $[0, 1]$. For instance, to extend boolean conjunction, a wide class of operators called t–norms is at our disposal: a t–norm is any symmetric, associative, increasing $[0, 1]^2 \to [0, 1]$ mapping T satisfying $T(1, x) = x$ for every $x \in [0, 1]$. Common t–norms include the minimum and the product in $[0, 1]$, but also the Łukasiewicz t–norm T_W which has several desirable properties (see e.g. [11]) and which is defined by, for x, y in $[0, 1]$,

$$T_W(x, y) = \max(0, x + y - 1) \tag{1}$$

Another prominent contribution of fuzzy set theory is the ability to perform *approximate reasoning*. In particular, we may summarize flexible, generic knowledge in a fuzzy rulebase like[1]

IF X is A_1 and Y is B_1 THEN Z is C_1
IF X is A_2 and Y is B_2 THEN Z is C_2
. . .
IF X is A_n and Y is B_n THEN Z is C_n

where X, Y and Z are variables taking values in the respective universes U, V and W, and where for i in $\{1, \ldots, n\}$, A_i (resp. B_i and C_i) is a fuzzy set in U (resp. V and W). Our aim is then to deduce a suitable conclusion about Z for every specific input of X and Y. Numerous approaches exist to implement this, with varying levels of sophistication; for our purposes, we used the conceptually simple and very efficient Mamdani method [9], that uses real numbers as inputs and outputs. It can be seen as a four–step process:

1. Given the observed values u of X and v of Y, we calculate for the ith rule its activation level $\alpha_i = \min(A_i(u), B_i(v))$.
2. We "cut off" C_i at level α_i, i.e. we compute $C_i'(w) = \min(\alpha_i, C_i(w))$ for w in W. We thus obtain n individual conclusions.
3. The C_i''s are aggregated into the global inference result C' by means of $C'(w) = \max_{i=1}^{n} C_i'(w)$, for w in W.
4. Finally, a *defuzzification* method is used to transform the result into a crisp value of W; this can be, for instance, the center of gravity of the area below the mapping C' (center–of–gravity method, COG).

Another way of looking at Mamdani's method, is as a flexible way to interpolate an unknown, underlying (possibly very complex) mapping from $U \times V$ to W by means of linguistic labels.

---

[1] This can of course be generalized to an arbitrary number of variables in the antecedent.

4   Fuzzy Ants

Our algorithm is in many ways inspired by the algorithm of Monmarché [10]. We will consider however only one ant, since the use of multiple ants on a non–parallel implementation has no advantages[2]. Instead of introducing several passes, our ant can pick up one item from a heap or an entire heap. Which case applies is governed by a model of division of labour in social insects by Bonabeau et al. [2]. In this model, a certain stimulus and a response threshold value are associated with each task a (real) ant can perform. The response threshold value is fixed, but the stimulus can change and represents the need for the ant to perform the task. The probability that an ant starts performing a task with stimulus s and response threshold value θ is given by

$$T_n(s; \theta) = \frac{s^n}{s^n + \theta^n} \qquad (2)$$

where n is a positive integer[3]. We will assume that $s \in [0, 1]$ and $\theta \in \,]0, 1]$.

Let us now apply this model to the problem at hand. A loaded ant can only perform one task: dropping its load. Let s_{drop} be the stimulus associated with this task and θ_{drop} the response threshold value. The probability of dropping the load is then given by

$$P_{drop} = T_{n_i}(s_{drop}; \theta_{drop}) \qquad (3)$$

where $i \in \{1, 2\}$ and n_1, n_2 are positive integers. When the ant is only carrying one item n_1 is used, otherwise n_2 is used. An unloaded ant can perform two tasks: picking up one item and picking up all the items. Let s_{one} and s_{all} be the respective stimuli and θ_{one} and θ_{all} the respective response threshold values. The probabilities for picking up one item and picking up all the items are given by

$$P_{pickup\_one} = \frac{s_{one}}{s_{one} + s_{all}} \cdot T_{m_1}(s_{one}; \theta_{one}) \qquad (4)$$

$$P_{pickup\_all} = \frac{s_{all}}{s_{one} + s_{all}} \cdot T_{m_2}(s_{all}; \theta_{all}) \qquad (5)$$

where m_1 and m_2 are positive integers.

The values of the stimuli are calculated by evaluating fuzzy IF–THEN rules as explained below. We assume that the objects that have to be clustered belong to some set U, and that E is a binary fuzzy relation in U, which is reflexive (i.e. $E(u, u) = 1$, for all u in U), symmetric (i.e. $E(u, v) = E(v, u)$, for all u and v in U) and T_W–transitive (i.e. $T_W(E(u, v), E(v, w)) \leq E(u, w)$, for all u, v and w in U). For u and v in U, $E(u, v)$ denotes the degree of similarity between the

---

[2] Note, however, that the proposed changes do not exclude the use of multiple ants.

[3] In fact, this is a slight generalization which was also used in [12]; in [2] only the case were $n = 2$ is considered.

items u and v. For a non-empty heap $H \subseteq U$ with centre[4] c in U, we define the average and minimal similarity of H, respectively, by

$$avg(H) = \frac{1}{|H|} \sum_{h \in H} E(h, c) \qquad\qquad min(H) = \min_{h \in H} E(h, c) \qquad (6)$$

Furthermore, let $E^*(H_1, H_2)$ be the similarity between the centres of the heap H_1 and the heap H_2.

4.1 Dropping Items

The stimulus for a loaded ant to drop its load L on a cell which already contains a heap H is based on the average similarity $A = avg(H)$ and an estimation of the average similarity between the centre of H and items of L. This estimation is calculated as $B = T_W(E^*(L, H), avg(L))$, which is a lower bound due to our assumption about the T_W–transitivity of E and can be implemented much more efficiently than the exact value. If B is smaller than A, the stimulus for dropping the load should be low; if B is greater than A, the stimulus should be high. Since heaps should be able to grow, we should also allow the load to be dropped when A is approximately equal to B. Our ant will perceive the values of A and B to be Very High (VH), High (H), Medium (M), Low (L) or Very Low (VL). The stimulus will be perceived as Very Very High (VVH), Very High (VH), High (H), Rather High (RH), Medium (M), Rather Low (RL), Low (L), Very Low (VL) or Very Very Low (VVL). These linguistic terms can be represented by fuzzy sets in $[0, 1]$. The rules for dropping the load L onto an existing heap H are summarized in Table 1(a).

Table 1. Fuzzy rules for inference of the stimulus for (a) dropping the load, (b) picking up a heap.

A \ B	VH	H	M	L	VL
VH	RH	H	VH	VVH	VVH
H	L	RH	H	VH	VVH
M	VVL	L	RH	H	VH
L	VVL	VVL	L	RH	H
VL	VVL	VVL	VVL	L	RH

(a)

A \ B	VH	H	M	L	VL
VH	VVH	-	-	-	-
H	M	VH	-	-	-
M	L	RL	H	-	-
L	VVL	VL	L	RH	-
VL	VVL	VVL	VVL	VL	M

(b)

---

[4] We do not go into detail about how to define and/or compute the centre of a heap, as this can be dependent on the kind of data that needs to be clustered.

4.2 Picking up Items

An unloaded ant should pick up the most dissimilar item from a heap if the similarity between this item and the centre of the heap is far less than the average similarity of the heap. This means that by taking the item away, the heap will become more homogeneous. An unloaded ant should only pick up an entire heap, if the heap is already homogeneous. Thus, the stimulus for an unloaded ant to pick up a single item from a heap H and the stimulus to pick up all items from that heap are based on the average similarity $A = avg(H)$ and the minimal similarity $M = min(H)$. The stimulus for picking up an entire heap, for example, can be inferred using the fuzzy rules in Table 1(b).

4.3 The Algorithm

During the execution of the algorithm, we maintain a list of all heaps. Initially there is a heap, consisting of a single element, for every item in the dataset. Picking up an entire heap H corresponds to removing a heap from the list. At each iteration our ant acts as follows. If the ant is unloaded, a heap from the list is chosen at random; the probabilities for picking up a single element and for picking up all elements are given by formulas (4)–(5). The case where H consists of only 1 or 2 items, should be treated separately (i.e. without using fuzzy rules). If the ant is loaded, a new heap containing the load L is added to the list of heaps with a fixed probability. Otherwise, a heap H from the list is chosen at random; the probability that H and L are merged is given by formula (3). The case where H consists of a single item, should be treated separately.

For evaluating the fuzzy rules, we used a Mamdani inference system with COG as defuzzification method. All response threshold values were set to 0.5. The other parameters are discussed in the next section.

5 Evaluation

We assume that the n objects to be clustered are characterized by m numerical attributes, i.e. $U = \{u_1, \ldots, u_n\}$ with $u_i \in \mathbb{R}^m$, $i = 1, \ldots, n$. To compute the similarity between vectors, we use the fuzzy relation E in U defined by, for u_i and u_j in U,

$$E(u_i, u_j) = 1 - \frac{d(u_i, u_j)}{d^*(U)} \tag{7}$$

where d represents Euclidean distance and $d^*(U)$ is (an estimation of) the maximal distance between objects from U. It can be proven that E is indeed reflexive, symmetric and T_W–transitive. To evaluate the algorithm, we compare the obtained clusters with the correct classification of the objects. For u in U, let $k(u)$ be the (unique) class that u belongs to and $c(u)$ the heap u was put in after algorithm execution. Following Monmarché [10], we define the classification error F_c by

$$F_c = \frac{1}{|U|^2} \sum_{1 \leq i,j \leq n} \epsilon_{ij} = \frac{2}{|U|(|U| - 1)} \sum_{1 \leq i < j \leq n} \epsilon_{ij} \tag{8}$$

with

$$\epsilon_{ij} = \begin{cases} 0 & \text{if } (k(u_i) = k(u_j) \wedge c(u_i) = c(u_j)) \vee (k(u_i) \neq k(u_j) \wedge c(u_i) \neq c(u_j)) \\ 1 & \text{otherwise} \end{cases}$$
(9)

As an important benefit, this evaluation criterion strongly penalizes a wrong number of clusters [10]. As test cases for evaluating our algorithm, we took the "Wine", "Iris" and "Glass" datasets from the UCI Machine Learning Repository [1]. The "Wine" and "Iris" dataset consist of three classes; the "Glass" dataset consists of two main classes, each of which can be further split up into 3 subclasses. Table 2 shows the effect of changing the parameter n_1. For each dataset, the average value for F_c over 50 runs is shown; $(m_1, m_2, n_2) = (5, 5, 20)$ is kept constant. The results were obtained after 10^6 iterations of the algorithm. This reveals that $n_1 = 10$ is a good choice, and moreover small changes in the value of n_1 have little impact on the result. Similar conclusions can be drawn for the other parameter values[5]. The table also contains the results Monmarché obtained with his two–phase algorithm for these datasets. Clearly, for both "Wine" and "Glass" our results are a significant improvement. We also remark that the classification error of the "Glass" dataset was computed w.r.t. the 2 main classes, while Monmarché considered the 6 subclasses. Our algorithm always identifies the 2 main classes, while Monmarché's fails to identify either the 6 subclasses or the 2 main classes in a reliable way.

Initial experiments with artificial datasets suggest that for a dataset of size n, cn (with c an appropriate integer constant) is a good choice for the number of iterations. The corresponding execution time of the algorithm is approximately proportional to $n \log_2 n$, while rule evaluation happens in linear time.

Table 2. Influence of n_1 on F_c.

	$n_1 = 5$	$n_1 = 10$	$n_1 = 15$	$n_1 = 20$	Monmarché
Wine	0.50	0.13	0.14	0.16	0.51
Iris	0.17	0.16	0.17	0.17	0.19
Glass	0.16	0.12	0.13	0.14	0.40

6   Concluding Remarks

We have presented a clustering algorithm, inspired by the behaviour of real ants simulated by means of fuzzy IF-THEN rules. Like all ant-based clustering algorithms, no initial partitioning of the data is needed, nor should the number of clusters be known in advance. The machinery of approximate reasoning from fuzzy set theory endows the ants with some intelligence. As a result, throughout the whole clustering process, they are capable to decide for themselves to pick up either one item or a heap. Hence the two phases of Monmarché's original

---

[5] Due to limited space, we omit the corresponding data.

idea are smoothly merged into one, and k-means becomes superfluous. Initial experimental results with artificial datasets indicate good scalability to large datasets. Outliers in noisy data are left apart and hence do not influence the result, and the parameter values are observed to be dataset-independent which makes the algorithm robust.

Acknowledgments

Martine De Cock and Chris Cornelis would like to thank the Fund for Scientific Research – Flanders for funding their research.

References

1. Blake, C.L., Merz, C.J.: UCI Repository of Machine Learning Databases. University of California, available at:
   http://www.ics.uci.edu/~mlearn/MLRepository.html (1998)
2. Bonabeau, E., Sobkowski, A., Theraulaz, G., Deneubourg, J.L.: Adaptive Task Allocation Inspired by a Model of Division of Labor in Social Insects. *Working Paper* **98-01-004**, available at http://ideas.repec.org/p/wop/safiwp/98-01-004.html (1998).
3. Deneubourg, J.L., Goss, S., Franks, N., Sendova–Franks, A., Detrain, C., Chrétien, L.: The Dynamics of Collective Sorting Robot–Like Ants and Ant–Like Robots. From Animals to Animats: Proc. of the 1st Int. Conf. on Simulation of Adaptive Behaviour (1990) 356–363.
4. Handl, J., Meyer, B.: Improved Ant-Based Clustering and Sorting in a Document Retrieval Interface. Proc. of the 7th Int. Conf. on Parallel Problem Solving from Nature (2002) 913–923.
5. Hölldobler, B., Wilson, E.O.: The ants. Springer-Verlag Heidelberg (1990).
6. Kanade, P.M., Hall, L.O.: Fuzzy Ants as a Clustering Concept. Proc. of the 22nd Int. Conf. of the North American Fuzzy Information Processing Soc. (2003) 227–232.
7. Lučić, P.: Modelling Transportation Systems using Concepts of Swarm Intelligence and Soft Computing. PhD thesis, Virginia Tech. (2002).
8. Lumer, E.D., Faieta, B.: Diversity and Adaptation in Populations of Clustering Ants. From Animals to Animats 3: Proc. of the 3th Int. Conf. on the Simulation of Adaptive Behaviour 501-508 (1994).
9. Mamdani, E.H., Assilian, S.: An Experiment in Linguistic Synthesis with a Fuzzy Logic Controller. Int. J. of Man-Machine Studies **7** (1975) 1–13.
10. Monmarché, N.: Algorithmes de Fourmis Artificielles: Applications à la Classification et à l'Optimisation, PhD thesis, Université François Rabelais (2000).
11. Klement, E.P., Mesiar, R., Pap, E.: Triangular norms, Kluwer Academic Publishers, Dordrecht (2002).
12. Ramos, V., Muge, F., Pina, P.: Self-Organized Data and Image Retrieval as a Consequence of Inter-Dynamic Synergistic Relationships in Artificial Ant Colonies. Soft Computing Systems: Design, Management and Applications (2002) 500-509.
13. Zadeh, L.A.: Fuzzy sets. Information and Control **8** (1965) 338–353.

How to Use Ants for Hierarchical Clustering

Hanene Azzag[1], Christiane Guinot[2], and Gilles Venturini[1]

[1] Laboratoire d'Informatique de l'Université de Tours
École Polytechnique de l'Université de Tours - Département Informatique
64, Avenue Jean Portalis, 37200 Tours, France
Phone: +33 2 47 36 14 14, Fax: +33 2 47 36 14 22
hanene.azzag@etu.univ-tours.fr, venturini@univ-tours.fr
http://www.antsearch.univ-tours.fr/webrtic
[2] C.E.R.I.E.S.
20 rue Victor Noir, 92521 Neuilly sur Seine Cedex
christiane.guinot@ceries-lab.com

Abstract. We present in this paper, a new model for document hierarchical clustering, which is inspired from the self-assembly behavior of real ants. We have simulated the way ants build complex structures with different functions by connecting themselves to each other. Ants may thus build "chains of ants" or form "drops of ants". The artificial ants that we have defined will similarly build a tree. Each ant represents one document. The way ants move, disconnect or connect themselves depends on the similarity between these documents. The result obtained is presented as a hierarchical structure with a series of HTML files with hyperlinks.

1 Introduction

Many web systems and applications often require the use of a clustering algorithm. For instance, one can define a portal site as a hierarchical partitioning of a set of documents (the documents contain information from different subject) which will repeat the following property: at each node (or category), the subcategories are similar to their mother, but they are as much dissimilar to each others.

The major problem of the automatic construction of portal sites is the automatic definition of this hierarchy of documents which, in actual systems, must be given by a human expert [3][9][15][13]. If we want to work with an important number of documents, or if we wish to let the computer autonomously do all this work, then standard approaches are useless.

Many researchers in computer science have been inspired by real ants [7] and have defined artificial ants paradigms for dealing with optimization or machine learning problems [2]. Therefore, several studies involve ants behavior and data clustering [6] [12] [10][4]. In this paper we deal with a new biological model based on artificial ants which as far as we know, has never been used before to solve computer science problems. We model the ability of ants to build structures by assembling their bodies together[16] in order to discover, in a distributed and

M. Dorigo et al. (Eds.): ANTS 2004, LNCS 3172, pp. 350–357, 2004.
© Springer-Verlag Berlin Heidelberg 2004

unsupervised way, a tree-structured organization of the document set as in actual system of portals sites.

The remainder of this paper is organized as follows: in section 2, we give an overview of the biological model that we have use and we present the details of the AntTree algorithm. Its properties and the results obtained will be given in section 3. Section 4, concludes and describes future evolutions of this method.

2   AntTree Algorithm

2.1   Biological Inspiration

Real ants provide a stimulating self-organization model for the clustering problem. In this paper we consider a new biologically observed behavior: ants are able to build mechanical structures by a self-assembling behavior. Biologists observe for instance the formation of drops of ants [17], or the building of chains of ants [11]. These types of self-assembly behavior have been observed with *Linepithema humiles* Argentina ants and African ants of gender *Oecophylla longinoda*. The goal of drop structures built by *Linepithema humiles* is today still obscure. This ability has been recently experimentally demonstrated [17]. The drop can sometimes fall down. For *Oecophylla longinoda* ants, it can be observed that two types of chains are built: for crossing an empty space, and for building their nest [11]. In both cases, crossing chains and building chains, these structures disaggregate after a given time.

From these self-assembly behaviors, we can extract properties that will constitute the framework of our algorithm:

- ants build this type of live structures starting from a fixed support (stem, leaf,...),
- ants can move on this structure whilst it is being currently built,
- ants can cling anywhere on the structure because every position can be reached. Nevertheless, in the formation of chains for example, ants will preferably fix themselves at the end of the chain, because they are attracted by gravity or by the object to reach,
- the majority of ants which constitute the structure can be blocked without any possibility of displacement. For example, in the case of a chain of ants, this corresponds to ants placed in the middle of the chain,
- some ants (a number generally much more reduced than for blocked ants considered in the previous point) are connected to the structure but with a link which they maintain by themselves, they can thus be detached (disconnected) from the structure whenever they want. In the case of a chain of ants, this corresponds to the ants placed at the end of the chain,
- we can observe a phenomenon of growth but also of decrease of the structure.

2.2   Artificial Ants Behavior

For our algorithm called AntTree. Each ant $a_i, i \in [1, N]$ represents one document d_i to cluster. An ant a_i is moving over the support denoted by a_0 or over an other ant denoted by a_{pos} (see the ants colored in gray on figure 1).

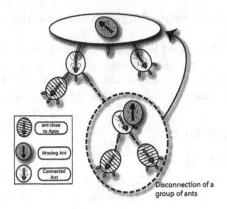

Fig. 1. Disconnection of ants

The similarity measure used (denoted by $Sim(i, j)$) is based on cosine measure [14] which encodes each text d_i and d_j as a vector of word count. We have used a common weighting scheme, i.e. *tf-idf* (term frequency - inverse document frequency) to represent these vectors. *tf* denotes the word count of the document and *idf* denotes the inverse document frequency (document frequency is the number of documents which contain the considered word).

The main principles of our deterministic algorithm are the followings: at each step, an ant a_i is selected in a sorted list of ants (we will explain how this list is sorted in the following section) and will connect itself or move according to the similarity with its neighbors. While there is still a moving ant a_i, we simulate an action for a_i according to its position (i.e. on the support or on another ant). In the following we consider that a_{pos} denotes the ant or the support over which the moving ant a_i is located, and a^+ is the ant (daughter) connected to a_{pos} which is the most similar to a_i (see figure 1).

For clarity, consider now that ant a_i is located on an ant a_{pos} and that a_i is similar to a_{pos}. As will be seen in the following, when an ant moves toward another one, it means that it is similar enough to that ant. So a_i will become connected to a_{pos} provided that it is dissimilar enough to ants connected to a_{pos}. a_i will thus form a new sub-category of a_{pos} which is as dissimilar as possible from the other existing sub-categories. For this purpose, let us denote by $T_{Dissim}(a_{pos})$ the lowest similarity value which can be observed among the daughters of a_{pos}. a_i is connected to a_{pos} if and only if the connection of a_i decreases further this value. The test that we perform consists in comparing a_i to the most similar ant a^+. If these two ants are dissimilar enough ($Sim(a_i, a^+) < T_{Dissim}(a_{pos})$), then a_i is connected to a_{pos}, else it is moved toward a^+.

Since this minimum value $T_{Dissim}(a_{pos})$ can only be computed with at least two ants, then the two first ants are automatically connected without any tests. This may result in "abusive" connections for the second ant. Therefore the second ant is removed and disconnected as soon as a third ant is connected (for this latter ant, we are certain that the dissimilarity test has been successful). When this second ant is removed, all ants that were connected to it are also dropped, and all these ants are placed back onto the support (see figure 1). This algorithm can thus be stated, for a given ant a_i, as the following ordered list of behaviour rules:

R_1 (no ant or only one ant connected to a_{pos}): a_i connects to a_{pos}
R_1' (2 ants connected to a_{pos}, and for the first time):
1. Disconnect the second ant from a_{pos} (and recursively all ants connected to it)
2. Place all these ants back onto the support a_0
3. Connect a_i to a_{pos}
R_2 (more than 2 ants connected to a_{pos}, or 2 ants connected to a_{pos} but for the second time):
1. let $T_{Dissim}(a_{pos})$ be the lowest dissimilarity value between daughters of a_{pos} (i.e. $T_{Dissim}(a_{pos}) = Min\ Sim(a_j, a_k)$ where a_j and $a_k \in$ {ants connected to a_{pos}}),
2. If a_i is dissimilar enough to a^+ ($Sim(a_i, a^+) < T_{Dissim}(a_{pos})$) Then a_i connects to a_{pos}
3. Else a_i moves toward a^+

When ants are placed back on the support, they may find another place where to connect using the same behavior rules. It can be observe that, for any node of the tree, the value $T_{Dissim}(a_{pos})$ is only decreasing, which ensures the termination and convergence of the algorithm.

One must notice that no parameters or predefined thresholds are necessary for using our algorithm: this is one major advantage, because ants based methods often need parameter tuning, as well as clustering methods which often require a user-defined similarity threshold.

3   Results

3.1   Testing Methodology

We have evaluated AntTree on 4 databases which have from $N = 258$ to 1025 texts (see table 1). The *Reuters* databases contains 1025 texts extracted from the reuteurs21578 database (8653 texts). Since clusters in the Reuters database can be very small (two or three texts for instance), we have selected some of the

Table 1. Description of used databases (see text for more explanation)

Databases	Size (# of documents)	Size (Mb)	# of classes
Reuters	1025	4.05	9
CE.R.I.E.S.	258	3.65	17
Database 1	319	13.2	4
Database 2	524	20	7

largest clusters only. The *CE.R.I.E.S.* database [5] contains 258 texts dealing with human skin (the *CE.R.I.E.S.* is a laboratory funded by Chanel). *Database 1* consists of web pages from different scientific topics (73 about scheduling, 84 about pattern recognition, 81 about TcpIp network, 94 about vrml courses). *Database 2* consists of web pages with general topics (55 about c++ courses, 82 about the Danone food company, 86 about IEEE, 90 about cinema, 50 about the Le Monde newspaper, 63 about the sfr phone company, 101 about medicine category from Google's directory). *Database 1* and *Database 2* are extracted from the web.

The real classes of documents (C_r) are of course not given to the algorithms. They are used in the final evaluation of the obtained partitioning.

The evaluation of the results is performed with the number of found clusters C_f, with the purity P_r of clusters (percentage of correctly clustered document in a given cluster), and the classification error measure, denoted by Ec, this measure represents the proportion of document couples which are not correctly clustered, i.e. in the same real cluster but not in the same found cluster, and vice versa.

We have tested three methods: AntTree Disc (the algorithm presented in this paper), AntTree Stoch [1] a previous stochastic version and AHC which is an efficient clustering method [8] (we have used the Ward criterion for cutting the dendrogram).

3.2   Analyzing the Results

Results are presented in tables 2 and 3. As far as the number of classes is concerned, AntTree Disc and Stoch obtain the best results compared to AHC, except for Database 1 where the number of classes is small. The explanation is that AHC often get a lower number of classes than the other algorithms. The

Table 2. Comparative results obtained between AntTree$_{Disc}$, AntTree *stoch* and AHC (C is the number of clusters, P is the purity, and Ec a classification error)

Databases	C_r	AntTree *Disc*			AHC			AntTree *Stoch*		
		Ec	C_f	P_r	Ec	C_f	P_r	$Ec[\sigma_{Ec}]$	$C_f[\sigma_{C_f}]$	$P_r[\sigma_{P_r}]$
Reuters	9	0.22	10	0.47	0.21	5	0.50	0.35 [0.004]	12 [0.00]	0.40 [0.007]
CERIES	17	0.26	7	0.27	0.21	7	0.30	0.15 [0.001]	17 [0.00]	0.37 [0.012]
Database 1	4	0.16	9	0.84	0.09	7	0.82	0.29 [0.011]	7 [0.00]	0.68 [0.012]
Database 2	7	0.06	9	0.87	0.23	3	0.52	0.10 [0.006]	8 [0.00]	0.80 [0.009]

Table 3. Prossesing Time: seconds

Databases	AntTree *Disc*	AHC	AntTree *Stoch*
Reuters	0.81	120	1.51
CERIES	0.03	4	0.04
Database 1	0.08	6	0.12
Database 2	0.25	25	0.34

automatic cutting procedure of the dendogram consider the largest value of the Ward criterion.

Purity values are comparable for AHC and AntTree Disc, except for Database 2 where AHC does not perform well. AntTree Disc performs better than AntTree Stoch for two databases. The definition of thresholds is one drawback of the stochastic method. If these thresholds are well tuned (by hand), then the results are very good. So obviously, the automatic threshold computation in AntTree Disc is efficient. Disconnecting the ants and placing them back onto the support also seems to increase the performances.

The classification error results lead us to the following conclusions: results of AHC and AntTree Disc are comparable (equal for two databases, and one algorithm better than the other for the two other databases). AntTree Stoch does not perform as well as the two other methods.

In table 3, we give the computation time needed once the similarity measure has been computed. AnTree was programmed in C++ and AHC in Visual Basic and the tests were performed on a standard PC (Pentium 2GHz, 512Mo). Both AntTree algorithms clearly outperform AHC. This is due to the fact that AntTree exploits the tree-structure very well and avoids exploring the whole tree when connecting a new ant. It is known for instance that the sorting algorithms based on trees (such as heap sorting for instance) have a lower complexity than other methods. When one consider the whole portal site generation, a run approximatively takes 1 second, including the text extraction from the HTML pages, the display of the tree and the generation of the HTML output as explained in the next section.

Portal Site Generation. When all of texts have been clustered in a tree, then it is easy to generate the corresponding portal site. The hierarchy of documents is represented in our actual implementation as a directory tree with indentation. The tree is encoded in a database in a few seconds and the generation of HTML files is dynamic using PHP language. Figure 2 gives an example of the portal home page obtained for *Database 2* (524 documents) and in figure 3 we have integrated a search engine based on a word index which is automatically generated in the database.

4 Conclusion

In this paper we have described a new algorithm directly inspired from the ants self-assembly behavior and its application to the automatic generation of portals sites. The results obtained are extremely encouraging and the main perspective of this work is to keep on studying this promising model. As future work, we are currently including the automatic extraction of keywords in order to automatically annotate the sub-categories of a given node. Another direction of research is to perform a sampling of the database when it contains many documents (e.g. more than 3000 pages for instance): the texts which have not been used for learning the tree are assigned to a category by following the path with

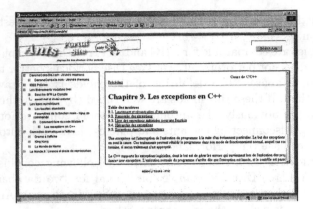

Fig. 2. A typical portal site generated from the *Database 2*

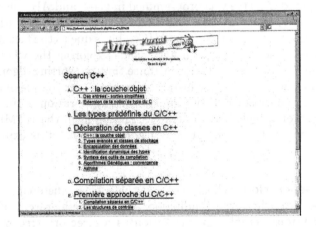

Fig. 3. Searching through the portal site using "C++" term

highest similarity. This procedure is on again very fast. We also want to generalize AntTree to the generation of graphs (and not just trees). We could generate hypertexts with the same self-assembly principles.

References

1. N. Azzag, H. Monmarché, M. Slimane, G. Venturini, and C. Guinot. Anttree : a new model for clustering with artificial ants. In *IEEE Congress on Evolutionary Computation*, Canberra, Australia, 08-12 December 2003.
2. E. Bonabeau, M. Dorigo, and G. Theraulaz. *Swarm Intelligence: From Natural to Artificial Systems*. Oxford University Press, New York, 1999.
3. D. Filo and J. Yang. Yahoo!, 1997.
4. S. Goss and J.-L. Deneubourg. Harvesting by a group of robots. In Varela, editor, *Proceedings of the First European Conference on Artificial Life*, pages 195–204, Paris, France, 1991. Toward a Practice of Autonomous Systems.

5. C. Guinot, D. J.-M. Malvy, F. Morizot, M. Tenenhaus, J. Latreille, S. Lopez, E. Tschachler, and L. Dubertret. Classification of healthy human facial skin. Textbook of Cosmetic Dermatology Third edition (to appear), 2003.

6. J. Handl, J. Knowles, and M. Dorigo. On the performance of ant-based clustering. In A. Abraham, M. Köppen, and K. Franke, editors, *Design and application of hybrid intelligent systems, Proceedings of the Third International conference on Hybrid Intelligent Systems (HIS'03)*, volume 104 of *Frontiers in Artificial intelligence and Applications*, pages 204–213. IOS Press, Amsterdam, The Netherlands, 2003.

7. B Hölldobler and E-O Wilson. *The Ants*. Springer Verlag, Berlin, 1990.

8. A.K. Jain and R.C. Dubes. *Algorithms for Clustering Data*. Prentice Hall Advanced Reference Series, 1988.

9. R. Kumar, P. Raghavan, S. Rajagopalan, and A. Tomkins. On semi-automated web taxonomy construction. In *WebDB*, Santa Barbara, May 2001.

10. P. Kuntz, D. Snyers, and P. Layzell. A stochastic heuristic for visualising graph clusters in a bi-dimensional space prior to partitioning. *Journal of Heuritics*, 5(3), October 1999.

11. A. Lioni, C. Sauwens, G. Theraulaz, and J.-L. Deneubourg. The dynamics of chain formation in oecophylla longinoda. *Journal of Insect Behavior*, 14:679–696, 2001.

12. E.D. Lumer and B. Faieta. Diversity and adaptation in populations of clustering ants. In *Proceedings of the Third International Conference on Simulation of Adaptive Behaviour*, pages 501–508, 1994.

13. Andrew K. McCallum, Kamal Nigam, Jason Rennie, and Kristie Seymore. Automating the construction of internet portals with machine learning. *Information Retrieval*, 3(2):127–163, 2000.

14. Gerard Salton and Christopher Buckley. Term weighting approaches in automatic text retrieval. In *information processing and management*, volume 25, pages 513–523, 1988.

15. Mark Sanderson and W. Bruce Croft. Deriving concept hierarchies from text. In *Research and Development in Information Retrieval*, pages 206–213, 1999.

16. Christian Sauwens. *Étude de la dynamique d'auto-assemblage chez plusieurs espèces de fourmis*. Thèse de doctorat, Université libre de bruxelles, 2000.

17. G. Theraulaz, E. Bonabeau, C. Sauwens, J.-L. Deneubourg, A. Lioni, F. Libert, L. Passera, and R.-V. Solé. Model of droplet formation and dynamics in the argentine ant (linepithema humile mayr). *Bulletin of Mathematical Biology*, 2001.

Inversing Mechanical Parameters of Concrete Gravity Dams Using Ant Colony Optimization

Mingjun Tian and Jing Zhou

Dalian University of Technology, Dalian 116024, China
tianyuan@student.dlut.edu.cn

Abstract. Parameters inverse model of concrete gravity dams is usually formulated as an optimization problem with continuous variables, thus inverse procedures require optimization of an objective function. Ant colony optimization (ACO) is a novel technique proposed mainly for solving discrete optimization problems. In this paper, we attempt to present an adaptation and application of the ACO technique to the parameters inverse of concrete gravity dams. For this purpose, we firstly discretized the search space of mechanical parameters so that a parameters inverse problem was transformed into a combinatorial optimization problem. Secondly, according to the characteristics of the parameters inverse model, we adapted the ACO to the parameters inverse by redefining trail intensity and visibility. Lastly, we gave an example to test the adapted ant colony optimization (AACO) and the results show the AACO can optimize the inverse model validly and efficiently.

1 Introduction

The deformation analysis of a concrete gravity dam is usually based on material parameters. But for complex reasons, such as structure aging, inferior construction quality and etc., design parameters given in design stage always vary from current actual parameters distinctly. Meanwhile, material parameters are difficult and even impossible to measure accurately because of cost, scale and apparatus. An attractive procedure for obtaining mechanical parameters in recent years has been through inverse modelling.

There are mainly three types of inverse methods: inversive solving method, atlas method and direct (i.e.,optimal) method [10]. Because of its special advantages, the optimal method is becoming more and more used in parameters inverse field. In this method, inverse procedures usually require optimization of an objective function. Many methods (for example, Levenber-Marquardt method [12], Gauss-Newton method [13], Powell method, Rosenbork method) are proposed to obtain optimal values of parameters from observed data. However, these optimization algorithms have the defects of being dependent on initial guess of parameters and prone to premature convergence to local minima.

The ACO is a novel simulating evolutionary algorithm and it focuses on discrete optimization problems. For example, the travelling salesman problem

M. Dorigo et al. (Eds.): ANTS 2004, LNCS 3172, pp. 358–365, 2004.
© Springer-Verlag Berlin Heidelberg 2004

(TSP)[7], quadratic assignment problem (QAP)[11] and job-shop scheduling problem (JSP)[3]. And the ACO has been proved to be an efficient optimization tool for hard combinatorial optimization problems [4].

In this paper, we attempt to present an adaptation and application of the ACO to the parameters inverse of concrete gravity dams.

2   Ant Colony Optimization

The first ACO meta-heuristic, called ant system [2,9], was inspired by studies of the behavior of real ants. The works of Colorni et al. [2], Dorigo et al. [8,6, 5] offer detailed information on the workings of the algorithm and the choice of the values of the various parameters.

For parameters inverse problems could be transformed into combinatorial optimization problems similar to the TSP, we first briefly review the ACO's benchmark application to the the the TSP [2].

Given a set of n towns, the TSP can be stated as the problem of finding a minimal length closed tour that visits each town once. Let $s_i(t)$ $(i = 1, 2, \ldots, n)$ be the number of ants in town i at time t and $m = \sum_{i=1}^{n} s_i(t)$ be the total number of ants. Each ant is forced to visit each town only once in a tour by the control of a tabu list and it lays a substance called trail on each visited edge when it completes a tour. Tabu$_k$ is defined as the dynamically growing vector that contains the tabu list of the k-th ant.

Let d_{ij} be the distance between i-th and j-th town and $\tau_{ij}(t)$ be the trail intensity on edge (i, j) at time t. We call visibility η_{ij} the quantity $1/d_{ij}$. The transition probability from town i to town j for the k-th ant at time t is as follows

$$p_{ij}^k(t) = \begin{cases} \dfrac{[\tau_{ij}(t)]^\alpha (\eta_{ij})^\beta}{\sum_{k \in \text{allowed}_k} [\tau_{ik}(t)]^\alpha (\eta_{ik})^\beta} & \text{if } j \in \text{allowed}_k \\ 0 & \text{otherwise} \end{cases} \tag{1}$$

Where allowed$_k = \{N - \text{tabu}_k\}$, and N is a set of n towns; α and β parameters control the relative importance of trail versus visibility.

After each ant cycle, trail intensity on each edges is updated as follows

$$\tau_{ij}(t + n) = \rho \tau_{ij}(t) + \Delta \tau_{ij}(t) \tag{2}$$

Where ρ is a coefficient such that $(1 - \rho)$ represents the evaporation of trail between time t and $t + n$, and it must be set to a value less than 1; and

$$\Delta \tau_{ij} = \sum_{k=1}^{m} \Delta \tau_{ij}^k \tag{3}$$

Where $\Delta \tau_{ij}^k$ is the quantity per unit of length of trail substance laid on edge (i, j) by the k-th ant between time t and $t + n$, and it is given by

$$\Delta \tau_{ij}^k = \begin{cases} \frac{Q}{L_k} & \text{if } k\text{-th ant select edge } (i,j) \text{ between } t \text{ and } t+n \\ 0 & \text{otherwise} \end{cases} \quad (4)$$

Where Q is a constant and L_k is the tour length of the k-th ant.

3   Application of the ACO to Parameters Inverse

3.1   Problem Description

The most commonly used parameters inverse model in direct method for concrete dams is as follows

$$\begin{cases} \text{Min} & T(P) = \sum_{i=1}^n (a_i - a_i^*)^2 \\ s.t. & [K_P]\{A\} = \{R\} \\ & d_j \leq p_j \leq u_j \ (j = 1, 2, \ldots, k) \end{cases} \quad (5)$$

Where $T(P)$ is the objective function; P is the mechanical parameters vector; n is total number of observed points; a_i^* is the i-th observation datum on the i-th observed point and a_i is the corresponding computational datum obtained by the second equation of the inverse model; $[K_P]$ is the stiffness matrix correlative to P; $\{A\}$ is nodal displacement vector; $\{R\}$ is equivalent nodal force vector; p_j is the j-th parameter to be inversed, while u_j and d_j are the up and low boundaries of it respectively; k is the total number of parameters to be inversed. It should be noted that the values a_i depend on P via $[K_P]$.

In the inverse procedure, $\{R\}$ is usually known beforehand. For a given (trial) P, the corresponding $[K_P]$ is assembled. Then $\{A\}$ is figured out in the second equation of the inverse model by using the finite element method (FEM). And the computational information on the observed points is selected out and transmitted into the right hand side of the objective function definition so that the corresponding objective function value can be obtained. Also, the objective function value is stored and the mechanical parameter vector P is revalued. This process is iterated until the stop criteria is satisfied.

3.2   Discretization of Search Space

The inverse model (5) is an optimization problem with continuous variables. In order to apply the ACO to treat the inverse model, we have to transform it into a combinatorial optimization problem by discretizing search space of mechanical parameters to be inversed.

We adopt the discretization approach proposed in paper [1]. Suppose P_{inv} is the vector of the parameters to be inversed, which is a subset of P, and S is the search space of P_{inv}. Without loss of generality, S can be described as the hyperparallelepiped

$$S = \{p_i \mid d_i \leq p_i \leq u_i; i = 1, 2, \ldots, k\} \quad (6)$$

To discretize S, interval $[d_i, u_i]$ is divided into a number, say h_i, of segments. Let each segment be represented by the value at the middle of the segment. Thus there will be $A = h_1 h_2 \ldots h_k$ possible pathways through S. Let h_{ij} be the segment j of the parameter p_i in S.

An example of search space discretization is shown in Fig. 1. In which P_{inv} is a 4 dimension vector and each parameter interval is divided into 4 segments. Thus there are totaly 256 possible pathways through the search space. And the 1-th, 3-th and 252-th pathways are shown in Fig. 1.

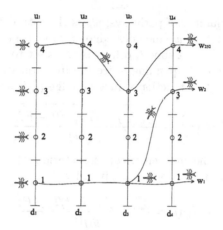

Fig. 1. Discretization of search space

Based on the discretization of search space, the inverse model is transformed into a combinatorial optimization problem. To solve it is to find a pathway, which corresponds to a set of input parameters, to minimize the objective function.

3.3   Adaptation of the Ant Colony Optimization

It is shown in formula (4) that the tour length L_k is essential for the update of trail intensity. However the pathway in the inverse model is only a virtual concept without length. For the objective of solving the inverse model is to find a pathway that can minimize the objective function, we define L_k as follows

$$L_k = T_k(P) + g \tag{7}$$

Where $T_k(P)$ is the corresponding objective function value of the current pathway selected by the k-th ant; g is a constant to avoid the outflow of $\Delta\tau_{ij}^k$, and it can be given by

$$g = Q/M \tag{8}$$

Where M is a pre-defined constant, which denotes the maximum value that $\Delta\tau_{ij}^k$ could obtain (When $T_k(P) = 0$) in a step. In this paper, M is set to 100.

Since the visibility between town i and town j is defined as $\eta_{ij} = 1/d_{ij}$, the distance between town i and town j should be known beforehand to calculate η_{ij}. However, distance between two towns is meaningless in a discretization search space. Hence it is necessary to redefine the visibility.

In order to redefine the visibility, we compute standard deviations of objective function values on edges. Let $g_{ij}(t)$ be the total number of pathways involving edge (i, j) at time t

$$g_{ij}(t) = \sum_{k=1}^{A} f_{w_k}(ij, t) \tag{9}$$

Where A is total number of pathways; w_k is the k-th pathway; $f_{w_k}(ij, t)$ is a identifier denoting whether w_k includes edge (i, j) at time t. $f_{w_k}(ij, t)$ is set to 1 if w_k includes edge (i, j) at time t, otherwise it is set to 0.

Let $T_{w_k}(P)$ be the corresponding objective function value of w_k. The standard deviation of objective function values on edge (i, j) at time t is given by

$$\sigma_{ij}(t) = \sqrt{\frac{\sum_{k=1}^{A} f_{w_k}(ij, t) \left[T_{w_k}(P) - EX_{ij}(t) \right]^2}{g_{ij}(t)}} \tag{10}$$

Where $EX_{ij}(t)$ is the mathematical expectation of all objective function values on edge (i, j) at time t, and it is given by

$$EX_{ij}(t) = \frac{\sum_{k=1}^{A} f_{w_k}(ij, t) T_{w_k}(P)}{g_{ij}(t)} \tag{11}$$

The value of $\sigma_{ij}(t)$ reflects the sensitivity of the edge (i, j) with respect to other edges. We can prevent algorithm from premature convergence by giving more ants choose those more sensitive edges with higher probability. Thus we redefine the visibility η_{ij} as follows

$$\eta_{ij}(t) = \sigma_{ij}(t) \tag{12}$$

Based on the above discussion, the transition probability from town i to town j for the k-th ant at time t is redefined as follows.

$$p_{ij}^k(t) = \begin{cases} \frac{[\tau_{ij}(t)]^\alpha [\sigma_{ij}(t-n)]^\beta}{\sum_{k \in \text{allowed}_k} [\tau_{ik}(t)]^\alpha [\sigma_{ik}(t-n)]^\beta} + h & \text{if } j \in \text{allowed}_k \\ 0 & \text{otherwise} \end{cases} \tag{13}$$

Where h is a constant ensuring the edges with zero transition probability also have slim chances to be selected by ants.

After being adapted according to formula (7) to (13), the adapted ant colony optimization (AACO) can be easily applied to parameters inverse.

4   Example

A segment of Fengman Dam is taken as an example to test the validity and efficiency of the AACO. The dam body and foundation rock are totally divided into three regions, as schematically shown in Fig. 2.

We inverse the elastic modulus of each region based on observed displacements. The loads taken account are water pressure and material weights.

It is known that the density and Poisson ratio of the foundation rock are 2700kg/m^3 and 0.25 respectively, the corresponding parameters for the concrete of dam body are 2400kg/m^3 and 0.18, while the water head upstream is 56m.

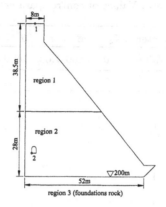

Fig. 2. Material regions of a dam

Suppose the elastic modulus' actual values are all known. And these values are listed in the second row of Table 1. In Table 1, E_1, E_2 and E_3 represent the elastic modulus of region 1, region 2 and foundation rock respectively.

Table 1. Actual and inverse values of material elastic modulus

Parameter	E_1	E_2	E_3
Actual value/GPa	25.00	31.00	20.00
Inverse value/GPa	24.98	30.93	19.87

Based on these actual values of elastic modulus, we can figure out displacements of the dam using FEM. And the displacements on observed points (point 1 and point 2, as shown in Fig. 2) are defined as observed data.

To test the AACO, we suppose the elastic modulus of each region are unknown and we inverse their values based on the observed data.

The initial parameter ranges are shown in Table 2.

In the discretization of the search space, the parameter ranges of E_1 and E_2 are both divided into 100 segments while range of E_3 is divided into 150 segments. The values of control parameters in the AACO are shown in Table 3.

The stop criteria used in this example is that the ant cycle times reach its maximum number, which is predefined as 100. The inverse results are shown in

Table 2. Initial parameter ranges

Unknown parameter	E_1	E_2	E_3
Low boundary/GPa	1.00	1.00	10.00
Up boundary/GPa	36.00	36.00	90.00

Table 3. Values of control parameters

Control parameter	Q	m	α	β	ρ	g	h
Value	1.00	100.00	1.00	1.00	0.80	0.01	0.01

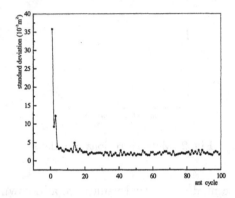

Fig. 3. Standard deviations of objective function values in each ant cycle

the third row of Table 1. The standard deviations of objective function values in each ant cycle are shown in Fig. 3.

Table 1 illustrates that the difference between the actual and inverse parameters is slight. Fig. 3 shows in the early cycles the AACO identifies good parameters that are subsequently refined in the rest of the cycles. Since the standard deviation never drops to zero, we are assured that the algorithm can contain the diversity of the search pathways and avoid premature convergence.

5   Conclusions

In this paper, we presented an adaptation and application of the ACO to the parameters inverse of concrete gravity dams. We adopted discretization method to transform a parameter inverse problem into a combinatorial optimization problem. In the adaptation of the ACO, we redefined the trail intensity and visibility according to the characteristics of the inverse model. The results obtained in this research have shown that the AACO could solve the inverse model validly and efficiently. This asserted the ACO as a new and effective way to inverse mechanical parameters of concrete gravity dams.

References

1. Abbaspour, K. C., Schulin, R., van Genuchen, M. Th.: Estimating unsaturated soil hydraulic parameters using ant colony optimization. Advances in Water Resources. **24** (2001) 827-841
2. Colorni, A., Dorigo, M., Maniezzo, V.: Distributed optimization by ant colonies. Proceeding of the First European Conference on Artificial Life, Paris, France, F, Varela. and P, Bourgine. (Eds), Elsevier Publishing. (1992) 134-142
3. Colorni, A., Dorigo, M., Maniezzo, V., Trubian, M.: Ant system for job-shop scheduling. Belgian Journal of Operations Research, Statistics and Computer Science. **34** (1994) 39-53
4. Dorigo, M., Di Caro, G., Gambardella, L. M.: Ant algorithms for discrete optimization. Artificial Life. **5(2)** (1999) 137-172
5. Dorigo, M., Di Caro, G.: The Ant Colony Optimization meta-heuristic. In D, Corne., M, Dorigo. and F, Glover., editors, New Ideas in Optimization. McGraw-Hill. (1999) 11-32
6. Dorigo, M., Gambardella, L, M.: Ant colonies for the traveling salesman problem. BioSystems. **43** (1997) 73-81
7. Dorigo, M., Maniezzo, V., Colorni, A.: Positive feedback as a search strategy. Technical Report 91-016, Dipartimento di Elettronica, Politecnico di Milano, IT. (1991)
8. Dorigo, M., Maniezzo, V., Colorni, A.: The Ant system: Optimization by a colony of cooperationg agents. IEEE Trans Syst Man Cybern-Part B. **26(1)** (1996) 29-41
9. Dorigo, M.: Optimization, learning and natural algorithms. Ph.D. Thesis, Politecnico di Milano, Italy. (1992)
10. Feng, ZL., Lewis, RW: Optimal estimation of in-situ round stress from displacement measurements. Int. J. Numer. Anal. Meth. Geomech. **11** (1987) 397-408
11. Gambardella, L, M., Taillard, E, D., Dorigo, M.: Ant colonies for the QAP. Journal of the Operational Research Society. **50(2)** (1999) 167-176
12. Okabe, T., Hayashi, K.: Inverse of drilling-induced tensile fracture data obtained from a single inclined borehole. Int. J. Rock Mech. Min. Sci. **35(6)** (1998) 747-758
13. William, WGY.: Aquifer parameter identification with optimum dimension in parameterization. Water Resources Research. **17(3)** (1981) 664-672

Large Pheromones: A Case Study
with Multi-agent Physical A*

Ariel Felner[1,2], Yaron Shoshani[2],
Israel A. Wagner[3,4], and Alfred M. Bruckstein[4]

[1] Dept. of Information Systems Engineering
Ben-Gurion University of the Negev Beer-Sheva, 84104, Israel
felner@bgumail.bgu.ac.il
[2] Dept. of Computer Science, Bar-Ilan University, Ramat-Gan, 52900, Israel
[3] IBM Haifa Labs, MATAM, Haifa 31905, Israel
[4] Dept. of Computer Science, Technion, Haifa, 32000, Israel

Abstract. Physical A* (PHA*) and its multi-agent version MAPHA*
[3, 4] are algorithm that find the shortest path between two points in
an unknown real physical environment with one or many mobile agents.
Previous work assumed a complete sharing of knowledge between agents.
Here we apply this algorithm to a more restricted model of communication which we call *large pheromones*, where agents communicate by
writing and reading data at nodes of the graph that constitutes their
environment. Unlike small pheromones where only a limited amount of
data can be written at each node, the *large pheromones* model assumes
no limitation on the size of the pheromones and thus each agent can
write its entire knowledge at a node. We show that with this model of
communication the behavior of a multi-agent system is almost as good
as with complete knowledge sharing.

1 Introduction

This paper introduces the notion of *large pheromones* as a model of communication and global knowledge sharing in multi-agent systems. With *pheromones*,
agents communicate by writing and reading data at the nodes of the graph that
constitutes their environment (e.g. [7, 8]). Unlike pheromones in previous work
where only a limited amount of data can be written in each node, in the *large
pheromones* model, there is no restriction on the amount of data that can be
written in the the nodes and thus each agent can write its entire knowledge in a
node. We apply this model of communication to the multi-agent physical A* algorithm (MAPHA*) which is the multi agent version of Physical-A* (PHA*) [3,
4]. These algorithms modify the A* algorithm to find shortest paths in physical
environments with mobile agents that move around the environment and explore unknown territories. These algorithms are designed to minimize the travel
effort of the agents. We will show that increasing the amount of data that can
be stored at each of the pheromones dramatically reduces the travel effort of the
agents. With maximal usage of this model with unlimited size of pheromones the

behavior of a multi-agent system is almost as good as with complete knowledge sharing between the agents.

2   Physical A*

The A* algorithm [5] is a common method for finding a shortest path in graphs that have exponential number of nodes (like combinatorial puzzles). A* keeps an *open-list* of generated nodes and expands them in a best-first order according to a cost function of $f(n) = g(n) + h(n)$, where $g(n)$ is the distance traveled from the initial state to n, and $h(n)$ is a heuristic estimate of the cost from node n to the goal. $h(n)$ is *admissible* if it never overestimates the actual cost from node n to the goal. A* was proved to be admissible, complete, and optimally effective [1]. Therefore, any other algorithm claiming to return the optimal path must expand at least all of the nodes that are expanded by A* given the same houristic h [1]. An A* expansion cycle is usually carried out in constant time as it takes a constant amount of time to retrieve a node from the open-list and to generate all its neighbors by applying domain-specific operators to the expanded node. Thus the time complexity of A* can be measured in terms of the number of generated nodes.

Physical A* (PHA*) modifies A* to find the shortest path in much smaller graphs which correspond to a real physical environment. Consider a mobile agent who needs to find a shortest path between two physical locations and assume that only a very small portion of the environment graph is known to the agent. Since A* is optimally effective, the mobile agent needs to activate the A* algorithm on this physical graph. For this type of graph, however, we cannot assume that expanding a node from the open list takes constant time. Many of the nodes and edges of this graph are not known in advance. Therefore, to expand a node that is not known in advance, a mobile agent must first travel to that node in order to explore it and learn about its neighbors. The cost of the search in this case is the cost of moving an agent in a physical environment, i.e., it is proportional to the distance traveled by the agent. PHA* expands all the mandatory nodes that A* would expand and returns the shortest path between the two points but is designed to minimize the traveling effort of the agent by intelligently choosing the next assignment of the traveling agent. Note that since small graphs are considered here, we can omit the actual computation time and focus only on the travel time of the agent.

Unlike ordinary navigation tasks the purpose of the agent in PHA* is not to reach the goal node as soon as possible, but rather to explore the graph in such a manner that the shortest path will be retrieved for future usage. On the other hand, our problem is not an ordinary exploration problem where the entire graph should be explored in order for it to be mapped out. See [3, 4] for more comparison to other algorithms and problems.

An example for a real application can be the following scenario. A division of troops is ordered to reach a specific location. The coordinates of the location are known. Navigating with the entire division through unknown hostile territory until reaching its destination is unreasonable and inefficient. Instead, one may have a team of scouts search for the shortest path for the division to pass through. The scouts explore the terrain and report shortest path for the division to move along in order to reach its destination. PHA* is an algorithm designed to help these scouts.

PHA* works in two levels. The high level (which invokes the low level as a subroutine), acts like a regular A* search algorithm: at each cycle it chooses the best node from the open-list for expansion. Nodes are evaluated according to a heuristic function $h(n)$, which, in our case, is the Euclidean distance between n and the goal node. If the node chosen by the high level has not been explored by the agent, the low level, which is a navigation algorithm, is activated to bring the agent to that node and explore it. After a node has been explored by the low level it is expandable by the high level. If the chosen node has already been explored, or if its neighbors are already known, then it is readily expandable by the high level without the need to send the agent to visit that node.

In [3, 4] a number of navigation algorithms for the low level are presented. They all attempt to navigate to the target via unexplored nodes. Thus, while navigating through unknown parts of the graph, the agent might visit new nodes that have not been explored yet and explore them on the fly. This may save the need to travel back to those nodes at a later time, should they be selected for expansion by the high-level algorithm. See [3, 4] for a comprehensive description and all technical details of these ideas and algorithms.

2.1 MAPHA*: Multi-agent Physical A*

In [3, 4], PHA* was generalized to the Multi-agent Physical A* (MAPHA*) where a number of agents cooperate in order to find the shortest path. The task is that these agents should explore the necessary portion of the graph, i.e., the A* nodes as fast as possible. The assumption in [3, 4] was that each agent can communicate freely with all the other agents and share data at any time. Thus any information gathered by one agent is available and known to all of the other agents. This framework can be obtained by using a model of a centralized supervisor that moves the agents according to the complete knowledge that was gathered by all of them. Another possible model for complete knowledge-sharing is that each agent broadcasts any new data about the graph to all the other agents.

MAPHA* also uses a two level framework. The high level chooses which nodes to expand, while the low level navigates the agents to these nodes. Since complete knowledge sharing is assumed there is one central high level which activates A* and distributes the agents to different tasks. Suppose that we have p available agents and We would like to distribute these p agents to the nodes from the front of the open lest as efficiently as possible. This is done by distributing more agents to nodes in the front of the window, (i.e. with a relatively small f-value) but on the other hand give high priority to assigning an agent to a relatively close-by node. See [3, 4]

3 Communication Models and Pheromones

There are many models for communication in multi agent systems. As described above, the most trivial model is complete knowledge sharing where any new discovery of an agent is immediately shared with all the other agents. Other models restrict the level of communication. Some models allow broadcasting or message exchanging between agents but restrict the size or frequency of message. Many times, a penalty cost is associated with each message.

In nature, ants and other insects communicate and coordinate by leaving trails of odor, i.e., chemical *pheromones*, on the ground. Inspired by nature, A famous model for communicating in a multiagent system is that of ant-robotics, (e.g. [7,8]). In this model, information is spread to other agents via *"pheromones"*, i.e., small amounts of data that are written by an agent at various places in the environment (e.g. nodes in the graph), and can be later used or modified by other agents visiting that node. In our ant-inspired model we assume that the information network is a graph, and the role of a pheromone is taken by a memory area on each node, that our search a(ge)nts can read and modify. This paradigm suggests a distributed group of one or more lightweight autonomous agents that traverses the environment in a completely autonomous and parallelized way. Data is spread by the agents via these pheromones, which together serve as a distributed shared memory.

Usually, it is assumed that a very small amount of data (no more than a few bytes) can be written in each pheromone. It turns out, however, that despite these severe limitations on pheromone size, such agents are able to cooperate and achieve goals like covering a faulty graph [8], finding an Euler cycle in a graph [9] and solving various combinatorial optimization problems [2]. A small sized pheromone can only include local data and is not very suitable for problems such as finding a shortest path in a graph where global data sharing is needed. In the sequel we consider the effect of using larger pheromones, i.e. storing more data in the nodes, on the efficiency of multiagent search.

4   Large Pheromones

We suggest a new model of communication which we call *large pheromones*. Unlike conventional pheromones, we cancel the restriction on the amount of data that can be stored in each node, and consider the effect of this increased storage on the performance of the search algorithm. At the extreme, we assume that an agent can write its entire knowledge base (e.g. a complete list of nodes and edges known to that agent) at each of the nodes. With today's hardware capabilities and computer architecture this is a reasonable assumption. With pheromones, we already assume that each agent has the necessary hardware devices to allow reading and writing data in the environment. We also assume that there is a storage device in each of the nodes. Given that a storage device is installed in each node it is not reasonable to limit the size of the storage device as with current technology memory is very cheap. For example, it is not un realistic to assume, say, one megabyte of memory at a node which can store a graph of tens of thousands of nodes. We can also assume that the time to read and write data from the large pheromones can be omitted when considering the traveling time of the agents. This additional data storage in the *large pheromones* paradigm can help the agent to solve the problem faster and much more efficiently. However, large memory is not always available, e.g. in a system of nanorobots within a hostile environment, where only a very limited use of the environment is possible. Hence, we also consider a more modest memory capacity of the storage devices in the nodes. In that case, given the limited memory capacities and the entire knowledge base, we will address the question of selecting the most relevant portion of the knowledge for storing at the nodes.

4.1 Spreading Data with Large Pheromones

With such large capacities of memory in each node, we present the following communication paradigm between the agents in general and in exploring unknown environments in particular. Each agent maintains a database with a partial graph that is known to it. Similarly, each node holds a database with a partial graph that is 'known' to it, i.e., knowledge that was written to it by the agents. Whenever an agent reaches a node, it merges the data known to it with the data that is written in that node. The agent then writes the combined data in that node and updates its own database according to the new data that was obtained. We call this the *data-merge* operation. In this way data will be spread out very fast as long as agents visit many nodes in many areas of the graph and perform *data-merge* operations at all the nodes that they visit. For example, assume that agent A visits node n and write its knowledge in that node. After a while, agent B visits node n. Agent B will read the information in the node and will merge it with its own knowledge. After merging the knowledge, both agent B and node n hold the information gathered by both agent A and agent B.

If we assume a limited memory capacity of the nodes then the data-merge operation has two stages. First, the agent merges its own data with the data that is written in the node. Then the agent erases the previous pheromone and applies a selection algorithm to determine the most informative portion of the data for writing in the node. This selection algorithm is of course domain dependent.

5   MAPHA* with Large Pheromones

we no present our new version for MAPHA* where the large pheromones model is employed. When the large pheromones model is employed there is no centralized entity and each agent activates the high-level A* on its own, based on the partial knowledge of the graph that is known to it at any point of time. Thus each agent keeps its own open-list of nodes and chooses to expand the best node from **that** open-list. If, neighbors of that node are not known to the agent, then, as with single agent PHA*, the agent will navigate to that target node with the help of a low-level navigation algorithm. When the large pheromones paradigm is employed then at each node that a navigating agent visits, it performs a *data-merge* operation. Its own knowledge is added to the node and knowledge about nodes that were not known to the agent is now learned by the agent. This might have a positive effect on the open-list and the high-level A* that is activated by this agent. For example, suppose that agent A choose to expand node t from front of the open-list, and this node was not yet explored by agent A. Thus, agent A should now navigate to that node. On its way to t, agent A visits node n. Suppose that another agent, B, already explored node t and later wrote that data in node n while visiting it. When agent A reaches node n and performs a data-merge operation, it learns about node t. Therefore, agent A does not need to continue the navigation to node t and that node can be expanded by agent A immediately.

If we only assume a limited memory capacity of the nodes then after the agent merges its data with data from the pheromone it will have to decide which are the most relevant nodes to write back to the pheromone. We have tried many variants and found out that the best performance was achieved by writing data about nodes that are closest to the current node.

Note that PHA* as well as MAPHA* with large pheromones are deterministic algorithms that are designed to work only in a static environments where the structure of the graph is stable throughout the search. If the graph is dynamic and changes during the search then a probabilistic approach would probably be a better choice.

6   Experiments

We have implemented the algorithms described above and performed experiments on Delaunay graphs [6] which are derived from Delaunay triangulations. Delaunay graphs simulate Roadmap graphs which are real physical graphs.

Figure 1 illustrates the time elapsed as a function of the number of agents that were used to find the shortest path with different versions of MAPHA* on Delaunay graph with 500 nodes. Since we assume constant speed, the time is reported as the distance traveled by the agents until the solution was found. Every

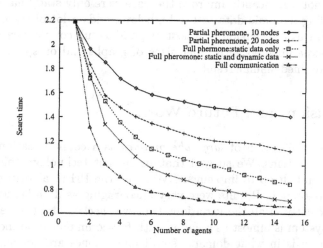

Fig. 1. MAPHA* with large pheromones, 500 nodes

data point (here and in all the other experiments) corresponds to an average of 250 different pairs of initial and goal nodes, that were picked at random. There are 5 curves in the figure. The bottom curve corresponds to the best version of MAPHA* with full communication from [3,4] and is used as a benchmark. Other curves show the overall time cost of different versions of MAPHA* with large pheromones. The top two curves show the results where we only assumed a limited memory capacity at the nodes. In particular, in the top curve. 10 nodes were allowed to be written and the second curve allowed 20 nodes to be written at each pheromone. As explained above, the nodes selected to be memorized are those closest to the current node. The rest of the curves assume unlimited data capacity and thus the entire graph can be written at each node. The third curve, "Full pheromones: static data only" shows the case where only static data about the graph was used. Finally, the forth curve, "Full pheromone: static and dynamic data" uses the most powerful pheromone available i.e. unlimited data capacity of the pheromones and both static data about the graph as well as dynamic data about behaviors of other agents.

The results show a clear phenomenon. As the pheromone becomes larger and includes more knowledge a significant improvement in the overall time is obtained. This parametric effect is achieved even though no explicit control is forced by a centralized supervisor.

The figure clearly shows that all the versions that use the large pheromones paradigm with unlimited memory capacity keep most of the potential of the full knowledge sharing paradigm. Their performance is rather close to the performance of the full communication version. This means that with large pheromones, data is spread to other agents rather fast and it is almost as good as full communication and full knowledge sharing. This is true even for the simple version which includes static data only.

Dynamic data, with knowledge about routes, tasks and decisions of other agents further improved the performance. It seems that using it only in the low level did not significantly improve the case were only static data was used. However, adding dynamic data to the high level adds a lot of strength to this algorithm and this last version is almost as good as full communication model.

We have experimented with other sizes of graphs and on sparse and dense graphs and obtained similar results.

7   Conclusions and Future Work

We introduced the notion of *large pheromones* as a communication paradigm in multi-agent systems. We showed that in current technology this paradigm is reasonable and cheap to implement. We used the PHA* algorithm as a test case for this paradigm. Results were very encouraging as data is indeed spread out quite efficiently in all the variations that we checked and the behavior of a multi-agent system is almost as good as with full-communication model.

The question is in what domains small pheromones are not sufficient and large pheromones are needed. We believe that large pheromones are needed in any domain where global data from different areas of the environment is critical to the decision making of all agents at all times. In that case smaller pheromones will not do the job. Our problem of activating A* in a physical environment is an example for this. Since A* expands nodes in a global best-first search order local data is not enough for this as shown in figure 1.

For another example, consider a team of fireman agents that have to extinguish fire. The general geometrical structure of the fire is very important as it might cause agents to move to different locations. A counter-example might be a group of agents who are trying to explore and map unknown territories. Whenever an agent reaches a new node, it learns new valuable information. Thus, when locally realizing that there is a nearby unexplored area, moving to that area is always beneficial. Knowledge of other areas of the environments is not so crucial at every point of time. Similarly, the works in [8, 9, 2] need local data to improve their efficiency and thus very small pheromones were enough.

Future work will proceed by applying this paradigm to other problems. Indeed we are currently implementing these ideas to multi-agent fire detecting. Preliminary results look promising. Also, A mathematical analysis of spreading data should be figured out, providing a better insights and bounds on achievable performance.

References

1. R. Dechter and J. Pearl. Generalized best-first search strategies and the optimality of A*. *Journal of the Association for Computing Machinery*, 32(3):505–536, 1985.
2. M. Dorigo, V. Maniezzo and A. Colorni, *The Ant System: Optimization by a Colony of Cooperating Agents*, IEEE T-SMC, Part B, 26(1):29-41
3. A. Felner, R. Stern, A. Ben-Yair, S. Kraus, and N. Netanyahu. Pha*: Finding the shortest path with a* in unknown physical environments. *Accepted to JAIR*, 2003.
4. A. Felner, R. Stern, and S. Kraus. PHA*: Performing A* in unknown physical environments. In *Proceedings of the First International Joint Conference on Autonomous Agents and Multi-Agent Systems*, pages 240–247, Bologna, Italy, 2002.
5. P. E. Hart, N. J. Nilsson, and B. Raphael. A formal basis for the heuristic determination of minimum cost paths. *IEEE Transactions on Systems Science and Cybernetics*, SCC-4(2):100–107, 1968.
6. A. Okabe, B. Boots, and K. Sugihara. *Spatial Tessellations, Concepts, and Applications of Voronoi Diagrams*. Wiley, Chichester, UK, 1992.
7. A. Wagner and A. M. Bruckstein. ANTS: Agents, networks, trees, and subgraphs. *Future Generation Computer Systems Journal*, 16(8):915–926, 2000.
8. V. Yanovski, I. A. Wagner, and A. M. Bruckstein. Vertex-ant-walk: A robust method for efficient exploration of faulty graphs. *Annals of Mathematics and Artificial Intelligence*, 31(1-4):99–112, 2001.
9. V. M. Yanovski, I. A. Wagner, A. M. Bruckstein, *A distributed ant algorithm for efficiently patrolling a network*, Algorithmica (2003) 37: 165-186.

Near Parameter Free Ant Colony Optimisation

Marcus Randall

Meta-heuristic Search Group
Bond University, QLD 4229, Australia
Phone: +61 7 55953361
mrandall@bond.edu.au

Abstract. Ant colony optimisation, like all other meta-heuristic search processes, requires a set of parameters in order to solve combinatorial problems. These parameters are often tuned by hand by the researcher to a set that seems to work well for the problem under study or a standard set from the literature. However, it is possible to integrate a parameter search process within the running of the meta-heuristic without incurring an undue computational overhead. In this paper, ant colony optimisation is used to evolve suitable parameter values (using its own optimisation processes) while it is solving combinatorial problems. The results reveal for the travelling salesman and quadratic assignment problems that the use of the augmented solver generally performs well against one that uses a standard set of parameter values. This is attributed to the fact that parameter values suitable for the particular problem instance can be automatically derived and varied throughout the search process.

1 Introduction

Meta-heuristic search strategies, including tabu search, simulated annealing, GRASP and ant colony optimisation (ACO), invariably require a set of parameters in order to solve combinatorial optimisation problems. These parameters directly impact on the performance of the solver and as such, researchers and practitioners will often "hand tune" parameter values before the application of the production meta-heuristic or use a set of values that have been found to be traditionally "good" by other researchers (i.e., a standard set).

Relatively little research has been conducted into either the analysis of parameter values or the ways in which they can be automatically derived or tuned by meta-heuristics themselves. In this paper, ant colony optimisation is examined as it is an optimisation framework that has been successfully applied to a range of combinatorial optimisation problems [3, 7]. ACO represents a group of constructive meta-heuristics (often coupled with local search) that use collective intelligence present in insect colonies. The reader is referred to Dorigo and Gambardella [5] and Dorigo and Di Caro [3] for an overview and background of ACO. In this work, ant colony system (ACS) [5] is used as it is a robust and reliable technique.

In regards to ACO studies in which parameters have been analysed, Colorni, Dorigo and Maniezzo [2], Dorigo and Gambardella [4], Dorigo, Maniezzo

M. Dorigo et al. (Eds.): ANTS 2004, LNCS 3172, pp. 374–381, 2004.

and Colorni [6], Maniezzo and Colorni [9] and Shmygelska, Aguirre-Hernández and Hoos [13] have each compared and contrasted various parameter values on particular problems (most notably the travelling salesman problem (TSP) and quadratic assignment problem (QAP)) in order to derive suitable parameter sets.

In terms of automatic parameter adaptation, Ingber [8] developed an implementation known as Adaptive Simulated Annealing in which parameter values are changed in a systematic manner throughout the search process. Pilat and White [10] in their work used concepts from genetic algorithms in order to evolve solution parameters for a constituent ACO technique, ant colony system. This was developed as a "Meta ACS-TSP" that would run standard ACS within a genetic algorithm that evolves solution parameters (a computationally expensive exercise). Using this technique, they were able to suggest alternative good parameter values to Dorigo and Gambardella [5] for ACS and the TSP.

The approach adopted within uses standard mechanisms of ant colony system [5] to modify and to determine appropriate parameter values while problems are being solved. Therefore, it is conceptually simple to integrate this approach into an ant colony implementation (moreso than other search techniques, particularly iterative meta-heuristics). Another advantage is that ant based techniques *learn* appropriate values for particular problem instances (without the researcher/practitioner having to derive these manually). Additionally, it does not add a significant computational overhead to the native algorithm (unlike, for instance, that of Pilat and White [10]). The results show that good quality solutions are achieved for a range of TSP and QAP problem instances. The remainder of the paper is organised as follows. The extensions that allow ACS to evolve its own parameter values (using aspects of the native algorithm) are given in Section 2. Computational experiments, using benchmark TSP and QAP problem instances, are reported in Section 3. Finally, the future directions of this work and conclusions are outlined in Section 4.

2   Evolving ACO Parameter Values

ACS can use the same mechanics for generating solutions to evolve appropriate values for its parameters. Within ACS, the core parameters (apart from the number of ants) are as follows. These are introduced and described in greater detail by Dorigo and Gambardella [5].

- q_0: This parameter determines whether the greedy or probabilistic form of component selection equation is used by an ant at each step of the algorithm. A low value will more likely result in the use of the probabilistic form of the equation (and vice versa).
- ρ: The local pheromone updating factor.
- γ: The global pheromone updating factor.
- β: Is the relative importance placed on the visibility heuristic.

The standard ACS algorithm is augmented at each iteration by allowing each ant to select a value for each parameter before commencing the selection of the solution components. Thus each ant maintains its own parameter values and in turn uses these to adapt the parameter values[1]. Additionally, a separate pheromone matrix is kept so that the system may learn appropriate parameter values. Selection of a particular value is based exclusively on its pheromone value, as there is no heuristic analogue (such as the distance measure for the TSP) that will determine the quality of a particular parameter value.

Local as well as global pheromone updates are used to adjust the parameter pheromone levels. The γ (global pheromone decay) value of the ant that returns the best quality solution is used in these equations.

Each parameter must be given a suitable range in which its values can lie. It must be noted that the setting of the parameter bounds is not quite the same as setting parameter values. The mechanics of the self-adaptation are designed to quickly identify suitable regions and values for the problem being solved. Due to the nature of the ACS equations, the parameters q_0, ρ, γ are bound between the constant values of 0 and 1. Dorigo and Gambardella [5] have specified that sensible values for β solving the TSP and QAP (as minimisation problems), range between -5 and -1. The initial value of each parameter is chosen as the halfway point in its range. The parameter pheromone matrix is specified as $v(i,j)$, where i represents a particular parameter ($1 \le i \le V$), j is the range of values and V is the number of parameters (four in this case). Naturally j is discretised and is bounded between 1 and P. P is a constant the defines the granularity of parameter values. The parameter division, w_i, is chosen using analogues for the solution component selection equations for $v(i,j)$. The actual value of each parameter is then calculated according to Equation 1.

$$p_i = l_i + \frac{w_i}{P}(u_i - l_i) \qquad 1 \le i \le V \tag{1}$$

Where:

p_i is the value of the ith parameter,
l_i is the lower bound value of the ith parameter,
w_i is the discretised division chosen for the ith parameter and
u_i is the upper bound value of the ith parameter.

In terms of computational overhead, the use of this scheme adds little burden to the existing ACS search process. As the number of parameters and parameter divisions is fixed, the overall worst case complexity of the ant algorithm remains unaffected.

3   Computational Experiments

The computing platform used to perform the experiments is a 2.6GHz Red Hat Linux (Pentium 4) PC with 512MB of RAM[2]. Each problem instance is run across ten random seeds. Two groups of experiments are performed, one using

---

[1] It would be also be possible to allow only one ant to modify parameter values.
[2] The experimental programs are coded in the C language and compiled with gcc.

a control strategy (referred to as *control*) and the other using the solver that evolves its own parameter values (referred to as *evolveparam*). The control strategy simply allows the user to manually specify these values. In this case they have been chosen as $\{\beta = -2, \gamma = 0.1, \rho = 0.1, q_0 = 0.9\}$ because these values have been found to be robust by Dorigo and Gambardella [5]. The number of ants, m, and the number of parameter value divisions, P, remain constant at 10 and 20 respectively in these experiments for both *control* and *evolveparam*.

3.1   Algorithm Implementation

The implementations for the TSP and QAP differ slightly due to their respective problem definitions. Each problem, however, requires solutions to be permutations. The objective functions for the TSP and QAP are given by Equations 2 and 3 respectively.

$$\text{Minimise} \sum_{i=1}^{N-1} d(x(i), x(i+1)) + d(x(N), x(1)) \tag{2}$$

Where:

$d(i, j)$ is the distance between cities i and j,
$x(i)$ is the ith city visited and
N is the number of cities.

$$\text{Minimse} \sum_{i=1}^{M} \sum_{j=1}^{M} a(i, j) b(y(i), y(j)) \tag{3}$$

Where:

$a(i, j)$ is the distance between locations i and j,
$b(i, j)$ is the flow between facilities i and j,
$y(i)$ is the facility placed at location i and
M is the number of facilities/locations.

The main differences between the two problems can be characterised by the pheromone representation, visibility heuristic and the transition operators used within the local search phase. The pheromone used for the TSP is given by $\tau(i, j)$ where i and j both represent cities (consistent with Dorigo and Gambardella [5]). For the QAP, $\tau(k, l)$ is used in which k represents the location and l is the facility assigned to the location (in accordance with Maniezzo and Colorni [9]).

For the visibility heuristic, the TSP uses the distance measure between the current city and the potential next city. For the QAP, there are a number of choices for this heuristic, such as the use of a range of approximation functions and not using them at all [14]. The definition used here is given by Equation 4.

$$\eta(w, j) = \sum_{i=1}^{w-1} a(i, w) b(y(i), j) \tag{4}$$

Where:

w is the current location and

j is the the potential facility to assign to $y(w)$.

The local search phase is performed by each ant at every iteration of ACS. The transition operators used for the TSP and QAP are inversion and 2-opt respectively. These operators have been found by Randall and Abramson [11] to provide good performance for these problems. For each operator, the entire neighbourhood is evaluated at each step of the local search phase. The phase is terminated when a better solution cannot be found, guaranteeing a local minimum.

3.2   Problem Instances

Twelve TSP and QAP problem instances are used to test both the effectiveness of the *control* and *evolveparam* strategies. These problems are from TSPLIB [12] and QAPLIB [1] respectively and are given in Table 1.

Table 1. Problem instances used in this study. "Size" for the TSP and QAP is recorded in terms of the number of cities and facilities/locations respectively

Name	Size	Best-Known Cost	Name	Size	Best-Known Cost
hk48	48	11461	nug12	12	578
eil51	51	426	nug15	15	1150
st70	70	675	nug20	20	2570
eil76	76	538	tai25a	25	1167256
kroA100	100	21282	nug30	30	3124
bier127	127	118282	tai35a	35	2422002
d198	198	15780	ste36a	36	9526
ts225	225	126643	tho40	40	240516
gil262	262	2378	sko49	49	23386
pr299	299	48191	tai50a	50	4941410
lin318	318	42029	sko56	56	34458
pcb442	442	50778	sko64	64	48498

3.3   Results

The results are grouped in terms of the problem type. Tables 2 and 3 each show the results of the *control* and *evolveparam* strategies. A particular run of the ACS solver is terminated when a maximum of 3000 iterations of the algorithm has elapsed. This should give the ACS solver sufficient means to adequately explore the search space of these problems. In order to report the results, non-parametric descriptive statistics are used throughout. This is because the distribution of results is highly non-normal. The cost results are reported as the relative percentage deviation (RPD) from the best known solution cost. This is calculated as $\frac{E-F}{F} \times 100$ where E is the result cost and F is the best known cost. The runtime is recorded as the number of CPU seconds required to obtain the best solution within a particular run.

The results reveal that overall, *evolveparam* produces very good results compared to *control* on the QAPs and the smaller TSPs. While it sometimes finds

Table 2. The results of the *control* and the *evolveparam* strategies on the TSP instances. Each result is given by a percentage difference (RPD) between the obtained cost and the best known solution. Note that "Min", "Med" and "Max" represent the minimum, median and maximum respectively

Problem	control						evolveparam					
	Cost			Runtime			Cost			Runtime		
	Min	Med	Max	Min	Med	Max	Min	Med	Max	Min	Med	Max
hk48	0	0.08	0.08	0.04	1.29	16.32	0	0.04	0.08	0.25	3.41	45.02
eil51	0.47	2	2.82	0.08	0.49	40.69	0	0.23	3.52	0.07	9.75	26.42
st70	0.15	1.33	2.07	36.39	43.48	87.56	0	0.89	4.15	8.59	34.9	144.02
eil76	0.19	1.3	2.42	0.08	70.23	114.73	0	0	1.12	2.52	24.58	73.21
kroA100	0	0	0.54	8.67	34.58	192.17	0	0.42	6.32	2.45	195.93	587.38
bier127	0.32	0.72	1.87	58.64	253.21	855.28	0.51	3.63	10.45	5.13	64.14	1248.73
d198	0.16	0.33	0.6	154.53	1723.34	2422.52	0.44	1.69	6.59	10.05	1975.71	4378.37
ts225	0.63	1.15	1.93	513.65	3019.9	5484.59	1.9	3.53	10.84	611.94	9935.43	
gil262	0.63	2.02	2.65	404.07	1674.22	2726.53	0.59	2.78	6.31	121.42	5566.07	19385.71
pr299	0.42	0.92	2.68	10139.87	10794.69	13470.37	0.57	3.36	14.62	23.33	3209.58	19323.87
lin318	1.39	1.92	3	10388.72	14185.36	16090.43	2.19	6.27	15.24	36.11	6020.23	27762.95
pcb442	3.11	3.53	4.39	26903.59	38445.9	57383.08	2.33	12.1	13.13	63.4	199.06	52065.77

Table 3. The results of the *control* and the *evolveparam* strategies on the QAP instances

Problem	control						evolveparam					
	Cost			Runtime			Cost			Runtime		
	Min	Med	Max	Min	Med	Max	Min	Med	Max	Min	Med	Max
nug12	0	0	0	0	0.02	0.14	0	0	0	0	0	0.04
nug15	0	0	0	0	0.04	0.38	0	0	0	0	0	0.01
nug20	0	0	0	0.02	0.11	2.54	0	0	0	0.02	0.11	7.12
tai25a	0.4	0.63	0.89	2.56	14.45	47.33	0	0.65	1.66	0.09	21.17	50.07
nug30	0	0.07	0.39	1.59	26.79	100.71	0	0.07	0.39	0.27	11.7	32.19
tai35a	0.9	1.34	1.58	84.35	154.21	239.59	0.73	1.41	2.34	1.52	123.51	242.63
ste36a	0	0.39	0.8	33.76	117.47	246.01	0	0	1.6	0.32	63.84	220.34
tho40	0.01	0.16	0.38	16.57	167.27	340.83	0	0.19	1.13	1120.75	1265.13	1275.84
sko49	0.05	0.09	0.25	1.24	6.78	13.38	0.05	0.11	0.35	14.74	113.35	480.7
tai50a	1.87	2.21	2.42	68.91	1060.15	1191.32	1.91	2.18	2.75	1.3	187.79	802.48
sko56	0.02	0.24	0.55	178.62	734.91	1768.55	0	0.18	0.6	42.67	287.06	903.22
sko64	0	0.25	0.6	890.83	2362.08	3553.67	0	0.12	0.58	8.12	298.59	1039.06

better solutions on the large TSPs (such as pcb442) than *control*, frequently its average behaviour is not as good as *control*. Inspection of the runtimes for both strategies shows there is no consistent advantage of using one approach over another (i.e., *evolveparam* does not take considerably longer to run than *control*).

It is interesting to note that the parameter values produced by *evolveparam* often resemble those of Dorigo and Gambardella [5]. The characteristic values of each parameter, in terms of both problems, are summarised as follows:

- q_0: The values were generally between 0.8 and 0.9 for the TSP with the value 0.85 being frequently encountered. However, for the QAPs, lower values, between 0.2 and 0.5 were often evolved.
- β: Generally the range of values was between -2 and -2.5 (with -2.4 being common) for the TSP, while the QAPs tended towards values between -1 and -1.5. Occasionally for both problems, the values would be in the range -4.5 to -5.

- ρ and γ: Values for both parameters were generally in the range of 0.1 to 0.3. Occasionally, larger values (such as 0.85) would be produced for both TSP and the QAP.

A characteristic of *evolveparam* for the larger TSPs is that the values tend to converge to a stable set within approximately a hundred iterations of the algorithm. However, the results for all the QAPs indicated that parameter values were more likely to undergo changes throughout the search process. This correlation between increased performance and the dynamic variation of parameter values warrants further investigation. It would seem appropriate therefore to apply an explicit diversification strategy to the selection of parameter values to counter any premature convergence, particularly for TSPs.

4   Conclusions

Parameter tuning for meta-heuristic search algorithms can be a time consuming and inexact way to find appropriate parameter values to suit various classes of problems. An alternative approach has been explored in this paper in which the algorithmic mechanics of ACS are used to produce and constantly refine suitable values while problems are being solved. The results for the TSP and QAP instances used to test this notion suggest that its performance, in terms of solution costs and runtimes, is comparable to a standard implementation in which values from Dorigo and Gambardella [5] are used. In fact the performance, in terms of objective cost, is often an improvement over the control strategy. This may be attributed to the new solver's ability to tailor parameter values to the problem instance being solved. It is important to note, however, that the parameter values produced by the solver often resemble those of Dorigo and Gambardella [5]. An interesting exercise would be to compare *evolveparam* with a normal ACS that uses optimised parameters for each problem instance to further test its performance.

A concern remains that, at times, the parameter values converge rapidly. This was particularly so for the large TSPs. However, parameter values tended to vary more considerably on the QAPs. This coincided with an increased performance of *evolveparam* strategy over *control*. An extension to this work will be to apply some form of intensification/diversification strategy so that a greater range of values can be explored and/or limiting the pheromone trail values (as done in $\mathcal{MAX} - \mathcal{MIN}$ Ant System [15]). After these extensions are carried out, a more quantitative analysis of the parameter values (and the way in which they change throughout the search process) can be sensibly undertaken. Additionally, the issue of dynamic colony sizes (i.e., varying the value of m, the number of ants) is in the process of being investigated.

References

1. R. Burkard, S. Karisch, and F. Rendl. QAPLIB - A quadratic assignment problem library. *Journal of Global Optimization*, 10:391–403, 1997.

2. A. Colorni, M. Dorigo, and V. Maniezzo. An investigation of some properties of an "Ant algorithm". In *Palallel Problem Solving from Nature Conference (PPSN 92)*, pages 509–520, Brussels, Belgium, 1992. Elsevier Publishing.
3. M. Dorigo and G. Di Caro. The ant colony optimization meta-heuristic. In D. Corne, M. Dorigo, and F. Glover, editors, *New Ideas in Optimization*, pages 11–32. McGraw-Hill, London, 1999.
4. M. Dorigo and L. Gambardella. A study of some properties of Ant-Q. In *Proceedings of PPSN IV - Fourth International Conference on Parallel Problem Solving From Nature*, Berlin, Germany, 1996. Springer-Verlag.
5. M. Dorigo and L. Gambardella. Ant Colony System: A cooperative learning approach to the traveling salesman problem. *IEEE Transactions on Evolutionary Computation*, 1(1):53–66, 1997.
6. M. Dorigo, V. Maniezzo, and A. Colorni. The ant system: Optimization by a colony of cooperating agents. *IEEE Transactions on Systems, Man and Cybernetics - Part B*, 26(1):1–13, 1006.
7. T. Hendtlass and M. Randall. A survey of ant colony and particle swarm meta-heuristics and their application to discrete optimisation problems. In *Proceedings of the Inaugual Workshop on Artificial Life*, pages 15–25, Adelaide, Australia, 2001.
8. L. Ingber. Simulated annealing: Practice versus theory. *Computer Modelling*, 18:29–57, 1993.
9. V. Maniezzo and A. Colorni. The ant system applied to the quadratic assignment problem. *IEEE Transactions on Knowledge and Data Engineering*, 11(5):769–778, 1999.
10. M. Pilat and T. White. Using genetic algorithms to optimize ACS-TSP. In M. Dorigo, G. Di Caro, and M. Sampels, editors, *Third International Workshop on Ant Algorithms, ANTS 2002*, volume 2463 of *Lecture Notes in Computer Science*, pages 282–287, Brussels, Belgium, 2002. Springer-Verlag.
11. M. Randall and D. Abramson. A general meta-heuristic solver for combinatorial optimisation problems. *Journal of Computational Optimization and Applications*, 20:185–210, 2001.
12. G. Reinelt. TSPLIB - A traveling salesman problem library. *ORSA Journal on Computing*, 3:376–384, 1991.
13. A. Shmygelska, R. Aguirre-Hernandez, and H. Hoos. An ant colony optimization algorithm for the 2D HP protein folding problem. In M. Dorigo, G. Di Caro, and M. Sampels, editors, *Third International Workshop on Ant Algorithms, ANTS 2002*, volume 2463 of *Lecture Notes in Computer Science*, pages 40–52, Brussels, Belgium, 2002. Springer-Verlag.
14. T. Stützle and M. Dorigo. ACO algorithms for the quadratic assignment problem. In D. Corne, M. Dorigo, and F. Glover, editors, *New Ideas in Optimization*, pages 33–50. McGraw-Hill, London, 1999.
15. T. Stützle and H. Hoos. The $\mathcal{MAX} - \mathcal{MIN}$ Ant System and local search for combinatorial optimization problems. In S. Voss, S. Martello, I. Osman, and C. Roucairol, editors, *Meta-Heuristics: Advances and Trends in Local Search Paradigms for Optimization*, pages 313–329. Kluwer Academic Publishers, Boston, MA, 1998.

Particle Swarm Optimization Algorithm for Permutation Flowshop Sequencing Problem

M. Fatih Tasgetiren[1], Mehmet Sevkli[2],
Yun-Chia Liang[3], and Gunes Gencyilmaz[4]

[1] Fatih University, Department of Management
Buyukcekmece, 34500 Istanbul, Turkey
ftasgetiren@fatih.edu.tr
[2] Fatih University, Department of Industrial Engineering
Buyukcekmece, 34500 Istanbul, Turkey
msevkli@fatih.edu.tr
[3] Yuan Ze University, Department of Industrial Engineering and Management
No 135 Yuan-Tung Road, Chung-Li, Taoyuan County, Taiwan 320, R.O.C
ycliang@saturn.yzu.edu.tw
[4] Istanbul Kultur University, Department of Management
E5 Karayolu Uzeri, Sirinevler, Istanbul, Turkey
g.gencyilmaz@iku.edu.tr

Abstract. This paper presents a particle swarm optimization algorithm (PSO) to solve the permutation flowshop sequencing problem (PFSP) with makespan criterion. Simple but very efficient local search based on the variable neighborhood search (VNS) is embedded in the PSO algorithm to solve the benchmark suites in the literature. The results are presented and compared to the best known approaches in the literature. Ultimately, a total of 195 out of 800 best-known solutions in the literature is improved by the VNS version of the PSO algorithm.

1 Introduction

In a PSO algorithm, members of entire population are maintained through the search procedure so that information is socially shared among individuals to direct the search towards the best position in the search space. Each member is called *particle*, which moves around in the multi-dimensional search space with a velocity. Two variants of the PSO algorithm are developed, namely, PSO with a global neighborhood, and PSO with a local neighborhood. According to the global neighborhood, each particle moves towards its best previous position and towards the previous position of best particle in the whole swarm, called *gbest* model. On the other hand, according to the local variant, called *lbest*, each particle moves towards its best previous position and towards the previous position of best particle in its restricted neighborhood (Kennedy et al. [7]). Since PSO was first introduced by Kennedy and Eberhart [3,7], it has been applied to a wide range of applications in [1,6,10,15]. Tasgetiren et al. [19] developed a PSO algorithm for the single machine total weighted tardiness problem, where the

M. Dorigo et al. (Eds.): ANTS 2004, LNCS 3172, pp. 382–389, 2004.

smallest position value (SPV) rule, borrowed from the *random key representation* of Bean et al. [2], is presented to convert a position vector to a job permutation. Following the successful application above, this paper aims at employing PSO in solving the PFSP with makespan criterion.

In the PFSP, given the processing times p_{jk} for job j on machine k, and a job permutation $\pi = \{\pi_1, \pi_2, ..., \pi_n\}$, n jobs ($j = 1, 2, ..., n$) will be sequenced through m machines ($k = 1, 2, ..., m$) using the same permutation. For $n/m/P/C_{max}$ problem, $C(\pi_j, m)$ denotes the completion time of job π_j on machine m. The calculation of completion time for n-job m-machine problem is given as follows:

$$C(\pi_1, 1) = p_{\pi_1,1} \tag{1}$$

$$C(\pi_j, 1) = C(\pi_{j-1}, 1) + p_{\pi_j,1} \qquad j = 2, .., n \tag{2}$$

$$C(\pi_1, k) = C(\pi_1, k - 1) + p_{\pi_1,k} \qquad k = 2, .., m \tag{3}$$

$$C(\pi_j, k) = max\{C(\pi_{j-1}, k), C(\pi_j, k - 1)\} + p_{\pi_j,k} \; j = 2, .., n \; k = 2, .., m \tag{4}$$

Then the PFSP is to find a permutation π^* in the set of all permutations Π such that

$$C_{max}(\pi^*) \leq C(\pi_n, m) \;\; \forall \pi \in \Pi \tag{5}$$

Since the PFSP is NP-Hard [14], many heuristic algorithms have been presented in the literature. Some applications of constructive algorithms are presented in [4,9] while modern heuristics methods are presented in [5,11,12,13,16,17]. This paper is organized as follows. Section 2 gives the methodology of the proposed PSO algorithm, and computational results of test problems are shown in Section 3. Finally, Section 4 summarizes the concluding remarks.

2   PSO Algorithm for PFSP

We follow the *gbest* model of Kennedy et al. [7] with the inclusion of SPV rule in the algorithm. Pseudo code of the PSO algorithm is given in Figure 1.

The basic elements of PSO algorithm are summarized as follows:

Particle: X_i^t denotes the ith particle at iteration t, and is represented by n number of dimensions as $X_i^t = [x_{i1}^t, x_{i2}^t, ..., x_{in}^t]$, where x_{ij}^t is the position value of the ith particle with respect to the jth dimension.

Population: pop^t is the set of ρ particles in the swarm at iteration t, i.e. $pop^t = [X_1^t, X_2^t, ..., X_\rho^t]$

Permutation: π_i^t, denotes the permutation of jobs, which can be defined as $\pi_i^t = [\pi_{i1}^t, \pi_{i2}^t, ..., \pi_{in}^t]$, where π_{ij}^t is the assignment of job j of the particle i in the permutation π_i^t at iteration t with respect to the jth dimension.

Particle Velocity: V_i^t denotes the velocity of particle i at iteration t. It can be defined as $V_i^t = [v_{i1}^t, v_{i2}^t, ..., v_{in}^t]$, where v_{ij}^t is the velocity of particle i at iteration t with respect to the jth dimension.

Inertia Weight: w^t is a parameter to control the impact of the previous velocities on the current velocity.

Initialize parameters
Initialize population
Find permutation
Evaluate
Do {
        Find the personal best
        Find the global best
        Update velocity
        Update position
        Find permutation
        Evaluate
        Apply local search or mutation(optional)
} While (Termination)

Fig. 1. PSO Algorithm with Local Search for PFSP

Personal Best: P_i^t represents the best position of the particle i with the best fitness until iteration t. In a minimization problem with the objective function $f(\pi_i^t \leftarrow X_i^t)$, where π_i^t is the corresponding permutation of particle X_i^t, the personal best of the ith particle is obtained such that $f(\pi_i^t \leftarrow P_i^t) \leq f(\pi_i^{t-1} \leftarrow P_i^{t-1})$ To simplify, we denote the fitness function of the personal best as $f_i^p = f(\pi_i^t \leftarrow P_i^t)$. For each particle, the personal best is then defined as $P_i^t = [p_{i1}^t, p_{i2}^t, ..., p_{in}^t]$ where p_{ij}^t is the position value of the ith personal best with respect to the jth dimension.

Global Best: G^t denotes the best position of the globally best particle achieved so far in the whole swarm, which can be obtained such that $f(\pi^t \leftarrow G^t) \leq f(\pi_i^t \leftarrow P_i^t)$ for $i = 1, 2, ..., \rho$. To simplify, we denote the fitness function of the global best as $f^g = f(\pi^t \leftarrow G^t)$. The global best is then defined as $G^t = [g_1^t, g_2^t, ..., g_n^t]$ where g_j^t is the position value of the global best with respect to the jth dimension.

Termination Criterion: It is a condition that the search process will be terminated. It might be a maximum number of iteration or maximum CPU time to terminate the search.

Table 1. Solution Representation of Particle X_i^t

dimension(j)	1	2	3	4	5	6
Position x_{ij}^t	1.8	-0.99	3.01	-0.72	-1.20	2.15
Velocity v_{ij}^t	3.89	2.94	3.08	-0.87	-0.20	3.16
Job $\quad \pi_{ij}^t$	5	2	4	1	6	3

To construct a direct relationship between the problem domain and the PSO particle, we present n number of dimensions for n number of jobs. The particle $X_i^t = [x_{i1}^t, x_{i2}^t, ..., x_{in}^t]$ corresponds to the position values for n number of jobs in the PFSP. Table 1 illustrates the solution representation of particle where the position vector is converted to the job permutation according to the SPV rule in Tasgetiren et al. [19].

Initial position values are established between 0.0 and 4.0 randomly whereas initial velocities are generated between -4.0 and 4.0 randomly. Velocity values are restricted to some range, namely $v_{ij}^t = [v_{min}, v_{max}] = [-4.0, 4.0]$. Since the objective is to minimize the makespan, the fitness function is given as $f(\pi_i^t \leftarrow X_i^t) = C_{max}(\pi_n, m)$. To simplify, $f(\pi_i^t \leftarrow X_i^t)$ will be denoted as f_i^t.

The complete computational flow of the PSO algorithm is given as follows:

Step 1. *Initialization*: Set $t=0$, ρ=twice the number of dimensions. Generate ρ particles randomly as $\{X_i^0, i = 1, 2, ..., \rho\}$ where $X_i^0 = [x_{i1}^0, x_{i2}^0, ..., x_{in}^0]$. Generate the initial velocities randomly as $\{V_i^0, i = 1, 2, ..., \rho\}$ where $V_i^0 = [v_{i1}^0, v_{i2}^0, ..., v_{in}^0]$. Apply the *SPV* rule to find the permutation $\pi_i^0 = [\pi_{i1}^0, \pi_{i2}^0, ..., \pi_{in}^0]$ for $i = 1, 2, ..., \rho$. Evaluate each particle i using the objective function $f_i^0, i = 1, 2, ..., \rho$. For each particle i, set $P_i^0 = X_i^0$, where $P_i^0 = [p_{i1}^0 = x_{i1}^0, p_{i2}^0 = x_{i2}^0, ..., p_{in}^0 = x_{in}^0]$ together with its best fitness value, $f_i^p = f_i^0$ for $i = 1, 2, ..., \rho$. Find the best fitness value among the whole swarm such that $f_l = min\{f_i^0\}$ for $i = 1, 2, ..., \rho$, with its corresponding particle X_l^0. Set global best to $G^0 = X_l^0$ with its fitness value $f^g = f_l$

Step 2. *Update iteration counter*: $t = t + 1$

Step 3. *Update inertia weight*: $w^t = w^{t-1} \times \alpha$ where α is decrement factor

Step 4. *Update velocity*: $v_{ij}^t = w^{t-1}v_{ij}^{t-1} + c_1 r_1(p_{ij}^{t-1} - x_{ij}^{t-1}) + c_2 r_2(g_j^{t-1} - x_{ij}^{t-1})$ where c_1 and c_2 are social and cognitive parameters, r_1 and r_2 are uniform random numbers between (0, 1).

Step 5. *Update position*: $x_{ij}^t = x_{ij}^{t-1} + v_{ij}^t$

Step 6. *Find permutation*: Apply the SPV rule to find the permutation $\pi_i^t = [\pi_{i1}^t, \pi_{i2}^t, ..., \pi_{in}^t]$

Step 7. *Update personal best*: Each particle is evaluated by using the permutation to see if personal best will improve. That is, if $f_i^t < f_i^p$, $i = 1, 2, ..., \rho$, then personal best is updated as $P_i^t = X_i^t$ and $f_i^p = f_i^t$ for $i = 1, 2, ..., \rho$.

Step 8. *Update global best*: Find the minimum value of personal best. That is, $f_l^t = min\{f_i^p\}$, $i = 1, 2, ..., \rho$ and $l \in i; i = 1, 2, ..., \rho$. If $f_l^t < f^g$, then the global best is updated as $G^t = X_l^t$ and $f^g = f_l^t$.

Step 9. *Local search or mutation*: Five percent of the population is mutated at each iteration t, where the insert operator is applied to the position values of randomly chosen particles. Mutation is not used for PSO with VNS since VNS is employed instead.

Step 10. *Termination criterion*: If the number of iteration exceeds the maximum number of iteration, or maximum CPU time, then stop, otherwise go to step 2.

The local search in this paper is applied to the permutation π^t of the global best solution G^t at each iteration t *only once*, and based on the *insert+interchange* variant of the VNS method presented in [8]. For simplicity, PSO with the SPV rule is denoted as PSO_{spv}, and PSO with an extra VNS as PSO_{vns}. Pseudo code of the local search is given in Figure 2.

```
s₀ = πᵗ permutation of global best;
η = rnd(1,n);  κ = rnd(1,n);  η ≠ κ
s=insert(s₀,η,κ);
loop=0;
do{
        kcount=0; max-method=2;
        do {
            η=rnd(1,n); κ=rnd(1,n); η ≠ κ
            if (kcount=0) then s₁=insert(s, η, κ)
            if (kcount=1) then s₁=interchange(s, η, κ)
            if (f(s₁) < f(s)) then{
                    kcount=0;
                    s = s₁;}
            else  {kcount++;}
        while (kcount<max-method)
loop++;
while (loop<n*(n-1)
if (f(s) ≤ f(πᵗ)){
        πᵗ = s;
        repair(Gᵗ ); }
```

Fig. 2. Pseudo Code of VNS Local Search Employed

3   Experimental Results

The PSO algorithm is coded in C and run on an Intel P4 2.6 GHz PC with 256MB memory. In addition, a traditional GA is also coded in C to compare the performance of two population based methods. In GA, population size is twice the number of jobs, permutation representation is used, crossover and mutation probabilities are taken as 1.0 and 0.05 percent respectively. For the mate selection, one individual is selected randomly and the other one by tournament selection with size of 2 to carry out the two-cut crossover. Again the tournament selection with size of 2 is used for constructing the population for next generation. GA has also used the insert operator as a mutation scheme. For the PSO parameters, c_1, c_2, w^0, and α are taken as 2, 2, 0.9, and 0.975 respectively. Inertia weight is never decreased below 0.40. The solution quality is measured with the percent relative increase in makespan (Δ) with respect to the upper bounds provided by Taillard [18] or to the best known solutions provided by Watson et al. [20].

For the Taillard's benchmarks, we first compare the performance of PSO_{spv} with GA to see the impact of the SPV rule on the solution quality. Then the results for PSO_{vns} are discussed to show how the local search has improved the solution quality. To do so, we run each instance for 500 iterations and 10 replications are conducted for each instance to obtain the statistics. Table 2 shows the results of GA, PSO_{spv} and PSO_{vns}. In terms of both Δ_{avg} and Δ_{std}, PSO_{spv} produced slightly better results than GA. But GA is much faster than PSO_{spv}, especially for the larger problems because PSO_{spv} consumes more

CPU time to update all the dimensions at each iteration. Since the VNS local search is computationally expensive, we run the Taillard's 20 job problems for 100 seconds, 50 job problems for 200 seconds, 100 job problems for 300 seconds, 200 job problems for 400 seconds, and five replications are conducted for each instance to obtain the statistics for the PSO_{vns} algorithm. From Table 2, it is obvious to see that both Δ_{avg} and Δ_{std} are significantly reduced by the PSO_{vns} algorithm at the expense of increased CPU times.

Table 2. Performance of PSO on Taillard's Benchmark Suite

	GA			PSO_{spv}			PSO_{vns}		
	Δ_{avg}	Δ_{std}	t_{avg}	Δ_{avg}	Δ_{std}	t_{avg}	Δ_{avg}	Δ_{std}	t_{avg}
20x5	3.13	1.86	0.28	1.71	1.25	0.19	0.28	0.49	0.19
20x10	5.42	1.72	0.31	3.28	1.19	0.23	0.70	0.46	0.38
20x20	4.22	1.31	0.45	2.84	1.15	0.30	0.56	0.34	0.72
50x5	1.69	0.79	1.46	1.15	0.70	1.40	0.18	0.22	2.35
50x10	5.61	1.41	1.55	4.83	1.16	1.58	1.04	0.64	5.20
50x20	6.95	1.09	2.03	6.68	1.35	2.10	1.71	0.48	10.21
100x5	0.81	0.39	6.54	0.59	0.34	10.39	0.11	0.17	17.75
100x10	3.12	0.95	6.88	3.26	1.04	10.87	0.67	0.33	39.42
100x20	6.32	0.89	8.57	7.19	0.99	12.76	1.28	0.39	79.23
200x10	2.08	0.45	32.13	2.47	0.71	81.81	0.43	0.22	292.35
200x20	5.10	0.60	36.88	7.08	0.81	89.77	0.87	0.36	571.08
Mean	4.04	1.04	8.83	3.73	0.97	19.22	0.71	0.37	92.63

Since the VNS local search has a significant impact on the solution quality, we conducted further runs for the recent benchmark problems provided by Watson et al. [20]. We only present the results for their random (RN) and narrow random problems (NR), where 100 problems are generated for 20, 50, 100 and 200 jobs with 20 machines, and consider only two tabu search algorithms, namely, $TABU_{sh}$, and $TABU_{ns}$ where $TABU_{sh}$ stands for for tabu search with shift operator, and $TABU_{ns}$ for tabu search with critical path move operator. They allocated five minutes of CPU time to each of the 20x20 and 50x20 problems, and ten minutes of CPU time to each of the 100x20 and 200x20 problems, and conducted five replication for each instance. The best solution found by any algorithm is reported as the best-known solution which can be found in *http://www.cs.colostate.edu/sched/generator*. See Watson et al. [20] for the details. We also allocated the same CPU times and five replications are conducted for each instance. We present only the comparison of PSO_{vns} with their tabu search algorithms. The results for the Watson's benchmark suite are given in Table 3. When comparing PSO_{vns} with $TABU_{sh}$ and $TABU_{ns}$, $TABU_{ns}$ is the clear winner since 765 out of 800 best-known solutions (n_{bnw}) are obtained by $TABU_{ns}$ with a very low mean percent relative increase in makespan of $\Delta_{avg} = 0.05\%$. On the other hand, PSO_{vns} produced better results than $TABU_{sh}$ in terms of both the number of best-known solutions (n_{bnw})

Table 3. Performance of PSO_{vns} on Watson's Benchmark Suite

	$TABU_{sh}$		$TABU_{ns}$		PSO_{vns}		
	n_{bnw}	Δ_{avg}	n_{bnw}	Δ_{avg}	n_{bnw}	Δ_{avg}	n_{imp}
20x20-RN(1,99)	99	0.09	96	0.04	90	0.09	1
20x20-NR(45,55)	99	0.05	98	0.05	97	0.01	0
50x20-RN(1,99)	16	0.72	87	0.20	58	0.36	49
50x20-NR(45,55)	36	0.09	86	0.05	89	0.01	59
100x20-RN(1,99)	0	1.42	100	0.00	31	0.52	24
100x20-NR(45,55)	2	0.16	98	0.05	71	0.02	59
200x20-RN(1,99)	0	1.38	100	0.00	18	0.95	2
200x20-NR(45,55)	0	0.16	100	0.00	2	0.11	1
Mean	252	0.51	765	0.05	456	0.26	195

and the mean percent relative increase in makespan (Δ_{avg}) since PSO_{vns} is able to find the 456 out of 800 best-known solutions (n_{bnw}) with a mean percent relative increase in makespan of $\Delta_{avg} = 0.26\%$ whereas $TABU_{sh}$ is able to find 252 out of 800 best-known solutions(n_{bnw}) with the mean percent relative increase in makespan of $\Delta_{avg} = 0.51\%$. In addition, PSO_{vns} is able to improve a total of 195 out of 800 best-known solutions provided by Watson et al. [20]. New best-known solutions with their permutations can be found in *http://www.fatih.edu.tr/~ftasgetiren/pso.html*. It should be noted that we used approximately six times faster machine than the one used by Watson et al. [20].

4   Conclusions

To the best of our knowledge, this is the first reported application of the PSO algorithm to the Taillard's benchmarks for the PFSP with makespan criterion in the literature. In summary, the results presented in this paper are very encouraging and promising for the application of the PSO algorithm to combinatorial optimization problems. It should be noted that extensive use of the *insert+interchange* variant of the VNS local search had a significant impact on the the solution quality. However, a total of 195 out of 800 best-known solutions provided by Watson et al. [20] is improved by the VNS version of the PSO algorithm. From Table 2, we can also conclude that GA with the VNS local search could be producing similar results as PSO_{vns}. For the future work, this paper may be extended to the inclusion of the critical path move operator in the VNS structure to solve the Taillard's benchmarks for the best-known solutions provided by Nowicki and Smutnicki [11], Reeves and Yamada [13], and Grabowski and Wodecki [5].

References

1. Abido, M.A.: Optimal power flow using particle swarm optimization, Electrical Power and Energy Systems, 24, (2002), pp.563-571.

2. Bean, J.C.: Genetic algorithm and random keys for sequencing and optimization, ORSA journal on computing,6(2), (1994), pp.154-160
3. Eberhart, R.C., Kennedy, J.: A new optimizer using particle swarm theory, Proc. of the Sixth International Symposium on Micro Machine and Human Science, Nagoya, Japan, (1995), pp.39-43.
4. Framinan, J.M., Leisten, R.: An efficient constructive heuristic for flowtime minimisation in permutation flow shops, Omega, 31, (2003), pp.311-317
5. Grabowski, J., Wodecki, M.: A very fast tabu search algorithm for the permutation flowshop problem with makespan criterion, Computers and Operations Research, 31(11), (2004), pp.1891-1909.
6. Hu, X., Eberhard, R.C.,Shi, Y.: Swarm Intelligence for Permutation optimization: A case study of n-Queens problem,Proc. of the 2003 IEEE Swarm Intelligence Symposium,Indianapolis,IN, (2003), pp.243-246
7. Kennedy, J., Eberhart, R.C., Shi, Y.: Swarm intelligence, Morgan Kaufmann, San Mateo, (2001), CA
8. Mladenovic, N., Hansen, P.: Variable Neighborhood Search, Computers and Operations Research, 24, (1997), 1097-1100.
9. Nawaz, M., Enscore Jr, E., Ham, I.: A heuristic algorithm for the m-machine, n-job flow shop sequencing problem, Omega, 11(1), (1983), pp.91-95.
10. Onwubolu, G.C., Clerc, M.: Optimal operating path for automated drilling operations by a new heuristic approach using particle swarm optimisation, International Journal of Production Research, 42(3), (2004), pp.473-491.
11. Nowicki, E., Smutnicki, C.: A fast tabu search algorithm for the permutation flowshop problem, European Journal of Operational Research, 91, (1996), pp.160-175.
12. Osman, I., Potts, C.: Simulated annealing for permutation flow shop scheduling, Omega, 17(6), (1989), pp.551-557.
13. Reeves, C., Yamada, T.: Genetic algorithms, path relinking and the flowshop sequencing problem, Evolutionary Computation, 6, (1998) pp.45-60.
14. Rinnooy Kan, A.H.G: Machine scheduling problems: Classification, complexity and computations, Nijhoff, The Hague, (1976).
15. Salman, A., Ahmad, I., Al-Madani, S.: Particle swarm optimization for task assignment problem, Microprocessors and Microsystems, 26, (2003), pp.363-371.
16. Stützle, T.: Applying iterated local search to the permutation flowshop problem", Technical Report, AIDA-98-04, (1998), Darmstad University of Technology, Darmstad, Germany.
17. Stützle, T.: An ant approach to the flowshop problem,In Proc. of EUFIT'98, (1998), pp.1560-1564.
18. Taillard, E.: Benchmarks for basic scheduling problems, European Journal of Operational Research, 64, (1993), pp.278-285.
19. Tasgetiren, M.F., Sevkli, M.,Liang, Y.-C., Gencyilmaz, G.: Particle swarm optimization algorithm for single machine total weighted tardiness problem, Proc. of the 2004 Congress on Evolutionary Computation(to appear).
20. Watson, J.P., Barbulescu, L., Whitley, L.D., Howe, A.E.: Contrasting structured and Random Permutation Flowshop Scheduling Problems: Search space Topology and Algorithm Performance, ORSA Journal of Computing, 14(2), (2002), pp.98-123

Search Bias in Constructive Metaheuristics and Implications for Ant Colony Optimisation

James Montgomery[1,*], Marcus Randall[1], and Tim Hendtlass[2]

[1] Faculty of Information Technology, Bond University, QLD 4229, Australia
{jmontgom,mrandall}@bond.edu.au
[2] School of Information Technology, Swinburne University, VIC 3122, Australia
thendtlass@swin.edu.au

Abstract. Constructive metaheuristics explore a tree of constructive decisions, the topology of which is determined by the way solutions are represented and constructed. Some solution representations allow particular solutions to be reached on a greater number of paths in this construction tree than other solutions, which can introduce a bias to the search. A bias can also be introduced by the topology of the construction tree. This is particularly the case in problems where certain solution representations are infeasible. This paper presents an examination of the mechanisms that determine the topologies of construction trees and the implications for ant colony optimisation. The results provide insights into why certain assignment orders perform better in problems such as the quadratic and generalised assignment problems, in terms of both solution quality and avoiding infeasible solutions.

1   Introduction

An implicit assumption when using any metaheuristic is that it offers relatively unbiased access to all parts of the solution space, provided that deliberate bias towards "good" solutions is removed. That is, if search decisions are made in an undirected fashion (i.e., randomly) then each solution has approximately equal probability of being found. Of course, all common metaheuristics *are* biased towards solutions that appear promising. The neighbourhood in which constructive metaheuristics search forms a tree of constructive decisions, or *construction tree*. The nature of this construction tree can introduce a bias to the search. Constructive metaheuristics such as Ant Colony Optimisation (ACO), which use previously generated solutions to learn appropriate features to include in future solutions, operate essentially randomly during the early stages of a run. Thus, any bias that affects the undirected construction of solutions may reduce the effectiveness of the learning mechanism employed.

Research is this area is largely restricted to the work of Blum [2] (previously published in Blum and Sampels [3, 4] and Blum, Sampels and Zlochin [5]). These studies have provided an investigation of *model bias*, or the bias introduced by

---

* Corresponding author.

M. Dorigo et al. (Eds.): ANTS 2004, LNCS 3172, pp. 390–397, 2004.

the interaction between a particular pheromone representation and problem constraints. The focus on how frequently particular pheromone values are updated for a given problem–pheromone combination does not fully take into account underlying biases in the constructive approach that may unfairly advantage some solutions.

This paper investigates how a chosen solution representation and construction mechanism combine to bias constructive metaheuristics. The interaction between these biases and different pheromone representations is not discussed, although these issues are addressed by Montgomery [12]. Thus, it is largely complementary to the work of Blum [2]. Section 2 considers the underlying sources of bias that act on a constructive heuristic, while Section 3 summarises the experimental work undertaken to investigate this issue. Section 4 discusses some implications of these findings for the ACO approach.

2   Constructive Metaheuristics and Bias

Constructive metaheuristics take an empty solution ($\varnothing$) and successively add *solution components* to build a complete, typically feasible, solution to the problem at hand. The nature of the solution components depends on the problem specification. For instance, in the travelling salesman problem (TSP), solution components are typically cities (see e.g., [8]), which are successively added to $\varnothing$ to produce a complete solution (i.e., a permutation of the cities). Hence, these metaheuristics explore a tree of constructive decisions, or construction tree, where the root corresponds to $\varnothing$ and leaves correspond to complete (or infeasible partial) solutions. At the heart of such metaheuristics is the constructive algorithm (hereafter denoted by $\mathcal{A}$) used to define solution components and the mapping from sequences of solution components (i.e., paths in the construction tree) to solutions. The construction tree defined by $\mathcal{A}$ is denoted by $\mathcal{T}_{\mathcal{A}}$. A distinction must be made between a sequence of solution components $\mathfrak{s}$ and the solution to which it corresponds $s = X_{\mathcal{A}}(\mathfrak{s})$, where $X_{\mathcal{A}}$ is a mapping from sequences to solutions. The set of solutions corresponding to a solution is denoted by $X_{\mathcal{A}}^{-1}(s)$.

Five common combinatorial optimisation problems are used throughout this paper to illustrate bias in constructive metaheuristics: TSP [8], subset problems such as the multiple knapsack problem (MKP) [9], quadratic assignment (QAP) [11], generalised assignment (GAP) [10], and permutation scheduling problems such as the job and open-shop scheduling problems (JSP and OSP respectively) [3]. The constructive algorithm used in each is that used most commonly in the respective ACO algorithms for these problems.

When constructive decisions are undirected two primary sources of bias may be identified: *representation* bias and *construction* bias. The term *undirected* is used to indicate that the constructive algorithm in question makes each constructive decision probabilistically using a uniform random distribution over the available choices at each step. However, it is assumed that constraints are still enforced such that at each step no options are available that would make the

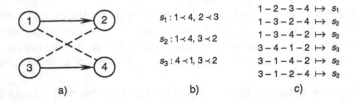

$s_1 : 1 \prec 4, 2 \prec 3$

$s_2 : 1 \prec 4, 3 \prec 2$

$s_3 : 4 \prec 1, 3 \prec 2$

$1-2-3-4 \mapsto s_1$
$1-3-2-4 \mapsto s_2$
$1-3-4-2 \mapsto s_2$
$3-4-1-2 \mapsto s_3$
$3-1-4-2 \mapsto s_2$
$3-1-2-4 \mapsto s_2$

a)   b)   c)

Fig. 1. Adapted from Blum and Sampels [4]. a) A small JSP instance; directed arcs indicate required order of operations within each job, dashed lines indicate operations that require the same machine. b) The three solutions to this problem described in terms of the relative order of operations that require the same machine, where $i \prec j$ indicates i is processed before j. c) The six possible solution representations an ACO algorithm may produce and the solutions to which they correspond.

partial solution infeasible. All constructive algorithms discussed in this section are considered to be undirected.

The nature of the problem representation used may allow distinct solutions to be represented in multiple ways. In many problems, the number of representations per solution is not uniformly distributed. Consequently, in ACO, some solutions will be overrepresented in the representation space in which ants search. For instance, solutions to many machine scheduling problems such as the JSP and OSP are represented as permutations of the operations to be scheduled. As solutions are uniquely described in terms of the relative order of operations that require the same machine (or that are part of the same job in the OSP), some operations may be exchanged in a permutation without changing the solution represented. Consider the JSP depicted in Fig. 1. There are three distinct solutions to this problem, yet six feasible representations. Of these, four correspond to solution s_2, which accordingly appears to have a $66\frac{2}{3}\%$ probability of being discovered by an undirected search, twice that expected if each distinct solution could be found with equal probability.

Definition 1. *A constructive process $\mathcal{A}$ applied to a given combinatorial optimisation problem is said to have a* representation bias *if there exist two solutions s_1 and s_2 such that $|X_{\mathcal{A}}^{-1}(s_1)| \neq |X_{\mathcal{A}}^{-1}(s_2)|$.*

Fig. 2 depicts the possible paths an ant may take to produce feasible solutions to the JSP described in Fig. 1 when representing solutions as permutations of the operations. The probability of choosing a particular component at a given node in the tree is inversely proportional to the number of alternative components at that node. If there are no infeasible sequences defined by $\mathcal{A}$, then the degree of branching at each level in the tree will be uniform within that level. This is the case, for example, in the TSP and QAP, where all permutations of cities or facilities represent feasible solutions. In problems where some solution representations correspond to infeasible solutions, the degree of branching within each level will not be uniform, as in the JSP. Consequently, solutions found on paths with less branching are more likely to be discovered than those on paths with

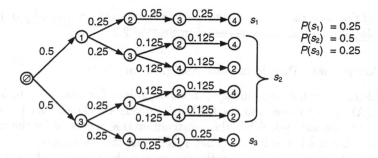

Fig. 2. Construction tree for small JSP instance (see Fig. 1 for problem description). Arcs are labelled with the probability of their being traversed given an undirected construction process. End points are labelled with the solution represented by that path. Aggregate solution probabilities appear on the right.

more branching. Hence, decisions that push the solution closer to the boundaries of feasible space lead to solution representations with a higher probability of being found. This constitutes a *construction bias*.

Definition 2. *A construction tree T_A has a* construction bias *if there exist two nodes in T_A such that their heights are equal yet their degrees are not equal.*

The two biases interact. For instance, in the JSP depicted in Fig. 1, although solution s_2 has $66\frac{2}{3}\%$ of all solution representations, Fig. 2 shows it has only a 50% probability of being found. This is because the distribution of paths in the tree corresponding to each representation determines the actual likelihood of constructing the solutions represented.

An interesting form of construction bias exists in problems where sequences are of variable length – shorter paths typically have fewer branching points and consequently have an elevated probability of being traversed. For example, in subset problems such as the MKP, feasible solutions are of varying lengths. Analysis of the construction tree for small instances confirms that a sequence's probability is inversely proportional to its length. However, such problems also have a representation bias (assuming sequences are built from the individual items that make up the subset) where each solution of size k is represented by $k!$ sequences in the construction tree. Further analysis reveals that the representation bias in these problems will always dominate.

In highly constrained problems such as the GAP partial sequences may be constructed that cannot be completed to produce a feasible solution. In the absence of backtracking these must then be abandoned[1]. Given such infeasible partial solutions are by definition found on shorter paths than feasible solutions, they have an elevated probability of being discovered. Furthermore, the more constrained a problem, the shorter will be the paths that lead to infeasible solutions. This issue is of particular interest in assignment problems, where the

---

[1] Some algorithms admit infeasible solutions, often relying on an accompanying local search to transform these into feasible solutions (e.g., Lourenço and Serra [10]).

nature of the constructive algorithm used can reduce the probability of reaching infeasible solutions, an issue discussed in Section 2.2.

2.1   Assignment Problems and Assignment Order

In problems involving assignment of *items* to *groups* (i.e., facilities to locations in the QAP, jobs to agents in the GAP) there exists a choice over the order in which items are assigned. While the construction tree is defined by the constructive algorithm used to solve the TSP, MKP and JSP, in assignment problems the choice of assignment order partly determines the topology of the tree. The solutions represented remain unchanged. Solutions that share much of their respective paths in the construction tree under one assignment order may diverge much earlier under another. In effect, the assignment order determines the constructive neighbourhood in which solutions are found.

In problems with no representation or construction bias, such as the QAP, different assignment orders will not alter solutions' respective probabilities. However, they will change which solutions are neighbours (in the construction tree) and hence may produce differing results in directed algorithms such as ACO. ACO algorithms for the QAP have taken a variety of approaches to determining assignment order, including selecting items randomly [15] and predetermining an order such that facilities with high flow requirements are assigned early with the intention they are assigned relatively central locations [11].

2.2   Assignment Order and Infeasible Space

In problems which do have infeasible solution representations, different assignment orders not only redistribute solutions in the construction tree, but may also alter their respective probabilities. Decisions that take partial solutions closer to infeasibility increase the probability of the (possibly infeasible) solutions to which they correspond. Consequently, the earlier such decisions are made, the greater the increase in probability. By altering the assignment order such decisions may be moved to any level in the construction tree, thereby altering their respective solutions' probabilities. In problems where feasible solutions are not guaranteed, choosing an appropriate assignment order is thus a significant issue.

A good construction tree topology for these problems is one in which the probability of discovering a feasible solution is maximised. However, finding an assignment order that produces such a tree is non-trivial. A static order fixes the topology of the construction tree and so also fixes the probability of infeasible representations. In contrast, a dynamic random order allows a range of construction trees to be used at various times in the algorithm and will likely be superior to an arbitrary static order. More commonly, assignment orders are chosen heuristically, such as in one ACO algorithm for the GAP developed by Randall [13].

Assigning highly constrained items early will likely produce trees with shorter paths leading to infeasible solutions, which consequently have an elevated probability relative to the proportion of paths they represent. However, using such an assignment order the total number of paths leading to infeasible solutions is reduced, as decisions that lead to infeasibility are consolidated nearer the root of the construction tree. Although a small number of infeasible paths does not guarantee a high probability of reaching a feasible solution our empirical studies show that trees with fewer infeasible paths typically have a higher probability of reaching a feasible solution than those with more infeasible paths. Indeed, a commonly used static assignment order assigns highly constrained items (e.g., jobs with high resource requirements in the GAP) early [10, 14]. Empirical testing reveals that simple heuristics for determining a static assignment order are often not the best. Accordingly, Costa and Hertz [6] consider a number of static and dynamic assignment orders for the graph colouring problem based on heuristics related to the principle of assigning more highly constrained items early.

3   Experimental Investigation

This section presents a summary of our empirical investigation of bias in five common combinatorial problems: Symmetric TSP [8], MKP [9], group-shop scheduling problem (GSP) (a generalisation of the JSP and OSP) [3], QAP [11] and GAP [10][2].

Complete exploration of the construction trees for small instances was performed, collecting data concerning the existing distribution of costs in each instance and the expected probability of solutions given an undirected constructive algorithm (referred to as $\mathcal{A}_{undir}$)[3]. Results were confirmed by comparing them against the frequency with which solutions were found by an implementation of $\mathcal{A}_{undir}$. The two analyses confirmed that the TSP and QAP have no inherent bias. In the MKP, only on the most trivial instance studied did a small solution show an elevated bias over others, further suggesting that in this problem the representation bias dominates.

The GSP and JSP show a mixture of representation and construction biases, with further analysis revealing that high probability sequences correspond to under-represented solutions and vice versa. This is an interesting property of these problems when solutions are represented as permutations of the operations, whereby those sequences with the highest probability are also those that are least able to be perturbed without changing the solution represented (see Montgomery [12] for full details).

Large problem instances were studied for all problems except the TSP using $\mathcal{A}_{undir}$[4]. On these instances, the effects of any underlying bias could not be observed, as $\mathcal{A}_{undir}$ can only take a sample from the very large space of sequences and solutions to these instances.

---

[2] Full results are available on request from the corresponding author.

[3] Small instances consisted of up to 14 cities in the TSP, 20 items in the MKP, 11 operations in GSP, 12 locations in the QAP, and 5 agents, 20 jobs in the GAP.

[4] Large instances consisted of up to 100 items in the MKP, 225 operations in GSP, 256 locations in the QAP, and 10 agents, 60 jobs in the GAP.

3.1   Assignment Order in the QAP and GAP

Given that the QAP has no inherent bias, changing the assignment order should have no impact on the frequency with which different solutions are found using $\mathcal{A}_{undir}$. Applying $\mathcal{A}_{undir}$ with a range of assignment orders to various QAP instances confirmed this.

In contrast, construction trees for all but the smallest contrived GAP instances contain infeasible partial solutions, the number and distribution of which is determined by the assignment order. Although the effects of representation and construction bias are not evident in large MKP and GSP instances, the effects of different assignment orders were observed in large GAP instances. Analysis was made of the construction trees for every static assignment order for a trivial instance with 3 agents and 8 jobs (adapted from a 5 agent, 15 job instance from the gap1 problem set, available from the OR-Library [1]). Across most assignment orders, the probability of producing infeasible solutions was elevated above what would be expected given the total number of infeasible paths in the construction tree, in some cases by more than 35%. This is in line with predictions made in Section 2.2. The probability of reaching a *feasible* solution when the number of paths leading to infeasible solutions is minimal was found to be better than the probability under the worst assignment order (13% versus 3%). However, the highest probability (34%) was shown under an another assignment order in which the number of such paths was low, but not minimal. Under this assignment order, the probability of producing infeasible solutions was *below* what would be expected given their number. Furthermore, the best assignment order did not have jobs in non-increasing order of constrainedness, which had a 16% probability of producing feasible solutions.

Sampling of randomly generated static assignment orders for instances from the gap1 (5 agents, 15 jobs) and gap2 (5 agents, 20 jobs) problem sets [1] reveals that very few assignment orders can produce a relatively high probability of finding feasible solutions. Comparison with the construction tree produced by a static assignment order in which jobs appear in non-increasing order of constrainedness indicates that it typically produces a higher probability of finding feasible solutions than 90% of alternative static orders for gap1 instances (although on one instance it was better than only 24% of sampled orders) and 80% of some gap2 instances. Interestingly, several different dynamic assignment orders based on the same heuristic of assigning more highly constrained jobs early did not produce a higher probability of feasibile solutions in general.

4   Implications for ACO

The nature of the constructive process is such that solution representation and problem constraints may introduce a bias into any search process that uses it. Constructive metaheuristics such as ACO use heuristic information and accumulated pheromone information to adapt their searches towards promising solutions. Additionally, almost all current ACO algorithms use a local search procedure to improve solutions produced by the ACO algorithm [7]. Each of these serves to counteract any underlying bias in the constructive process.

In assignment problems such as the QAP and GAP, the assignment order can alter the distribution of solutions within the construction tree. This may impact on the performance of ACO.

Infeasible space is a problem for any metaheuristic, but may be particularly so for constructive techniques as infeasible solutions individually have an elevated probability of being found. In problems involving assignment, the assignment order can alter the number and distribution of infeasible solutions in the construction tree. Heuristically determined static assignment orders can reduce the probability of reaching an infeasible solution, but may not necessarily produce the best construction tree.

References

1. J. E. Beasley. OR-library: Distributing test problems by electronic mail. *J. Oper. Res. Soc.*, 41:1069–1072, 1990.
2. C. Blum. *Theoretical and practical aspects of ant colony optimization.* PhD dissertation, Université Libre de Bruxelles, 2004.
3. C. Blum and M. Sampels. Ant colony optimization for fop shop scheduling: A case study on different pheromone representations. In *Proceedings of CEC 2002*, pages 1558–1563, 2002.
4. C. Blum and M. Sampels. When model bias is stronger than selection pressure. In *Proceedings of PPSN-VII*, volume 2439 of *Lecture Notes in Computer Science*, pages 893–902. Springer-Verlag, Berlin, 2002.
5. C. Blum, M. Sampels, and M. Zlochin. On a particularity in model-based search. In *Proceedings of GECCO 2002*, pages 35–42, New York, 2002.
6. D. Costa and A. Hertz. Ants can colour graphs. *J. Oper. Res. Soc.*, 48:295–305, 1997.
7. M. Dorigo and L. M. Gambardella. Ant colony system: A cooperative learning approach to the traveling salesman problem. *IEEE Trans. Evol. Comput.*, 1(1):53–66, 1997.
8. M. Dorigo, V. Maniezzo, and A. Colorni. The ant system: Optimization by a colony of cooperating agents. *IEEE Trans. Sys. Man Cyb. B*, 26(1):1–13, 1996.
9. G. Leguizamón and Z. Michalewicz. A new version of ant system for subset problems. In *Proceedings of CEC 99*, pages 1459–1464, 1999.
10. H. R. Lourenço and D. Serra. Adapative search heuristics for the generalized assignment problem. *Mathware Soft Comput.*, 9(2):209–234, 2002.
11. V. Maniezzo and A. Colorni. The ant system applied to the quadratic assignment problem. *IEEE Trans. Knowledge Data Eng.*, 11(5):769–778, 1999.
12. J. Montgomery. Search bias in constructive metaheuristics and implications for ant colony optimisation. Technical Report TR04-04, Faculty of Information Technology, Bond University, Australia, 2004.
13. M. Randall. Heuristics for ant colony optimisation using the generalised assignment problem. In *Proceedings of CEC 2004*, Portland, OR, USA, 2004.
14. C. Solnon. Ants can solve constraint satisfaction problems. *IEEE Trans. Evol. Comput.*, 6(4), 2001.
15. É. D. Taillard and L. M. Gambardella. Adaptive memories for the quadratic assignment problem. Technical Report IDSIA-87-97, IDSIA, 1997.

Task Oriented Functional Self-organization of Mobile Agents Team: Memory Optimization Based on Correlation Feature

Sorinel Adrian Oprisan

Department of Psychology, University of New Orleans
New Orleans, LA 70148
soprisan@uno.edu

Abstract. We developed a new optimization algorithm for multiagent coordination based on indirect and unsupervised communication. The mobile agents team task is simply searching and collecting "food items". The global coherent behavior is emergent, meaning that despite the fact that agents have no global map of the environment and do not directly communicate with each other they coordinate their behavior to achieve a global "goal". The coordinated response of the agents is the result of indirect communication via local changes in the environment. Each agent records the encountered objects in a memory register and by appropriate weighting of local perception the agent tries to estimate the global spatial distributions of the objects in the environment. The range of spatial and temporal indirect coupling among the agents is controlled via a "memory radius". We developed an optimized an algorithm that adapts the "memory radius" according to environment changes to minimize the computational time required to achieve the "goal" (piling the objects of the same kind together). Our optimization procedure is based on the correlation feature of the emergent pattern. The maximum speed of feature decreases leads to an optimized dependence of the "memory radius" on simulation time step. We derived also an analytic relationship between the "memory radius" and the time step based on the intermediate steady-state assumption. Numerical simulations confirmed that our analytic relationship coincides with the numerical optimization criterion based on the correlation feature.

1 Introduction

The capability of physical and chemical systems to generate order from chaos is usually termed *self-organization.* According to Farley and Clark *A self-organizing system is a system that changes its basic structure as a function of its experience and environment* [24].

Autonomous robotic systems receive a significant attention in last few years due to potential applicability to space missions, operations in hazardous environment, and military operations [20, 23]. Large distance or the robustness and versatility of tasks limed the applicability of traditional central controller or hierarchical methods. As a result, research in autonomously navigation promoted

M. Dorigo et al. (Eds.): ANTS 2004, LNCS 3172, pp. 398–405, 2004.
© Springer-Verlag Berlin Heidelberg 2004

the swarm-based robotics approach as a new and promising paradigm. A swarm of simple robots may have the advantages of the flexible, reliability, and fault-tolerance intrinsic to complex system. Controlled randomness or fluctuations in robots behavior, far from being harmful, may enhance the swarm's ability to explore new behaviors and new solutions. In addition, the emergent coherent behavior at the team level based on communication between the mobile agents via environmental cues reduces significantly the direct communications between agents. Central control is not well suited to deal with a large number of agents because the required resources scales proportionally to the square of the agents number, and because the failure of the central controller implies failure of the whole system.

We studied the emergent behavior at the team level for a simple task: search and collect "food-items" randomly distributed in a foraging area, and sort them in disjoint piles. We also studied different quantitative measures of organizational degree, and applied them to characterize the emergent behavior.

2   Computational Model of Stochastic Functional Self-organization

We modeled the dynamics of mobile agents by cellular automata (CA) under the following assumptions: (1) The environment was a two-dimensional rectangular lattice. The lattice sites are occupied by *objects* denoted with the letters a, b, c and so on. A free site is occupied by a ϕ-type object. (2) The agents that change the shape of the environment, called *robots* or *robots-like-ants (RLAs)*, perform a random walk like motion. The RLAs have the capability to distinguish between different object types, and to carry them around. (3) When a RLA reaches a new site it decides if there are conditions to put down the carrying object and to pick up the existing one. The swapping condition is

$$f_\alpha \geq f_\beta, \tag{1}$$

where f_α is the weighted frequency of the carried α-type object and f_β is the weighted frequency of the encountered β-type object. To compute the occurrence frequencies, the RLAs are endowed with memory registers [16,17]. The current implementation of our stochastic functional self-organization algorithm used separate registers for each object-type. The register content is a binary string with the following structure

$$s_{\alpha,\tau} : u_{\alpha,1} u_{\alpha,2} \ldots u_{\alpha,\tau}, \tag{2}$$

where

$$u_{\alpha,i} = \begin{cases} 1 \text{ if an } \alpha \text{ type object was encountered at step } i, \\ 0 \text{ otherwise,} \end{cases} \tag{3}$$

The following conservation rules are the result of (2) and (3): $\sum_{i=1}^{\tau} u_{\alpha,i} = n_\alpha$, for any $\alpha = \overline{1,T}$, where n_α is the total number of α-type object encountered and $\sum\limits_{\alpha=1}^{T} u_{\alpha,i} = 1$, for any $i = \overline{1,\tau}$, where T is the total number of distinct object-types.

Instead of using the binary strings stored in the registers, we associated to every object-type, at any instant τ, a real number called the weighted frequency

$$f_\alpha(\tau) = \frac{\sum\limits_{i=1}^{\tau} w(i)u_{\alpha,i}}{\sum\limits_{i=1}^{\tau} w(i)}, \tag{4}$$

where $w(i)$ is an appropriate weighting function. The weighting function we choose is

$$w(i) = \frac{1}{r^{i-1}}. \tag{5}$$

The above equation indicates that if $r \gg 1$, then the contribution of τ-th step (with $\tau \gg 1$) to present decision is not significant, and the RLA has a *short-type memory*. The limit case $r = 1$ corresponds to an infinite length memory. The case when $r < 1$ exacerbates the contribution of the τ steps with $\tau \gg 1$, and diminishes the contribution of the most recently ones [1, 13]. In our simulations the control parameter, called "memory radius" r was always greater than unit.

We previously found that to achieve the goal (sorting the objects in distinct piles according to their type) in shortest time the memory radius r should depend on the cluster dimension (i.e. aggregation stage) [16, 17]. Our numeric simulations showed that clustering process occurs for any value of the "memory radius" $r > 1$ and that the speed with which the system reaches his sorted steady state depends sensibly on r (see Fig.1).

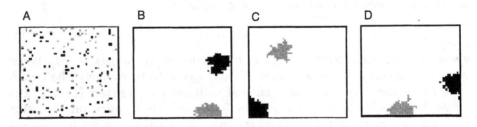

Fig. 1. Sorting time dependence on the memory radius. The initial configuration contains two object-types. The lattice dimensions were 50×50 with periodic boundary conditions. For each object-type, 100 objects were randomly distributed in the lattice. Sorting was performed by 30 RLAs, and the piling process was stopped when each object type was contained in a single connected cluster (A). The average time to reach final sorting strongly depends on the memory radius. For $r = 1.1$ stable clusters occurred after $2.5\,10^5$ steps (B) while for $r = 1.3$ they occurred after $6.5\,10^5$ steps (C), and for $r = 1.5$ after 10^6 steps (D).

2.1   Global Measure of Organizational Degree

The above described functional self-organization algorithm (see [1, 13, 16, 17] for details) requires a well-defined global cost-function to decide when the final aggregation stage was reached. We used a global measure of aggregation stage based on the texture analysis. Let $p(i, j)$ denote the normalized matrix of relative frequencies with which two cells, separated by distance d, occur on the image, one with gray tone i and the other with gray tone j. The matrix of gray-tone spatial-dependence frequencies depends on angular relationship between the neighboring cells. For simplicity, we refer here only to the horizontal gray-tone spatial-dependence matrix. The correlation feature gives global information about gray-tone linear dependence in the image, and is defined by

$$Correlation = \frac{1}{\sigma_x \sigma_y} \left(\sum_{i=1}^{N_s} \sum_{j=1}^{N_g} ijp(i,j) - \mu_x \mu_y \right), \tag{6}$$

where μ_x, μ_y are the mean and σ_x, σ_y are the standard deviations of the p_x and respectively p_y vectors defined by $p_x(i) = \sum_{j=1}^{N_g} p(i,j), \quad p_y(j) = \sum_{i=1}^{N_g} p(i,j)$. The noisy texture do not has any correlation between pixels and, therefore, the correlation feature has a low value.

3   Results

3.1   Intermediate Steady-States Hypothesis. Analytic Relationship Between Memory Radius and Aggregation Stage

A transition from a cluster containing p objects to p+1 objects at time step n is described by simultaneous conditions

$$f_\alpha^n < f_\beta^n, \quad f_\alpha^{n+1} > f_\beta^{n+1}, \tag{7}$$

where the first condition states that at time step n object type α cannot replace the object type β, whereas the second condition shows that at the time step $n + 1$ α-type could substitute β-type object. In a two object-types environment (black and white image) and a very long memory record ($\tau \to \infty$) the above conditions give

$$\frac{r(2 - r)}{2(r - 1)} < f_\alpha^n < \frac{r}{2(r - 1)}. \tag{8}$$

We assume that the aggregation process takes place progressively: starting with a random distribution of the objects, the RLAs put together, in a first stage, only two-object clusters. When this stage is completed, they form only three-object clusters and so on. This *intermediate steady-states* assumption is the bases of our quantitative relationship between the memory radius and the time dependent characteristic features. Based on above derived relationship, we concluded that the transition from p-objects clusters steady state to $(p + 1)$-objects clusters required $r \in (2^{1/(p+1)}, 2^{1/p})$.

To asses the relationship between the feature and the computational time we used the sum of the mean-free distance between clusters. For example $\lambda^{1\to 2} = \frac{\sqrt{N}}{2} \sum_{i=\frac{N_\alpha}{2}}^{N_\alpha} \frac{1}{\sqrt{4i-1}-1}$, where the superscript $1 \to 2$ design the initial and, respectively, final aggregation steady states. The proportionality constant depends on the successive steps required to visit the same cluster in order to transport it to its nearest neighbor. The generalized mean free path that takes into account the multiplicity of the stochastic paths is

$$\lambda^{k \to k+1} = (2k + 1)\frac{\sqrt{N}}{2} \sum_{i=\frac{N_\alpha}{k+1}}^{\frac{N_\alpha}{k}} \frac{1}{\sqrt{4i-1}-1}. \tag{9}$$

Substituting the sum in (9) by $\sum_{i=\frac{N_\alpha}{k+1}}^{\frac{N_\alpha}{k}} \frac{1}{\sqrt{4i}}$, which is appropriate if $N_\alpha \gg 1$, and using Euler formula $\sum_{i=a}^{b} f(i) \approx \int_a^b f(x)dx + \frac{1}{2}(f(a) + f(b)) + \frac{1}{12}(f'(b) - f'(a)) + \dots$, one gets

$$\lambda^{k \to k+1} \approx (2k + 1)\frac{\sqrt{N}}{2} \sum_{i=\frac{N_\alpha}{k+1}}^{\frac{N_\alpha}{k}} \frac{1}{\sqrt{4i}} \approx 2N(2k + 1)\frac{\sqrt{k+1} - \sqrt{k}}{\sqrt{k(k+1)}},$$

which simplifies to $k \approx \frac{2N}{\lambda}$. We also proved that in order to built a k-objects cluster the memory radius satisfies $r \in (2^{1/(k+1)}, 2^{1/k})$. Therefore, the last two relations give

$$r \propto \exp\frac{\lambda \ln 2}{2N}. \tag{10}$$

The above relationship agrees with our numerical findings.

3.2   Comparison with the Numerical Results

The quantitative feature called correlation is sensitive to aggregation stage and offers a quantitative meaning of this fuzzy concept (see Fig.2).

We found that it is computationally advantageous to start with a high value of the memory radius r. As a result, we obtained steep decrease of the feature, and unfortunately a rapid slowing-down of the algorithm, too. To avoid slowing-down the object's clustering we appropriately changed the "memory radius" every iteration. The optimal relationship between the memory radius value and the time step was found by monitoring the dynamics of the slop of the correlation feature. We assumed that optimality (the shortest computational time) could be achieved if the "memory radius" value at every time step ensures the

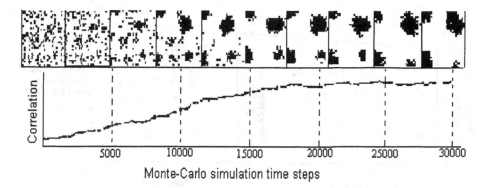

Fig. 2. Qualitative and quantitative aggregation degree measures Snapshots of the lattice every 5000 time-steps (upper panel) and the corresponding value of the correlation feature (lower panel). The environment was a rectangular 100×100 lattice, 10 % black objects' concentration, 40 RLAs, and a constant memory radius $r = 1.1$. The initial and final values of the correlation feature were analytically evaluated and used to quantify the distance between the current and the final aggregation stage.

fastest change in the feature (maximum slope). Therefore, we performed numerical experiments using different values of the "memory radius" starting with he same random initial configuration. For each "memory radius" value r the correlation feature versus time step n was computed (see lower panel in Fig.2, and the maximum value of its slope was found from the first derivative of the correlation feature. The maximum slope of the correlation feature versus "memory radius" is shown in Fig.3A. The relationship between the "memory radius" and the corresponding value of time step that ensures the maximum possible slope of the correlation feature is shown in Fig.3B.

The above computationally derived relationship agrees with our theoretical predicted relationship (10).

4 Conclusions

We proposed and studied a realistic computational model of self-organizing mobile agents team based on a first order recurrent memory function [1, 13, 16–18]. The present study addresses the problem of parameter optimization, particularly, its correlation length or *memory radius* dependence on the simulation time step. Based on *the intermediate steady-state* hypothesis we derived a theoretical time-dependent "memory radius" that leads to a minimum aggregation time. The intuitive idea behind our approach is that each two-, three-, four-, etc. clusters are metastable and the inherent stochastic behavior of the robot-like-ants (RLA) is the mechanism that drives the systems from intermediate (metastable) steady states to a final (stable) steady state. We found an analytical expression for the relationship between the "memory radius" and the simulation time based on the intermediate steady-states assumption.

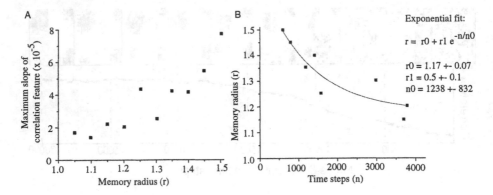

Fig. 3. Maximum slope of the correlation feature depends on the value of the memory radius r (panel A). The environment is a rectangular 100×100 lattice, with 10 % black objects' concentration, and 40 RLAs. To reduce the computational time, the value of the memory radius must ensure the maximum slop of the feature at every time step. **"Memory radius" (r) versus simulation time step (n) for maximum slope of the correlation feature.** (panel B) The best fit of the "memory radius" r dynamics is an exponential decaying function.

The analytical result was compared against experimental findings. For this purpose, we used the correlation feature as the quantitative measure that characterized the cluster aggregation stage. For a fix value of the "memory radius" we computed the maximum slope of the correlation feature and the corresponding time step at which that maximum slope was found. Our main assumption was that the minimum aggregation time could be attained if at every time step the "memory radius" maximizes the correlation feature slope. We found that the analytically derived optimization criterion and correlation feature-based numerical criterion are the same. This way we validated our theoretical assumption of intermediate steady-state aggregation mechanism, and provided a reliable numerical optimization criterion for functional self-organization algorithm.

Our study also pointed to limitations of the functional self-organization algorithm that must be addresses. For instance, stagnation is due to the lack of global knowledge at the agent level, and leads to a deadlocks (lack of progress in task completion). Secondly, we implicitly assumed in this study that all RLAs were identical units. In real applications of robotic teams we should consider variability in RLAs capabilities to correctly recognize and classify objects.

Above described mechanism of functional self-organization involves no learning. We are working to include learning capabilities to RLAs by allowing them to classify new object types and sort them.

References

1. Amarie D., S.A. Oprisan, M. Ignat, Phys. Lett. A **254**, 112 (1999).
2. Blacher S., F. Brouers, R. van Dyck, Physics A **197**, 516 (1993).

3. Blacher S., F. Brouers, R. Fayt, P. Teyssie, J. of Polymer Science: Part B: Polymer Physics **31**, 655 (1993).
4. Boccara N. , E. Goles, S. Martinez, P. Picco ed., *Cellular Automata and Cooperative Systems*, Kluwer Academic, Dorderecht, 1993.
5. J. von Neumann, in *Essays on Cellular Automata*, edited by A. Burks (Univ. of Illinois Press, 1970).
6. Deneubourg J. L., J.M. Pasteels, J.C. Verhaeghe, J. Theor. Biol. **105**, 259 (1983).
7. Deneubourg J. L., D. Fresneau, S. Goss, J.-P. Lachaud, I.M. Pasteels, Experientia Supplementum **54**, 177 (1987).
8. Deneubourg J. L., S. Goss, Ethology, Ecology and Evolution **1**, 295 (1989).
9. Deneubourg J. L., S. Gross, G. Sandini, F. Ferrari, P. Dario, in *Proc. Japan - USA Symposium on Flexible Automaton, 1990*.
10. Deneubourg J. L., S. Gross, N. Franks, A. Sandova-Franks, C. Detrian, L. Chretien, in *Proc. 1st Int. Conf Simulation of Adaptive behavior* (MIT Press, Cambridge, 1991, p. 356.
11. Deneubourg J. L., Insectes Sociaux **24**, 117 (1977).
12. Frisch U., B. Hasslacher, Y. Pomeau, Phys. Rev. Lett. **56**, 1505 (1986).
13. Giuraniuc C.V., S.A. Oprisan, Phys. Lett. A **259**, 334 (1999).
14. Haralick R., K. Shanmugan, I. Distein, IEEE Trans. System Man and Cybern. **3**, 610 (1973).
15. Neumann von J. , in *Theory of Self-Reproducing Automata*, edited by A.W. Burks (Univ. of Illinois Press, Urbana,1966).
16. Oprisan S.A., V. Holban, B. Moldoveanu, Phys. Lett. A **216**, 303 (1996).
17. Oprisan S.A., J. Phys. A: Math & Gen. *31*, 8451 (1998).
18. Oprisan S.A., A. Ardelean, P.T. Frangopol, Bioinformatics **60**, 1 (2000).
19. Perdang J.M., A. Lejeune, *Cellular Automata* (World Scientific, Singapore, 1993).
20. Rodin E.Y, S.M. Amin, in *Proc. Intelligent Symposium on intelligent control* (Arlington, VA, 1989), p. 366.
21. Toffoli T., Physica **10 D**, 117 (1984).
22. Tsallis C., Stariolo D. A., Physica A **233**, 395 (1996).
23. Unsal C., M.S. thesis, Virginia Politechniq Insitute, 1993.
24. Yovits M.C, G.T. Jacobi, G.D. Goldstein, *Self-organizing systems* (McGregor and Werner, Washington, 1962).

Towards a Real Micro Robotic Swarm

Ramon Estaña, Marc Szymanski, Natalie Bender, and Jörg Seyfried

Institute for Process Control and Robotics(IPR)
Kaiserstr. 12, Universität Karlsruhe(TH), D-76128 Karlsruhe, Germany
{estana,szymanski,nbender,seyfried}@ira.uka.de
http://microrobotics.ira.uka.de

Abstract. This paper introduces the I-SWARM project (Intelligent Small World Autonomous Robots for Micro-manipulation). This project aims at the development and production of a very large-scale artificial swarm (VLSAS) composed of several hundred micro-robots with a proposed size of $2 \times 2 \times 1$ mm. This will be the first realisation of a swarm with such a large number of robots. The extremely small size of the robots will impose severe limitations on their sensory and computational capabilities which is to be compensated by collective behaviour and emerging swarm effects. This paper presents an overview over one faces in the realisation of such a swarm based on extremely miniaturised robots. Further a new concept for an on-board ego-positioning system is proposed and some initial concepts for simulation and task planning in such a VLSAS are presented.

1 Introduction

Today, technology is still far away from the first "artificial ant" which would integrate all capabilities of these comparatively simple, yet highly efficient swarm building insects. Up to now, in classical micro robotics, highly integrated and specialised robots have been developed, which are able to perform micromanipulations controlled by a central high-level control system.

Realisations of small robot groups of some (i.e. 10, 20 or sometimes 50) robots are capable to mimic some aspects of such social insects, however, the employed robots are usually huge compared to their natural counterparts, and very limited concerning perception, manipulation and cooperation capabilities. In order to go beyond the current state of the art in the realization of robotic swarms, the European Commission has granted funding to the I-SWARM project. In this project, a swarm with mass-produced micro robots is planned with a number significantly higher than 50 robots: a swarm of several hundred up to 1,000 robots is envisioned.

This paper describes the underlying concepts of this project, the hardware restrictions one is facing in the realization of such a swarm, and work currently being carried out in the predecessor project being the basis for the realization of such a swarm. The latter project will establish a small cluster of (up to five) micro robots equipped with on-board electronics, sensors and wireless power

M. Dorigo et al. (Eds.): ANTS 2004, LNCS 3172, pp. 406–413, 2004.

supply, while the I-Swarm project aims at technological breakthroughs to facilitate the mass-production of micro robots, which can then be employed as a "real" swarm. These robot clients will all be equipped with limited, pre-rational on-board intelligence. The swarm will consist of a huge number of heterogeneous robots, differing in the type of sensors, manipulators and computational power. Such a robot swarm is expected to perform a variety of applications, including micro assembly, biological, medical or cleaning tasks.

2   Swarm Intelligence

The I-SWARM project will utilise approaches from swarm intelligence. Swarm intelligence (SI) is a property of systems of unintelligent agents of limited individual capabilities exhibiting collectively intelligent behaviour [2]. SI is based on the principles underlying the behaviour of natural systems consisting of many agents.

An agent represents an entity with sensor properties which undertakes simple processing in order to perform an action. Intelligent behaviour arises through indirect communication between the agents, this is known as the principle of stigmergy [3].

The application of SI principles to multi-robot systems has three advantages scalability, self organisation and robustness [10]. Scalability relates to agents and their control architecture. Self organisation is achieved by agents being dynamically added, removed or reallocated to different tasks without explicit reorganisation. Finally, robustness is possible through the design of simple agents.

To understand the principles of the SI we analyse the dynamics of swarm systems first on a microscopic level. This level contains the state diagram of the robot controller. The macroscopic level builds on the microscopic level.

On the microscopic level, we analyse the state-to-state transitions depending on the interaction of a robot with another teammate and with its environment. Each robot is represented by its own hybrid automat.

The macroscopic level places the whole robotic team with states which can be represented by the average of teammates at a certain time step in a particular state. The environment can be considered as a passive, shared resource whose modifications are generated by parallel actions of the robots or as active resource by involving this for task execution.

3   Navigation and Ego Positioning

Beside the very important issues regarding swarm behaviour, the navigation is also a necessary point within ant-like agents to think about. Positioning systems for small micro robots has been established successfully in [4], [6] and [5]. But within those micro agents, there has been plenty of space and power to assemble sensors for navigation on-board, and furthermore, calculation power was provided by external computers connected to a global camera for robot control, and a control software observed the whole scene.

But now, thinking towards very small agents in the size of ants, this is not longer possible. The robots should navigate by themselves with on-board possibilities. This is the motivation for the following thoughts regarding those issues.

Using the experiences out of the projects mentioned before, there could be a solution found by inverting and adapting the principles used therein. Therefore, an ego positioning using a so-called moiré-based navigation principle could be used.

As realised within [6], this kind of navigation is very precise. To realise this kind of positioning, we have to imagine a robot equipped with a very small 32 x 32 cell CCD-chip on its back (see Fig. 1). Assembled onto the surface of the CCD is a filter working as a semi-transparent Moiré-mark. We assume that the beamer is also able to project a so called **Moiré based potential field** onto the working platform. This potential field is working as a stationary Moiré grid firstly. Together with the circular Moiré filter on the robot's CCD-chip, a Moiré effect is caused directly on the CCD-chip surface. Analysing these Moiré effect by extracting the Moiré points enables the position calculation on-board by calculating the center position of the assembled Moiré filter grid related to the Moiré potential field.

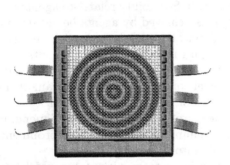

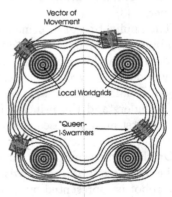

Fig. 1. Moiré-based Ego positioning robot

Fig. 2. Moiré Potential Field projected with a beamer

Sadly, solving this problem like described above makes a higher effort necessary in on-board computing. To avoid this disadvantage, we will use pre-solved, fixed wired equation solutions on-board. By preparing the equation system and leading it to a system which is already solved to "Zero" before, we can calculate a vector which points to the center to the potential field. These centers are the centers of the so called **Local World Grid**, like showed in Fig.2.

This method can not only be used during initialisation, but also for guiding the swarm, because the beamer is able to "drive" the local world grids over the working plate, so that at least the swarm can be guided from outside to find the working places on the platform. This supports the synergetic effect of the swarm behaviour itself.

Using this more intelligent agent together with the Moiré Potential Field, will lead to a scenario showed in Fig. 4. Herein, we can see the small robot clients walking along the area of the potential field using both, their tactile sensors for communication under each other and the potential field for navigation.

4 Sensing and Communication

One major task is involving sensors on the robots which act for different issues:

- Detecting obstacles
- Detecting other robots
- Detecting parts to handle
- Also usable for communication

As we will see, sensing and communication within these total new small world are nearly the same. Using light for communication is nearly impossible, because the light energy on the receiving side is much lower than usually used in remote controls: the transmitting power on the robots is about 15μW, and the light force at the receiving photo diode is well below 0.01 *lux*. Additionally, the light noise of the environment will disturb the weak light signals arriving at the robot.

As we know already, power is the main restriction for all on-board activities (we assume about 150μW overall power). For this, we have to use sensors with the following characteristics:

- very low energy consumption
- very lightweight
- easy to assemble
- very sensitive
- possibilities regarding communication

One way could be using small piezo foil strips. We will use strips with a size of $0,2mm * 0.05mm * 1mm$. The strips are fixed at the robot. Because of their tiny size, we have some advantages:

- easy to bend because of the material property; important because of the low robot weight (about $50mg$) and his small actuation forces
- very sensitive regarding bending
- good response characteristics
- can also be used as an actuator; small amplitudes $(0,05mm)$ with $1..2$ V supply voltage are thinkable

Within the scope of usage in the micro world, we can focus our sight onto three different working principles. Within this, we assume both, low energy receiving and transmitting signals:

- low frequency mode $\rightarrow$ detecting obstacles
- passive resonance mode $\rightarrow$ detecting parts
- active resonance mode $\rightarrow$ used for communication

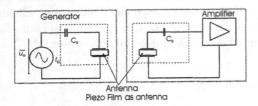

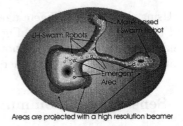

Areas are projected with a high resolution beamer

Fig. 3. Piezo strip used as communication tool

Fig. 4. Light trails together with local world grids

While the first two operation modes are easy to realise, the third and most auspicious mode is the **Active Resonance mode**. Within this, the operating point of the frequency generator is equal to the Eigenfrequency of the piezo strip. The complex source resistance value of the generator is the same as the complex load of the piezo stripe (Fig. 3). This requires a bi-directional signal handling. It can be established by using a bi-directional analogue multiplexer. One robot is the transmitter, the other robot is the receiver. Both robots switch periodically between both modes. They are not synchronised. As soon as one robot receives a signal within the defined spectra from another robot, a communication protocol is invoked, and the conversation can start.

Because this is a very sensitive measurement method, the piezo strips does not have to be in contact, because the electro magnetic antenna characteristics of the strip will be used. Estimated calculations gives us a maximum distance of about six Millimeters of communication distance.

5 Simulation Issues

Matarić discovered in her work [8] significant differences between simulated experiments and experiments performed on real robots. These differences arose from problems simulating each robot's sensors and actuators. Thinking about a swarm consisting of over hundreds up to thousand agents it may be unpredictable within a simulation what is going to happen in the real world. Especially regarding the strict constraints we encounter as a matter of size of the artificial ant and the micro scale effects caused by the robots' and objects' sizes. From those constraints not only the problem of handling and sensing objects in the micro world arises. But also the problem of implementing the software on physically restricted reprogrammable integrated circuits like FPGAs or microcontrollers with very limited capabilities. Not every behaviour model that was proposed up to now will fit physically on the chip.

Being restricted to a low computational and perceptional power on-board the robots and therefore often being unable to cope with unusual situations, it is important to design behaviours that are statistically stable[1] related to the

---

[1] The probability P_g for achieving the goal g over a set of runs must be almost one, $P_g \approx 1$.

subgoals and the main goal of the swarm. Facing these restrictions on-board-online learning as proposed by Mataric to overcome those problems may not be possible or will at least be hard to achieve.

Due to the lack of work on very large-scale artificial swarms, it is not foreseeable if those discrepancies between simulations and the real world may be scaled to a point, where a statistical forecast of the overall swarm behaviour is not possible. Otherwise in the lucky case one maybe reach a critical number of agents where such simulation problems will statistically not be relevant anymore. The answers to those arising questions are part of the research goals of the I-SWARM project.

From this follows that at the beginning of the project where no VLSAS is available there are two concepts for the sensor simulation to study such effects. On the one hand side we have poor sensor/actuator models and on the other hand, we have sensor/actuator models as exact as possible derived from experiments and FEA of small world effects. This division leads to a software architectural problem. For the poor sensor model it is possible to run a whole simulation on one host. Considering this from the view of distributed computing one is able to perform multiple simulations on multiple computers. But for the latter case it may be not possible to run one simulation on one host in an acceptable time. Therefore we need a well designed distributed parallel computing architecture for the simulation of one run.

Another fact that must be mentioned are the coherences between the robot hardware and the resulting models. the literature[1, 7] that show the tight connections between hardware and behaviour and vice versa. Often it is not quite clear if the hardware was adapted to proposed models or if the models were adapted to the hardware. In our case we know the stringent hardware and software constraints but we are not really able to fulfil the needs of both. This will be compensated by an iterative development process of both using a well designed multi-agent system simulation (MASS) tool that covers hardware and software simulation aspects.

6 Swarm Task Planning

As mentioned above, on-board-online behaviour learning may be not feasible due to the strict hardware constraints. The same applies for the manual behaviour design which can be applied for up to twenty agents but not for a VLSAS that ought to achieve several subtask, in order to fulfil the goal. The knowledge about the subtasks must be distributed over the whole swarm. It can be argued that all agents may have the same rules that lead to the fulfilling of the subtasks as it is shown in several papers. But this homogenous approach may need a complex set of behaviours and suitable sensor data processing that is too large to be implemented in our hardware. Therefore, developing a swarm that is heterogeneous by behavioural means is necessary. The allocation of behaviours to classes and the distinction how many different classes of behaviour arrangements are needed and how to distribute them over the swarm is a very hard optimisation problem with several degrees of freedom (DoF). Additional DoFs may be different start configurations or different parameterisations of the sensor/actuator models to state only a few more than mentioned above.

This leads to the following shortly outlined swarm task planning process that relies on a well defined simulation. Figure 5 shows a coarse sketch of the proposed architecture.

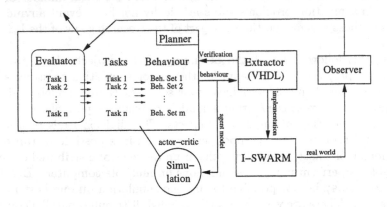

Fig. 5. Coarse sketch of the proposed architecture

Like in general centralised robotics it is useful to break up the task into several subtasks. This is done by the planner. Those subtasks must be identifiable by a swarm observing process called the observer. These subtasks are used in the optimisation process by the evaluator to evaluate the success of the swarm and the individuals. After evaluation, the planner computes a new set of behaviour sets and verifies their feasibility regarding the hardware through the extractor. If the generated behaviour set can be implemented on a micro-robot, the simulation will be started. The information one gets from the simulation will lead to an optimisation and adaption of the planner and therefore to a new set of behaviour sets. This process iterates until a set of behaviour sets is statistically stable enough to be implemented on the real swarm. One can say that the simulation acts as an "actor-critic" for the planner.

Using a simulation makes it also possible to implement "on-board-online" learning inside the virtual agents in order to search for a fixed behaviour policy that can be transfered to the robots. The "on-board-online" learning for instance can be done with behaviour based reinforcement learning as done by [8, 9]. Other optimisation techniques e.g. evolutionary algorithms, SVM or ANN need to be evaluated and may be combined inside the planner.

As we can see is it important to verify the set of behaviour sets on a broad range of different parameterised simulation runs to grant statistical stability for the swarm. Performing the "on-board-online" learning inside the virtual agents requires the use of a sensor/actor simulation as exact as possible in order to learn behaviour policies that can be transfered to the micro-robot.

Acknowledgement

This work is beeing performed at the Institute for Process Control and Robotics (IPR, head: Prof. H. Wörn). It is supported by the European Union within the "Beyond Robotics" Proactive Initiative - 6th Framework Programme: 2003-2007 (I-SWARM project, Project Reference: 507006).

References

1. R. Beckers, O. E. Holland, and J. L. Deneubourg. From local actions to global tasks: Stigmergy and collective robotics. In R. A. Brooks and P. Maes, editors, *Proceedings of the 4th International Workshop on the Synthesis and Simulation of Living Systems (Artificial Life IV)*, pages 181–189, Cambridge, MA, USA, July 1994. MIT Press.
2. G. Beni and J.Wang. Swarm intelligence in cellular robotic systems. *Proceedings of the NATO Advanced Workshop on Robots and Biological Systems*, 1989.
3. E. Bonabeau, M. Dorigo, and G. Theraulaz. *Swarm intelligence: from natural to artifical systems.* New York: Oxford Univ. Press, 1999.
4. R. Estaña and H. Wörn. Moiré-based positioning system for micro robots. In *SPIEs Int. Conference on Optical measurement Systems for Industrial Inspection III*, pages 431–442, Munich, Germany, 2003.
5. R. Estaña and H. Wörn. Moiré-based positioning system for small micromanipulation units. In *OPTO2004 Conference, Methods and Signal Processing*, 2004. (accepted).
6. J. S. et al. The MiCRoN EU-project, ist2001-33567, 2001.
7. R. C. Kube and E. Bonabeau. Cooperative transport by ants and robots. *Robotics and Autonomous Systems*, Volume 30(Issue 1/2):85–101, 2000. ISSN: 0921-8890.
8. M. J. Matarić. *Interaction and Intelligent Behavior.* PhD thesis, Massachusetts Institute of Technology, May 1994.
9. M. J. Matarić. Learning to behave socially. In *Proceedings of the third international Conference on Simulation of adaptive behavior*, volume 3 of *From animals to animates*, pages 454–462, April 1994.
10. T. White and B. Pagurek. Artificial life, adaptive behavior, agents, application oriented routing with biologically-inspired agents. *In Proceedings of the Genetic and Evolutionary Computation Conference (GECCQ-99)*, July 1999.

A Hybrid Ant Colony System Approach for the Capacitated Vehicle Routing Problem

Lyamine Bouhafs, Amir Hajjam, and Abderrafiaa Koukam

Laboratoire Systèmes et Transports, Equipe Systèmes Multi-Agents
Université de Technologie de Belfort-Montbéliard
90010 Belfort cedex, France
{Lyamine.Bouhafs,Amir.Hajjam,Abder.Koukam}@utbm.fr
http://www.utbm.fr/set

1 Introduction and Problem Definition

In this paper we propose a hybrid approach for solving the capacitated vehicle routing problem (CVRP). We combine an Ant Colony System (ACS) with a Savings algorithm and, then, we improve solutions by a local search heuristic. The CVRP is a class of well-known NP-hard combinatorial optimization problem, which can be formally defined as a complete graph $G = (V, E)$ where $V = \{0, \ldots, n\}$ is a set of vertices and E is a set of arcs [4]. The vertex $\{0\}$ represents the depot and the other vertices represent customers. The cost of travel between vertices i and j is denoted d_{ij} and represents the distance or the travel time. We assume that costs are symmetric(i.e. $d_{ij} = d_{ji}$), and an unlimited fleet of identical vehicles, each of capacity $Q > 0$, is available. Each customer i has a demand q_i, with $0 < q_i \leq Q$. Each customer must be served by a single vehicle and no vehicle can serve a set of customers whose demand exceeds its capacity. The task is to find a set of vehicle routes of minimum cost, where each vehicle used leaves from and returns to the depot. In the following, we explain our algorithm then, we give results and conclusions.

2 The Hybrid Ant Colony System for the CVRP

Our approach is based on a combination of Ant Colony System [3] and Savings algorithm [1, 5]. The Savings algorithm is the basis of most commercial software tools for solving vehicle routing problems in industrial applications [2]. It is initialized with the assignment of each customer to a separate tour. After that for each pair of customers i and j the following parametrical Savings measure is calculated: $\gamma_{ij} = d_{i0} + d_{0j} - g.d_{ij} + f.|d_{i0} - d_{0j}|$ where d_{ij} represents the distance between locations i and j, the index 0 denotes the depot, and g and f represent two parameters of the Savings heuristic. Thus, the values γ_{ij} contain the Savings of combining two customers i and j on one tour as opposed to serving them on two different tours.

The probabilistic rule used to construct routes is as follow. Ant k positioned on node i chooses the next customer j to visit with probability $p_k(i, j)$ given in

M. Dorigo et al. (Eds.): ANTS 2004, LNCS 3172, pp. 414–415, 2004.
© Springer-Verlag Berlin Heidelberg 2004

Equation (1). With F_k is the set of feasible customers that remain to be visited by ant k.

$$p_k(i,j) = \begin{cases} arg\ max\{(\tau_{ij})^\alpha.(\eta_{ij})^\beta.(\gamma_{ij})^\lambda\} & \text{if } q \leq q_0, j \in F_k \\ \dfrac{(\tau_{ij})^\alpha.(\eta_{ij})^\beta.(\gamma_{ij})^\lambda}{\sum_{u\ in F_k}(\tau_{uj})^\alpha.(\eta_{uj})^\beta.(\gamma_{uj})^\lambda} & \text{if } q > q_0, j \in F_k \\ 0 & \text{otherwise} \end{cases} \tag{1}$$

Where q is a random number uniformly distributed in $[0,\ldots,1]$ and q_0 is a parameter($0 \leq q_0 \leq 1$) which determines the relative importance of exploitation against exploration. τ_{ij} is the pheromone associated with arc (i,j), η_{ij} is the inverse of the distance between customer i and j, and γ_{ij} is the Savings function which leads to better results. The parameters α, β and λ determine the relative importance of the trails, distance and the Savings heuristic, respectively. The routes are improved by a local search heuristic(2-opt-heuristic). The rule of updating pheromone trail is the same like in [3].

3 Results and Conclusions

We have tested our approach on some CVRP benchmark instances available at http://neo.lcc.uma.es/radi-aeb/WebVRP. Preliminary tests of the algorithm show competitive results. New best solutions are found by our approach for some benchmarks. In this paper we have shown the possible improvements to Ant Colony System approach for the CVRP through the use of a problem specific heuristic, namely the Savings algorithm. The computational study performed shows the performance of our new approach. New best solutions are found for some benchmark problems in the literature. Furthermore, our approach is competitive with other meta-heuristics such as Tabu Search and Simulated Annealing.

References

1. Bullnheimer, B., Hartl, R. F. and Strauss, Ch.: An improved ant system algorithm for the vehicle routing problem. Annals of Operations Research **89** (1999) 319-328
2. Doerner, K, Gronalt, M., Hartl, R.F., Reiman, M., Strauss, Ch. and Stummer, M.: SavingsAnts for the Vehicle Routing Problem, in: Cagnoni et al. (Eds.): Application of Evolution-ary Computing, Springer LNCS 2279, Berlin/Heidelberg, (2002) 11-20.
3. Dorigo, M. and Gambardella, L.: Ant Colony System: A Cooperative Learning Approach to the Travelling Salesman Problem. IEEE Transactions on Evolutionary Computation, 1(1)(1997) 53-66
4. Letchford, A.N., Eglese, R.W., and Lysgaard, J.: Multistars, Partial Multistars and the Capacitated Vehicle Routing Problem (2001). Technical Report available at http://www.lancs.ac.uk/staff/letchfoa/pubs.htm
5. Paessens, H.: The savings algorithm for the vehicle routing problem, Eur. J. Oper. Res. **34** (1988) 336-344

A Swarm-Based Approach for Selection of Signal Plans in Urban Scenarios

Denise de Oliveira[1], Paulo Roberto Ferreira Jr.[1],
Ana L.C. Bazzan[1], and Franziska Klügl[2]

[1] Instituto de Informática, UFRGS, C.P. 15064, 91501-970, P. Alegre, RS, Brazil
[2] Dept. of Art. Int., Univ. of Würzburg, Am Hubland, 97074 Würzburg, Germany

1   Introduction

This paper presents a swarm approach to the problem of synchronisation of traffic lights in order to reduce traffic jams in urban scenarios. Other approaches for reducing jams have been proposed. A classical one is to coordinate or synchronise traffic lights so that vehicles can traverse an arterial *in one direction*, with a specific speed, without stopping. Coordination here means that if appropriate signal plans are selected to run at the adjacent traffic lights, a "green wave" is built so that drivers do not have to stop at junctions. This approach works fine in traffic networks with defined traffic flow patterns like for instance morning flow towards downtown and its similar afternoon rush hour. However, in cities where these patterns are not clear, that approach may not be effective. This is clearly the case in big cities where the business centres are no longer located exclusively downtown.

In [1] a multi-agent based approach is described in which each traffic light is modelled as an agent. This approach makes use of techniques of evolutionary game theory and has several benefits. However, payoff matrices have to be explicitly formalised by the designer of the system.

Therefore, an approach is presented here in which each crossing with a traffic light behaves like a social insect that grounds its decision-making on mass recruitment mechanisms found in social insects. The signal plans are seen as tasks to be performed by the insect without any centralised control or task allocation mechanism. Stimuli are provided by the vehicles that, while waiting for their next green phase, continuously evaporate some "pheromone".

We have simulated the flow of vehicles in an arterial and its vicinity under different situations. The results show that the swarm-based approach is more flexible: traffic lights adapt to the current flow of vehicles by selecting the appropriate signal plan, thus reducing the density in the arterial. As a measure of effectiveness we seek to optimise a measure of the density (vehicles/unit of length) in the roads. This is shown briefly in the next section due to space limitation. The extended version of this paper can be found at http://www.inf.ufrgs.br/~mas/traffic/siscot/downloads.html.

M. Dorigo et al. (Eds.): ANTS 2004, LNCS 3172, pp. 416–417, 2004.

2 Model of Task Allocation in the Traffic Scenario and Results of the Simulations

Our approach is based on the model of Theraulaz and colleagues for division of labour in colonies of social insects [2]. These concepts are used in our approach in the following way: each traffic light has different tendencies to execute one of its signal plans, according to the environment stimulus and particular thresholds. The liberated pheromone dissipates in a pre defined rate in time and its intensity indicates the vehicle flow in the street section. The traffic light stimulus is the average of the accumulated pheromone of all its lanes.

The accumulated pheromone in one lane, $d_{i,t}$ (for lane i at time t) is given by $d_{i,t} = \dfrac{\sum\limits_{i=0}^{n} \beta^{-i}(d_{i,t})}{\sum\limits_{i=0}^{t} \beta^{i}}$, where n is the time-window size, and β the pheromone dissipation rate of the lane. While the vehicles are waiting for the green light they remain releasing pheromone so the amount of pheromone increases.

The stimulus s of the plan j is thus: $s_j = \sum\limits_{i=0}^{n}((1 - \alpha)d_{in_{i,t}} + \alpha d_{out_{i,t}})\Delta t_i$ where n is the number of phases of the signal plan j, $d_{in_{i,t}}$ and $d_{out_{i,t}}$ are as defined above (accumulated pheromone trail in the incoming and outgoing lanes in phase i at time t), Δt_i is the time fraction of the phase i, and α is a constant employed to set different priorities to the input and output lane densities. A higher time interval indicates a phase change priority in the plan. Individuals may change task because high levels of stimulus related to a direction exceed their response threshold.

The scenario is part of a real network situated in the city of Porto Alegre (Brazil). The main street or arterial has eight traffic lights, each with two possible plans: signal plan 1 gives priority to the main direction (west-east) and it is synchronised with the adjacent traffic lights in this direction, while plan 2 is not synchronised. We have performed several comparisons: without any coordination between traffic lights, with fixed coordination, and with the swarm approach. The swarm approach achieves the best result when we use an exponential success function. Our approach was able to perceive the difference in traffic pattern (flow) and to adapt the traffic lights to priorise the higher flow. A total lack of synchronisation among the agents shows the highest densities levels.

References

1. A. L. C. Bazzan. Evolution of coordination as a metaphor for learning in multi-agent systems. In G. Weiss, editor, *DAI Meets Machine Learning*, number 1221 in LNAI, pages 117–136. Springer–Verlag, Berlin Heidelberg New York, 1997.
2. G. Theraulaz, E. Bonabeau, and J. Deneubourg. Response threshold reinforcement and division of labour in insect societies. In *Proceedings of the Royal Society of London B*, volume 265, pages 327–332, 2 1998.

Ant Colony Behaviour as Routing Mechanism to Provide Quality of Service

Liliana Carrillo, José L. Marzo, Lluís Fàbrega, Pere Vilà, and Carles Guadall

Institut d'Informàtica i Aplicacions (IIiA), Universitat de Girona (UdG)
Campus Montilivi, Av. Lluís Santaló s/n, Edifici P-IV, 17071 Girona, Spain
{lilianac,marzo,fabrega,perev,cguadall}@eia.udg.es

1   Introduction

Quality of Service (QoS) guarantees must be supported in a network that intends to carry real-time and multimedia traffic effectively. The effort of satisfying the QoS different requirements of these applications has resulted in the proposals of several QoS-based frameworks, such as Integrated Services (IntServ) [1], Differentiated Services (DiffServ) [2], and Multi-Protocol Label Switching (MPLS) [6].

The support of QoS services is underpinned by QoS Routing (QoSR), known as the task of determining paths that satisfy QoS constraints [4]. There exist many QoSR algorithms and also some first applications of ant-inspired algorithms [8]. In this paper, we propose a QoSR scheme called *AntNet-QoS*. Taking inspiration from Ant Colony Optimization (ACO) ideas, and in particular from the AntNet routing algorithm [5] (which was designed for best-effort IP networks), AntNet-QoS is intended to provide routes under QoS constraints in a DiffServ network, as well as under traditional best-effort IP routing conditions.

In the DiffServ model [2], QoS conditions to be satisfied are specified in Service Level Agreements (SLA) [7]. The SLA specify what Classes of Traffic (CoT) will be provided, what guarantees are needed for each class, and how much data will be sent for each class. Using the DiffServ model a variety of CoT can be defined based upon the DiffServ Code Point (DSCP) field [2]. DiffServ allows for a maximum of 14 different CoT. In AntNet-QoS, we will consider four of these: an Expedited Forwarding (EF) class to transport real-time traffic, two Assured Forwarding (AF_1 and AF_2) classes for traffic with different flavors for losses, and, as usual, a Best Effort (BE) class for traffic with no QoS requirements.

In the following, we describe the general characteristics of the algorithm and briefly discuss our future plans for implementation and testing.

2   AntNet-QoS – General Description

AntNet-QoS (more details are presented in [3]) is a QoSR algorithm based on the use of different classes of ant-like mobile agents for the different CoT with different QoS requirements. In AntNet-QoS, four different and independent classes of ant agents are used to sample and find paths that can provide the QoS associated to the specific CoT managed by its specific class of ant. The scheme we

M. Dorigo et al. (Eds.): ANTS 2004, LNCS 3172, pp. 418–419, 2004.

propose follows the DiffServ model, and in summary operates as follows: When a data session arrives at a node with certain parameters of requested QoS, defined in the DSCP field in the IP header of a packet, forward ants are generated reactively for sampling paths (like in AntNet), and to find a QoS-feasible path depending on the requirements of the CoT it belongs to. When an ant finds a path that satisfies the QoS required, a backward ant returns to the source to inform about the existing path, updating at each node along the return path two structures: the *QoS data model* and the *routing table*. In the case of Dynamic SLAs resources, it also has to interact with a signaling protocol (e.g. RSVP) to reserve the resources under the hop-by-hop assumption.

The QoS data model contains statistics collected by ants about residual bandwidth, delay, jitter and packet losses. The entries in the routing tables are pheromone variables that represent the estimated probability of getting the required QoS through each next hop for each forwarding class of data.

After paths are set up, the algorithm maintains them proactively: it periodically launches ants to monitor the QoS routes for each CoT and automatically balance the load over the network. This behavior is necessary to maintain the high quality needed for the EF class (with requirements of low delay and low jitter).

3 Future Work

The next step in this research is to program the scheme into a network simulator (e.g. using NS-2 or QualNet) and perform a series of realistic simulations. This will allow us to compare its performance with current state-of-the-art algorithms and give us better insight as to where improvements are possible.

References

1. Braden R., Clark D. and Shenker S. Integrated Services in the Internet Architecture: An Overview. RFC 1633, Internet Engineering Task Force, June 1994.
2. Carlson M., Weiss W., Blake S., Wang Z., Black D., and Davies E. An Architecture for Differentiated Services. RFC 2475, Internet Engineering Task Force, Dec. 1998.
3. Carrillo L., Marzo J.L., Vilà P., Fàbrega L. and Guadall C. A Quality of Service Routing Scheme for Packet Switched Networks based on Ant Colony Behavior. Accepted in SPECTS 2004, San José, California (EEUU). July 25 - 29, 2004.
4. Crawley, Nair R., Rajagopalan B., Sandick H. A Framework for QoS-based Routing. RFC 2386, Internet Engineering Task Force, August 1998.
5. Di Caro G. and Dorigo M. AntNet: Distributed Stigmergetic Control for Communications Networks. Journal of Artificial Intelligence Research 9. Pp. 317-365. 1998.
6. Rosen E., Viswanathan A. and Callon R. Multiprotocol Label Switching Architecture. RFC 3031, Internet Engineering Task Force, January 2001.
7. Xiao X. and Ni L.M. Internet QoS: a Big Picture. IEEE Network, pp. 8-18, March 1999.

Applying Ant Colony Optimization to the Capacitated Arc Routing Problem

Karl F. Doerner[1], Richard F. Hartl[1], Vittorio Maniezzo[2], and Marc Reimann[3]

[1] Department of Management Science, University of Vienna, Austria
{Karl.Doerner,Richard.Hartl}@univie.ac.at
[2] Department of Computer Science, University of Bologna, Italy
maniezzo@csr.unibo.it
[3] Inst. for Operations Research, Swiss Federal Inst. of Tech., Zurich, Switzerland
Marc.Reimann@ifor.math.ethz.ch

1 Extended Abstract

The Capacitated Arc Routing Problem (CARP) is a prototypical optimization problem asking a fleet of vehicles to serve a set of customer demands located on the arcs of a network. The problem is closely related to Vehicle Routing Problem (VRP), and in fact every CARP instance can be transformed into an equivalent VRP instance using a graph which has a number of nodes twice the number of customer arcs of the original CARP graph [1] plus one.

This report extends results previously reported [2] on our ACO algorithm for the CARP. Our algorithm is a Rank based Ant System algorithm which works directly on the CARP graph: ants decide which arcs to traverse in order to serve the next required arc. In determining the visibility we exploit shortest path information and take into account the orientation of traversing a required arc. As opposed to that, the pheromone information only deals with the sequence of compulsory arcs on each route, ignoring arc orientation. This leads to efficient data structures for both the construction graph and the pheromone information.

Heuristic information: The heuristic information is based on a shortest path distance matrix between every pair of nodes being endpoints of required arcs. Each arc (ij) can be traversed starting from node i or from node j. As the distance from the current node k to the two endpoints i and j of arc (ij) will in general be different, we have two heuristic values for each compulsory arc (ij).

Pheromone information: In the pheromone information a *node to arc* encoding is chosen using a matrix where the rows correspond to the nodes and the columns correspond to the required arcs. During the construction phase an ant positioned on a node, an endpoint of the just visited compulsory arc, will use the pheromone information to decide which arc should be visited next, and $\tau_k(ij)$ denotes the trail level on the connection traversing arc (ij) after node k.

The algorithm put forth in this work hybridizes ACO with local search, in order to fast achieve good quality solution. The local search routines used in this work are standard in the CARP literature and perform the following operations: Route breakup, Move of an arc, Swap of two arcs, Route crossover.

M. Dorigo et al. (Eds.): ANTS 2004, LNCS 3172, pp. 420–421, 2004.
© Springer-Verlag Berlin Heidelberg 2004

Table 1 presents the results of five runs on each of the 23 well known DeArmon benchmark instances, compared with the TSCS of Greistorfer (2003), *Carpet* of Hertz et al. (2000) and the Memetic Algorithm (MA) of Lacomme et al. (2004). For each instance we show the number of nodes, the number of arcs, a Lower Bound (LB) and the results of the algorithms. Our Ant System seems to produce solutions superior to TSCS and *Carpet* as measured by the average and worst deviation from the LB, but is outperformed by the MA. The Ant System is robust with respect to solution quality, we obtain an average worst case deviation of 0.72 % with a worst result of 4.62 %. These results are obtained at an expense of 202 seconds on average, a large number in comparison with the runtimes reported for the other approaches. The large averages come mainly from runs where the Ant System does not find the best solution but slowly converges to a poor one. We are currently working on eliminating this effect.

Table 1. Detailed computational results for the DeArmon instances

				Ant System				TSCS (1 run)		Carpet (1 run)		MA (1 run)	
				best	avg.	worst	time	best	time	best	time	best	time
Probl.	Nodes	Arcs	LB	(in %)	(in %)	(in %)	(in s)	(in %)	(in s)	(in %)	(in s)	(in %)	(in s)
1	12	22	316	0	0	0	0.0	0	1.7	0	17.1	0	0.0
2	12	26	339	0	0	0	24.6	0	13.2	0	28.0	0	0.4
3	12	22	275	0	0	0	0.4	0	1.0	0	0.4	0	0.1
4	11	19	287	0	0	0	0.0	0	3.3	0	0.5	0	0.0
5	13	26	377	0	0	0	10.6	0	19.3	0	30.3	0	0.1
6	12	22	298	0	0	0	2.0	0	12.7	0	4.6	0	0.2
7	12	22	325	0	0	0	0.0	0	0.1	0	0.0	0	0.1
10	27	46	344	2.33	2.56	2.91	1070.2	1.16	166.3	2.33	330.6	1.74	0.7
11	27	51	303	3.30	3.96	4.62	936.0	5.94	116.0	4.62	292.2	0	7.1
12	12	25	275	0	0	0	21.6	0	16.3	0	8.4	0	0.1
13	22	45	395	0	1.82	3.04	636.0	0	21.6	0	12.4	0	1.3
14	13	23	450	1.78	1.78	1.78	16.2	1.78	1.0	1.78	111.8	1.78	0.1
15	10	28	536	0.75	1.34	1.49	81.4	1.49	4.7	1.49	13.1	0	7.4
16	7	21	100	0	0	0	0.0	0	0.3	0	2.6	0	0.1
17	7	21	58	0	0	0	0.0	0	0.3	0	0.0	0	0.0
18	8	28	127	0	0	0	61.8	0	16.5	0	9.2	0	0.1
19	8	28	91	0	0	0	0.0	0	0.1	0	0.0	0	0.1
20	9	36	164	0	0	0	3.2	0	5.4	0	1.5	0	0.1
21	8	11	55	0	0	0	0.0	0	0.0	0	1.1	0	0.0
22	11	22	121	0	0	0	94.6	0	67.9	0	51.5	0	0.3
23	11	33	156	0	0	0	180.8	0	68.7	0	6.1	0	0.2
24	11	44	200	0	0.4	1	860.0	0	40.1	0	18.3	0	3.4
25	11	55	233	0.86	1.37	1.72	645.8	0.86	1.7	0.86	186.3	0	51.2
Avg				0.39	0.58	0.72	202.0	0.49	25.1	0.48	49.0	0.15	3.2

References

1. R. Baldacci and V. Maniezzo. Exact methods based on node routing formulations for arc routing problems. Technical Report ULBCS-2004-10, Department of Computer Science, University of Bologna, 2004.
2. K. Doerner, R. Hartl, V. Maniezzo, and M. Reimann. An ant system metaheuristic for the capacitated arc routing problem. In *Proceedings of the Fifth Metaheuristics International Conference (MIC 2003)*. 2003.

Dynamic Optimization Through Continuous Interacting Ant Colony

Johann Dréo and Patrick Siarry*

Université Paris XII Val-de-Marne, Laboratoire d'Étude et de Recherche en
Instrumentation Signaux et Systèmes (LERISS), 94010 Créteil, France
{dreo,siarry}@univ-paris12.fr

1 Extended Abstract

In recent past, optimization of dynamic problems has evoked the interest of the researchers in various fields which has resulted in development of several increasingly powerful algorithms. Unlike in static optimization, where the final goal is to find the fixed global optimum, in dynamic optimization the aim is to find and follow the evolution of the global optimum during the entire optimization time.

Dynamic optimization has been recently studied under the name of "real-time optimization" (RTO) mainly in the field of model-based real-time optimization of chemical plants [3] and optimization of "non-stationary problems" or "dynamic environments" [1]. There are many fields that deal with dynamic optimization, our aim is to focus on the use of metaheuristics for the optimization of difficult continuous dynamic problems. Only a few metaheuristics have been studied on such problems, mainly genetic algorithms, particle swarm optimization, etc. Ant colony algorithms have been employed on discrete dynamic problems but not on continuous ones.

We propose in this paper a new algorithm inspired from the behaviour of ant colonies. The so-called "Dynamic Hybrid Continuous Interacting Ant Colony" ($DHCIAC$) algorithm is population-based, articulates itself around the notions of heterarchy and communication channel (see [2] for an exact definition of these terms), and uses simplex algorithm as local search. Two versions of this algorithm are proposed, using two different approaches for optimizing dynamic problems.

To test this algorithm, we have worked out a set of dynamic benchmark problems, that attempts to cover a large scale of different dynamic difficulties. We have determined twelve characteristics that can be used: continuity, convexity, symmetry of the search space, symmetry of the function, quadratic nature, modality (i.e. presence of one or several optima), periodicity, number of local optima, number of dimensions, constraints, optimum position and the shape of the basin of attraction. One example of problem, extracted from this set, is presented in Figure 1.

We can distinguish two main components of a dynamic optimization problem: the first one can be named the structure of the objective function, and the second one is the evolution of this structure. Respectively called the *structure function*,

---

* Corresponding author.

M. Dorigo et al. (Eds.): ANTS 2004, LNCS 3172, pp. 422–423, 2004.
© Springer-Verlag Berlin Heidelberg 2004

$D(x, t) = S(x, T(t))$, with:

- *Periodical* time function: $T(t) = 8 \cdot \cos(t + 0.5)$,
- *Morrison* structure function:

  $D(x_1, x_2, t) = \sum_{i=1}^{5} \min\left(-H_i + R_i \cdot \left((x_1 - P_{i,1})^2 + (x_2 - P_{i,2})^2\right)\right)$

  with: $-10 < x_1, x_2 < 10$, $H = \{1, 3, 5, (7 + T(t)), 9\}$, $R = \{1, 1, 1, 1, 1\}$ and

  $P = \left\{ \begin{array}{cccc} 2.5 & -2.5 & -5 & T(t) & -4 \\ -2.5 & 2.5 & 5 & T(t) & -4 \end{array} \right\}$,

- Target: optimum #4 position and value,
- Characteristics: continuous, symmetric search space, multi-modal, convex optimum attraction basin.

Fig. 1. Example of a dynamic benchmark problem: O(P+V)P (variations of the Optimum Position and Value, Periodically).

and the *time function*, these components are connected through the *variation target*. The proposed benchmark set is based on these notions.

The *DHCIAC* algorithm uses two communication channels to gather informations about the problem to be optimized: a *stigmergic channel*, calling upon the spots of pheromone, deposited within the search space, and a *direct channel*, implemented in the form of messages exchanged between pairs of ants. One additional characteristic of *HCIAC* is its hybridization with the Nelder-Mead simplex local search.

Two different versions of the *DHCIAC* algorithm have been developed so far: $DHCIAC_{find}$ and $DHCIAC_{track}$. Indeed, there exist two different strategies to find a global optimum when optimizing a dynamic problem: find the optimum very quickly as soon as it changes or follow the evolution of the optimum. Thus, $DHCIAC_{find}$ is aimed at searching new optima appearing during the optimization time, so that the correct parameter set can show a "diversification" behaviour. On the contrary, $DHCIAC_{track}$ tries to follow an evolving optimum, so that the better behaviour is an "intensification" one.

We have found that the two different strategies are best suited for distinct classes of dynamic problems, namely value-based problems for $DHCIAC_{find}$ and position-based problems for $DHCIAC_{track}$. Moreover, *DHCIAC* seems to be particulary useful for slow dynamic problems with a lot of local optima where the values and the positions of the optima are changing.

The efficiency of DHCIAC is demonstrated through numerous tests, conducted involving one new benchmark platform.

References

1. J. Branke. *Evolutionary Optimization in Dynamic Environments*, volume 3 of *Genetic Algorithms and Evolutionary Computation*. Kluwer, 2001.
2. J. Dréo and P. Siarry. Continuous Interacting Ant Colony Algorithm Based on Dense Heterarchy. *Future Generation Computer Systems*, 2004. to appear.
3. Q. Xiong and A. Jutan. Continuous optimization using a dynamic simplex method. *Chemical Engineering Science*, 58(16):3817–3828, 2003.

Dynamic Routing in Traffic Networks Using AntNet

Bogdan Tatomir[1,2], Ronald Kroon[1], and Leon Rothkrantz[1,2]

[1] Delft University of Tehnology, Mekelweg 4, 2628 CD Delft, The Netherlands
{B.Tatomir,R.Kroon,L.J.M.Rothkrantz}@ewi.tudelft.nl
[2] DECIS Lab, Delftechpark 24, 2628 XH Delft, The Netherlands

1 State of the Art

Road traffic is getting busier and busier each year. Everyone is familiar with traffic congestion on highways and in the city. And everyone will admit that it is a problem that affects us both economically as well as mentally. Furthermore finding your way in an unknown city can be very difficult even with a map. Navigation systems like CARiN can help in such cases. These systems display the route to be followed when the user has set his destination. Most current systems are based on static information. The latest versions are also able to use congestion information to avoid trouble spots. But such information is only available for highways and not in a city. This paper addresses the dynamic routing of traffic in a city. We want to set up a routing system for motor vehicles that guides them through the city using the shortest way in time, taking into account the load on the roads. Furthermore we want the routing system to be distributed, for more robustness and load distribution. Applied in packet switched networks, the Ant-based algorithms have proven to be superior to other distributed routing algorithms. In this paper we will apply a variant of such an algorithm (AntNet), to a traffic network in a city.

2 Research

To route traffic dynamically through a city we need dynamic data about the state of the traffic in the city. The vehicles, themselves, can provide the system with information about the path they followed and the time it took them to cover it. The current technology enables to fix the position of a vehicle with an accuracy of a few meters. That position can be communicated to the system along with the covered route. We will use this information as dynamic data. But of course the model is open for additional types of dynamic data. The information from the vehicles is handled by a separate part of the routing system, called the timetable updating system. This subsystem takes care that the information is processed for use by the ant-based algorithm. This way one vehicle drives a certain route and sends its performance to the routing system. Another vehicle is able to use that information to choose the shortest route.

M. Dorigo et al. (Eds.): ANTS 2004, LNCS 3172, pp. 424–425, 2004.

The most important problem of this research is solved by the timetable updating system and the route finding system. These two subsystems together form the routing system (see figure 1). The reason why we need the timetable updating system is the following. The route finding system needs information about the state of the network. This information is provided by the timetable updating system. It is the time it takes to cover a road. Vehicles send information about their covered route to the timetable updating system. From that information this system computes the travel-times for all roads and stores it in the timetable in the memory. Besides the timetable also a history of measurements is stored in the memory.

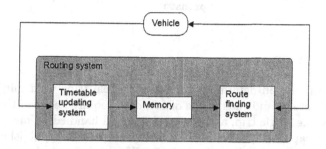

Fig. 1. Design of the Routing system

The route finding system uses the information in the timetable to compute the shortest routes for the vehicles. When the vehicle requests route information, the route finding system sends this information back to the vehicle. We used the idea of emergent behavior of natural ants to build the routing tables in the traffic network of a city, i.e. the composition of the roads and their intersections.

This network is represented by a directed graph. Each node in the graph corresponds to an intersection. The links between them are the roads. Mobile agents, whose behavior is modelled on the trail laying abilities of natural ants, replace the ants. The agents move across this virtual traffic network between randomly chosen pairs of nodes. As they move, pheromone is deposited as a function of the time of their journey. That time, provided by the timetable updating system, is influenced by the congestion in the real traffic. They select their path at each intermediate node according to the distribution of the simulated pheromone at each node. Each node in the network has a probability table for every possible final destination. The tables have entries for each neighbouring node that can be reached via one connecting link. The probability of the agents choosing a certain next node is the same as the probability in the table. The probability tables only contain local information and no global information on the best routes. Each time an agent visits a node the next step in the route is determined. This process is repeated until the agent reaches its destination. Thus, the entire route from a source node to a destination node is not determined beforehand.

First Competitive Ant Colony Scheme
for the CARP

Philippe Lacomme[1], Christian Prins[2], and Alain Tanguy[1]

[1] Blaise Pascal University, Laboratory of Computer Science (LIMOS)
UMR CNRS 6158, Campus des Cézeaux, 63177 Aubiere Cedex
{lacomme,tanguy}@sp.isima.fr
[2] University of Technology of Troyes, Laboratory for Industrial Systems Optimization
12, Rue Marie Curie, BP 2060, F-10010 Troyes Cedex, France
prins@utt.fr

1 Introduction

The CARP consists on determining a set of vehicle trips with minimum total cost. Each trip starts and ends at the depot, each required edge of the site graph is serviced by one single trip, and the total demand handled by any vehicle does not exceed the capacity Q. The most efficient metaheuristics published so far are a sophisticated taboo search method (CARPET) of Hertz et al. [1] and a hybrid genetic algorithm proposed by Lacomme et al. [2]. In the beginning, no collective memory is used and ants use only heuristic information. Pheromone deposition is proportional to the fitness that can be defined for minimization objectives as the inverse of the solution quality or solution cost. Local search can also be applied to increase the convergence rate. The process is iterated until a lower bound is reached or a maximal number of iterations is carried out.

2 Ant Colony Proposal

The network is stored as a directed internal graph using two opposite arcs per edge and one dummy loop on the depot. The nodes are dropped out and an arc index list is used. In any solution, each trip is stored as a sequence of required arcs with a total cost and a total load. Shortest paths between tasks are not stored. The cost of a trip is the collecting costs of its required arcs plus the traversal costs of its intermediate paths. The solutions are giant tours with no trip delimiters sorted in increasing cost order. They are split into solutions regarding the vehicle constraints, using the Split procedure [2]. The ant system framework consists in the following steps:

1. generation of solutions by powerful constructive heuristics for the CARP;
2. generation of solutions by ants according to pheromone information;
3. application of a local search to the ant solutions with a fixed probability;
4. updating the pheromone information;
5. iterate step 2 to 4 until the lower bound or some completion criteria are reached.

M. Dorigo et al. (Eds.): ANTS 2004, LNCS 3172, pp. 426–427, 2004.

The population is composed of elitist ants and non-elitist ants. The elitist ants tend to favor the convergence of the algorithm and the non-elitist ones attempt to control the diversification process. Whatever the solution cost found by a non-elitist ant, it replaces the previous one. For the elitist ants, the solution is replaced only if it is more promising. To decrease the probability of being captive in a local minimum, the pheromone is erased when n_s iterations have been performed without improvement. The contribution level of the pheromone update depends on the quality of the solution. The local search scheme [2] dedicated to the CARP is applied with a certain probability to improve solution. The split procedure [2] is applied to get the solution cost.

3   Numerical Evaluation and Concluding Remarks

In this section, we present numerical experiments for the proposed Ant Colony Scheme compared to the best methods for the CARP including CARPET and the Genetic Algorithm [2]. The benchmark has been performed using the well-known instances of DeArmon, Eglese, Belenguer and Benavent. Results (see Table 1) are given for three restarts of BACO (Best Ant Colony Optimization).

Table 1. Experimental results

BACO	DeArmon inst. CARPET	GA	Belenguer/Benavent inst. CARPET	GA	Eglese inst. CARPET	GA
Better	3	0	17	0	16	0
Equal	20	20	15	27	0	3
Worse	0	3	2	7	9	21

The Ant Colony scheme is competitive with the best methods previously published providing high quality solutions in rather short computational time. It outperforms the CARPET algorithm and competes with the Genetic Algorithm for small and medium scale instances. The computational time is acceptable but the Ant Colony scheme can not compete, for a computational point of view, with the Genetic Algorithm. Further researches are required to increase the convergence rate.

References

1. A. Hertz, G. Laporte and M. Mittaz: A Tabu Search Heuristic for the Capacitated Arc Routing Problem. Operations Research, vol. 48, no. 1, pp. 129-135, 2000
2. P. Lacomme, C. Prins and W. Ramdane-Cherif: Competitive genetic algorithms for the Capacitated Arc Routing Problem and its extensions. Lecture Notes in Computer Science, E.J.W. Boers et al. (Eds.), LNCS 2037, pp. 473-483, Springer-Verlag, 2001

Hypothesis Corroboration in Semantic Spaces with Swarming Agents

Peter Weinstein, H. Van Dyke Parunak, Paul Chiusano, and Sven Brueckner

Altarum Institute
3520 Green Court, Suite 300, Ann Arbor, Michigan 48105, USA
{peter.weinstein,van.parunak,paul.chiusano,sven.brueckner}@altarum.org

1 The Ant CAFÉ Prototype

Our poster describes the architecture and innovative Swarm Intelligence algorithms of the Ant CAFÉ system that extracts and organizes textual evidence that corroborates hypotheses about the state of the world to support Intelligence analysts.

Intelligence analysts try to connect clues gleaned from massive data measured in petabytes (10^{15}), which also contains complex interconnectivity and heterogeneity at all levels of form and meaning. In the investigative process, analysts construct hypotheses (tentative assertions about the world) and then submit them to systems that find and organize evidence to corroborate the hypotheses (with some degree of persuasiveness). We represent hypotheses as graphs of concepts at varying levels of abstraction. Finding evidence requires matching edges of the hypotheses graphs against document text. Organizing evidence means joining pieces of evidence according to the template provided by the hypothesis.

Our Ant CAFÉ system combines multiple Swarm Intelligence [2] mechanisms in a single system, utilizing both digital pheromones and sematectonic stigmergy. These processes include clustering of text to yield an orderly space; identifying relations in text to yield matches; and assembly of matches into structures that instantiate hypotheses. These processes are guided by a concept map [1] that represents an analyst hypothesis about the state of the world, essentially acting as a template for evidence assembly. In most systems stigmergy occurs in environments whose topology is simple and of low dimensionality (often mapping directly to a two or three-dimensional physical space). In Ant CAFÉ's evidence assembly, stigmergy occurs in "manifestation set" (or mset) space, where each manifestation set is a directed acyclic graph of WordNet synsets [3] that specialize a concept associated with either a node or edge in the concept map.

In evidence assembly our goal is clarity, a measure that quantifies the degree of understandability of a set of assemblies (the system's response to an investigation at some point in time). In high clarity solutions, a few assemblies stand out, they are coherent, and they are well differentiated from each other. As a global metric, clarity must emerge from the local behavior of evidence assembly agents.

M. Dorigo et al. (Eds.): ANTS 2004, LNCS 3172, pp. 428–429, 2004.

2   Preliminary Results

In evidence assembly, matches produced by relation identification self-organize into instantiations of the concept map. A match contains three bindings, associating an edge of the concept map to text elements that provide evidence for that relation. In the Most Likely Collisions (MLC) algorithm, bindings move within manifestations sets, looking to join matches that have proximate bindings.

To test whether evidence assembly using MLC works as intended, we generated artificial populations of matches that vary with respect to the quality of potential solutions. In particular, the Hidden Solutions data includes matches generated by perturbing hypothetical solutions that include a synset from each manifestation set associated with the concept map.

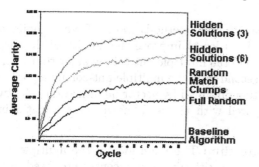

Fig. 1. Average clarity

Figure 1 shows the average clarity achieved across 90 runs for each type of data as the runs progress. The more orderly the data, the greater the clarity achieved. These results indicate that the clarity of a set of evidence assemblies is a global property that emerges from local decisions about the semantic proximity of pairs of match bindings. Thus swarming evidence assembly can scale to handle massive data. [4] is a full version of this paper.

Acknowledgments

This study was supported and monitored by the Advanced Research and Development Activity (ARDA) and the National Geospatial-intel-ligence Agency (NGA) under Contract Number NMA401-02-C-0020. The views, opinions, and findings contained in this report are those of the authors and should not be construed as an official Department of Defense position, policy, or decision, unless so designated by other official documentation.

References

1. J. W. Coffey, R. R. Hoffman, A. J. Cañas, and K. M. Ford. A Concept Map-Based Knowledge Modeling Approach to Expert Knowledge Sharing. In Proceedings of IASTED International Conference on Information and Knowledge Sharing, 2002.
2. H. V. D. Parunak. 'Go to the Ant': Engineering Principles from Natural Agent Systems. Annals of Operations Research, 75:69-101, 1997.
3. G. A. Miller. WORDNET: An On-Line Lexical Database. Int. Journal of Lexicography, 3(4):235-312, 1990.
4. P. Weinstein, H. V. D. Parunak, P. Chiusano, S. Brueckner. Agents swarming in Semantic Spaces to Corroborate Hypotheses. Altarum Institute, Ann Arbor, MI, 2004. www.altarum.net/~vparunak/AAMAS04AntCAFE.pdf.

Mesh-Partitioning
with the Multiple Ant-Colony Algorithm

Peter Korošec[1], Jurij Šilc[1], and Borut Robič[2]

[1] Jožef Stefan Institute, Computer Systems Department, Ljubljana, Slovenia
[2] University of Ljubljana, Faculty of Computer and Information Science, Slovenia

1 Extended Abstract

We present two heuristic mesh-partitioning methods, both of which build on the multiple ant-colony algorithm in order to improve the quality of the mesh partitions. The first method augments the multiple ant-colony algorithm with a multilevel paradigm, whereas the second uses the multiple ant-colony algorithm as a refinement to the initial partition obtained by vector quantization. The two methods are experimentally compared with the well-known mesh-partitioning programs.

The Multiple Ant-Colony Algorithm: The main idea of the multiple ant-colony algorithm (MACA) for k-way partitioning was recently proposed in [6, 7] and based on the metaheuristics developed by Dorigo et al. [3]. There are k colonies of ants that are competing for food, which in this case represents the vertices of the graph (the elements of the mesh). Eventually, ants gather food to their nests, i.e., they partition the mesh into k submeshes.

Multilevel Algorithm M-MACA. An effective way to speed up and globally improve any partitioning method is the use of multilevel techniques [2]. The basic idea is to group vertices together to form clusters that define a new graph. This procedure is applied until the graph size becomes small enough. Each step is followed by a successive refinement of the graph (using the MACA).

Hybrid Algorithm H-MACA. Another way of using the MACA in the mesh-partitioning problem is to use the MACA as a refinement of some initial partitioning. We chose the vector quantization method [8] as the algorithm to obtain the initial partition.

Experimental Results: We present and discuss the results of the experimental evaluation of our M-MACA and H-MACA algorithms, and make comparisons with the well-known partitioning programs Chaco 2.0 with the multilevel Kernighan-Lin global partitioning method [4], k-Metis 4.0 [5], p-Metis 4.0 [5], and MultiLevel refinated mixed Simulated Annealing and Tabu Search (MLSATS) [1]. The test graphs used in our experiment were acquired from the Graph Collection Web page at the University of Paderborn (FEM2 benchmark suite).

The results, which are given in terms of edge-cut, are shown in Table 1. Here we must mention that for both the M-MACA and the H-MACA the imbalance

M. Dorigo et al. (Eds.): ANTS 2004, LNCS 3172, pp. 430–431, 2004.
© Springer-Verlag Berlin Heidelberg 2004

was kept inside 0.2%. Note that the Chaco and the p-Metis are well-balanced while the imbalance for the k-Metis and MLSATS was 3% and 5%, respectively. Because both the M-MACA and the H-MACA are stochastic algorithms we ran both algorithms 32 times on each graph and calculated the average edge-cut value and its standard deviation.

Table 1. Comparison of the algorithms for $k = 4$

Graph	Algorithm					
	Chaco	k-Metis	p-Metis	MLSATS	M-MACA	H-MACA
3elt	258	250	252	199	286.5 ± 39.3	253.0 ± 33.8
3elt_dual	130	125	120	123	145.5 ± 21.2	180.4 ± 11.3
airfoil1	182	190	179	200	228.4 ± 32.1	204.0 ± 9.2
airfoil1_dual	111	84	84	64	122.8 ± 17.8	117.5 ± 4.6
big	416	385	405	367	505.2 ± 76.1	428.7 ± 4.6
big_dual	219	171	196	205	270.8 ± 39.0	238.1 ± 4.5
crack	457	478	458	401	450.5 ± 48.0	379.5 ± 3.7
crack_dual	228	200	201	211	211.7 ± 30.7	169.5 + 3.1
grid1	48	49	40	39	41.2 ± 4.1	38.2 ± 0.4
grid1_dual	37	38	35	35	35.5 ± 0.6	35.0 ± 0.0
grid2	106	114	121	92	114.0 ± 14.0	101.6 ± 1.9
grid2_dual	99	112	91	88	111.0 ± 16.5	97.0 ± 0.2
netz4504	66	61	62	48	57.8 ± 7.2	51.2 ± 1.7
netz4504_dual	54	50	49	50	58.1 ± 7.6	46.2 ± 1.4
U1000.10	200	100	108	177	174.1 ± 30.1	123.5 ± 4.8
U1000.20	664	566	515	460	826.3 ± 126.9	535.5 ± 15.6
ukerbe1	82	72	64	70	88.0 ± 15.5	61.9 ± 1.2
ukerbe1_dual	56	63	51	59	73.6 ± 11.2	48.0 ± 0.2
whitaker3	439	407	424	448	453.2 ± 47.9	388.0 ± 2.2
whitaker3_dual	251	223	210	216	243.7 ± 32.8	199.1 ± 2.3

The multilevel and hybrid methods were quite similar in terms of producing the best results. The only difference was in the standard deviation of the results, which was in favor of the hybrid method.

References

1. R. Banos, C. Gil, J. Ortega, and F.G. Montoya, Multilevel heuristic algorithm for graph partitioning, Lect. Notes Comp. Sc. 2611 (2003) 143–153.
2. S.T. Barnard, H.D. Simon, A fast multilevel implementation of recursive spectral bisection for partitioning unstructured problems, Concurrency-Pract. Ex. 6 (1994) 101–117.
3. M. Dorigo, G. Di Caro, and L.M. Gambardella,Ant algorithms for discrete optimization, Artif. Life 5 (1999) 137–172.
4. B. Hendrickson, R. Leland, The Chaco User's Guide, Version 2.0, Technical Report SAND95-2344, Sandia National Laboratories, Albuquerque, NM, 1995.
5. G. Karypis, V. Kumar, Multilevel k-way partioning scheme for irregular graphs, J. Parallel Distr. Com. 48 (1998) 96–129.
6. P. Korošec, J. Šilc, B. Robič, An experimental study of an ant-colony algorithm for the mesh-partitioning problem, Parallel Distr. Com. Pract. 5 (2002) 313–320.
7. A.E. Langham, P.W. Grant, Using competing ant colonies to solve k-way partitioning problems with foraging and raiding strategies, Lect. Notes Comput. Sc. 1674 (1999) 621–625.
8. Y. Linde, A. Buzo, R.M. Gray, An algorithm for vector quantizer design, IEEE Trans. Commun. 28 (1980) 84–95.

Author Index

Lecture Notes in Computer Science

For information about Vols. 1–3052

please contact your bookseller or Springer-Verlag